Eponym Dictionary of Odonata

Bo Beolens

Whittles Publishing

Published by
Whittles Publishing Ltd.,
Dunbeath,
Caithness, KW6 6EG,
Scotland, UK

www.whittlespublishing.com

© 2018 Bo Beolens

ISBN 978-184995-365-8

Print production managed by Jellyfish Solutions

Contents

Dedication

This book is dedicated to my son Ashley Crombet-Beolens, his partner Zoe and their sons, Owen and Toby and their daughter Bo (my namesake). Ash began this project with me and has a continuing interest in Odonata, is more knowledgeable about them than I and takes beautiful photographs of them too.

Cover Image: A courting pair of Neurobasis kaupi (Great Blue metalwing) from Sulawesi, a species named after Johann Jakob Kaup. Artwork by Albert Orr (q.v.)

Foreword

In the 19th and early 20th centuries, the heyday of insect collecting in the United Kingdom and many European countries, Odonata (an insect order comprising dragonflies and damselflies, often simply called 'dragonflies') were largely neglected in favour of butterflies, larger moths and beetles. No doubt the major reason for this prejudice was that most dragonfly species completely lose their bright colours after death and are unattractive when pinned in elegant mahogany cabinets. However, during the last 25 years or so, observing, photographing and recording Odonata has become an increasingly common hobby among nature lovers, and they are also becoming the research topic for a growing number of scientists.

The interest in these fabulous insects has been facilitated by the ease with which superb photographs can be taken using digital cameras, a technology in its infancy just 15 years ago. Also during this period new standards have been set in entomological art by several illustrators. There has thus been a spate of colourful guide books enabling identification of dragonflies and damselflies in many parts of the world, often without even capturing a specimen. They have become 'the birdwatchers' insects' and unlike yesteryear many present enthusiasts have come from this, rather than an entomological background. It is natural therefore that Bo Beolens, an avid birdwatcher, also known by his cognomen, 'The Fat Birder', should produce this fine book on the eponyms in dragonflies similar to his earlier books on the eponyms of the vertebrate species: *Whose bird? Men and women commemorated in the common names of birds*, *The eponym dictionary of birds*, *The eponym dictionary of mammals*, *The eponym dictionary of amphibians*, *The eponym dictionary of reptiles* and *Sharks: an eponym dictionary*.

There is another reason why dragonflies are a natural topic for this kind of book. Eponyms, i.e. scientific names based on those of individual persons, are very common in Odonata nomenclature. Of the nearly 8,600 names given to species and subspecies of extant Odonata (including those not in current use, technically known as synonyms), nearly one quarter are eponyms named after an individual person or a group of people. Moreover, this practice seems to be accelerating. Of about 1150 names published after 1 January 1995, as many as 41 percent were eponyms. There are several reasons for this. Taxonomy is becoming increasingly a social activity, with multiple authors participating in the naming of species, rather than a single authority labouring away in solitude. Also at present there are more Odonata collectors and taxonomists, residing in or visiting tropical countries where new species can still be found, than at any previous point in history. From the time of Linnaeus (1758) to the end of 1994, a total of about 330 scientists had described or co-described new Odonata taxa, but this number is now approaching 570. Thus, the number of individuals deserving recognition is rapidly increasing. Another recent phenomenon has been to sell naming rights, to obtain funding for taxonomic research. This is a form of crowd-funded research where particularly generous donors are honoured with an eponym of their choosing, and most choose their own name or that of their loved ones. Moreover, few modern taxonomists have even basic knowledge of Latin and still less of Greek, and thus find it difficult to formulate a philologically correct descriptive species epithet. Naming a species after a person is an easy and convenient solution and for technical reasons it is also easier to manage such names later, should they be moved to a different genus or different gender.

Over 1,330 individuals have been immortalized in these species names, and a small number also appear in genus names. This book tells us something of their stories. The biographical accounts on the life and achievements of these 'eponymees' vary considerably in length and content. This is no reflection of the importance and merit of the person in question. Individuals honoured come from all levels of society, from plebs to princes, with much disparity in available information. Many collected the type specimens, or are fellow odonatologists, museum curators or other colleagues and benefactors. Family members, especially wives, children and grandchildren often have names dedicated to them, sometimes even on their first birthday. But among them are also royal family members, famous explorers, scholars, poets, musicians and other persons of note whom the authors have wished to honour for various reasons. The book also covers names based on ethnic groups, fictional and mythological characters and names which structurally resemble eponyms, but in fact are named after location (toponyms) or have some other meaning. I warmly congratulate Bo Beolens on his achievement and am confident that this book will offer enjoyable reading to anyone interested in dragonflies and curious to know the secrets of their names.

Matti Hämäläinen
Espoo, Finland

Acknowledgements

This project started out as a 'Father & Son' project between my son, Ashley Crombet-Beolens and myself. He took the most comprehensive list of Odonata we could find and identified those that were, or seemed to be, eponyms. He also began to find the identity of some of those people and write brief biographies. However, we had not progressed very far before the exigencies and vicissitudes of a full-time job, an external degree course and a growing family meant that he found less and less of his time could be devoted to the project and reluctantly withdrew. It seems to me, therefore, befitting to his contribution that this volume be dedicated to him and his partner and children.

As I continued with the project and made good progress, at some stage I asked Dr Albert Orr the prominent Australian odonatologist and lepidopterist, to provide his own entry and he began helping me with further advice and support. Part of this was to introduce me to his friend and colleague Dr Matti Hämäläinen whom most readers of this volume will know as an authority on Odonata taxonomy with expert knowledge of the Asian fauna, especially of Caloptera damselflies. It happens that he also has had a long interest in eponyms and the history of odonatology.

He initially gave me help, support and advice, more than anyone else before or since. Indeed, he was so enthusiastic about the project that he devoted a lot of time to identifying people and researching them for the biographies. So much time in fact that he (and Bert Orr) felt they should be credited as co-authors. We all agreed on this and continued with the research. However, as we progressed it became clear that Matti in particular wanted to produce a slightly different book to me, using some different conventions. I felt strongly that this volume should reflect in style, content and conventions the other eponym dictionaries I have worked on. Reluctantly, and without rancour, Matti and Bert withdrew. Nevertheless, both have continued to help. In Matti's case this help has been extensive and this volume would be far less comprehensive and less accurate had he not continued to offer support and bring to bear his expertise.

Matti used the information which he had gathered for his own publication, originally suggested by Bert: *Catalogue of individuals commemorated in the scientific names of extant dragonflies, including lists of all available eponymous species-group and genus-group names*, which appeared as volume 80 of the *International Dragonfly Fund – Report* (2015). A revised edition was published as volume 92 in the same series (2016). Matti wrote in his introduction as follows, "*I have prepared the following catalogue in a spirit of cooperation* [with Bo Beolens' eponym dictionary project]. *I hope that it will serve as a useful reference, facilitating more detailed biographic accounts of the individuals listed, as planned by Bo Beolens, as well as complementing that publication when it appears.*" Indeed, his catalogue (and its later additions available online) has greatly facilitated my work. After all, I am not even a zoologist, let alone an entomologist, merely an amateur wildlife enthusiast and writer. Matti has been incredibly generous in his contribution of time. This spring he has again spent well over one hundred working hours acting as unpaid editor of my manuscript, ensuring the accuracy of the (particularly scientific) species names and greatly expanding the contents of the individual entries of most odonatologists (past and present) from his personal knowledge or through his correspondence with his fellow odonatologists. He has also provided much relevant information for numerous other person entries. Without his extensive contribution, this volume would be far less comprehensive and less accurate. I thank him for his generosity, support and continuing help, and, especially, from preventing me carrying forward errors. Needless to say, all remaining errors remain my own responsibility.

Particular thanks are due to several other people who have given me a great deal of time and information:

First among these is Dr Florian Weihrauch, editor of *Odonatologica,* who has winkled out some hard to find articles, helping me confirm or identify some of the people after whom Odonata have been named and, as he would put it, add 'salt' to the thinner entries. He has brought to the project his own expertise and his extensive library as well as the journal's resources. Moreover, he seems to know anyone who is anyone in the world of odonatology and has introduced me to many other prominent entomologists and others who have been very open and kind in their support and sharing of their knowledge too.

Mike Watkins, my friend and co-author on a number of books, has also helped me find some people I was having difficulty identifying, or details of some elusive people, often thanks to his grasp of several European languages. Thanks Mike!

Several people whose brief biographies appear in this book not only helped with their own entries but those of relatives, friends or colleagues, so my sincere thanks are also due to Professor Dr Bastiaan Kiauta, Dr Oleg Kosterin, Dr Stephen A Marshall, Dr Dennis Paulson, Dr Akihiko Sasamoto, Graham Vick and K D P Wilson in particular. Dr Klaas-Douwe Dijkstra kindly provided me with his lists of the English common names of African and Madagascan species, many of which have not yet been published elsewhere.

Albert Orr generously offered the use of his handsome artwork, depicting courtship in *Neurobasis kaupi,* as a cover illustration gratis – an illustration of an eponymous species by another person appearing in this book. This was based on his observations of the species while a participant on 'Project Wallace', the 1985 Royal Entomological Society Expedition to North Sulawesi, Indonesia.

Many of the other people that have entries appearing in this book, too many to individually name, were kind enough to fill in gaps, or correct their own entries and sometimes entries of friends or colleagues.

So, this is not only an ongoing project as information comes to light or new species are described, but a collaborative effort which we all hope will prove a resource to odonatologists and zoologists in general as well as the general reader.

Index & Bibliography

As this is an alphabetical volume, and because wherever possible I have given cross references, I have not provided an index; it would be redundant. Nor have I provided an extensive bibliography. Many of the entries are drawn from webpages, CVs and the personal notes of many of those mentioned. The process of tracking down the individuals and cross-checking their facts has been a very long-winded one.

It started with a scientific name and the etymology given by the species account author (barring those who didn't bother to say who was honoured) – although some original papers proved really hard to get hold of. Listing every etymology would more than double the length of this book.

The next stage was to write an entry based on published facts about that person. Wherever possible, the living people themselves were consulted. In some cases, close friends or relatives gave us further information even more so with those who were recently deceased.

The next most useful sources were a few (acknowledged) experts in the field, odonatologists who have extensive networks of their fellow entomologists. Of course, they may have consulted books and papers as well as exercised their own memories, but as those sources were 'once removed' they are not listed either.

After this was a long trawl through everything online. This may sound straight forward, but it is actually often really tough to find and serendipity played a role in finding things in seemingly unrelated sources. Even then a lot of people honoured in species names turn out to be unsung heroes appearing in no other literature. This is particularly true of local or commercial collectors employed by travellers in colonial times.

So listed below are just those works I had a hand in, as often I was able to use previous research on individuals after whom many different taxa types are named.

Beolens, B. and Watkins, M. 2003. *Whose Bird?* London, Christopher Helm A. and C. Black.

Beolens, B., Watkins, M. and Grayson, M. 2009. *The Eponym Dictionary of Mammals.* Baltimore, Johns Hopkins University Press.

Beolens, B., Watkins M. and Grayson, M. 2011. *The Eponym Dictionary of Reptiles.* Baltimore, Johns Hopkins University Press.

Beolens, B., Watkins, M. and Grayson, M. 2013. *The Eponym Dictionary of Amphibians.* Pelagic.

Beolens, B., Watkins, M. and Grayson, M. 2014. *The Eponym Dictionary of Birds.* Christopher Helm.

Watkins, M. and Beolens, B. 2015. *Sharks : An Eponym Dictionary.* Pelagic.

Hämäläinen, M. 2016. Catalogue of individuals commemorated in the scientific names of extant dragonflies, including lists of all available eponymous species-group and genus-group names – Revised edition. *International Dragonfly Fund – Report* 92: 1-132.

Introduction

One might call this section 'how to use this book' as I use some abbreviations and have adopted some conventions that I and my fellow authors found useful in our previous eponym dictionaries.

So far as I know, no one has set out to bring together in book form scientific eponyms and I developed a style with my fellow authors which we think works, so I have used that format in this book. So the layout etc. should be familiar to anyone who has looked at our previous efforts, and there is much in those introductions that apply here too.

The book is in the form of a dictionary so entries are alphabetical by the person named in a common or scientific name of Odonata, whether that be genus, species or sub-species.

I use the name honoured *almost* in the form it appears in the name; 'almost' because I leave out the (usually Latin) declensions. So, where *roberti* appears you will find the listing under **Robert**, when the genitive *robertae* appears I list it under **Roberta**. If a whole name is used such as *robertgoodfellowi* I list it as the proper name **Robert Goodfellow**. If your Latin is as sparse as mine beware some binomials can fool you when a masculine name ends in an 'a', it is sometimes rendered incorrectly.

When, as sometimes occurs, an individual is honoured differently in a number of different species such as *roberti* in some and *robertgoodfellowi* in others there will be an entry for each. One entry will give the full biographical details, and end with (see…) and the other will be much briefer and end with (see…). Nicknames, clever allusions, etc., are treated in the same way unless there are no entries under the real name, in which case the full biography will be listed under that nickname without declension but the biography will start with the full name and dates (if I have them).

Where an entrant has family connections etc. with other entrants each will end with (Also see…) so the reader can see where the connections are. When, in the text, another entrant is mentioned this will be acknowledged with (q.v.) and where an entrant appears as an author of a species under another entrant their name will be **emboldened**. I only do this for the first such use of any entry and do not embolden names in taxonomic synonyms which often appear in square brackets beneath eponymous species.

Each entry also follows a conventional form with names and dates, university education and career with dates and institutions, expeditions, collecting trips, major or relevant publications and finally a hint at other taxa that also carry the person's name. Of course, some entries are much fuller than others and may contain more information such as how the author referred to the honoured person in their etymology.

Inevitably there are some educated guesses where no etymology is given and there is no living person or knowledgeable contemporary to consult. If there is no supporting evidence known to me I have tried to select a likely candidate. In some cases, this is obvious as the name is the same as someone close to the author, maybe a work colleague, mentor or family member.

Some are virtually impossible ever to know and this is especially true where there is no etymology and the name used is a forename. There are a few examples where the etymology is omitted because the person wished to remain anonymous or the author didn't want the recipient to be sure that they were being honoured. Fortunately, these are few.

Abbreviations

The following abbreviations may be used in the lists of species and the general texts.

A.	Alternative Common Name
S	Synonym
JS	Junior Synonym of
Orig.	Original Scientific Name
AMNH	American Museum of Natural History, New York
BMNH	British Museum of Natural History, London
CSIRO	Commonwealth Scientific and Industrial Research Organisation
IRD	Institut de Recherche pour le Développement
MNHN	Muséum Nationale d'Histoire Naturelle, Paris
MRAC	Musée Royale de l' Afrique Centrale, Tervuren, Belgium
ORSTOM	Office de la Recherche Scientifique et Technique Outre-mer
P	Priority common name*
RMNH	Rijksmuseum van Natuurlijke Histoire, Leiden, Netherlands
USNM	Smithsonian Institution (National Museum of Natural History)
UN	United Nations

* A number of leading odonatologists are struggling to bring order to common (English) names of species to cut through the confusion caused by differing names. To this end there is increasing agreement of which common names should be used. Throughout I have adopted all those recommendations that I am aware of, EXCEPT where an eponymous common name is available. In these cases, I have added the priority name below not marked 'A' for alternative, but 'P' for priority. The book is, after all, about eponymous names.

Conventions

All entries have titles that are in emboldened normal font or italics. Those in ordinary text are real people after whom the species listed were named.

Bold italicised entries are those referring to places (toponyms) or peoples, generic forenames, mythical and fictional people that might appear to be eponyms but are not (or at least not directly as some may be named after something that is named after a person such as Victoria, which is a place named after Queen Victoria and the species is named for the former as a location, not directly the latter). The toponyms may be rivers, geographical areas, towns, mountains or other physical features. Usually those named after peoples are endemic to those people's area or were first found there.

I also italicise entrants that are historical figures (usually from classical times of Roman or Greek history). They are eponyms, but do not so much honour the person as use their fame or characteristic similarly to the use of fictional or mythical characters. Thus, **Julius** derives from Julius Caesar but doesn't honour him in the way that a loved one, colleague or great man of entomology might be honoured.

It is usually the case that the others are named because of some characteristic of that fictional or historical figure or the translation of the generic forename. So, for example, a mythical person who was very beautiful has been used as a name for a particularly beautiful dragonfly, or a historical figure might have been very aggressive and the Odonata is also, and so on. The historical figures are not being 'honoured' in the way truly eponymous ones are.

Species are listed first in taxonomic order in that first are eponymous genera, then species and lastly subspecies, with a gap between each category. Within each category species are listed in date order and within each year alphabetically by scientific name.

Brackets are not used in the way of a normal taxonomic convention. I use them merely to show alternative names and synonyms below the listed species.

Authors' names in the species lists are emboldened if they have an entry of their own in that name elsewhere in the dictionary (i.e. if the only entry they have is under, say, a forename, it is not emboldened, but a reference may be made at the end of the entry in brackets).

A few entries are listed under the name of the person honoured, but that name may only be partly referred to in the scientific name. Alternatively, a few are listed under the honorific which is then explained. For example, some forenames are Latinised to their equivalent, A good example is listed as 'Eliseva' for Ellis Bettina Grootveldas, Elisabeth is the root of both given names of Ellis and Bettina and the name 'eliseva' (noun in apposition) is derived from the original Hebrew equivalent of Elizabeth, the name Elisheba or Elisheva. There are, inevitably, some clever jokes hidden in names where, for example, a surname is translated into Latin too.

On a more mundane level, journals are placed between single quotes rather than in italics which is the publishing convention; this is to avoid adding to the already oft-used italic setting.

The dictionary attempts to list all names named after an individual person that have been published up to 1st June 2017.

A

Aaron

Coral-Fronted Threadtail *Neoneura aaroni* **Calvert**, 1903

Samuel Francis (Frank) Aaron (1862–1947) was an American (Texas born) naturalist who was Custodian of Insects in the entomological section at the Academy of Natural Sciences of Philadelphia (1884–1885). He was Economic Entomologist at the Philadelphia Commercial Museum (1896–1906) and published widely on entomology, having described numerous species new to science. He collected with his half-brother Eugene Murray-Aaron* and they wrote together: *List of a collection of diurnal Lepidoptera from southern Texas* (1885) published in 'Papilio', which his brother edited. It was during this trip that Frank collected the threadtail that Calvert studied nearly two decades later and named after him. The American Entomological Society published his first entomological paper (1883). He was also a writer of popular nature articles for newspapers and magazines and wrote for the *Radio Boys* young adult series of books. Philip Calvert wrote his obituary in the 'Entomological News'. He is remembered in the names of a variety of insects. Calvert wrote: *"The material before me was probably collected by Mr. S. F. Aaron in the vicinity of Corpus Christi in 1884."*

*Sources regularly confuse the half-brothers and their achievements. Eugene Murray-Aaron (1852–1940) was an entomologist who was the first editor of the 'Entomological News' (1890). In addition to entomology he had diverse professional interests including natural history, geography, cartography, travel, human biology, cycling and government. He travelled extensively throughout the Caribbean, Central, and South America (1880s–1890s). His major natural history books were: *Butterfly Hunters in the Caribbees* (1884) and *The New Jamaica* (with E M Bacon, 1890). He was also the last Editor of 'Papilio' (1884), the journal of the New York Entomological Club. He was Geographic Editor for the George F Cram Company (1899–1917). His passion in entomology focused on skippers (*Hesperiidae*) and Aaron's Skipper, is named after him.

Abbot

Regal Darner *Aeschna abboti* **Hagen**, 1863
[JS *Coryphaeschna ingens* (Rambur, 1842)]

John Abbot (1751–1841) was an English-born American entomologist and ornithologist and a talented illustrator. He studied drawing and engraving under the engraver Jacob Bonneau (1741–1786) and was consequently encouraged by a group of naturalists from the Royal Society to go to Virginia (1773–1775) to study, draw and collect natural history specimens. Afterwards (1776) he settled in Georgia. There he produced thousands of insect illustrations and several sets of bird drawings. Many are now in the collection of the BMNH and others at Harvard and other US Universities. His only published work, as co-author, was *The Natural History of the Rarer Lepidopterous Insects of Georgia* (1797) although examples of his work appeared in other later publications.

Abbott

Eastern Yellowwing *Allocnemis abbotti* **Calvert**, 1892
[Orig. *Disparoneura abbotti*]
Little Skimmer *Orthetrum abbotti* Calvert, 1892
Flatwing sp. *Ortholestes abbotti* Calvert, 1894
[JS *Hypolestes triniatitis* (Gundlach, 1888)]
Tiger sp. *Gomphidia abbotti* **EB Williamson**, 1907
Rare Basker *Urothemis abbotti* **Laidlaw**, 1927

Dr William Louis Abbott (1860–1936) was a student, naturalist and collector. Initially qualifying as a physician at the University of Pennsylvania and working as a surgeon at Guy's Hospital in London, he decided not to pursue medicine but use his private wealth to engage in scientific exploration. As a student (1880) he had collected in Iowa and Dakota and in Cuba and San Domingo (1883), in the company of Joseph Krider son of the taxidermist John Krider. He went to East Africa (Kenya and Tanganyika) (1887) spending two years there as well as visiting Haiti. He studied the wildlife of the Indo-Malayan region (1889–1891), using his Singapore-based ship 'Terrapin', and made large collections of mammals from South-east Asia for the National Museum of Natural History (Smithsonian) in Washington D.C. He switched to Siam (1897) and spent 10 years exploring and collecting in and around the China Sea. He provided much of the Kenya material in the Smithsonian and was the author of *Ethnological Collections in the United States National Museum from Kilima-Njaro, East Africa* in their report (1890/91). He returned to Haiti and San Domingo (1917) exploring the interior and discovering more new species. He retired to Maryland but continued his study of natural history all his life. Amongst other taxa eighteen birds, two mammals, two amphibians and two reptiles are named after him.

Abuna

Firetail sp. *Telebasis abuna* **GH Bick** & **JC Bick**, 1995
Wedgetail sp. *Acathagrion abunae* **Leonard**, 1977

Named after the type locality, Abuna, Rhondonia state, Brazil

Acantha

Dragonlet sp. *Erythrodiplax acantha* **Borror**, 1942

Acantha was a nymph in Greek mythology. In one version of her story, Acantha refused Apollo's advances and scratched his face when he tried to rape her. Apollo then turned her into a bush.

Acco

Sylvan sp. *Coeliccia acco* **Asahina**, 1997
[JS *Coeliccia pyriformis* Laidlaw, 1932]

Clubtail sp. *Asiagomphus xanthenatus acco* Asahina, 1996

Dr Akiko ('Akko') Saito – see **Akiko (Saito)**

Acosta

Bannerwing sp. *Thore acostai* **Navás**, 1924
[JS *Polythore vittata* (Selys, 1869)]

Father José Acosta SJ (1540–1600) was a Spanish missionary and naturalist in Latin America. He entered a seminary (1553) in his hometown of Medina del Campo, Spain, as did three of his brothers. He lectured in theology at Ocana (1569) and was then sent to Lima, Peru (1570). There he was almost immediately sent across the Andes with the Spanish Viceroy. He was later (1583) sent to Mexico before returning to Spain (1587). He became Professor of Theology there at the Roman College (1594). He has been called the 'Pliny of the New World' because his book: *Historia natural y moral de las Indias* was the first detailed description of the geography and culture of Latin America, especially Peru. He also studied altitude sickness in the Andes, analysed earthquakes, volcanoes, tides, and meteorology, local culture and natural history.

Actaeon

Shadowdamsel sp. *Drepanosticta actaeon* **Laidlaw**, 1934

Actaeon was a hero in Greek mythology. When he lost favour with the gods he was turned into a stag.

Adam

Adam's Shadowdamsel *Drepanosticta adami* **Fraser**, 1933
[A. Adam's Forest Damselfly, Orig. *Ceylonosticta adami*]
Adam's Gem *Libellago adami* Fraser, 1939

This is a toponym as they are named after Adam's Peak (Sri Pada), Sri Lanka's holy mountain where both species are found. The names of the mountain differ according to different religious traditions, such as Buddhists seeing it as Buddha's footprint from a very large (5'7" × 2'6") impression looking like a footprint. Each religion has attributed it to a different deity or prophet. In Arabic tradition, they record it as being the solitary footprint of Adam (the first man according to the Bible, Koran, etc.) where he stood on one foot for a thousand years of penance. An Arab tradition tells that when Adam was expelled from heaven, God put him on the peak to make the shock less terrible – Ceylon being that place on earth closest to and most like heaven.

Adams, C C

Amber-winged Dancer *Argia adamsi* **Calvert**, 1902

Charles Christopher Adams (1873–1955) was an American zoologist from Bloomington, Illinois. Illinois Wesleyan University awarded his first degree and Harvard his doctorate. He was an assistant entomologist at Illinois State Laboratory of Natural History (1896–1898) then Curator of the University of Michigan Museum (1903–1906). He was then employed at the Cincinnati Society of Natural History and the Museum of the University of Cincinnati (1906–1907). He became associate professor at the University of Illinois (1908), Assistant Professor of Forest Zoology at the New York State College of Forestry, Syracuse University (1914) and then full Professor (1916). He was the first Director (1919) of the Roosevelt Wild Life Forest Experiment Station in the Adirondacks, managed by the New York State College of Forestry and was instrumental in the development of the Allegheny State Park in Western New York. He left the Roosevelt Station (1926) to become the Director of the New York State Museum until retirement (1943). Among his publications was: *Guide to the*

3

Study of Animal Ecology (1915). He collected the holotype in Panama. He lent his collection to Calvert to study, who wrote in the etymology: "*...Named after Mr. C. C. Adams, of Bloomington, Illinois, whose collection of Argia, lent with the greatest liberality, has been an important source of knowledge for the present work.*"

Adams, E E

Adam's Emerald Dragonfly *Archaeophya adamsi* **Fraser**, 1959
[A. Horned Urfly]

Ernest E Adams collected the holotype. He is described as "*...the well-known collector from Edungalba in central Queensland*". He was a cattle owner as evidenced by this newspaper account: *When mustering on Mourangee, Mr Ernest Adams met with an unusual accident. Galloping after a beast in thick, tall cane grass his horse, in jumping over a fallen tree, collided with a bullock, which had been lying in the cane grass behind the tree, but which at that moment rose to its feet. In the impact horse, rider and beast found themselves on the ground together. Mr Adams received injuries to the right arm and shoulder, necessitating medical attention.* His specimens are in the Forestry Service Collection. I believe this to be the same individual who donated various documents to the Central Queensland University. Ernest E Adams' grandfather arrived in Central Queensland about 1876. Subsequently, members of the family occupied a number of grazing and farming properties, including Merion, Mourangee, Honeycombe, Bellwood, and Glencoe. The material consists of Diaries (1912–1919, 1944–1947, 1961), station accounts (various periods 1926–1965), rainfall records (1934–1993), correspondence (various dates 1925–1963), and miscellaneous papers and publications.

Adela

Duskhawker sp. *Gynacantha adela* **Martin**, 1909

Adela is a female given name. Martin did not provide any etymology.

Adelais

Pond Damselfly sp. *Libellula adelais* Fourcroy, 1785
[JS some European Coenagrionidae species]

Based on the French woman's name Adelaide. (See **Cecilia**)

Adoniram

Archtail sp. *Nannophlebia adonira* **Lieftinck**, 1938

Adoniram is a character from the bible who was in charge of conscripted timber cutters during the building of King Solomon's temple.

Aethra

The Genus *Aethriamanta* **Kirby**, 1889
Adjudant sp. *Aethriamanta aethra* **Ris**, 1912

Aethra was a daughter of King Pitheus of Troezen, and mother to Theseus in Greek mythology.

Afonso

Pond Damsel sp. *Leptagrion afonsoi* **Machado**, 2007

Dr Afonso Pelli (b.1965) is a Brazilian biologist and limnologist and Machado's former student. He was a biologist at the Universidade Federal do Triângulo Mineiro (1988) and collected the type specimen (1988) in the Serra do Caraça reserve. He has written (1968–2011) a number of entomology papers in connection with aquatic insects in the state of Minas Gerais, such as, with F A R Barbosa: *Insetos coletados em Salvinia molesta Mitchell (Salviniaceae), com especial referência às espécies que causam dano à planta, na lagoa Olhos d'Água, Minas Gerais, Brasil* (1968).

Aglaé

Common Bluetail *Agrion aglae* **Fonscolombe**, 1938
[JS *Ischnura elegans* (Vander Linden, 1820)]

Aglaé is a French given name for an unspecified woman.

Aglaia

Archtail sp. *Nannophlebia aglaia* **Lieftinck**, 1948
Wiretail sp. *Selysioneura aglaia* Lieftinck, 1953

A name from the Greek mythology meaning 'shining one' and can relate to several figures. She was the youngest of the Charities, the three daughters of Zeus, the goddess of beauty. Most other references are to nymphs or people named after the goddess.

Aguesse

Aguesse's Sprite *Pseudagrion aguessei* **Pinhey**, 1964

Pierre-Charles Aguesse (b.1929) is a French research scientist. He was educated in Algeria, then at the Faculty of Science at Nancy University, which awarded his bachelor's degree. He gained his doctorate (1961) from the Sorbonne for his thesis entitled: *Contribution a l'etude ecologique des Zygoptere de Camarge* (1963). He wrote, among other works: *Les Odonates de l'Europe occidentale du Nord de l'Afrique et des îles Atlantiques* (1968). He has described over 20 new dragonly species, most of them from Madagascar.

Ailsa

Threadtail sp. *Protoneura ailsa* **Donnelly**, 1961

Ailsa Donnelly (née MacEwen) (b.1933) is the wife of the describer, T W Donnelly (q.v.) and has regularly participated in his field trips.

Aino

Skimmer Genus *Aino* **Kirby**, 1890
[JS *Nannothemis* **Brauer**, 1868]
Sparkling Jewel *Aristocypha aino* **Hämäläinen**, **Reels** & **Zhang**, 2009

Aino is a nymph from the Kalevala – a 19th-century work of epic poetry compiled by Elias Lönnrot from Finnish and Karelian oral folklore and mythology. W F Kirby, who translated the Kalevala into English – *Kalevala: The Land of Heroes* (1907) named (1890) a new dragonfly genus and species *Aino puella* ('Aino girl'), inspired by this charming and tragic Kalevala character. Sadly, Kirby's name was later synonymised with *Nannothemis bella* (Uhler, 1857), thus disqualifying *Aino* as a genus group name. It was reintroduced to odonatology as a species name, largely as a tribute to Kirby who was first to list the known odonate fauna of Hainan (1900.)

Akiko (Saito)

Sylvan sp. *Coeliccia acco* **Asahina**, 1997
[JS *Coeliccia pyriformis* Laidlaw, 1932]
Clubtail ssp. *Asiagomphus xanthenatus acco* Asahina, 1996

Dr Akiko ('Akko') Saito is a Japanese entomologist who is Curator at the Zoology Department of the Natural History Museum & Institute, Chiba Province, Japan. She collected when taking part in three Japanese expeditions to Vietnam (1994–1995). She also collected on an expedition to the Kamchatka Peninsula and North Kuril Islands (1995) and in Sulawesi, Indonesia (2001). Among other papers she wrote: *Cerambycid beetles (Coleoptera, Cerambycidae) from northern Vietnam – A new species of the genus Pidonia (Lepturinae)* (1995) and co-wrote: *Check-list of Longicorn-beetles of Taiwan* (1992), *Records of some Cerambycidae (Insecta: Coleoptera) from the Northern Mariana Islands, Micronesia* (2007). At least one beetle *Stenandra saitoae*, that she collected in Sulawesi is also named after her. (See also **Acco**).

Alan

Scissortail sp. *Microgomphus alani* **Kosterin**, 2016

Alan Michael Andrews (b.1957) from Neath, Wales, is a friend of the describer, Oleg Engelsovich Kosterin. He was a restaurateur in Cambodia and diving instructor in Thailand, at the time of this publication he is a schoolteacher in Thailand. He was instrumental in the clubtail being found as he quite unexpectedly gave Kosterin a great deal of very efficient help. It started as a casual conversation in the restaurant when Kosterin asked him when the rains would start and told him about his collecting trip looking for new species of dragonfly. The following morning Andrews took Oleg to a really excellent remote area where they found the eponymous paratypes.

Albarda

Dragonhunter sp. *Sieboldius albardae* **Selys**, 1886

Forktail ssp. *Macrogomphus parallelogramma albardae* Selys, 1878
[Orig. *Macrogomphus albardae*]

Johan Herman Albarda (1826–1898) was a Dutch lawyer and amateur naturalist who was Deputy Prosecutor at Leeuwarden. His main interests were ornithology and entomology. Moreover, he was a friend of Selys (q.v.) and had entomologist contacts in England including Eaton and McLachlan (q.v.). He published around 13 scientific papers and described six new species of dragonflies, four of which are valid such as: *Zyxomma obtusum*. There is a brief biography written about him by D T E van der Ploeg entitled: *A Lawyer as Amateur Biologist – Herman Albada 1826–1998* (1985). He is also honoured in the name of a Chrysopid by McLauglin.

Albert (Kennedy)

Paiute Dancer *Argia alberta* **Kennedy**, 1918

Albert Hamilton Kennedy was the father of the species author Professor Clarence Hamilton Kennedy (q.v.) who collected the holotype of the new dancer species at Laws, California, USA, naming it after his father.

Albright

Five-Striped Leaftail *Phyllogomphoides albrighti* **Needham**, 1950
[Orig. *Gomphoides albrighti*]

Paul N ('Philip') Albright (1901–1999) worked for the Public Health Department of San Antonio, Texas, USA. He is mentioned in Needham's *A Manual of the Dragonflies of North America* (1954). He collected several specimens of this new species in Texas (1948 & 1949).

Albuquerque

Threadtail sp. *Epipleoneura albuquerquei* **Machado**, 1964

Professor Dr Dalcy de Oliveira Albuquerque (1918–1982) was an entomologist. He was Director, Museu Paraense Emílio Goeldi (1962–1968) and Museu Nacional, Rio de Janeiro (1972–1976). He started the Brazilian national collection of Diptera in Rio de Janeiro (1944). He went on many expeditions within Brazil and other South American countries. He wrote, among other papers: *Quinta nota sobre os tipos de Macquart (Diptera – Muscidae), existentes no Museu Nacional de História Natural de Paris e descrição de uma espécie nova, proveniente do Chile* (1951). A reptile is named after him.

Alcathoe

Duskhawker sp. *Gynacantha alcathoe* **Lieftinck**, 1961

In Greek mythology Alcathoe was a daughter of Minyas.

Alcebiades

Skimmer sp. *Elasmothemis alcebiadesi* **Santos**, 1945
[Orig. *Dythemis alcebiadesi*]

Dr Alcebíades Marques was one of the Brazilian zoologists of the Estação Experimental de Caça e Pesca, in Pirassununga. He helped and facilitated Santos during his stay at the station. He was director of the experimental fish farming station (1943–1948) and then appointed Administrator of the Pisciculture division there (1948). He took part in an expedition himself (1946) organized by Insituto Oswaldo Cruz.

Alcestis

Iridescent Flutterer *Rhyothemis alcestis* **Tillyard**, 1906
[JS *Rhyothemis braganza* Karsch, 1890]

Alcestis was a princess, daughter of Pelias, King of Iolcus in the Greek mythology, known for her love of her husband.

Alcyone

Aquamarine Bannerwing *Cora alcyone* **Selys**, 1873
[JS *Cora marina* Selys, 1869]
Midget sp. *Mortonagrion alcyone* **Laidlaw**, 1931
[Orig. *Agriocnemis alcyone*]

In Greek mythology Alcyone was the daughter of Aeolus. She was turned into a kingfisher, and many kingfisher species are of the genus Alcyon.

Alecto

Tigertail sp. *Synthemis alecto* **Lieftinck**, 1953

In Greek mythology Alecto was one of the Furies, the daughter of Gaea. Her job was to castigate moral crimes such as anger.

Alexander

Small Dragonhunter *Sieboldius alexanderi* **Chao**, 1955
[Orig. *Hagenius alexanderi*]

Dr Charles Paul Alexander (1889–1981) was an American entomologist. He attended Cornell University (1909), which awarded his BSc (1913) and his PhD (1918). He became an entomologist at Kansas University (1917–1919) then at the University of Illinois (1919–1922). He then became Professor of Entomology at the Massachusetts Agricultural College. Most interested in Diptera he described over 11,000 species and genera of flies! There is no etymology in the original Chinese description. However, Chao attended Massachusetts University (1948–1951) and named the species after the Professor there. (Also see **Margarita (Alexander)**)

Alexia

Archtail sp. *Nannophlebia alexia* **Lieftinck**, 1933

In Greek mythology, Alexis was a young man of Ephesus, beloved of Meliboea.

Alfonso

Spreadwing sp. *Lestes alfonsoi* **González-Soriano** & Novelo-Gutiérrez, 2001

Alfonso Gonzáles Figueroa (b.1988) is the son of the senior author (q.v.) who dedicated it to him "...*for his companionship and help during the collecting trips*".

Alfur

Flatwing sp. *Argiolestes alfurus* **Lieftinck**, 1956
Reedling sp. *Indolestes alfurus* Lieftinck, 1960

Named after the Alfur people in the Maluku Islands. Both species were found in these islands.

Alicia

Skimmer sp. *Elasmothemis aliciae* **González**-Soriano & Novelo-Gutiérrez, 2006

Alicia Rodríguez Palafox (1963–2003) was a Mexican entomologist specialising in wasps. She graduated with a biology degree from the University of Guadalajara (1985). She undertook her masters at UNAM (1990) and registered for her PhD there (1998) but did not complete it. She worked from 1991 to 1993 as an academic technician for a researcher at the Faculty of Sciences of UNAM and from 1998 to 1999 as an academic technician at the Chamela Biology Station, looking after the Station's entomological collection. The authors said in their etymology: "*We dedicate this species to our beloved and unforgettable friend Alicia Rodriguez Palafox, an enthusiastic and dedicated student of the Vespidae (Hymenoptera).*" She died of ovarian cancer and was survived by her husband Felipe A Noguera (who wrote her obituary in *Folia Entomológica Mexicana*) and son Filipin (b.1999). She wrote or co-wrote at least a dozen papers. She is commemorated in the names of c. twenty-five other insects, mostly beetles.

Allen, E

Small Reedling *Indolestes alleni* **Tillyard**, 1913
[Orig. *Austrolestes alleni*]

E Allen collected insects of various types around Cairns, Queensland, Australia. Most of the specimens noted as holotypes are now in BMNH and other major institutions. There was a collector called Edmund Allen also collecting in northern Queensland at the same time, but I cannot be 100% sure that they are one and the same as most references are to Mr E Allen rather than Edmund. Other documents infer that Edmund was a professional collector yet it seems more likely that E Allen was an amateur making his own collection, but sharing unusual specimens with well-known entomologists of the day. The author wrote: "*I have dedicated this species to its discoverer, Mr. E. Allen, to whom I am indebted for much valuable help in the form of careful collecting of Odonata in the Cairns district.*" He had collected other Odonata that Tillyard described such as *Lestoidea conjuncta*, upon which Tillyard comments "*A unique male taken by my friend, Mr. E. Allen of Cairns*" that Allen collected in Kuranda, Queensland, January 1908. He also has at least two beetles named after him.

Allen, G M

Rapids Clubtail *Gomphus alleni* **Howe**, 1922
[JS *Gomphus quadricolor* Walsh, 1863]

Dr Glover Morrill Allen (1879–1942) was a collector, curator, editor, librarian, mammalogist, ornithologist, scientist, taxonomist, teacher and writer. He was Librarian at the Boston Society of Natural History (1901–1927). He was employed to oversee the mammal collection at the Museum of Comparative Zoology, Harvard (1907) having been awarded his PhD (1904). He was Curator of Mammals (1925–1938) and then Professor of Zoology (1938–1942). He was keen on all vertebrates, particularly birds (editing *Auk* 1939–1942) and mammals (President of the American Society of Mammalogists 1927–1929). He made many collecting trips (1903–1931), variously to Africa, including the Harvard African Expedition to Liberia (1926), Australia, the Bahamas, Brazil, Labrador and the West Indies. He wrote a great many scientific papers and articles and a number of books. Early works include: *The Birds of Massachussetts* (1901) where he notes taking a specimen of Passenger Pigeon (1904), maybe the last recorded in the wild. He collected the clubtail holotype in New Hampshire (1907). Three birds are named after him.

Allen (Davies)

Coralleg sp. *Rhipidolestes alleni* **Wilson**, 2000

Professor David Allen Lewis Davies (1923–2003). (See **Davies**)

Alluaud

Indian Ocean Fineliner *Teinobasis alluaudi* **Martin**, 1896
[Orig. *Telebasis alluaudi*]
Eastern Whiskerleg *Paracnemis alluaudi* Martin, 1903
Highland Hooktail *Paragomphus alluaudi* Martin, 1915
[Orig. *Onychogomphus alluaudi*]

Charles A Alluaud (1861–1949) was an explorer, entomologist, botanist and naturalist who collected for the NMNH, Paris. He came from a wealthy family, living in a chateau where the painter Corot was a frequent visitor; his father was President of the Royal Porcelain Factory, Limoges. Left a fortune by his parents he was able to travel extensively (1887–1930) often with his wife Jeanne, including scientific expeditions to the Seychelles and Madagascar (1892–1893) and Kenya (1911–12) with René Jeannel (1952) (q.v.) where he collected the type specimens, as well as to Sudan (1905–1906), Tanzania (1908–1909), Morocco (1919–1924), the Ivory Coast & Niger (1930–1931), Tunisia (1935–1936), and the Canary Islands. His 24 voyages seem to have been to increase his understanding of bio-geography. He was elected President of the Entomological Society of France (1899 & 1914) and of the French Zoological Society (1909 and honorary president 1926). He published 165 'notes' on his scientific work (1886–1942) all of which were brought together in a biography of him by Jeannel (1952). Dozens of plants and animals are named after him including four reptiles, a fish and an amphibian, but particularly Coleoptera.

Alma

Odalisque *Epallage alma* **Selys**, 1879
[JS *Epallage fatima* Charpentier, 1840]

Alma is a female given name. Selys named his species from a female specimen from Persia.

Almeida

Dasher sp. *Micrathyria almeidai* **Santos**, 1945

Romualdo Ferreira D'Almeida (1891–1969) was a former colleague of Dr Santos, whose scientific jubilee was being celebrated that year (1945). He was interested in butterflies from an early age and spent every spare penny on his collection. He became a civil servant (1917) as a cleaner, then mailman (1921) working afternoons, leaving the mornings free to pursue his entomological interests. He was ignored by the scientific community in Brazil despite publishing 24 papers in French and German (from 1913) until introduced to Dr Travassos of the Oswaldo Cruz Institute (1933). He arranged for him to work in the mail room at the institute where he could use the library and other facilities and soon gave up delivering mail. Here he wrote revisions of the E*urema* genus and *Actinote* genus. Under the Institute's auspices he also took part in his first collecting expedition (1937). He worked there (ostensibly delivering mail but actually studying butterflies) until appointed (1940–1944) as an assistant in the zoology department of the Secretary of Agriculture for São Paulo State under Dr Oliverio Pinto, then transferred to Rio de Janeiro National Museum (1944–1967), which meant he could continue working on his own collection which eventually amounted to over 27,000 specimens. He wrote 112 scientific papers; perhaps his most significant work was: *Catálogo dos Papilionidae americanos* (1966). Olaf Mielke (q.v.) wrote his obituary.

Alvarenga

Damselfly sp. *Austrotepuibasis alvarengai* **Machado** & **Lencioni**, 2011

Colonel Moacyr (Moacir) Alvarenga (1914–2010) was a Brazilian Air Force officer as well as being a zoologist and entomologist; a coleopterist expert on *Erotylidae* who was associated with the Museu Nacional, Rio de Janeiro. One of his collecting companions and writing collaborators was Bokermann, whose obituary he wrote (1995). Dr José H. Leal, the Director and Curator of the Bailey-Matthews Shell Museum, said he was "...*a true old-school naturalist, during my tenure as a young graduate student at Museu Nacional in Rio, I was lucky to share many trips and social occasions with him and his late son, Luiz Carlos. Seu Moacir will be sorely missed.*" The dedication says: "...*Named in honour of the entomologist Moacir Alvarenga who collected the species (Brazil 1970) and also in recognition of his enormous contribution to the odonatological studies of senior author by gift of valuable Odonata specimens.*" Among the many taxa named after him is a dung beetle.

Alwis

Alwis's Shadowdamsel *Ceylonosticta alwisi* Priyadarshana & Wijewardana, 2016

Lyn De Alwis (1930–2006) was a zoologist who graduated from the University of Colombo, Sri Lanka, and joined (1955) the staff of the National Zoological Gardens at Dehiwela. He became the Director of the Department of Wildlife Conservation (1965–1969 & 1978–1983). His reputation as a zookeeper was such that Prime Minister Bandaranaike of Sri Lanka was persuaded by Singapore's premier, Lee Kwan Yew, to second Alwis to Singapore (1970–1972) to oversee the construction of their zoo. He founded the Young Zoologists Association in Sri Lanka. In the etymology of a reptile that honours him, he was thanked for "...*his initiative in igniting a research culture in the country leading to conservation of wildlife resources.*"

Amalia

Amalia Helicopter *Mecistogaster amalia* **Burmeister**, 1839
[Orig. *Agrion amalia*]

Amalia is a female forename. There was a St. Amalia in the sixth century, whose name probably is related to the Austrogoths royal family, the Amali, to which the famous Theoderic belonged. Burmeister evidently wanted the name to be suitable to the related species *lucretia* [l. female member of the Lucretii-clan], for which Drury had chosen a name from classical antiquity. In Rome's legend, a Lucretia had played a great role. Having been raped by the son of king Tarquinius Superbus she urged her husband and Brutus to take vengeance for her, then killed herself. Thus, the expulsion of the kings was initiated by which Rome became a republic. Burmeister's intention to follow Drury can be deduced from the fact that next to *amalia* he described a female *M. lucretia*, which he did not recognise, as *Agrion tullia* [l. female member of the Tullii-clan].

Amanda

Amanda's Pennant *Celithemis amanda* **Hagen**, 1861
[Orig. *Diplax amanda*]
Jewel ssp. *Rhinocypha tinca amanda* **Lieftinck**, 1938

Hagen does not say why he used this name – but in Latin it means 'she who must be loved' so he might have used it as a general term meaning 'Beloved Pennant'... as so often is the case a spurious apostrophe has sneaked in. Lieftinck surely had the same idea, since he uses words such as 'handsomely coloured', 'colour pattern... ...decidedly striking'.

Amata

Spreadwing sp. *Lestes amata* **Hagen**, 1862
[JS *Lestes concinnus* Hagen, 1862]
Superb Jewelwing *Calopteryx amata* Hagen, 1889
Leaftail sp. *Oligoaeschna amata* **Förster**, 1903
[Orig. *Jagoria amata*]

In Roman mythology Amata was wife of King Latinus. However, it could be that this is just Latin as 'amatus' means love/like, so they could just be species that the described or others liked.

Amazili

Amazon Darner *Anax amazili* **Burmeister**, 1839

Amazili is an Inca heroine in Jean Francois Marmontel's novel (1777) *Les Incas, ou la destruction de l'Empire du Pérou*. Amazili is also a character in the play by the German author August von Kotzebue (1761–1819) – *Die Sonnenjungfrau* (*The virgin of the sun*) (1791). No doubt Burmeister took his name from one (or both) of these sources. A genus of hummingbirds, Amazilia, is also called after Amazili.

Ambatoroa

Golden-tailed Sprite *Pseudagrion ambatoroae* **Aguesse**, 1968
Named after the type locality, Ambatoroa, Ile Sainte-Marie, Madagascar.

Ameeka

Satinwing sp. *Euphaea ameeka* **Van Tol** & Norma-Rashid, 1995

Ameeka Louise Thompson (b.1988) is the daughter of the discoverer, Professor David J Thompson. Jan van Tol kindly offered to name the species after her because her father set off for Brunei two days after she had started school. Ameeka subsequently received a first-class honours degree in biological sciences from Oxford University, an MPhil from Liverpool University and is currently (2016) in her final year of a graduate medicine course at Oxford University. She has frequently acted as a research assistant for her father's conservation work on UK dragonflies.

Amelia

Amelia's Threadtail *Neoneura amelia* **Calvert**, 1903
[A. Amelia's Firetail]
Aurora Bluetail *Agriocnemis amelia* **Needham**, 1930
[JS *Ischnura aurora* (Brauer, 1865)]

Amelia Catherine Calvert née Smith (1876–1966) was the wife (1901) of the describer of the threadtail P P Calvert (q.v.). The University of Pennsylvania awarded her BSc in biology (1899). She was the first author of the couple's joint book: *A year of Costa Rican Natural History* (1917).

Amphinome

Skimmer sp. *Oligoclada amphinome* **Ris**, 1919

In Greek mythology Amphinome is a Nereid, but also the wife of Aeson who, after Pelias had killed her husband and younger son, stabbed herself with a sword. As she lay dying she pronounced a curse against the king.

Amphion

Eastern Forktail *Amphiagrion amphion* **Selys**, 1876
[JS *Ischnura verticalis* (Say, 1840)]

In Greek mythology Amphion was the son of Zeus and Antiope.

Ampolomita

Violet Sprite *Pseudagrion ampolomitae* **Aguesse**, 1968

This is a toponym, named after the type locality Ampolomita in central Madagascar.

Amsel

Clubtail sp. *Gomphus amseli* **Schmidt**, 1961

Dr Hans Georg Amsel (1905–1999) was a German entomologist best known as a specialist in small butterflies. He was awarded a doctorate (1933) by Berlin University, his dissertation being: *The Lepidopteran fauna of Palestine. A zoogeographical-ecological-faunistic study*. He was in charge of the Entomology Department of the State Museum of Natural History, Karlsruhe (1955–1973) and even after retirement he was a long-term volunteer there. He laid the foundation of the internationally important collection of Microlepidoptera there. He initiated (1965) and jointly edited a 10-volume series on small butterflies in the Palearctic. He collected the type (1956) in Afghanistan and later undertook another expedition there (1966). He was also known for his rather controversial right-wing views which he wrote about in, for example: *Money and crime. A contribution to institutional thinking* (1965).

Amymone

Cruiser sp. *Macromia amymone* **Lieftinck**, 1952

In Greek mythology Amymone was a daughter of Danaus.

Ana

Dragonlet sp. *Erythrodiplax ana* Guillermo-Ferreira, Vilela, Del-Claro & Bispo, 2016

Ana Carolina Gonçalves Vilarinho (b.1984) is the wife of the first author Rhainer Guillermo-Ferreira (b.1985), who collected the type in Brazil. It was named for her "...*who went through a difficult moment at the time of the study.*"

Ana Clara

Threadtail sp. *Neoneura anaclara* **Machado**, 2005

Ana Clara Machado Tomelin (b.2002) is the granddaughter of the author (q.v.).

Ana Mia

Ana Mia's Shadowdamsel *Ceylonostica anamia* Bedjanič, 2010
[Orig. *Drepanosticta anamia*]

Ana Mia Bedjanič (b.2004) is the author's daughter. (See also **Bine** and **Mojca**, Matjaž Bedjanič's son and wife).

Andaman

Leaftail sp. *Oligoaeschna andamani* Chhotani, Lahiri & Mitra, 1983

This is a toponym, as the hawker is named after the Andaman Islands of India.

Ander

Longleg sp. *Anisogomphus anderi* **Lieftinck**, 1948

Dr Kjell Ernst Viktor Ander (1902–1992) was a Swedish entomologist. His PhD was awarded by the University of Lund (1939) and he became Docent (similar to Adjunct Professor) there (1939-1951). He was then a lecturer in Linköping. Among his written work is: *Die Insect Fauna des baltischen Bernstein nebst damit verknüpften zoogeographischen Problems* (1948). His main focus was grasshoppers, but he also had an interest in butterflies, bumblebees and Odonata. He sent the material (China, Hunan, G. Österlin leg.) to Lieftinck for study. He wrote several papers on Odonata, including: *Einige Odonaten aus aus Kashmir und Ostturkestan* (1944) and *Zur Verbreitung und Phänologie der boreoalpinen Odotanen der Westpaläarktis* (1940).

Anderson

Demoiselle sp. *Mnais andersoni* **McLachlan**, 1873

John Anderson (1833–1900) was a qualified physician who became Professor of Comparative Anatomy at the Medical School in Calcutta and Director of the Indian Museum there (1865). He joined an expedition to Burma and Yunnan in southwest China as naturalist (1867–1868) where he collected the type. A second expedition (1875) only collected in Burma. He wrote a zoological account of the two expeditions; the section on birds covers 233 species. R. Bowdler Sharpe gave assistance in the report's preparation by verifying the identifications. Anderson also wrote monographs of two whale genera. He was elected a Fellow of the Royal Society (1879). Four birds, three amphibians, three mammals and eight reptiles are also named after him.

Andersson

Stout Metalwing *Neurobasis anderssoni* **Sjöstedt**, 1926

Professor Johan Gunnar Andersson (1874–1960) was a Swedish geologist, polar explorer and archaeologist, affectionately called 'Kina-Gunnar' (i.e. China-Gunnar) by his compatriots. He worked as a geologist in China (1914–1924) and was involved in palaeontological and archaeological research. Among his greatest achievements was the discovery of the Neolithic Yangshao-culture in Henan with its unique painted pottery. He also played an important part in the discovery of Peking Man (*Homo erectus pekinensis*). Andersson's popular book *Children of the Yellow Earth* (1932/1934) describes his stay in China during which (1921) he collected the type. He founded the Museum of Far Eastern Antiquities in Stockholm. Several animal species have been named after him, including a Chinese fossil rhinoceros and an ostrich.

Andres

Hawker sp. *Andaeschna andresi* **Rácenis**, 1958
[Orig. *Aeshna andresi*]

Andrés Rácenis (1949–1970) was the 'dear son' of the author. (See also **Rácenis**)

Andromache

Pond Damsel sp. *Leptagrion andromache* **Hagen**, 1876

In Greek mythology Andromache was the wife of Hector.

Andromeda

Leaftail sp. *Phyllogomphoides andromeda* **Selys**, 1869
[Orig. *Cyclophylla andromeda*]

In Greek mythology Andromeda was the daughter of Cepheus and Cassiopeia.

Anduze

Forceptail sp. *Phyllocycla anduzei* **Needham**, 1943
[Orig. *Cyclophylla anduzei*]

Dr Pablo José Anduze (1902–1989) was a Venezuelan entomologist. He studied at the Maison de Melle, Gent, Belgium and the Saint George's College, Weybridge, England. His interest in arts and literature gave way to an increasing interest in science particularly medical entomology, bacteriology, parasitology and zoology. He then studied (1938) at the School of Agriculture and the Faculty of Veterinary Medicine at Cornell University (United States) and at the Gorgas Institute in Panama. On his return to Venezuela he became Head of the Department of Entomology of the Yellow Fever Prophylaxis Service (1940), then Head of Entomology, National Institute of Hygiene (1941). He was a member of the National Commission for the study of onchocerciasis and Chief Medical Entomologist at the Creole Petroleum Corporation (1946–1951). The Academy of Physics, Mathematics and Natural History made him a fellow (1947). He became head of the Commission of Zoology (1950) – the Franco-Venezuelan Expedition that determined the location of the sources of the Orinoco. He wrote about his experience as: *Shailili-ko*: the story of his search for the sources of the Orinoco, his scientific observations and the most comprehensive study ever done on the Yanomami people. He collected the holotype in Venezuela (1939).

Angel

Murray River Hunter *Austrogomphus angelorum* **Tillyard**, 1913
[Orig. *Austrogomphus angeli*]

Frank Milton Angel (1876–1969) and Sidney Percival Angel (1876–1966) of Adelaide were twin brothers who collected the type series in South Australia (1909). Their entomology collection is in the South Australia Museum.

Angela

The damselfly genus *Angelagrion* **Lencioni**, 2008

Damselfly sp. *Cyanallagma angelae* Lencioni, 2001

Angela Schmidt Lourenço Rodrigues (b.1967) is the wife of the author. He dedicated the species "*...for her continuous support and encouragement during my research*".

Angela

Intermediate Claspertail *Onychogomphus angela* **Martin**, 1915
[JS *Onychogomphus nigrotibialis* Sjöstedt, 1909]

Etymology not given, but Martin had also a manuscript name '*yvonna*' for a *Mnais* species. It is not clear whether these names are after people but the lack of a Latin ending suggests that they are not.

Angelina

Shadowdamsel sp. *Palaemnema angelina* **Selys**, 1860
Skimmer sp. *Libellula angelina* Selys, 1883
Yellow-fronted Sprite *Pseudagrion angelicum* **Fraser**, 1947
[JS *Pseudagrion camerunense* (Karsch, 1899)]

This is not an eponym, but a reference to the angelic qualities of the species.

Angelo

Sanddragon sp. *Progomphus angeloi* **Belle**, 1994
Threadtail sp. *Epipleoneura angeloi* Pessacq & **Costa**, 2010
Damselfly sp. *Psaironeura angeloi* **Tennessen**, 2016

Angelo Barbosa Montiero Machado (b.1934). Belle wrote in his etymology: "*I name this species in honour of my highly-esteemed friend Prof. Dr Angelo B.M. Machado, who has generously sent to me – two decennia long – numerous Brazilian Gomphidae for identification and eventual description. Thanks to his continuous attention and field work a large part of the gomphid fauna of Brazil could be disclosed.*" (See **Machado**)

Angelo Machado

Clubtail sp. *Cyanogomphus angelomachadoi* Pinto & De Almeida, 2016

(See Above, also see **Machado**)

Anna

River Bluet *Enallagma anna* **EB Williamson**, 1900

Anna Tribolet (1876–1950) was the wife (1902) of the author Edward Bruce Williamson (q.v.).

Anna

Limniad sp. *Amphicnemis annae* **Lieftinck**, 1940

This eponym was not explained in the etymology and there seems to be no family links or links to relatives of his collectors. So, it remains a mystery!

Anna Karl

Grabtail sp. *Lamelligomphus annakarlorum* Zhang, Yang & Cai, 2016

Anna Diehl (b.1942) and Dr Karl Schorr (1936–2012) were life partners, the dragonfly name is a joint dedication. Karl Schorr (uncle to Martin Schorr) (q.v.) was a physicist interested in mineralogy, but he was also a keen naturalist interested in fauna, especially bats. For many years, he also maintained an interest in dragonflies, collecting photographs of each of the German species. Anna was his companion, life partner and great love. Naming a dragonfly for them celebrates their great love.

Annaliese

Redspot sp. *Austropetalia annaliese* **Theischinger**, 2013

Annaliese Jones (b.2001) is the author's granddaughter.

Anna-Maija

Fineliner sp. *Teinobasis annamaijae* **Hämäläinen & Müller**, 1989

Anna-Maija Müller née Kaltula (b.1951) is the widow (m.1972) of the second author R. A. Müller (q.v.). She was born in Finland, but moved to Switzerland (1969).

Annandale

Fiery Gem *Libellago annadali* **Laidlaw**, 1903
[JS *Libellago aurantiaca*, Selys, 1859]
Bluetail sp. *Ischnura annandalei* Laidlaw, 1919
Goldenwing sp. *Cordulegaster annandalei* **Fraser**, 1924
[Orig. *Anotogaster annandalei*]
Shadowdamsel sp. *Drepanosticta annandalei* Fraser, 1924

Threadtail ssp. *Prodasineura verticalis annandalei* Fraser, 1921
[Orig. *Caconeura annandalei*]

Dr Thomas Nelson Annandale (1876–1924) was a Scottish zoologist. He went to India and became Deputy Superintendent of the Natural History Section (1904) then Director of the Indian Museum, Calcutta (1907). He started the Records and Memoirs of the Indian Museum journals and (1916) became the first Director of the Zoological Survey of India that he helped found. He wrote a number of scientific papers (1900–1930) including: *Fauna of the Chilka Lake: mammals, reptiles and batrachians* (1915). He was instrumental in establishing a purely zoological survey, not combined with anthropology, undertaking several expeditions, most notably the Annandale-Robinson Expedition that collected, including the types, in Malaya (1901–1902) and Burma. A mammal, six amphibians and four reptiles are named after him as well as other taxa.

Anne

Flashwing sp. *Vestalis anne* **Hämäläinen**, 1985

Since Selys had selected the genus name *Vestalis* after the ancient Roman Vestal virgins, the author wanted to select a woman's name for his new species. *Anne* was chosen for various reasons, allegedly because most other species names in this genus start with the letter *a*. No etymology is given in the original description but the author assured me it is NOT after any specific person.

Annika

Annika's Cascader *Zygonyx annika* **Dijkstra**, 2015

Dr Annika Hillers (b.1977) is an ecologist and conservationist currently working for the Wild Chimpanzee Foundation, based in Liberia. Her PhD was awarded (2008) by the University of Amsterdam after which (2009) she undertook post-doctoral research at the Museum für Naturjunde, Berlin and the Muséum National d'Histoire Naturelle, Paris. She then worked for the RSPB (2010–2017) in the Gola Rainforest NP, Sierra Leone. Among her c.35 written papers is: *A mix of community-based conservation and protected forests is needed for the survival of the endangered pygmy hippopotamus Choeropsis liberiensis* (2016). The etymology states it was: "*…named in honour of Dr Annika Hillers, who advanced research on the Gola rainforest ecosystem and enabled the author to make his contribution*".

Annina

Demoiselle sp. *Matrona annina* Zhang & **Hämäläine**n, 2012

Selys' genus name 'Matrona' refers to the Latin word for a respectable married woman. The authors selected the feminine name 'Annina' as it sounds well with the genus name. Annina is a diminutive form of 'Anna', which derives from the Hebrew 'Hannah' and the second author assured us it is NOT after any specific person.

Antia

Archtail sp. *Nannophlebia antiacantha* **Lieftinck**, 1963

This is a family name from Roman history.

Antigone

Pond Damselfly sp. *Antiagrion Antigone* **Ris**, 1928
Skimmer sp. *Diplacina antigone* **Lieftinck**, 1933

In Greek mythology Antigone was the daughter of Oedipus and Jocasta.

Antónia

Damselfly sp. *Igneocnemis antoniae* Gassmann & **Hämäläinen**, 2002
[Orig. *Risiocnemis antoniae*]

Dr Antónia Fraser Monteiro (b.1969) is Assistant Professor at Yale (2006). Edinburgh University awarded her PhD (1997) following which she undertook Postdoctoral work at Harvard University (1997–1998) and Leiden University (1998-2001). She became assistant professor at the University at Buffalo (2001–2006). Much of her work involves the wings of butterflies as evidenced in her many publications such as: *Butterfly wings: Colour patterns and now gene expression patterns* (1994). Dirk Gassman selected the name, writing: *"The new species is named after Dr. Antonia Monteiro (Leiden University), who was especially struck by the red legs of the male of the present species."*

Anu-Mari

Fairy Metalwing *Neurobasis anumariae* **Hämäläinen**, 1989

Anu-Mari Tanskanen née Hämäläinen (b.1988) is the describer's daughter – her father named the demoiselle in celebration of her first birthday, with the description being published on the same day. Anu-Mari received a BA degree from the University of Helsinki (2012), an MA (majoring in the Finnish language, 2015) and is presently working as a Finnish language teacher. She has accompanied her father on several dragonfly-hunting expeditions to Thailand. (See **Hämäläinen**)

Aphrodite

Blue Jewel *Chlorocypha aphrodite* **Le Roi**, 1915

In Greek mythology Aphrodite was goddess of love.

Apolinar Maria

Clubskimmer sp. *Nothemis apolinaris* **Navas**, 1915
[JS *Brechmorhoga rapax rapax* Calvert, 1898]
Spreadwing sp. *Lestes apolinaris* Navas, 1934

Brother Apolinar María (1867–1949) was a missionary Colombian monk and ornithologist. He was Director of the Institute La Salle in Bogota (1914). He collected the type in Colombia.

Apollo

Apollo Redspot *Phyllopetalia apollo* **Selys**, 1878

Apollo was an ancient Greek and Roman God.

Apomyius

Banner Clubtail *Gomphus apomyius* **Donnelly**, 1966

A name used by Zues at Olympia… when Heracles was offering a sacrifice to Zeus at Olympia, he was annoyed by hosts of flies, and in order to get rid of them, he offered a sacrifice to Zeus Apomyius, whereupon the flies withdrew across the river Alpheius. From that time, the Eleans sacrificed to Zeus under this name. Donnelly: *"The specific name, meaning, 'one who drives away flies', is very appropriate for this small but aggressive dragonfly".*

Arachne

Scarlet Spiderlegs *Planiplax arachne* **Ris**, 1912

Flatwing sp. *Megapodagrion arachne* **Rácenis**, 1959
[JS *Megapodagrion megalopus* (Selys, 1862)]

In Greek and Roman mythology Arachne was a great mortal weaver who boasted that her skill was greater than that of Athena, goddess of wisdom, weaving, and strategy.

Arara

Pond Damselfly sp. *Tuberculobasis arara* **Machado**, 2009

The Arara are a people who inhabit the municipality of Ji-Paraná.

Arakawa

Slim sp. *Nehalennia arakawai* **Matsumura**, 1931
[JS *Aciagrion migratum* (Selys, 1876)]

Although there is no etymology given it is evident that this species was named after a Mr Arakawa who co-operated with the author. He is from Iyo (Ehime prefecture, Shikoku Island), which is the type locality.

Aran

Grisette sp. *Devadatta aran* **Dow, Hämäläinen** & Stokvis, 2015

Reddish Aran (1963–2012) is one of the Kelabit people and the owner of a tourist lodge in Bario in the Tama Abu Range in Sarawak, an area sometimes referred to as the Kelabit Highlands. His first name was pronounced 'Raddish'. He accompanied the first author on his trip to the type locality. Very unfortunately he died an untimely death in a car accident outside the city of Miri.

Arba

Threadtail sp. *Disparoneura arba* **Krüger**, 1898
[JS *Prodasineura verticalis delia* Karsch, 1891]

Arba was a man mentioned in early, Old Testament verses of the Bible.

Archbold, J D

Clubskimmer sp. *Brechmorhoga archboldi* **Donnelly**, 1970
[Orig. *Scapanea archboldi*]

John Dana Archbold OBE (1910–1993) was a philanthropist, engineer, conservationist, businessman, yatchsman and agriculturalist, heir to Standard Oil who spent his childhood on an English dairy farm. His cousin was the zoologist Richard Archbold (1907–1976) (q.v.). He was educated AT Princeton University (1934) and the University of Geneva, Switzerland. He served in the US Navy (WW2) in Southeast Asia attached to the OSS. He donated (1979) to The Nature Conservancy a 950-acre tract of rainforest on Dominica that he had bought earlier and worked as a plantation (1935) and which they manage in conjunction with Clemson University. He was a Director Emeritus of The Nature Conservancy, a trustee emeritus of Syracuse University, and a fellow of the Virginia Museum of Fine Arts.

Archbold, R A

The Featherleg Genus *Archboldargia* **Lieftinck**, 1949

Richard A Archbold (1907–1976) was an independently wealthy American philanthropist who became a zoologist at the AMNH. He was educated privately then attended classes at Columbia University but never graduated. He was a member of the (1929–1931) Madagascar Expedition (funded by his father). He went on to finance and lead expeditions, particularly to New Guinea (1933–1934, 1936–1937 & 1938–1939). At times he was accompanied by, among others, R & G H H Tate (q.v.). He also set up a permanent research station at Lake Placid in Florida. Lieftinck wrote: "*The name I have chosen for this interesting genus, it will be observed, is an allusion to that of Mr. RICHARD ARCHBOLD, the leader of the American-Dutch expedition to the Snow Mountains of New Guinea.*" Eight birds, twenty-six insects, three spiders, a fish, an amphibian and a mammal are also named after him. (See **Richard**)

Arethusa

Archtail sp. *Nannophlebia arethusa* **Lieftinck**, 1948

In Greek mythology Arethusa was a nymph.

Ariel

Bluetail sp. *Ischnura ariel* **Lieftinck**, 1949
Fineliner sp. *Teinobasis ariel* Lieftinck, 1962

No etymology is given so I assume they are named after the folk legend and fictional fairy of Shakespeare's *The Tempest*.

Ariadne

Tigertail sp. *Synthemis ariadne* **Lieftinck**, 1975

In Greek mythology Ariadne was the daughter of Minos king of Crete and his queen Pasiphaë, daughter of Helios, the Sun-titan.

Ariken

Pond Damselfly sp. *Denticulobasis ariken* **Machado**, 2009

The Ariken Indian people inhabited the municipality of Ariquemes, Rondônia, and their descendants live now scattered throughout the region.

Arita

Duskhawker sp. *Cephalaeschna aritai* **Karube**, 2003
Skydragon sp. *Chlorogomphus aritai* Karube, 2013

Dr Yutaka Arita (b.1941) works as Honorary Professor at the Zoological Laboratory, Faculty of Agriculture, Meijo University, Nagoya, Japan. He (and Tetsup Miyashita) gave Karube material from Vietnam 'collected by natives' (2001–2002), which included the holotype. He has written or co-written twenty-eight papers and longer works naming forty-three taxa (1979–2016), mostly on Lepidoptera from: *A survey of the Japanese species of Anthophila*

Haworth and Eutromula Froelich and their early stages (Lepidoptera, Choreutidae) (1979) to *A remarkable new species of the genus Teinotarsina (Lepidoptera, Sesiidae) from Okinawa-jima, Japan (2016).*

Armageddon

Forest Jewel *Chlorocypha armageddoni* **Fraser**, 1940
[JS *Platycypha lacustris* (Förster, 1914)

Named after the place supposed to be mankind's nemesis.

Armstrong

Pond Damselfly sp. *Amorphostigma armstrongi* **Fraser**, 1926
Flutterer ssp. *Rhyothemis chalcoptilon armstrongi* Fraser, 1956
[JS *Rhyothemis chalcoptilon* Brauer, 1867]

Dr John Scaife Armstrong (1892–1977) was a physician who was New Zealand's Resident Medical Health Officer at the Government Hospital, Apia, Western Samoa. Here he collected the types (1923 & 1924). He was noted for his treatment for Yaws and (1924) reported on the history of Leprosy in Samoa.

Arnaud

Duskhawker sp. *Gynacantha arnaudi* **Asahina**, 1984

Dr Paul Henri Arnaud Jr (b.1924) is an American entomologist. He was a staff sergeant in the Corps of Engineers (1944–1946). San Jose State University awarded his first degree (1949) and Stanford University his masters (1950) and PhD (1961). After various posts as entomologist he became assistant curator (1964–1965), then associate curator (1965–1971) before becoming Curator of Entomology, California Academy of Sciences, San Francisco (1972–1995) and has since been Curator Emeritus. He also became Research Associate, Department of Entomology, National Museum of Natural History, Smithsonian Institution, Washington (2007). He made a number of collecting trips in California, Mexico, Japan and Switzerland and was on the Orca Expedition to Baja California and spent time at various museums and educational establishments in Europe, Canada and Japan. He has written more than 200 papers many describing over 100 new taxa. He is honoured in the names of many invertebrates.

Arnoult

Spotted Junglewatcher *Neodythemis arnoulti* **Fraser**, 1955

Jacques Arnoult (1914–1995) was an ichthyologist and herpetologist. He graduated in biology, agriculture and hydrobiology at Université de Toulouse. He was in charge of zoological research at the Institute of Scientific Research, Madagascar (1951) and became an assistant in the Department of Reptiles and Fish, Muséum National d'Histoire Naturelle, Paris (1954). He collected frequently in Africa and was notably successful in getting his live reptile specimens to breed in captivity. He was Director of the Aquarium, Monaco (1968–1981). His publications were mostly about fish such as *Fauna of Madagascar 10, Freshwater Fish* (1959), and latterly aquaria. A reptile is named after him.

Aroon

Skydragon sp. *Chlorogomphus arooni* **Asahina**, 1981

Aroon Samruadkit (d.2001) collected the holotype in the Khao Poh Ta Mountains in southern Thailand (1980). He was a very well-known butterfly collector working as a technician in the Entomology Section of the Department of Agriculture in Bangkok. He was one of the co-authors of *Field Guide to Butterflies of Thailand* (1977) illustrating 630 Thai butterflies and it is said that most of the specimens for the book were collected by him. He supplied Asahina with specimens including the holotype of at least three Odonata that Asahina described including this eponymous one. Three butterfly subspecies have been named after him.

Arria

Flutterer ssp. *Rhyothemis variegata arria* Drury, 1773
[Orig. *Libellula arria*]

Arria was an ancient Roman. According to Plinney the Younger, her husband Caecina Paetus was ordered by the emperor Claudius to commit suicide for his part in a rebellion, but was not capable of forcing himself to do so. Arria wrenched the dagger from him and stabbed herself, then returned it to her husband, telling him that it didn't hurt – '*Non dolet, Paete!*'.

Arses

Damselfly sp. *Palaiargia arses* **Lieftinck**, 1957

Arses was an ancient Persian King.

Arsinoe

Duskhawker sp. *Gynacantha arsinoe* **Lieftinck**, 1948
Skimmer sp. *Diplacina arsinoe* Lieftinck, 1953

Arsinoe was a character in Greek myth, one of the Nysiads or Minyades.

Artemis

Artemis Dasher *Micrathyria artemis* **Ris**, 1911
Emerald sp. *Procordulia artemis* **Lieftinck**, 1930

In Greek mythology Artemis was a daughter of Zeus.

Arthur (Wegner)

Clubtail sp. *Burmagomphus arthuri* **Lieftinck**, 1953
Duskhawker sp. *Gynacantha arthuri* Lieftinck, 1953

Dr Arthur M R Wegner (1894–1969) was a German born naturalised Dutch zoologist, entomologist and insect collector (1930s–1970s). He became a merchant in Indonesia (1932) but spent his free time studying nature. He built a private museum in Nongkojajar in the Tengar Mountains, East Java, exhibiting butterflies, birds, snakes and mammals that he had collected, and a small zoo and bee farm and applied for Dutch citizenship. Nevertheless, he was interned at the outbreak of war but during the Japanese occupation he lost all his personal belongings as well as the museum. He moved to Bogor, Java (1947) bringing his curator with him. He was Assistant Curator at Museum Zoologicum Bogoriense then Director (1955–1960) retiring to Ambon. Lieftinck worked in the same museum (1929–1954). Gilbert Church, in a paper of *Bufo melanostictus*, said of him as a collector (1959) *"...The enthusiasm and optimism of Mr A. M. R. Wegner, Zoological Museum, Bogor, is always a source of inspiration."* He collected the type series of both eponymous species (Burma 1950 & Sumba Island 1949) and took other trips such as one he led to North Maluku (1953). He wrote a number of papers such as: *On a collection of Rhopalocera from Panaitan Island, with description of a new species* (1953). A butterfly genus is named after him, which he collected in a cave he had discovered while insect collecting (1938) as well as other taxa, particularly insects.

Arthur (Wheeler)

Arthur's Midget *Mortonagrion arthuri* **Fraser**, 1942

Arthur Wheeler (b.1931) when aged only four years old, 'collected' the (male) holotype in his garden in Butterworth, Penang, Malaysia (November 1935). He was the son of Raymond Wheeler (q.v.). Fraser wrote that it was *"...surely a record in the annals of entomology."*

Asahina

Shadowemerald sp. *Macromidia asahinai* **Lieftinck**, 1971
Clubtail sp. *Sinogomphus asahinai* **Chao**, 1984
Damselfly sp. *Risiocnemis asahinai* Kitagawa, 1990
Cascader sp. *Zygonyx asahinai* **Matsuki** & **Saito**, 1995
Clubtail sp. *Perissogomphus asahinai* **Zhu**, Yang & **Wu**, 2007
Shadowdamsel sp. *Drepanosticta asahinai* **Sasamoto** & **Karube**, 2007
Emerald sp. *Procordulia asahinai* Karube, 2007
Shadowdancer sp. *Idionyx asahinai* Karube, 2011
Duskhawker sp. *Cephalaeschna asahinai* Karube, 2011
Phantomhawker sp. *Planaeschna asahinai* Karube, 2011
Clubtail sp. *Burmagomphus asahinai* **Kosterin**, Makbun & Dawwrueng, 2012

Four-Spotted Chaser *Libellula quadrimaculata asahinai* **Schmidt**, 1957
[A. Four-Spotted Skimmer]

Dr Syoziro Asahina (1913–2010) was a Japanese odonatologist, medical entomologist and an expert on taxonomy of oriental dragonflies and cockroaches. He graduated in zoology at the University of Tokyo (1938). Thereafter he joined the Research Institute of Infectious Diseases at Tokyo University. During WW2, he was obliged to work as a civil servant in Manchuria. After five years of unemployment he was appointed at the National Institute of Health, Tokyo (1950) and worked there until retirement (1979), most of the time as Chief of the Department of Medical Entomology. He did fieldwork in many Asian countries. His interest in dragonflies started in his school days and was intensified by a visit to Okinawa Island (May 1931) when he saw huge black-winged females of *Chlorogomphus brunneus* flying high overhead. Hokkaido University awarded his DSc (1953); his thesis *A morphological study of a relic dragonfly Epiophlebia superstes Selys (Odonata, Anisozygoptera)* was published as a book (1954). He founded (1958) the Japanese Society of Odonatology and the Society's journal, 'TOMBO' (Acta Odonatologica), the world's first scientific journal devoted to odonatology. Asahina's entomological bibliography includes nearly 1000 titles (1928–1998), most being short papers or notes on dragonflies of Japan and other East and South-East Asian countries. He also wrote a book on cockroaches: *Blattaria of Japan* (1991). He described c.150 species and subspecies of dragonflies and named six genera. In addition to Odonata c.40 species of several insect orders have been named after him. (Also see **Terue**)

Asato

Coralleg sp. *Rhipidolestes asatoi* **Asahina**, 1994

Clubtail ssp. *Stylogomphus ryukyuanus asatoi* Asahina, 1972

Susumu Asato (b.1947) is an amateur entomologist who collected the clubtail in Japan (1967). By profession he is an arts and cultural historian and archaeologist and has published a number of such papers. After graduating from Ryukyu University he worked in the private sector before a post in the Protection of Cultural Properties Division of Osaka (1979) then for their Board of Education (1988) rising to Chief of the Cultural Section (2003–2006) in Urasoe city. He then became (2006) a professor at Okinawa Prefectural University of Arts. He donated his Odonata collection to the Nago Museum, Okinawa. He wrote: *The Odonata of the Okinawa islands. 1. Odonata from Tokashiki-shima and iheya-shima* (1968).

Asclepiades

Gem sp. *Libellago asclepiades* **Ris**, 1916
[Orig. *Micromerus asclepiades*]

Asclepiades (c.124/129–40BC) was a Greek physician born at Prusa in Bithynia in Asia Minor and flourished at Rome, where he established Greek medicine.

Askew

Flatwing sp. *Celebargiolestes askewi* **Kalkman**, 2016

Dr Richard Robinson Askew (b.1935) is an English entomologist, formerly a Reader in Entomology at the Department of Zoology, Manchester University. His first-class degree in zoology was awarded by Kings College, Durham (1957); he then studied for his doctorate, which was awarded by the Hope Department of Entomology at Oxford University (1960). He took up a post as lecturer at Manchester University becoming reader (1972). His professional research has mostly been into the biology and taxonomy of parasitic Hymenoptera about which he has written around 150 papers. However, his spare time activity has been the study of Odonata. He has travelled widely including Japan, Indonesia, Siberia and the West Indies. He has written numerous works from the early 1960s through to the present day including a number of books such as: *Parasitic Insects* (1971), *The Dragonflies of Europe* (1988) and the co-written: *Butterflies of the Cayman* Islands (2008). He illustrated all his works including the 29 plates in his *Dragonflies of Europe*. In retirement, he lives in France and leads nature tours there.

Aspasia

Satinwing sp. *Euphaea aspasia* **Selys**, 1853

Aspasia was a Milesian woman who was famous for her involvement with the Athenian statesman Pericles. Very little is known about the details of her life. She spent most of her adult life in Athens, and she may have influenced Pericles and Athenian politics.

Astami

Midget sp. *Mortonagrion astamii* **Villanueva** & Cahilog, 2013

Shuaib Juaini Astami has been the Mayor of the Municipality of Balabac in the province of Palawan, Philippines for most of the last two decades (1998–2017). He was honoured for approving and facilitating the odonatological survey conducted in his jurisdiction. He owns Onuk Island, one of the 31 islands of Balabac.

Astarte

Cruiser sp. *Macromia astarte* **Lieftinck**, 1971

Astarte was a Greek name for a Mesopotamian goddess.

Astrape

Demoiselle sp. *Mnesarete astrape* **De Marmels**, 1989

In Greek mythology Astrape was a Goddess.

Astrid

Emerald sp. *Procordulia astridae* **Lieftinck**, 1935

Princess Astrid of Sweden, born Astrid Sofia Lovisa Thyra Bernadotte (1905–1935) became Queen of the Belgians (1934) having married (1926) Leopold III. Her youngest son, Albert was King of the Belgians until abdicating (2013). She died in a car accident (1935) in Switzerland. Her husband was driving and was only slightly injured; their chauffeur was sitting in the back and was unscathed. Tragically Astrid was pregnant and Albert aged just one. There is no etymology in the original description, but as Astrid died on 29 August 1935 and 'Treubia' 15(2) was printed in November 1935 I am confident this is who Lieftinck had in mind. Especially as in the paper the species is compared with *Procordulia leopoldi* – a species named after King Leopold III.

Athalia

Hawker sp. *Aeshna athalia* **Needham**, 1930
[Orig. *Aeschna athalia*]

Athalia is a feminine forename originating of a historical person Athaliah who was queen consort to King Jehoram of Judah, and later queen regnant of Judah for six years (842–837BC).

Athenais

Dasher sp. *Micrathyria athenais* **Calvert**, 1909

Athenais was a 4th century BC prophetess who told Alexander the Great of his allegedly divine descent.

Attenborough

Attenborough's Pintail *Acisoma attenboroughi* Mens, Schütte, Stokvis & Dijkstra, 2016

Sir David Frederick Attenborough (b.1926) is famous as a maker of wildlife television programmes. He studied natural sciences at Cambridge and joined a firm of publishers (1950), where he did not stay long before joining the BBC in the early days of its postwar television service. He has been associated with the BBC, first as an employee and later as a freelance journalist, virtually ever since. He rose high in the organisation's ranks, becoming controller of BBC2 and responsible for introducing colour television to Britain, yet his first love was not administration but photojournalism. He has made some of the most stunning series of nature programmes and produced excellent books to accompany them, such as *Life in Cold Blood* (2008). Two reptiles, an amphibian, a mammal and a bird and are also named after him, among other taxa.

Atkinson

Skydragon sp. *Watanabeopetalia atkinsoni* **Selys**, 1878
[Orig. *Orogomphus atkinsoni*]
Threadtail sp. *Elattoneura atkinsoni* Selys, 1886
[Orig. *Disparoneura atkinsoni*]
Oread sp. *Calicnemis atkinsoni* Selys, 1886
[JS *Calicnemia eximia* (Selys, 1863)]

William Stephen Atkinson (1820–1876) was a British entomologist who spent most of his working life in India. Trinity College, Cambridge awarded his degree (1843) following which he studied engineering, but was offered the post of Principal at Martiniere College in Calcutta. He became interested in moths and joined the Entomological Society (1857). He became Director of Public Instruction in Bengal (1860) and visited Darjeeling and Sikkim where he collected Lepidoptera. He was also a fine artist, painting many of the species he collected. His collection was purchased on his death and sent to the BMNH.

Attala

Black Pondhawk *Erythemis attala* **Selys**, 1857
[Orig. *Libellula attala*]
Shadowdamsel *Drepanosticta attala* **Lieftinck**, 1934

Attala (d.627) was a monk who became St Attala.

Âu Cơ'

Demoiselle sp. *Atrocalopteryx auco* **Hämäläinen**, 2014

Âu Cơ is a character in Vietnamese mythology; a young, beautiful mountain fairy who fell in love with Lac Long Quân (the Dragon Lord of Lac). They married and she gave birth to an egg sac from which hatched a hundred children known collectively as Bach Viet, the ancestors of the Vietnamese people. Âu Cơ is widely honoured as the mother of Vietnamese civilization. The author found this demoiselle species in northern Vietnam (June 2009).

Auca

Hawker sp. *Staurophlebia auca* **Kennedy**, 1937

This species was named after an indigenous tribe in Ecuador.

August

Flatwing sp. *Philogenia augusti* **Calvert**, 1924

August Busck (1870–1944) was a Danish entomologist who collected the type series in Panama (1911–1912). Ordrup College awarded his BA and the Royal University in Copenhagen his MA (1893). He became an American citizen after a trip to the World's Columbian Exposition in Chicago (1893). He opened a flower business in Charleston, Virginia but then worked for the US Department of Agriculture (1896–1940) at the Department of Entomology, NMNH. He collected widely in the USA and central America, particularly Panama. His friend, Lord Walsingham, invited him to England (1908) to help work on the *Biologia* volume dealing with the micro-Lepidoptera of Central America, the group for which he is best known as he described over 600 (mostly American) species. He published a great many papers such as: *A New American Tineina* (1900) and *Stenomidae* (1934).

Aurinda

Skimmer sp. *Garrisonia aurindae* Penalva & **Costa**, 2007

Aurinda Ramos Penalva (1925–1995) was the mother of the senior author Ruy Penalva de Faria Neto (b.1951).

Aurivillius

Horntail sp. *Tragogomphus aurivillii* **Sjöstedt**, 1900

Dr Per Olof Christopher Aurivillius (1853–1928) was a Swedish entomologist. His doctorate was awarded by Uppsala University (1880). He worked in the entomology department of the Natural History Museum in Stockholm (1881) becoming Director and also Professor there (1893). He specialised in Coleoptera, Lepidoptera and wasps. He was also long time Secretary of the Royal Swedish Academy of Science (1901). He wrote many papers, particularly on African Lepidoptera. He set up the National Entomological Institute (1890s). His brother was also a prominent naturalist and his son a zoologist. A bird is named after him.

Austen

Amberwing sp. *Perithemis austeni* **WF Kirby**, 1897
[JS *Perithemis bella* Kirby, 1889]
Giant Skimmer *Orthetrum austeni* WF Kirby, 1900
[Orig. *Thermorthemis austeni*]

Major Ernest Edward Austen (1869–1938) was a British entomologist. After education at Rugby School and the University of Heidelberg, he entered the service of the Trustees of the British Museum as Second Class Assistant (assistant keeper) in the entomological section of the Department of Zoology (1899), becoming Keeper (1927–1932). He was placed in charge of the Diptera to which he devoted his scientific life. He was with the first expedition of the Liverpool School of Tropical Medicine to Sierra Leone (1899) where he collected the skimmer, and was a member of the council as well as Vice-President of the Royal Society of Tropical Medicine and Hygiene. He collected the amberwing in Brazil (1896). He wrote around 150 papers and several longer works, most notably: *Handbook of the Tsetse Flies* (1923).

Austin

Austin's Shadowdamsel *Ceylononosticta austeni* **Lieftinck**, 1940
[Orig. *Drepanosticta austeni*]

G Douglas Austin was a government economic entomologist in Sri Lanka often deputed to plantations to study insect pest infestations, such as the nettle grub which infests tea plants. He was at the Tea Research Institute Substation at Passara, Ceylon (Sri Lanka). The author received from him "...*much courtesy and help during our tour*". He wrote a number of papers published in the 'Bulletin of the Department of Agriculture', Ceylon. The spelling of the binomial appears is incorrect, but cannot be amended.

Awamena

Highland Metalwing *Neurobasis awamena* Michalski, 2006

This is a tribute to the Awamena people of the Foi tribe, who inhabit the region in which the species was first collected.

Aztec

Aztec Glider *Tauriphila azteca* **Calvert**, 1906

Named after the native South American people and culture.

Azupizu

Shadowdamsel sp. *Palaemnema azupizui* **Calvert**, 1931

This is a toponym, it is a region of Peru.

B

Bacchus

Lesser Emperor *Anax bacchus* **Hagen**, 1867
[JS *Anax parthenope* (Selys, 1839)]
Darter sp. *Sympetrum baccha* **Selys**, 1884
[Orig. *Thecadiplax baccha*]

Bacchus is the Roman name for Dionysus, the god of wine and intoxication.

Bainbrigge

Duskhawker sp. *Gynacantha bainbriggei* **Fraser**, 1922

Thomas Bainbrigge Fletcher (1878–1950) (See **Fletcher, TB**)

Baker, C F

Grappletail sp. *Heliogomphus bakeri* **Laidlaw**, 1925
Giant Riverhawker sp. *Tetracanthagyna bakeri* **Campion**, 1928

Charles Fuller Baker (1872–1927) was an American naturalist, entomologist, botanist, agronomist and collector. He trained at the Michigan Agricultural College and was then employed as Assistant Entomologist at the Colorado Agriculture Experiment Station in Fort Collins. He collected widely in Brazil, Cuba, Malaysia, Mexico, Nicaragua, USA, Singapore and the Philippines when he moved there (1912) and became Professor and Dean of the College of Agriculture at Los Baños. He was also a staff member of the Botanic Gardens in Singapore and temporarily appointed acting Assistant Director (1917). Both species types were taken at Luzon. They, with others, were sent to BMNH and a hawker paratype to the Smithsonian (1928). He bequeathed his mycological herbarium collection to the Philippines National Herbarium, but it was destroyed during the Japanese occupation in WW2. Several plants are named after him including two genera.

Baker, S

Blue-fronted Citril *Ceriagrion bakeri* **Fraser**, 1941

Sir Samuel White Baker (1821–1893) was a British explorer, army officer, naturalist, big game hunter, engineer, writer and abolitionist. He held the title of Pasha in the Ottoman empire and Major-General in Egypt. He was Governor General (1869–1873) of the Equatorial Nile Valley (South Sudan and Northern Uganda). His most famous discovery was Lake Albert. He had an MA in civil engineering and also farmed in Ceylon (Sri Lanka) about which he wrote: *The Rifle and the Hound in Ceylon* (1853) and *Eight Years' Wanderings in Ceylon* (1855). Widowed at the age of 34 he went to Constantinople and Crimea (1856) and while acting as Royal Superintendent he constructed railways and bridges connecting the Danube with the

Black Sea. His first tour of Africa was exploring for the source of the Nile (1861) hoping to meet Speke and Grant near Lake Victoria, later discovering Lake Albert (1864). He was knighted (1866) and wrote: *The Albert N'yanza, Great Basin of the Nile, and Explorations of the Nile Sources* (1866) and *The Nile Tributaries of Abyssinia* (1867). He also spent time trying to suppress the slave trade (1869). Professor G D Hale Carpenter collected two male specimens of the citril from a rock pool in Sir Samuel Baker's old camp at Patiko, Gulu district and asked Fraser to name it after Baker.

Bal

Quarre's Fingertail *Gomphidia balii* **Fraser**, 1949
[A. Quarre's Tiger, JS *Gomphidia quarrei* (Schoudeten, 1934)]

Auguste M Bal was colonial administrator in the Bangala district of the former Belgian Congo (DRC). He collected the holotype there (1935). He was a keen photographer and amateur anthropologist, taking many photographs of local Mbuja and Ngombe people and collecting a few cultural objects which he sent to a number of museums, in particular the Belgian Royal Museum for Central Africa which has 142 of his photographs.

Balachowsky

Gabon Slim *Aciagrion balachowskyi* **Legrand**, 1982

Alfred Serge Balachowsky (1901–1983) was a Russian born French entomologist at the MNHN. During WW2, he spied in Paris for the British and was sent to a German concentration camp where he was put to work studying vaccines for typhus. He used the work to help some people, notably British SOE officers, to escape. He was a witness at the Nuremberg trials. He worked at Muséum National d'Histoire Naturelle (MNHN) in Paris and was President of the Société Entomologique de France (1948). He was a specialist of scale insects (Coccoidea) and some beetle families. Among his published works are: *Coléoptères Scolytides* in *Faune de France* (1949) and *Coléoptères* in *Entomologie appliquée a l'agriculture* (1963).

Balinsky

Balinsky's Sprite *Pseudagrion inopinatum* Balinsky, 1971

Boris Ivan Balinsky (1905–1997) was a Ukrainian and South African biologist, entomologist, embryologist and teacher and among the first to induce organogenesis in amphibian embryos. At 28 years of age he was a full professor and Deputy Director of the Institute of Biology, Kiev. A victim of soviet oppression under Stalin he remained under German occupation but fled to Poland after the war and then Germany and briefly Scotland working there on mice embryology. He then settled in South Africa (1949) where he became a lecturer at the University of Witwatersrand, then professor and finally took the chair of zoology (1954). He also had a long-term interest in entomology, particularly Plecoptera, Lepidoptera and Odonata. He described 14 new Odonata and around 180 Lepidoptera including 34 Genera. He collected odonates from all over South Africa amassing a collection of more than 4000 from 160 species which he later donated to the Transvaal Museum. His very first new discovery (Pinhey's Wisp) he took from the suburb of Johannesburg where he lived.

Ball

Westwern Stream Threadtail *Elattoneura balli* **Kimmins**, 1938

Dr Antoine Ball (1897–1981) was a minor member of the Belgian nobility and an entomologist at the Musée Royal d'Histoire Naturelle, Brüssels. He published papers on Psocoptera such as: *Les Psocidae de Belgique* (1926) and *Contribution a l'étude des Psocoptères* (1943).

Baltazar

Flatwing sp. *Luzonargiolestes baltazarae* Gapud & Recuenco-Adorada, 2001
[Orig. *Argiolestes baltazarae*]

Dr Clare Rilloraza Baltazar (b.1927) is an entomologist at the University of the Philippines, which awarded her BSA in entomology (1947). The University of Wisconsin awarded her Master of Science degree in economic entomology (1950) and her PhD in systematic entomology (1957). She has studied many insects, particularly Philippine-endemic Hymenoptera species. She has also discovered 8 genera and 1 subgenus of Hymenoptera, and 108 species of Philippine parasitic wasps. The Former President of the Philippines conferred her as a 'National Scientist' (2001). Among her publications are: *A Catalogue of Philippine Hymenoptera* (1966) and *Philippine Insects* (1980) the first book on insects to be published by a Philippine entomologist, because of which she has been called the 'Mother of Philippine Entomology'.

Baltodano

Cacao Shadowdamsel *Palaemnema baltodanoi* Brooks, 1989

Jorge Baltodano is a rancher in Costa Rica. Brooks honoured him because he had "*...very kindly allowed his ranch to be purchased by Guanacaste National Park.*"

Baluga

Junglehawker sp. *Indaeschna baluga* **Needham** & Gyger, 1937

The Baluga are an indigenous people in Central Luzon, where this species was found.

Bamptom

Angola Jewel ssp. *Chlorocypha crocea bamptoni* **Pinhey**, 1975

Ivan Bampton (1926–2010) was an amateur English ornithologist and self-taught botanist and entomologist. He and his family moved to Kenya (1953) to work for the East African Railways and Harbours as an engineer but stepped down (1969) to spend time working on birds with ornithologist John Williams. Shortly after he organised his first expedition to study and collect butterflies in southern Africa. He then (1972) undertook an expedition to Angola for the Allyn Museum in Florida. For many years thereafter he lived a nomadic life and went to different, mostly English-speaking, African countries in between times staying with his eldest son in Zimbabwe and later South Africa. However he returned each year to the UK to spend a month seeing his other children, grandchildren and sister and indulge his life-long passion for Newcastle FC. (On one such occasion he did a tree-climbing course so he could later study larvae living on the outer leaves of trees – he was 78 at the time!) For the last four decades of his life he travelled around Africa studying the plants and insects that interested him. He collected the Jewel in West Central Angola (October 1973). Pinhey wrote: *I take pleasure in naming this subspecies after Ivan Bampton who has extended his interests from ornithology to collect the early stages of many Lepidoptera and has also added some interesting Odonata and Heterocera to our collections.* At least a dozen butterflies are named after him and a series of butterflies named after the ladies in his life – his wife, daughters, daughter-in-law and granddaughter.

Banks

Northern Wiretail *Rhadinosticta banksi* **Tillyard**, 1913
[Orig. *Isosticta banksi*]

This is a toponym as it is named after Banks Island, Torres Strait, Australia. The island is, in turn, named after Sir Joseph Banks, the botanist on the HMS Endeavour during James Cook's voyage to Eastern Australia (1770). So, the species is an eponym, once removed.

Barbara (Moulds)

Mount Lewis Tigertail *Eusynthemis barbarae* **Moulds**, 1985
[Orig. *Choristhemis barbarae*]
Kimberley Hunter *Austrogomphus mouldsorum* **Theischinger**, 1999

Barbara J Moulds (b.1944) is the wife of the author Dr M S Moulds (q.v.). She assisted her husband in fieldwork over many years including an intensive survey of Odonata in remote areas of Cape York Peninsula, Australia (1974) and to the headwaters of the unexplored Jardine River (1978). Apart from collecting she also typed his manuscripts! (Also see **Moulds**)

Barbara (Watson)

Large Bluestreak *Lestoidea barbarae* **Watson**, 1967

Barbara Watson was the wife of the describer, Dr J A L Watson (1935–1993) (q.v.), whom he honoured with the dedication: *"…the species is named after my wife in recognition of her considerable assistance in field collecting over many years".*

Barber

Desert Forktail *Ischnura barberi* Currie, 1903

Herbert Spencer Barber (1882–1950) was associated with entomology in the Smithsonian Institute (1898–1950) where his large collection of insects and his numerous papers are housed. He was appointed Assistant Preparator of insects (1898) despite his minimal formal education and he worked under E A Schwarz (1898–1902). He collected the male holotype of the forktail in Arizona (1901). He was then employed by the US Department of Agriculture (1902–1904), but returned to the Museum (1904–1908). Then he was employed as an expert on beetles in the Division of Insect Identification in the Agriculture Department (1908–1950). During these years, while he worked mostly in the museum, he collected mainly insects in the US, Mexico and Guatemala and was an internationally recognized authority on lampyrid and chrysomelid bruchid beetles (now regarded as a subfamily of Chrysomelidae).

Baria

Forceptail sp. *Phyllocycla baria* **Belle**, 1987
Flatwing sp. *Heteragrion bariai* **De Marmels**, 1989

This is a toponym; both species are named after the type locality, Río Baria, Amazonas State, Venezuela.

Barbiellini

Malachite Darner *Remartinia barbiellina* **Navás**, 1911
[JS *Remartinia luteipennis* (Burmeister, 1939)]

Count Amadeu Amidei Barbiellini (1877–1955) was an agriculturalist in Brazil, and founded the Sao Paulo Society of Agriculture. He collected entomological specimens widely in Brazil, including the Odonata type (1910) and corresponded with all the leading entomologists of his times in North America and Europe. He sent a collection to the AMNH. The author wrote: "*I dedicate this species to our generous donor Count A A Barbiellini*". He is commemorated in the names of over 100 insect species, but did not describe any himself as an entomologist, rather he was a 'humble hunter and collector of flies and beetles'.

Baroni (Urbani)

Clubtail sp. *Davidius baronii* **Lieftinck**, 1977

Professor Dr Cesare Baroni Urbani (b.1943) is an Italian entomologist and taxonomist formerly Professor at the Institute for the Protection of Nature, University of Basle and Curator Naturhistorisches Museum in Basel, with a particular interest in ants. He was also at the Istituto di Zoologia, Universita di Siena. He collected the holotype in Bhutan. He has written a great many articles (1962 onwards) such as, with others: *Zoologische Expedition des Naturhistorischen Museum Basel in das Königreich Bhutan* (1973) and the more extensive: *The Zoogeography of Ants (Hymenoptera, Formicidae) in Northern Europe* (1977). In private life, he is a wine lover and collector of wine labels.

Baroalba

Blackwinged Threadtail *Nososticta baroalba* **Watson** & **Theischinger**, 1984

This is a toponym; Baroalba Creek Springs are north of Mount Cahill, Northern Territory, Australia.

Barrett

Comanche Dancer *Argia barretti* **Calvert**, 1902

Otis Warren Barrett (1872–1950) was an American entomologist and collector based in Tacubaya, Mexico and later at the Agricultural Experiment Station, Mayaguez, Puerto Rico. He collected the holotype in Linares, Nuevo Leon, Mexico. He has numerous insects named after him including a genus of cricket.

Bárta

Duskhawker sp. *Gynacantha bartai* **Paulson** & von **Ellenrieder**, 2005

Daniel Bárta (b.1969) is a Czech singer, songwriter and photographer, holder of several awards from the Czech Academy of Popular Music and been part of several popular music combos such as 'Sexy Dancers' and 'Illustratosphere'. He has also appeared in several films. In private life he is also a photographer of dragonflies (which is why he is honoured) having an exhibition (2008) in the Academia Gallery, Prague. He illustrated *Dragonflies of the Czech Republic* (2008). He accompanied D R Paulson to Explorer's Inn (2002) and pointed out this species to him, and "...*wielded a long-handled net with sufficient skill to catch two of the three specimens taken on that visit.*"

Bartels

Shadowdamsels sp. *Drepanosticta bartelsi* **Lieftinck**, 1937
Nighthawker sp. *Heliaeschna bartelsi* Lieftinck, 1940

Dr Max Bartels, Jr. (1902–1943) was one of the three sons of Max Eduard Gottlieb Bartels (1871–1936) the Dutch ornithologist, plantation owner and naturalist. He studied zoology at the University of Bern (1925–1929) and graduated cum laude, living there until returning to Indonesia (1932). Having independent means, he volunteered to do research for the Botanical Garden at Bogor, particularly researching mammals and publishing on them, but he also enjoyed big game hunting. He collected in Java (collecting the shadowdamsel 1935) and other parts of Indonesia with his brothers and father including Sumatra where he took the hawker type (1936). They were primarily ornithologists but collected all and any creatures they came across, much of which is still in the collection at RNMH. He was interned by the occupying Japanese forces (1942) and died when forced into labour on the notorious Burma Railway.

Bartenev

Cruiser sp. *Macromia bartenevi* **Belyshev**, 1973
[JS *Macromia amphigena fraenata* Martin, 1906]

Arctic Bluet ssp. *Coenagrion concinnum bartenevi* Belyshev, 1955
[JS *Coenagrion johanssoni* (Wallengren, 1904)]

Professor Alexandr Nikolaevich Bartenev (1882–1946) was a Russian entomologist and one of the most prominent Russian odonatologists. While a student at Moscow University he studied dragonflies in Belorussia, Lithuania, the eastern Ural foothills and Transcaucasia; publishing the results (1907–1909). After graduating he worked as a teacher of natural sciences at Tomsk Commercial College and studied dragonflies at Zoological Cabinet at Tomsk University. While there he also studied dragonflies in the surroundings and in former Khakasia and Transbaikalia. He moved to Warsaw (1909) and worked in Warsaw University. Due to WW1, Warsaw University and its staff were evacuated to Rostov-na-Donu (1915) to form Don University. Bartenev worked there first as the Keeper of Museum and later Professor, Vice-Rector and Rector (1920–1922). He returned to Moscow (1922) for administrative work in education. He defended his DSc dissertation (1925) and was elected Professor at the North Caucasian University; then was a professor at Krasnodar Medical Institute (1930–1933). He moved to Alma-Ata in Kazakhstan (1933) staying for the rest of his life. He wrote two volumes on *Libellulidae* (1915 & 1919) in the series *Faune de la Russie et des pays limitrophes*. In total, he published 102 scientific works, of which 80 were devoted to faunistics, taxonomy and biogeography of dragonflies. In these papers, he named nearly 70 new species or subspecies, of which around a dozen are now considered valid.

Bartolozzi

Leaftail sp. *Phyllogomphus bartolozzii* **Marconi**, Terzani & Carletti, 2001

Dr Luca Bartolozzi (b.1954) is Head of Zoology and Curator of Entomology at La Specola Zoological Museum, University of Florence where he graduated (1980) before becoming an employee (1981). He has collected and researched in much of Africa, Ecuador and Brazil, Madagascar, Malaysia, Israel, Armenia and eastern Russia. He has published many papers (from 1983) alone or with others including many descriptions of new species, in particular of beetles as this is his special area of interest. Others are taxonomy and faunistics of the Lucanidae and Brentidae. Many insects are named after him.

Basilewsky

Black Flasher *Aethiothemis basilewskyi* **Fraser**, 1954
Slender Jewel *Chlorocypha basilewskyi* Fraser, 1955
[JS *Stenocypha tenuis* (Longfield, 1936)]

Dr Pierre Basilewsky (1913–1993) was a Belgian entomologist of Russian origin specializing in Carabidae. He was awarded his diploma in agricultural engineering (1936) at the Institut agronomique de l'État in Gembloux, Belgium, and a further diploma in limnology and forestry (1938). His career was always focussed on the entomology of Africa, and he visited east, southeast and central Africa and some Atlantic islands. He wrote 68 works, published in 101 publications in 5 languages (1929–1992) such as: *Descriptions de coléoptères Carabidae nouveau d'Afrique et notes diverses sur des espèces déjà connues, VI* (1950) as well as a number of reports on the scientific results of some overseas expeditions for the Koninklijk Museum voor Midden Africa (Musée Royal de l'Afrique Centrale), Belgium where he was Curator of Entomology, then Director (1977–1978). He collected a series of both species in Ruanda-Burundi (March 1953).

Bastiaan

Firetail sp. *Telebasis bastiaani* **GH Bick** & **JC Bick**, 1996
Coralleg sp. *Rhipidolestes bastiaani* **Zhu** & **Yang**, 1998

Professor Dr Milan Boštjan (Bastiaan) Kiauta (b.1937) (See **Kiauta**)

Bates

Clubtail sp. *Zonophora batesi* **Selys**, 1869
Emerald sp. *Neocordulia batesi* Selys, 1871
[Orig. *Gomphomacromia batesi,* A. Shadowdragon]
Bannerwing sp. *Polythore batesi* Selys, 1879
[Orig. *Thore batesi*]
Tawny Pennant *Cannacria batesii* **WF Kirby**, 1889
[JS *Brachymesia herbida* (Gundlach, 1889)]
Pennant sp. *Idiataphe batesi* **Ris**, 1913
[Orig. *Ephidatia batesi*]

Henry Walter Bates (1825–1892) was an English explorer and naturalist, one of the giants of nineteenth century natural history. He was the first to describe animal mimicry for science. He was largely self-taught having left school at 13 and been apprenticed to a hosier in Leicester where he grew up, yet 10 years later published his first scientific paper in the 'Zoologist' (1848). He had joined the Mechanic's Institute and studied in their library and also collected insects in his spare time. He became friends with Wallace when Wallace took a teaching post at Leicester Collegiate School. He explored the Amazon (1848–1859), initially with A R Wallace where they collected 14,712 species, 8000 of which were new to science. Wallace left after four years and lost his collections to shipwreck, but Bates remained for eleven years and was said to have 'gone native'. He wrote up the findings of that expedition as: *The Naturalist on the River Amazons* (1863). After returning from the expedition he married (1863) and worked as Assistant Secretary of the Royal Geographical Society (1864). He was the first to describe animal mimicry based on his study of Amazonian butterflies, lending support to Charles Darwin's recently published theory of evolution, which he ardently supported. He died of bronchitis. His humble origins, and retiring nature meant that others were more famous, but few more important. Many of the taxa that he collected were not properly described for many years until 'rediscovered' in museum collections. A bird is also named after him.

Bauer

Lancet sp. *Skiallagma baueri* **Förster**, 1906
[JS *Xiphiagrion cyanomelas* Selys 1876]

F W Bauer was a friend of Förster who named the species after him when the former was Director of the German Middle School in São Paulo, the state where the only two specimens ever found in Brazil were thought to have been collected. Recently it has turned out that the specimens were in fact collected in the Oriental/Australian region.

Bayadére

The Genus *Bayadera* **Selys**, 1853
Duskhawker sp. *Gynacantha bayadera* Selys, 1891

Bayadére is the French version of the Portuguese word bailadeira, which refers to a Hindu dancing girl in Indian temples. Selys Longchamps (1853) introduced the genus-group name *Bayadera* for his Indian species *Epallage indica*.

Beadle

Variable Sprite ssp. *Pseudagrion sjöstedti beadlei* **Pinhey**, 1961
[JS *Pseudagrion sjoestedti* Förster, 1906]

Professor Leonard Clayton Beadle OBE (1905–1985) was a pioneer British limnologist. His lifelong interest in freshwater biology started when, as an undergraduate, he took part in an expedition to South America (1926). His first experience of Africa was when he participated in a Cambridge University expedition to the East African Lakes (1930). He made three further African trips; Algeria (1938) to work on oases and saline water, Uganda (1949) where he was head of the Zoology Department of Makerere University and a Trustee of Uganda Parks (1949–1966), and lastly again to Uganda on a Royal Society biological programme to Lake George. Among his written works are: *The Art of Science* (1955) and *The inland waters of tropical Africa* (1974) The etymology reads: "*I name this insect after Professor Beadle of Makerere College, in appreciation of my first visit to Uganda (May 1949) which was arranged by him.*"

Beata

Bannerwing sp. *Polythore beata* **McLachlan**, 1869
[Orig. *Thore beata*]
Clubtail sp. *Trigomphus beatus* **Chao**, 1954

Common Blue Jewel ssp. *Heliocypha perforata beatifica* **Fraser**, 1927
[Orig. *Rhinocypha perforata beatifica*]

Beata is a female forename, which comes from the Latin *beatus*, meaning 'happy' or 'blessed'. This name is derived from the same Latin word as the concept of beatification, of major importance in the Catholic religion. It does not appear to be to any particular individual.

Beatrix

Threadtail sp. *Nososticta beatrix* **Lieftinck**, 1949
[Orig. *Notoneura beatrix*]
Yellow-striped Flutterer ssp. *Rhyothemis phyllis beatricis* Lieftinck, 1942

Princess Beatrix Wilhelmina Armgard (Queen regnant 1980–2013) of the Netherlands (b.1938) became heiress presumptive (1948) at the time the threadtail was collected on the 3rd Archbold Expedition. No etymology is given for either, but the author has later confirmed that they were named after the princess.

Beaumont

Short-winged Shadowdamsel *Protosticta beaumonti* **KDP Wilson**, 1997

Jack Beaumont (1919–1993). Wilson wrote: "*…I am pleased to name this damselfly in honour of the late Jack Beaumont who greatly encouraged my interest in Odonata.*"

Becker

Threadtail sp. *Roppaneura beckeri* **Santos**, 1966

Professor Johann Becker (1932–2004) was a Brazilian zoologist, working first as a researcher then Curator at the National Museum of Rio de Janeiro. As a student (1951) he was admitted as apprentice in the Division of Insects, Museu Nacional (Rio de Janeiro). He graduated with an initial degree of BSc (1954) and became zoologist researcher and later a professor of that Museum being a specialist in genetics, evolution, and invertebrate palaeontology. He had collected insects, especially beetles, ever since he was a student and, after his death, his c.14,000 specimen entomological collection was presented to the Zoology Department of Universidade Estadual de Feira de Santana. An assassin bug, a bird and an amphibian are also named after him.

Becker and Olmiro Roppa, who was a preparator at the museum, collected this species in Minais Gerais (1963). The preparator was honoured with the genus eponym (see **Roppa**) and the professor with the species eponym. Not very democratic, but the name combination sounds better that way round.

Bedê

Emerald sp. *Lauromacromia bedei* **Machado**, 2005

Dr Lúcio Cadavel Bedê (b.1963) is currently (2017) Senior Biologist at Golder Associates Brazil, Meio Ambiente. His bachelor's degree (1987), masters (1992) and doctorate (2006) were all awarded by the Minas Gerais Federal University. He was previously (2002–2013) Project Manager at Conservation International – Atlantic Forest Program, Brazil, Belo Horizonte, Minas Gerais. He gave the holotype to the author, a single male specimen collected in a river at the 'cerrado' region of the State of Minas Gerais, Brazil (2004). His main focus is international conservation. Among his many published papers are the co-written: *Challenges and Opportunities for Biodiversity Conservation in the Brazilian Atlantic Forest* (2005) and *Two new genera and nine new species of damselflies from a localized area in Minas Gerais, Brazil (Odonata: Zygoptera)* (2016) written with Machado.

Bedford

Genus *Bedfordia* **Mumford**, 1942
[Homonym of *Bedfordia* Fahrenholz, 1936 in Phthiraptera; replaced with *Hivaagrion* Hämäläinen & Marinov, 2014]

This is only an eponym by one remove as the genus was named after the May Esther Bedford Fund, which, in turn was named after Mary (May) Esther Schiott née Bedford (d.1911), the late daughter of Edward T Bedford (1849-1931), Director of Standard Oil. She married (1906) Norwegian Naval Lieutenant Johannes Schiott.

Beebe

Pond Damselfly sp. *Bromeliagarion beebeanum* **Calvert**, 1948
[Orig. *Leptagrion beebeanum*]

Charles William Beebe (1877–1962) was an American ornithologist, zoologist, marine biologist, conservationist, naturalist, explorer and writer. He left Columbia University (1899) before gaining a degree (later he was granted honorary doctorates by Tiufts and Colgate Universities) and began his working life looking after the birds at the Bronx Zoo (New York) as Assistant Curator of the New York Zoological Park. For them he undertook a series of expeditions around the world documenting the world's pheasants. He became Curator of Ornithology, New York Zoological Society (1902–1918), and Director, Department of Tropical Research (1919). The trips led to his interest in marine biology and deep-sea exploration and he made a number of descents in the bathysphere including (1934) a then record descent of 923 metres (3,028 feet) off Nonsuch Island, Bermuda. He set up a camp (1942) at Caripito in Venezuela for jungle studies and (1950) bought 92 hectares (228 acres) of land in Trinidad and Tobago, which became New York Zoological Society's Tropical Research Station (Asa Wright Nature Centre). He married Helen Elswyth Thane Ricker (1900–1981) who wrote romantic novels (pen name Elswyth Thane). Much of his writings were popular books on his expeditions and his book: *The Bird, Its Form and Function* (1906) presented technical information about bird biology and evolution in a way that was accessible to the general public and made enough money to finance his later expeditions. He made various collecting trips to bring live birds back to the zoo including one to British Guiana (1924) where he collected the type. Perhaps his most outstanding ornithological work is the 4-volume: *A Monograph of the Pheasants* (1918–1922). He retired to Trinidad. A number of other taxa including several fish, a bird and two amphibians are named after him.

Beeson

Nighthawker sp. *Amphiaeschna beesoni* **Fraser**, 1922
[JS *Heliaeschna uninervulata* Martin, 1909
Margarita Jewel *Rhinocypha beesoni* Fraser, 1922
[JS *Heliocypha biforata* (Selys, 1859)]

Dr Cyril Frederick Cherrrington Beeson (1889–1975) was an English entomologist and forest conservator who worked in India. Interestingly he attended City of Oxford High School for Boys where his best friend was T E Lawrence (Lawrence of Arabia) who called him by his nickname 'Scroggs'. Together they cycled to every parish church in three counties making brass rubbings that they presented to the Ashmolean Museum. They also toured France by bike for two summers (1906 & 1907) visiting medieval castles. St John's College, Oxford awarded his degree in geology (1910) but he then changed disciplines to forestry and obtained a diploma. Oxford also awarded his MSc (1917) and DSc (1923). He was a Royal Army Medical Corps captain during WW1. He worked for the Imperial Forest Service (1911–1941) as a research officer, forest conservator and entomologist. He was appointed Forest Zoologist of India (1913) where he was associated with the development of the Forest Research Institute in Dehradun, India. When he retired (1941) he took up the study of antique clocks and wrote about them, although his first book was: *The Ecology and Control of the Forest Insects of India and the Neighbouring Countries* (1941). In the thirty years of forestry study he wrote around sixty papers on tropical forest insects. He then took another post as Director of the Imperial Forestry Bureau in Oxford (1945–1947). He was a founder member of the Antiquarian Horological Society (1953) and edited their journal (1959–1960). He also extended the clock collection of the Museum of the History of Science, Oxford. He wrote: *English Church Clocks 1280-1850: History and Classification* (1971). The chlorocyphid was collected by Beeson at Dehradun (1920) and the male aeshnid taken at dusk in Gahan, Burma (1921).

Bella

Elfin Skimmer *Nannothemis bella* **Uhler**, 1857
[Orig. *Nannophya bella*]
Amberwing sp. *Perithemis bella* **WF Kirby**, 1889
Riverking sp. *Pseudomacromia bella* **Lacroix**, 1920
[JS *Zygonoides lachesis* Ris, 1912]
Striped Flasher *Aethiothemis bella* **Fisher**, 1939
Skimmer-like Flasher *Cirrothemis bella* **Fisher**, 1939
[JS *Aethiothemis bequaerti* Ris, 1919]

Bella is Latin for beautiful.

Belladonna

Spreadwing sp. *Lestes belladonna* **MacLeay**, 1826

A long-forgotten name for an Australian damselfly species, the name means 'beautiful woman', but which actual beauty remains unknown.

Belle

Belle's Sanddragon *Progomphus bellei* Knopf & **Tennessen**, 1980
Clubtail sp. *Peruviogomphus bellei* **Machado**, 2005

Dr Jean Belle (1920–2001) was a Dutch odonatologist who collected in Latin America, particularly in Surinam with Dirk Geijskes. His appetite for Odonata was whetted when, as a boy in West Java, night watchmen at his father's office taught him how to collect dragonflies and roast and eat them. He moved to the Netherlands (1934) and studied maths at the university there, graduating (1947) and becoming a science and maths teacher in a secondary school. He married (1950) and he and his wife, who was also brought up in Indonesia, decided to emigrate to Suriname. The Odonata collection made by Dirk Geijskes (q.v.) in the museum in Paramaribo stimulated his interest, and Geijskes suggested he start collecting them too, which he did (1955), specialising in Gomphidae. He collected all over Suriname (1955–1965) before he and his family returned to the Netherlands and his teaching job. He started publishing accounts of dragonflies (1963) which he continued for the rest of his life. He was considering emigrating again, this time to Costa Rica, but changed his mind when his wife died (1987). His account of twenty-one new Suriname species (1973) doubled as his PhD thesis (1974); in all he described 110 and eight genera of new gomphidae and a further fourteen species and one genus of other Odonata. The Rijksmuseum van Natuurlijke, Leiden (RMNH), (now Naturalis Centre for Biodiversity) holds his collections which he sold to them (1987). His last, posthumously published (2001) paper was a checklist of the Odonata of Suriname.

Bellona

Coraltail sp. *Ceriagrion bellona* **Laidlaw**, 1915

Bellona was an ancient Roman goddess.

Belon

Genus *Belonia* **WF Kirby**, 1889
[JS *Libellula* Linnaeus, 1758]

Pierre Belon (du Mans) (1517–1564) (pen name Petrus Bellonius Cenomanus) was a French diplomat, writer, traveller and naturalist who was deeply interested in antiquity and extolled the virtues of the renaissance. He studied medicine in Paris being awarded the degree of doctor (1542) and then became the pupil of Valerius Cordus the botanist with whom he travelled to Germany. After this he undertook a scientific journey through Greece, Crete, Asia Minor, Egypt, Arabia and Palestine (1546–1549). He published an account of his travels as: *Observations* (1553). He undertook a second journey (1557) in northern Italy and its environs. He wrote widely on ichthyology, ornithology, botany, anatomy, architecture and Egyptology and his other works, often illustrated with anatomical drawings included: *Histoire naturelle des estranges poissons* (1551), *De aquatilibus* (1553), and *L'Histoire de la nature des oyseaux* (1555). He was murdered on a return journey to Paris. Although no etymology is given for this genus, in which Kirby placed a few North American *Libellula* species, it seems most likely that it was named after Pierre Belon.

Belyshev

Bluet sp. *Enallagma belyshevi* Haritonov, 1975
[JS *Enallagma circulatum* Selys, 1883]
Shadowdamsel sp. *Drepanosticta belyshevi* **Hämäläinen**, 1991
Sanddragon sp. *Progomphus belyshevi* **Belle**, 1991

Dr Boris Fedorovich Belyshev (1910–1993) was the most prolific Soviet odonatologist of his time. He was an expert on the dragonflies of Siberia. He published his first scientific paper on ornithology when only 16 years old and at 19 he became the President of the Siberian Ornithological Society. He was at Tomsk State University (1927–1930) and after graduating (1930) he enrolled in Leningrad University, but was arrested (1932) for his connections with foreign ornithologists. He was released, but was ordered to return to Siberia, where he worked for three years as a hunting manager in Tara Town, Omsk Province. In the course of Stalin's repressions, he was arrested again (1936) 'for counter-revolutionary activity' and imprisoned at Vorkuta, one of the most notorious forced labour camps of the Gulag. In the early 1940s he was released from the camp, but remained in exile working in a medical service fighting the most dangerous infections. Throughout this time, he continued publishing ornithological papers and developed an interest in dragonflies, to which he devoted the rest of his life. He was 'rehabilitated' (1953) and started to work in Biysk Museum of Local Lore. He lectured at the Biysk Institute of Pedagogics (1955–1958) and then became a researcher at the Siberian Institute of Physiology and Biochemistry of Plants of the USSR Academy of Science in Irkutsk (1959–1967). During this time, he also gained his DSc (1964) from Irkutsk University, his lengthy dissertation being on the: *Odonate fauna of Siberia*. He moved to Novosibirsk (1967) to work as an odonatologist at the Biology Institute of the Siberian Branch of the USSR Academy of Sciences at Novosibirsk. He wrote (from 1951) 182 odonatological papers and seven books. He named nearly 40 new odonate species and subspecies, of which only a few are presently ranked as valid taxa; among them is the Hawker species *Aeschnophlebia zygoptera*. His major contribution was the three-volume: *Strekozy Sibiri (Odonata)* [The dragonflies of Siberia (Odonata)] (1973–1974). He was a founder Member of Honour of the Societas Internationalis Odonatologica (1971).

Benken

Damselfly sp. *Palaiargia benkeni* **Orr**, **Kalkman** & Richards, 2014

Theodor Benken (b.1963) studied biology at the University of Freiburg. He is head of the division of Information technology at the media centre Baden-Württemberg in Karlsruhe. He was a founding member of the board of Schutzgemeinschaft Libellen in Baden-Württemberg, a dragonfly conservation organization, which he currently heads. He published several papers on Odonata of regional interest, and is co-author of the *Atlas of the Odonata of Germany* (2015). He was honoured for his generous support of Odonata research in New Guinea through the International Dragonfly Fund (IDF).

Bequaert

Skimmer-like Flasher *Aethiothemis bequaerti* **Ris**, 1919
Pincertail sp. *Nihonogomphus bequaerti* **Chao**, 1954

Professor Dr Joseph Charles Corneille Bequaert (1886–1982) was a Belgian botanist, entomologist and malacologist. He graduated with a doctorate in botany from the University of Ghent (1906) and worked for the colonial government in the Belgian Congo (1910–1915). He went to the US in 1916s later taking citizenship citizen (1921). He was a research assistant at AMNH, New York (1917–1922), then worked at Harvard (1923–1956), initially teaching entomology at Harvard Medical School, finally becoming Professor of Zoology, Museum of Comparative Zoology. In retirement, he became Professor of Biology, University of Houston (1956–1960) and Visiting Entomologist, University of Arizona. Among other works he co-wrote: *The Mollusks of the Arid Southwest* (1973). Among other taxa, a reptile and two amphibians are named after him.

Berawan

Shadowdamsel sp. *Telosticta berawan* **Dow** & **Reels**, 2012

Not a person but a people – the Berawan people live in the Mulu area of Sarawak where the type was discovered.

Berenice

Seaside Dragonlet *Erythrodiplax berenice* Drury, 1773
Flatwing sp. *Philogenia berenice* Higgins, 1901
Pin-tailed Flasher *Lokia berenice* **Fraser**, 1953
[JS *Aethiothemis erythromelas* (Ris, 1910)]

Berenice II (c.267–221 BC) was the wife of Ptolemy III of Egypt. According to a myth, she dedicated her hair to Aphrodite in return for her husband's safe return from an expedition. The hair was apparently then stolen from the temple where it had been placed, but it was said to have been carried to the heavens and put among the stars. While it is very probable that Drury named the Dragonlet after this person, there is no evidence either way on the other two species.

Berg

Darter sp. *Sympetrum bergi* **Grigoriev**, 1905
[JS *Sympetrum tibiale* (Ris, 1897)]

Lev (Leo) Semionovitch (Semenovich) Berg, (1876–1950) was a Russian academician, President of the Russian Geographical Society and a geographer and zoologist born in Bender, Moldavia. He established the foundations of limnology in Russia with his systematic studies on the physical, chemical, and biological conditions of fresh waters, particularly of lakes. His work in ichthyology was also noted in the palaeontology, anatomy, and embryology of Russian fish. He wrote: *Natural Regions of the USSR* (1950). He thoroughly investigated the lakes of the Russian Central Asia, including their fishes, and collected the dragonfly at Kirghizia (1903). It is a pity that he is most famous in his country for his theory of 'nomogenesis', an anti-Darwinian theory that still holds sway in some quarters. At least seven marine fish and some freshwater species are named after him.

Berla

Skimmer sp. *Nephepeltia berlai* **Santos**, 1950

Herbert Franzoni Berla (1912–1985) was a Brazilian ornithologist and entomologist who worked for the National Museum in Rio de Janeiro (MNRJ). He was admitted as a trainee (1932) at the Zoology Section of the MNRJ, where he learned the techniques of cataloguing and conservation of large zoological groups and became curator (1944). He made several collecting trips to Mato Grosso do Sul (1940) and Pernambuco (1946) then to many locations in Brazil over forty years. He collected the holotype of the skimmer in Brazil (April 1949) with Santos. He made a particular study of mites which live on birds and two birds are named after him.

Berland

Shadowdamsel sp. *Drepanosticta berlandi* **Lieftinck**, 1939
Cruiser sp. *Macromia berlandi* Lieftinck, 1941
Lancet sp. *Xiphiagrion berlandi* **Fraser**, 1951
[JS *Xiphiagrion cyanomelas* Selys, 1876]

Fineliner spp. *Teinobasis alluaudi berlandi* **Schmidt**, 1951
[JS *Teinobasis alluaudi* (Martin, 1896)]

Lucien Berland (1888–1962) was a French entomologist and arachnologist. He studied at the Lycée Charlemagne and the Sorbonne, graduating in natural sciences (1908). He joined the NMNH (1912) as an assistant in the entomology laboratory. He was seriously injured during WW1, but still managed to travel extensively in North and Sub-Saharan Africa. He became Director of the French Zoological Society (1952). Among his more than 200 papers was: *Les Araignées de l'Afrique Occidentale Française* (1941). The authors named these species after Berland, because as a museum curator he had arranged the material for them to study. At least six other species of various taxa are named after him.

Bernard

Batéké Sprite *Pseudagrion bernardi* Terzani & Carletti, 2001

Dr Bernardo Cecchi (b.1970) is an Italian entomologist particularly interested in Coleoptera. He works at Corpo Forestale dello Stato and previously worked at Museo di Storia Naturale dell'Università di Firenze where he also studied. The authors examined much material at the Entomology Section of the Department of Zoology, known as 'The Observatory' at the Museum of Natural History, University of Florence, which Cecchi was responsible for accumulating and studying. Among his published works is the co-written: *A Contribution to the Knowledge of the Coleoptera of the Aeolian Islands* (2006).

Bertha

Red-Veined Pennant *Celithemis bertha* **EB Williamson**, 1922

Bertha Paulina Currie (b.1876) was an American entomologist who worked for the Bureau of Entomology, US Department of Agriculture and as 'custodian of dragonflies' in the National Museum. Her brother Rolla Patteson Currie (1875–1960) was a 'neuropterist' (he was ordained later) and the first Museum Aid to be appointed at the NMNH (1894–1904) where he and Bertha (who was unpaid) began to form a collection of international scope. They collected 103 species of Odonata around Washington; only seven others have been found since. Rolla wrote ten papers on neuroptera and five on Odonata (1898–1918) and Bertha wrote several others including: *Gomphus parvidens, a new species of dragonfly from Maryland* (1917). She helped Williamson examine material at the USNM on several occasions.

Berthoud

Orange Streamcruiser *Hesperocordulia berthoudi* **Tillyard**, 1911

George Frederick Berthoud (1856–1936) was a Western Australian agriculturalist and horticulturalist. He collected botanical specimens (1874–1887) in Victoria and Queensland. He moved to Western Australia (1895) where (1896–1900) he began growing experimental crops on a railway reserve at Drake's Brook. He was then Director of the State Farm of Hamel, Waroona, WA, (1901–1910) and the State Forestry Nursery there (1911 onwards), which planted experimental crops. He collected entomological specimens (1895–1911) around the land he farmed. Tillyard wrote: *"Taken by my friend Mr G. F. Berthoud, to whom I dedicate this species"* and *"I am much indebted to Mr. G. Berthoud, of the State Farm, Hamel, for sending me a large number of specimens from Waroona (Murray District), thus linking together the northern and southern localities which I myself worked."*

Beschke

Flatwing sp. *Heteragrion beschkii* **Hagen**, 1862

Carl Heinrich Beské (sometime Bescké or Beschke) (1798–1851) was a German naturalist, particularly entomologist, who collected the holotype in Brazil, where he (1830s onwards) was resident in New Freiburg, Rio State as a taxidermist and natural history dealer. The town was something of a centre for exporting natural history material. (An English traveller recounted in 1821 that a member of the colony killed various toucans, parrots, woodpeckers and other birds with the aim of stuffing them.) Beschke collected the types of a number of birds. He sent material to many European museums and also sold directly to visiting naturalists such as Burmeister (q.v.) who explored the area (1850–1851).

Bhatnagar

Yellow Featherleg *Disparoneura bhatnagri* Sahni, 1965

[JS *Copera marginipes* (Rambur, 1842)]

Professor Dr S P Bhatnagar was the Head of Zoology Department of DSB College, Kumaun University, Nainital, India. Among his published works is the co-written *Dermaptera from Nainital* (1961).

Bianco

Emerald sp. *Neocordulia biancoi* **Rácenis**, 1970

Dr Jesús Maria Bianco (1917–1976) was a Venezuelan pharmacist, formerly (1963–1970) Rector of the Universidad Central de Venezuela, forced to resign by the military government. The author wrote: *"We dedicate this species to Venezuelan pharmacist, Dr. Jesús M. Bianco, who as the Rector of the Universidad Central de Venezuela for many years, has always defended the institutional autonomy of our university and comprehended the concerns and needs of its researchers."*

Bick

Firetail sp. *Telebasis bickorum* Daigle, 2002
Flatwing sp. *Heteragrion bickorum* Daigle, 2005

Dr George Herman Bick (1914–2005) and his wife Juanda Claire Bick (née Bonck) (1919–1999) were American entomologists and odonatologists. George graduated from Tulane University (1936) and gained his masters there (1938) and his PhD was awarded by Cornell University (1945) the year he and Juanda married. After US Navy service as a Lieutenant Commander in Australia, New Guinea and New York, he became Professor of Biology at Tulane, then Southwest Louisiana State University and eventually St Mary's College, Notre Dame, Indiana where a nature centre and trail are named after them, as is also an academic award for excellence in environmental biology. After retirement, he was a volunteer at the Arthropod Museum of the University of Florida. Juanda was an Instructor in Biology at Loyola University after gaining her MSc. They co-wrote a great many papers over more than six decades from the 1940s including: *A review of the genus Telebasis with descriptions of eight new species (Zygoptera: Coenagrionidae)* (1995). They jointly described twenty new damselfly species in the genera *Cora, Polythore, Philogenia* and *Telebasis.* They collected especially very widely in the USA. George died aged 91 of complications following a fall.

Bidayuh

Shadowdamsel sp. *Telostica bidayuh* **Dow** & **Orr**, 2012

Named for the Bidayuh people, the known range of the species lies in the Bidayuh heartland.

Biedermann

Gem sp. *Disparocypha biedermanni* **Ris**, 1916

Professor Dr Richard Biedermann-Imhoof (1865–1926), née Biedermann – he re-named himself (1906) – was a Swiss ornithologist and naturalist who had been a childhood friend of the author. He graduated in medical science at the University of Zurich (1889) and continued his studies in Berlin (1889–1890) and Kiel (1890–1893). Of independent means, he was able to devote himself to study and wrote ten ornithology papers (1896–1914). His collections were donated to various institutions including the zoological museums in Zurich and Berlin. He is credited with having collected in the Altai (1908) and the Indian sub-continent, but it seems more likely that he paid other collectors to travel on his behalf, which may account for his getting into financial difficulty (1917). Among his written work is: *Ornithologische Studien* (1908). Three birds are named after him as are a number of other taxa.

Billinghurst

Large Riverdamsel *Caliagrion billinghursti* **Martin**, 1901
[Orig. *Pseudagrion billinghursti*]

Francombe Lovett Billinghurst (1859–1937) was a bank manager in Alexandra, Victoria, Australia and he collected (1898–1900) over 40 species of Odonata, which he sent to Martin. Martin's types seem to be un-locatable but the specimens of *Caliagrion* that Tillyard collected from Billinghurst's Alexandra site just a few years later are in ANIC. Martin gives no etymology, but as the specimens were sent by Billinghurst there is no doubt who is honoured. Billinghurst himself wrote a paper regarding that collection although it probably contained more species than he recorded: *Some notes on the dragon-flies of the Alexandra district* (The Victorian Naturalist 1902).

Bine

Bine's Shadowdamsel *Ceylonosticta bine* Bedjanič, 2010
[Orig. *Drepanosticta bine*]

Bine Bedjanič (b.2005) is the son of the author Matjaž Bedjanič (**b.1972**).

Biolley

Yellow-lined Skimmer *Orthemis biolleyi* **Calvert**, 1906

Professor Dr Paul (Paulo A) Biolley (1862–1908) was a Swiss born teacher, naturalist and biologist who taught at the San José High School (1886–1889) and worked on botany and invertebrate zoology for the National Geographic Institute at the Museo Nacional in San Jose, Costa Rica, where he became Director (1904). He collected (1892–1906), and many of his spermatophytes are at the BMNH. He took part in the Hopkins-Stanford Galapagos Expedition (1898–1899) and visited the Cocos Islands where he collected; at least one insect collected there is also named after him. Among other works he wrote: *Elementos de historia natural* (1887) and *Costa Rica et son Avenir* (1889). He also has at least one plant, a cactus, named after him.

Bíró

Genus *Bironides* **Förster**, 1903

Archtail sp. *Nannophlebia biroi* Förster, 1900
[Orig. *Tetrathemis biroi*]

Dune Glider *Tramea loewi biroi* Förster, 1898
[JS *Tramea eurybia* Selys, 1878]

Dr Lajos Bíró (1856–1931) was a Hungarian traveller, collector, photographer, ethnographer, ornithologist and entomologist. He began studying theology in Budapest, but had to leave to seek work. He later enrolled in the Royal Hungarian University supported by the papers he wrote and a salary as a technician, but never finished his degree joining an expedition instead, so his only academic 'qualification' was the honorary doctorate he was awarded (1926) by University of Sciences, Szeged, Hungary. He joined the expedition to German New Guinea (1896–1901) where he collected both species types and sent back over 200,000 natural history specimens and ethnographic objects to European museums. He was in Singapore (1898), North Africa (1901–1903), Crete (1906) and Turkey (1925). He worked at the Budapest Natural History Museum (1903–1931) between trips. More than 200 species and genera, including two amphibians and a bird are named after him.

Blackburn

Blackburn's Hawaiian Damselfly *Megalagrion blackburni* **McLachlan**, 1883
Blackburn's Skimmer *Nesogonia blackburni* McLachlan, 1883
[Orig. *Lepthemis blackburni*]

Reverend Thomas Blackburn (1844–1912) was an English-born Australian entomologist. Interested in entomology from early youth, he and his brother started a journal (1816) 'The Weekly Entomologist' that ceased publication after two years, but he then became an early editor of 'Entomologist's Monthly Magazine'. He studied at the University of London (1866–1868) receiving a BA, but was later (1870) ordained as a Church of England priest serving a parish in England for six years. He was transferred (1876) to Hawai'i where he was chaplain to the Bishop of Honolulu. During his time there, he collected insects extensively on Oahu but also visiting other islands, sending specimens to BMNH and discovering, among others, 23 new beetles. McLachlan described five endemic damselflies and one dragonfly species from specimens collected in Hawai'i by Reverend Blackburn who also collected over the next few years describing four species himself including: *Megalagrion oahuense* and *Megalagrion koelense* (1884). He was posted to Port Lincoln, Australia (1882–1886) then to Woodville where he remained. After his arrival in Australia, his entomological studies were focused almost exclusively on coleoptera, specimens of which he collected throughout South Australia as well as on trips to the other states. He became the foremost Australian coleopterist, publishing descriptions of 3,069 Australian species. He became Honorary Curator of Entomology for the South Australian Museum in 1887.

Blanchard

Pond Damselfly sp. *Antiagrion blanchardi* **Selys**, 1876
[Orig. *Erythromma blanchardi*]

Charles Émile Blanchard (1819–1900) was a French zoologist and entomologist. At 14 he was allowed access to the MNHN where he became a préparateur (1838) and Assistant Naturalist (1841). He undertook a field trip to Sicily and published his first book: *Histoire des insects* (1845) followed by *Zoologie agricole* (1854–1856). He was appointed Professor of Natural History of Crustacea, Arachnida and Insects (1862–1894). Selys wrote: *"J'ai dedie cette espece au savant professeur Blanchard, qui a publie l'entomologie de grand ouvrage sur le Chili de M. Gay, a qi j'ai consacre l'espece voisine."* He wrote a number of significant papers on Chilean insects.

Blum

Bluetail sp. *Ischnura blumi* **Lohmann**, 1979
[JS *Ischnura isoetes* Lieftinck, 1949]

Dr J Paul Blum (b.1940) is a German herpetologist who made a number of important collections of frogs from New Guinea, as well as a series of dragonflies from the Snow Mountains (1976). He took part in the DFG interdisciplinary research project (1975, 1976 & 1979) in Irian Jaya, particularly in the Eipomek Valley (1979). While there he received his PhD thesis from the University of Freiburg. He also studied at Kiel University. He retired in 2005. He co-wrote: *Notes on Xenobatrachus and Xenorhina (Amphibia: Microhylidae) from New Guinea with description of nine new species* (1988). An amphibian is also named after him.

Bock

Satinwing sp. *Euphaea bocki* **McLachlan**, 1880

Carl Alfred Bock (1849–1932) was a Norwegian diplomat, naturalist and explorer. He served at the Norwegian-Swedish Consulate in Grimsby, UK (1869–1875) before moving to London. His first collecting trip for the Zoological Society of London was to Batavia (1878) and, on his return, he was commissioned by the Governor-General of the Netherlands East Indies to travel through and report on the interior of South-East Borneo (1879). Later (1881) he explored northern Siam. He wrote: *Journeyings in Sumatra* (1881) and more sensationally *The Headhunters of Borneo* (1882), his account of his observations on the 700-mile route from Tangaroeng to Bandjermasin. Highlights included his report on cannibalism among the Dayak people as well as his prolonged, but obviously unsuccessful efforts to locate a tribe of men with tails he had been told of. He also wrote: *Temples and Elephants: The Narrative of a Journey of Exploration Through Upper Siam and Lao* (1884). He was Vice-consul (1886) and Consul General (1893) in Shanghai. Later (1899–1900) he was Consul in Antwerp and (1900–1903) Consul General in Lisbon after which he left government service and settled in Brussels. He collected a single male of this damselfly at the Paio Mountains, Sumatra.

Bodkin

Clubtail sp. *Zonophora bodkini* **Campion**, 1920
[JS *Zonophora batesi* Selys, 1869]

Sir Gilbert Edwin Bodkin (1886–1955) was a biologist and zoologist with an interest in entomology. He was Government Economic Biologist, George Town, British Guiana (1912–1919). He later occupied a similar post in British occupied Palestine (1922–1929) at the Department of Agriculture, Mount Carmel, Haifa, and later still was Director of Agriculture, Mauritius (1937–1942) for which service he was knighted. Earlier he visited and reported on Trinidad, Barbados and Dominica (1911). He wrote a number of papers on economic insect problems such as: *Insects injurious to sugar cane in British Guiana, and their natural enemies* (1912). He found one female specimen of this new species in British Guiana (1915).

Bohart

Skimmer sp. *Tapeinothemis boharti* **Lieftinck**, 1950

Dr George Edward 'Ned' Bohart (1916–1998) was an internationally recognized expert in Apoidea taxonomy. He grew up interested in natural history and collected insects with his brother, Richard M 'Doc' Bohart (1913–2007) who also became a professor and an entomologist. University of California, Berkeley awarded his BSc, MSc and PhD (1947) – his thesis was: *Filth Inhabiting Flies of Guam*. In WW2 he was a navy medical entomologist assigned to malaria control and studied dipteran life cycles on decaying human corpses. He was an adjunct professor of biology and later professor at USU (Utah State University). He then worked as the pollination expert for the USDA-ARS Bee Biology and Systematics Lab, where he spent the rest of his career becoming director and retiring in 1973. He wrote over 140 papers and was still publishing 15 years later. He collected the holotype (and still the only known example) of the eponymous species (1945) that is probably endemic to the Florida Islands (Solomon Islands). Many other insects are named after him.

Bolívar

Genus *Bolivarides* **Martin**, 1907
[JS *Hadrothemis* Karsch, 1891]

Skimmer sp. *Diplacina bolivari* **Selys**, 1882

Ignacio Bolívar y Urrutia (1850–1944) was a Spanish naturalist, one of the founding fathers of Spanish entomology. After a law degree and doctorate in natural sciences he was offered a teaching post at the Central University of Madrid (1875) becoming Professor of Entomology (1877–1939). He also became Director of the Museum of Natural Sciences (1901-1934) and the Royal Botanical Garden of Madrid (1921–1930). During the Spanish Civil War, he was exiled to the south of France (1939) and moved from there to Mexico. Made an honorary doctor at Mexico's National University, he collaborated with scientists there for the five years before he died. He wrote more than 300 papers and books including: *Ortópteros de España nuevos o poco conocidos* (1873) and *Catálogo sinóptico de los ortópteros de la fauna ibérica* (1900). He described over one thousand species and 200 genera. He was also one of the founders (1871) of the 'Real Sociedad Española de Historia Natural'.

Bolton

Common Goldenring *Cordulegaster boltonii* **Donovan**, 1807
[Orig. *Libellula boltonii*]

Thomas Bolton (1722–1778) and his younger brother James (1735–1799) were keen naturalists. Whereas the more successful James was a very good illustrator and keen on mycology and botany, Thomas was more interested in ornithology and entomology, although he too was a competent botanist who collected for others across northern English counties including his native Yorkshire. Sources differ as to which brother collected the eponymous holotype later studied by Edward Donovan; the majority attributing it to Thomas. The two brothers contributed to the natural history section in: *The History and Antiquities of the Parish of Halifax in Yorkshire* (1775).

Boomsma

Firetail sp. *Telebasis boomsmae* **Garrison**, 1994

Tineke Boomsma (b.1954) is a Dutch entomologist. She discovered the species (1992) on the Gallon Jug Estate in Belize when undertaking a survey with Jan Meerman with whom she manages (since 1989) the Shipstern Nature Reserve, Belize. Among other papers she co-wrote, with Dunkle: *Odonata of Belize* (1996) and, with Meerman: *Biodiversity of the Shipstern Nature Reserve* (1993).

Borchgrave

Demoiselle sp. *Mnesarete borchgravii* **Selys**, 1869
[Orig. *Hetaerina borchgravii*]

Comte Paul Edmund Joseph de Borchgrave d'Altena (1827–1901) was a Belgian nobleman who was an aide to the king rising to Chef de Cabinet du Roi (Secretary) to King Leopold II (1895–1900). He collected around his home when he lived at Nova Friburgo, Rio de Janeiro, Brazil (1860s) during the time he was an ambassador there. He wrote a treatise on farming

Coypu! He was Selys' nephew and close friend, the son of Selys' stepsister Coralie, née Smits (1800–1854). After returning to Belgium he gave Selys many rare odonates including this one.

Borgmeier

Dasher sp. *Micrathyria borgmeieri* **Santos**, 1947

Father Thomas Borgmeier OFM (née Heinrich Fritz Hermann Borgmeier) (1892–1975) was a German born entomologist who spent most of his adult life in Brazil. He graduated in classics at the Bielefeld gymnasium and then went to Brazil (1910) to join the Franciscan Order of Friars (1911). He studied philosophy in Curitiba (1912–1914) and theology in Petropolis (1915–1918) when he was ordained. There he became interested in entomology in general and ants in particular. He met and befriended von Ihering (q.v.), the founder of Museu Paulista, Sao Paulo. He became Professor of Biblical Sciences in Pertopolis (1920–1924). His scientific papers (1920 on) included his first description of a new species (1922). He also became an adjunct research scientist at the National Museum in Rio de Janeiro, moving there in 1924. He became a Brazilian citizen (1927) and went to Sao Paulo as Assistant Entomologist (1928) at the newly founded Instituto Biologico. He returned to Rio (1933) becoming Head of the Entomological Section of the Instituto de Biologia Vegetal in the Botanical Gardens (1933–1941). He founded the international journal 'Revista de Entomologia' (1931) and edited it until it folded (1951). His studies were on the back burner (1940–1952) when he was Director of the Franciscan publishing house 'Vozes'. He spent the following twenty years devoted to entomology. He published 243 papers on entomology and described around 1100 insects. His beetle collection is in the Zoological Museum of the University of Sao Paulo.

Boris

Bulgarian Emerald *Somatochlora borisi* Marinov, 2001

Boris Milenov Marinov (b.1996) is the son of the author, Milen Georgiev Marinov (b.1968). The discovery of this previously unknown dragonfly species in Europe was a great surprise. He moved with his family to New Zealand (2008), but he is currently (2017) studying artificial intelligence at the University of Groningen, The Netherlands.

Borror

Skimmer sp. *Oligoclada borrori* **Santos**, 1945

Dr Donald Joyce Borror (1907–1988) was Professor of Entomology at the Ohio State University and was particularly interested in Odonata and bird song. Otterbein University awarded his BSc (1928) and Ohio State University his MSc (1930) and PhD (1935). He stayed on as an instructor becoming assistant professor (1946), associate professor (1948) and full professor (1959–1977). In between times he served in the Naval Reserves as a Lieutenant (1944–1945) where he was employed looking at malaria controls. He founded the Borror Laboratory of Bioacoustics at the university, which houses one of the largest collections of recorded animal sounds in the world; more than 30,000 recordings of over 1400 species. He wrote many papers such as: *The genus Oligoclada (Odonata)* (1931) and *A revision of the libelluline genus Erythrodiplax (Odonata)* (1942), in which he described a total of 25 new species. His longer works include: *An introduction to the study of insects* (1954) and *A Field Guide to Insects: America North of Mexico* (1970).

Bosq

Hawker sp. *Staurophlebia bosqi* **Navás**, 1927

Juan M Bosq was an Argentine collector (1920–1953), particularly of butterflies and beetles, but of all insects. He collected the holotype (1926) of the eponymous species. He collected in many locations such as the provinces of Corrientes (1920, 1949), La Plata (1920), Buenos Aires (1924, 1926, 1951), Entre Rios (1943), Misiones (1939), Santiago del Estero (1935, 1942), Parana (1939), and Jujuy (1924).

Botacudo

Dancer sp. *Argia botacudo* **Calvert**, 1909

The Botacudo are an indigenous people in eastern Brazil.

Bott

Collared Threadtail *Caconeura botti* **Fraser**, 1922
[JS *Prodasineura collaris* (Selys, 1860)]

John Richard Elton-Bott (1881–1942) was an English civil engineer and architect. He took (1901–1909) a managerial position in a rubber plantation in the Mergui district of Burma. He returned to England (1909–1912) but moved back to Burma, married a local woman and settled. There he collected several males of the eponymous species, as well as other new dragonfly species at King Island (1921). It seems he also collected insects in general. When WW2 broke out, he trekked to India but sadly died (1942) not long after arriving in Calcutta, due to the privations of the trek. (Also see **Elton**)

Böttcher

Sylvan sp. *Coeliccia boettcheri* **Schmidt**, 1951

Georg Böttcher (1890–1919) was an insect collector from Berlin. German coleopterologist Julius Moser (1863–1929) sent (1913) Böttcher to the Philippines to collect beetles for him. When WW1 broke out Böttcher could not return home, so had to stay in the Philippines. There he travelled extensively to various islands and collected many insects of different orders. When he was able to return to Berlin (1919), he died soon after being, as one source puts it, 'involved in communist affairs'. Böttcher had left his insect and ethnographical collections in Manila as security for the loans he had taken out to support himself. To retrieve the collection from Manila, Moser reimbursed his creditors with help of various foreign entomologists. Among them was Friedrich Ris (q.v.), who received about 1800 specimens of 135 Odonata species. This important collection is now in the Senckenberg Museum (SMF). This George Böttcher has often been confused with Dr Georg Böttcher (1865–1915), a German physician and entomologist working on Diptera.

Boucard

Desert Firetail *Telebasis boucardi* **Selys**, 1868
[JS *Telebasis salva* (Hagen, 1861)

Adolphe Boucard (1839–1905) was a French ornithologist, naturalist and trader who collected in Mexico and central America and spent c.40 years killing hummingbirds for science and the fashion trade. He lived in San Francisco (1851–1852) at the height of the 'gold rush'. He collected on expeditions into Mexico (1854–1867) including the eponymous species in (1867 or earlier). He moved to London (1890) where he set up a taxidermy company, *Boucard, Pottier & Co.*, but spent his later years (1894–1905) at his villa near Ryde on the Isle of Wight. He was author of: *The Hummingbird* (1891). He published a periodical 'The Hummingbird' (1891–1895) and wrote (1894) that *"...Now-a-days the mania of collecting is spread among all classes of society, and that everyone possess, either a gallery of pictures, aquarels, drawings, or a fine library, an album of postage stamps, a collection of embroideries, laces... ...and such like, a collection of humming-birds should be the one selected by ladies. It is as beautiful and much more varied than a collection of precious stones and costs much less..."* He wrote: *Travels of a Naturalist* (1894). A reptile and eight birds are also named after him.

Bouchard

Acuminate Snaketail *Ophiogomphus bouchardi* **Louton**, 1982
[JS *Ophiocomphus acuminatus* Carle, 1981]

Dr Raymond W Bouchard Jr (b.1944) is an American zoologist. His bachelor's degree was awarded (1967) by Massachusetts State College and his PhD (1972) by University of Tennessee. Bouchard discovered a number of crayfish when he was a postdoctoral fellow (1974–1975) at the Smithsonian, University of Tennessee. He was then (1976–1978) Assistant Professor of Biology, Department of Biology, University of North Alabama. He was (1986–2001) Senior Scientist and Director, Macroinvertebrate Section and then Senior Scientist & Director of the Invertebrate Zoology Sector of the Patrick Center for Environmental Research, Philadelphia Academy of Natural Sciences at Drexel University where his research focus is macroinvertebrates. When collecting crayfish (1971) he collected a single nymph of a female of the snaketail and succeeded in rearing it. It was put in the University of Tennessee collection where it remained until studied by his long-time friend and colleague Dr J A Louton (1976). [Louton raised others and this type series was described and the new species named in his paper: *A new species of Ophiogomphus (Insecta: Odonata: Gomphidae) from the western highland rim in Tennessee* (1982)]. Bouchard has published many scientific papers mostly on crayfish and other crustaceans. Some crayfish are named after him such as *Ascetocythere bouchardi* that he discovered.

Boudot

Boudot's Pincertail *Onychogomphus boudoti* Ferreira, 2014

Dr Jean-Pierre Boudot (b.1948) is a French entomologist and geologist who was a researcher at the French National Centre for Scientific Research (1977–2012). Among his 195 papers and symposia communications in the field of soil science, botany and entomology, 79 refer to odonatology. He has co-written books such as: *Les Libellules (Odonates) du Maroc* (1999), *Les Libellules de France, Belgique et Luxemburg* (2006), *The Atlas of the Odonata of the Mediterranean and North Africa* (2009) and *Cahier d'identification des Libellules de France, Belgique, Luxembourg & Suisse* (2014) and *The Atlas of European Dragonflies and Damselflies* (2015). He is also co-founder of the French Society of Odonatology.

Boumiera

Brownwater Skimmer *Orthetrum boumiera* **Watson** & Arthington, 1978

This is a toponym, the Aboriginal name for Brown Lake, North Stradbroke Island, Australia.

Boyer

The Hawker genus *Boyeria* **McLachlan**, 1896

Étienne Laurent Joseph Hippolyte Boyer de Fonscolombe (1772–1853) was an aristocratic French entomologist specialising in coleoptera, hymenoptera and pests. He was locked up (1793–1794) as a suspect during 'The Terror'. His mother described him as "...*a remarkable entomologist with intelligence, kindness, virtue and knowledge*". Always interested in natural history, he entrusted the management of his estates (1833) to his son-in-law so he could devote his time to entomology. He published most of his work in '*Mémoires de l'académie d'Aix*'. Much of his collection is in the MNHN in Paris but there are some of his Apoidea in the Hope Department of Entomology in Oxford. (Also see **Fonscolombe**)

Bradley

Fineliner sp. *Teinobasis bradleyi* **Kimmins**, 1957

Dr John David Bradley (1920–2004) was a British entomologist, primarily a lepidopterist. He left school at sixteen and became a laboratory assistant at the London School of Hygiene and Tropical Medicine (1936). He then joined the BMNH as a preparator (1938). During WW2, he served for six years in North Africa, Italy and Austria. When he returned to the BMNH he worked on micro-Lepidoptera which became his life focus. He collected the holotype on Guadalcanal Island during the BMNH Rennell Island Expedition (1953–1954), his wife Diana also collected on that trip. Still he retired to Somerset (1980) but continued with his study right up until his death. Among his 120 papers (1950–2000) were: *Microlepidoptera from Rennell and Bellona Islands* (1957) and *British Tortricoid moths. Cocliylidae and Tortricidae: Tortricinae* (1974). Despite his lack of a formal zoological education, an international friend and colleague encouraged him to submit his studies to the Charles University, Prague which awarded him a PhD.

Braganza

Iridescent Flutterer *Rhyothemis braganza* **Karsch**, 1890

Karsch erroneously thought that the single specimen of his new libellulid species came from Brazil. However, it later transpired that it was an Australian species. No etymology was given, but this name refers to the Portuguese royal house Braganza, and most likely the name was a dedication to Dom Pedro II (1825–1891), a member of Braganza House and the second and last ruler of the Empire of Brazil (1831–1889). He lost his throne in a coup on 15 November 1889, and two days later he and his family were sent to exile in Europe. Dom Pedro II was well known as a patron of arts and sciences.

Brauer

Emerald sp. *Antipodochlora braueri* **Selys**, 1871
[Orig. *Epitheca braueri*]

Skimmer sp. *Diplacina braueri* Selys, 1882
Bombardier *Lyriothemis braueri* **WF Kirby**, 1889
[JS *Lyriothemis cleis* Brauer, 1889]
Jewel sp. *Rhinocypha braueri* **Krüger**, 1898
[JS *Rhinocypha sumbana* Förster, 1897]
Archtail sp. *Nannophlebia braueri* **Förster**, 1900
[Orig. *Tetrathemis braueri*]
Siberian Winter Damsel *Sympycna braueri* **Bianchi**, 1904
[JS *Sympecma paedisca* (Brauer, 1877)]

Friedrich Moritz Brauer (1832–1904) was an Austrian entomologist who became Director of the Naturhistorisches Hofmuseum, Vienna. His interest in entomology was sparked by a gift of a small exotic insect collection which he quickly wanted to identify, leading him to contact curators, read around the subject and raise eggs (Neuroptera). Despite not initially passing entry qualification he was eventually allowed to enter The University of Vienna (1853) to study medicine. In the year he passed his medical exams (1861), he survived typhus and sought financial independence by taking a job as an assistant at what was to become the Natural History Museum at the University of Vienna where he worked for sixteen years and rose to Custodian of the Collections (1873). In the meantime, he had passed the second medical exam (1871) allowing him to pursue an academic career. He was appointed (1876) Curator of Entomology and later Professor of Zoology at the university (1878). He went on to become Director of the Zoological Museum (1898) by which time his health was failing. He first published in 1850 and went on to write many papers, particularly on diptera and neuroptera (then thought to encompass Odonata), being considered one of the foremost authorities on them in Europe. He wrote: *Neuroptera Austriaca* (1857) and his most famous work was: *System of Diptera* (1883); his last paper was a chapter on Diptera (1901). He established over twenty dragonfly genera and named about 100 new dragonfly species. (Also see **Leontine**)

Braulita

Limniad sp. *Sangabasis braulitae* Villanueva, 2005
[Orig. *Amphicnemis braulitae*]

Braulita Calingin Torayno (1921–2007) was the grandmother of the author Reagan J T Villanueva (b.1981). The author found the species in the Camiguin Island in the Philippines.

Bredo

Northern Fingertail *Gomphidia bredoi* **Schouteden**, 1934
[Orig. *Diastatomma bredoi*]
Sombre Cruiser *Macromia bredoi* Schouteden, 1934
[JS *Phyllomacromia melania* (Selys, 1871)]
Tiger Hooktail *Mesogomphus bredoi* Schouteden, 1934
[JS *Paragomphus serrulatus* (Baumann, 1898)]
River Dropwing *Trithemis bredoi* **Fraser**, 1953

Hans Joseph Anna Erich Richard Bredo (1903–1991) was the Belgian government entomologist who collected, including the types in the Congo (1930s). He was sent to the

Congo to advise the Belgian government on all locust matters (1930) and later was appointed Director of the International Red Locust Team (1949–1951). His major achievement was a scientific explanation for locusts swarming (i.e. a succession of environmental conditions affecting the behaviour of these usually non-gregarious insects). He wrote a number of scientific papers such as: *La lutte biologique et son importance econimique au Congo Belge* (1934). Two birds are also named after him.

Bréme

Long Skimmer *Libellula bremii* **Rambur**, 1842
[JS *Orthetrum trinacria* (Selys, 1841)]
Small Redeye *Agrion bremii* Rambur, 1842
[JS *Erythromma viridulum* Charpentier, 1840]

Ferdinando (François) Arborio Gattinara di Bréme, Marquis de Bréme, Duc de Sartirana Lomellina (1807–1869) was an Italian naturalist and entomologist specialising in beetles and flies. He was President of the Société Entomologique de France (1844). He collected the types of the species in Sicily. He wrote several entomological treatises including: *Essai Monographique et Iconographique de la Tribu des Cossyphides* (1842) and *Des Cossyphides* (1846).

Brevignon

Shadowdamsel sp. *Palaemnema brevignoni* Machet, 1990

Christian Brévignon (b.1955) is a French entomologist. Lalita Brévignon and Christian Brévignon collected insects in French Guiana (1989) (including the type), especially Lepidoptera. He is currently Directeur adjoint du Service régional de Guadeloupe de Méteo France and has co-written about the butterflies of the region: *Papillons des Antilles* (2003) and: *The Infratribe Strephonina in French Guiana* (2003). There are a number of butterflies among 84 new taxa he has described (1985–2015) and at least one is named after him.

Brewer

Flatwing sp. *Heteragrion breweri* **De Marmels**, 1989

Charles Brewer-Carías (b.1938) is a Venezuelan dentist (1960), the son of a British diplomat. After 20 years in practice he became a naturalist, explorer (known as the Humboldt of the twentieth century) and speleologist (1979). He has led over 200 expeditions to remote parts of Venezuela. His discoveries include the sink holes of Cerro Sarisariñama and the world's largest quartzite cave. He is currently out of favour with the Chavez regime as his brother, a lawyer, fled into exile after being openly critical of them. He was a cabinet minister (1979–1982) as the Minster for Youth Affairs & Sports. He is well known to the indigenous people of the Venezuelan highlands, as he fixed their teeth as well as exploring their territory. Thieves who broke into his home (2003) shot him, but he shot back and killed one. Despite some loss of flexibility in his shoulder he still explores. The quartzite cave is named after him, as are twenty-seven species including two amphibians, a bird, a reptile as well as many plants and other taxa.

Brian May

Flatwing sp. *Heteragrion brianmayi* **Lencioni,** 2013

Dr Brian Harold May CBE (b.1947) is an astrophysicist most famous as an outstanding guitarist, songwriter and conservationist, ..."*whose*," the author said, "...*marvellous sound and lyrics have enchanted the world for over four decades*'. He gained his PhD after his rock fame, at Imperial College (2007). Four new species were described in tribute to the 40th anniversary of the rock band Queen – one for each of its original members.

Brightwell

Rubyspot sp. *Hetaerina brightwelli* **W Kirby**, 1823
[Orig. *Agrion brightwelli*]

Thomas Brightwell (1787–1868) was an English nonconformist solicitor who became Mayor of Norwich (1837). He was a keen amateur natural historian, primarily a botanist and entomologist, although he also specialised in diatoms and protozoa. He made a collection of insects, especially Coleoptera, which he gave to the Norwich museum (1844). He wrote the religious book: *Notes on the Pentateuch* (1840) and the natural history book: *Sketch of a Fauna Infusoria for East Norfolk* (1848).

Brimley

Sandhill Clubtail *Gomphus brimleyi* Muttowski, 1911
[A Brimley's Clubtail, JS *Gomphus cavillaris* Needham, 1902]

Dr Clement Samuel Brimley (1863–1946) and his brother, Herbert Hutchinson Brimley (1861–1946) emigrated from England to North Carolina (1880) where the former collected the type (1910). He started collecting, preparing and selling natural history specimens. He worked for the Entomology Division, North Carolina Department of Agriculture (1919). His doctorate was an honorary Doctor of Laws conferred by University of North Carolina. He wrote a few papers on the dragonflies of North Carolina. The brothers co-wrote: *Birds of North Carolina* (1919). They both worked at the North Carolina Museum of Natural Sciences, which Herbert eventually took charge of in 1895. At least one amphibian is also named after Clement.

Brinck

Brinck's Shadowdamsel *Drepanosticta brincki* **Lieftinck**, 1971

Per Simon Valdemar Brinck (1919–2013) was a Swedish zoologist. The bulk of his collection is at the Museum of Zoology, Lund University where he was Professor of Zoology (1958–1986) and Director of the Department of Animal Ecology. His area of interest was entomology and astacology and he studied Coleoptera in particular. He was one of the organizers of the Lund University Expedition to Ceylon (1962) in conjunction with the Government of Ceylon (Sri Lanka) and UNESCO. Lieftinck studied the odonates collected during this expedition and dedicated this species to Brinck. He published over 230 papers and articles.

Brito

Skydragon sp. *Chlorogomphus brittoi* **Navás**, 1934

Brother João Heitor de Brito SJ (sometimes San Juan de Britto or John de Brito) (1647–1693) was a Jesuit saint whose father was Governor of Brazil. He became a novitiate (1662), studied at the College of Evora and University of Coimbra receiving holy orders (1673) and assigned to the missions of India in Malabar, where he became a panderam ascetic with beard and turban, mediator between the Brahmins and outcasts. He was caught and tortured in Madurai (1684) and returned to Portugal (1687) but soon went back to India, but he was once again captured and this time beheaded. Navás dedicated the species to him whom he described as a great apostle and martyr.

Broadway

Setwing sp. *Dythemis broadwayi* **WF Kirby**, 1894
[JS *Dythemis sterilis* Hagen, 1861]

Walter Elias Broadway (1863–1935) was an English botanist and naturalist who became Superintendent of the Botanical Gardens, Trinidad. He started working life as a gardener aged sixteen and after six years became a gardener at Kew, rising to sub-foreman. He was appointed as Assistant Superintendent at the Royal Botanic Gardens in Trinidad (1888) where he was in awe of the tropical plants and began learning all he could. He collected widely before taking a post in Granada as the Curator of the Botanical Gardens (1894–1904). In his decade there he collected specimens for sale. He returned to Trinidad (1906) and continued collecting there, then took a post on Tobago as a curator (1908–1915). He also collected in French Guyana (1921) and Venezuela (1922–1923). He was honoured as he had sent the specimens for Kirby to study.

Brookhouse

Barrington Flatwing *Austroargiolestes brookhousei* **Theischinger & O'Farrell**, 1986

Peter A Brookhouse (b.1955) is an Australian collector whose specimens are in the Australian National Insect Collection. He was very involved in collecting for the study the authors undertook and he collected for them over a number of years. He collected the type in New South Wales (1976).

Brosset

Yellow-winged Slim *Aciagrion brosseti* **Legrand**, 1982

Dr André Brosset (1926–2004) was a French zoologist. He was Assistant Director of the Museum of Natural History, Paris. He served in the French civil service in Morocco (1953–1959) and while there observed birds and collected in his spare time. He then worked in the consulate in Bombay (1959–1961) where bats were his spare time focus. He became Director of the Charles Darwin Research Station in the Galápagos Islands (1962–1963), continuing with his naturalism there and in Ecuador. When he returned to France he was asked to join the national centre for scientific research, CNRS (1963–1972) as a zoologist specialising in ecology and behaviour in Gambon where CNRS established a laboratory and asked him to take charge (1972–1982). He also visited French Guiana a number of times. He published numerous articles both in scientific journals and in popular science journals such as 'Nature' and was also a keen ornithologist. A mammal is also named after him.

Brown

Duskhawker sp. *Agyrtacantha browni* Marinov & **Theischinger,** 2012

Dr Rafe Marion Brown (b.1968) is an American zoologist and herpetologist who has described around 70 taxa. He studied for his BA (1994) and his MSc (1997) at Miami University, Ohio (Oxford, Ohio). The University of Texas at Austin awarded his PhD (2004) during which time he was also a curatorial assistant (1999–2001) and assistant instructor in biology (2000–2004). He also undertook postdoctoral research for a year (2004-2005) at the University of California, Berkeley and became an associate scientist of the Philippines' Silliman University (2003) at the same time. He now (2011–present) works as an Associate Professor at the Department of Ecology and Evolutionary Biology and Curator-in-charge of the Herpetology Division of KU's Biodiversity Institute, University of Kansas. His dissertation was: *Evolution of Eco-morphological Variation and Acoustic Diversity in Mate-Recognition Signals of Southeast Asian Forest Frogs (subfamily Platymantinae)* (2004). His research interests are biogeography and processes of diversification in island archipelagos; phylogenetic systematics, genomics, phylo-geography, population and speciation genetics; biodiversity conservation, systematics and the taxonomy of amphibians and reptiles of Southeast Asia. He has collected in the Philippines and Solomon Islands, where he took the type (2012). In his personal life, he loves rap and hip-hop music from the 1990s.

Brownell

Shadowdamsel sp. *Drepanosticta brownelli* Tinkham, 1938
[Orig. *Ceylonosticta brownelli*]

Chauncey Wells Brownell (1917–1995) was an American entomologist. He found two specimens of this species in Kwangtung (Guangdong), China (1935). Tinkham himself had collected one specimen of the same species a few days earlier in another location in Kwangtung.

Bruce (Kennedy)

Shadowdamsel sp. *Palaemnema brucelli* **Kennedy**, 1938

Bruce Albert Hamilton Kennedy (1928–2013) was the son of the author C H Kennedy (q.v.). The original description reads: "...*Brucelli = of Bruce. The smaller. Named for my [Clarence H. Kennedy's] son, Bruce Hamilton Kennedy, God-son of Edward Bruce Williamson, and in reference to the fact that brucelli is a close relative of [Palaemnema] brucei Calvert.*" Kennedy and E.B. Williamson collaborated on many projects involving Odonata during the 19th and early 20th century.

Bruce (Williamson)

Shadowdamsel sp. *Palaemnema brucei* **Calvert**, 1931

Edward Bruce Williamson of Bluffton Indiana collected a great deal of material described by Calvert. He was known for "...*his indefatigability as a collector, the minuteness of his discrimination of species and his generosity to other students.*" (See **Williamson**)

Bruch

Pond Damselfly sp. *Oxyagrion bruchi* **Navás**, 1924

Dr Franz Karl (Carlos) Bruch (1869–1943) was a German-born Argentine naturalist, entomologist and archaeologist. He was selected by the first director of the Museo de la Plata to organize its collections. He had a background as a professional photographer before becoming a professor at the University of La Plata and was a fine illustrator. He was particularly interested in ants and many of his papers were on them. He collected the type in Argentina (1920).

Brunnea

Southern Skimmer *Orthetrum brunneum* **Fonscolombe**, 1837
[Orig. *Libellula brunnea*]
Common Bluet *Agrion brunnea* **Evans**, 1845
[JS *Enallagma cyathigerum* (Charpentier, 1840)]
Satinwing sp. *Euphaea brunnea* **Selys**, 1879
[JS *Euphaea ochracea* Selys, 1859]
Giant Hawker *Tetracanthagyna brunnea* **McLachlan**, 1898

Brunnea is just Latin for brown.

Bryden

Percher-like Dropwing *Trithemis brydeni* **Pinhey**, 1970

John W Bryden was an agricultural entomologist (1960s–1970s) at the Mount Makulu Research Station near Chilanga. He planned and arranged the expedition to North Zambia, which included the author, who collected the holotype on reeds on the edge of the dam at Msamfu, Kasama (1969). He also took an interest in the fieldwork and helped to collect this new species. He was the author of a number of papers concerning agriculture such as: *Some Studies on Soils Pests Attacking Virginia Tobacco in Zambia* (1976).

Buch

Flatwing sp. *Philosina buchi* **Ris**, 1917

Pater A Buch SJ (b.1865) was a French Catholic missionary at the Mission in Ningpo, from the Vincentian order in China and Indo-china (1906–1952). He collected Lepidoptera specimens and sent them to many museums including the Field Museum. Buch sent the holotype to Ris from Fujian (1916). He collected the holotype of an eponymous snake.

Buchecker

Green Snaketail *Paragigma bucheckeri* Buchecker, 1876
[JS *Ophiogomphus cecilia* (Fourcroy, 1775)]

Heinrich Friedrich Buchecker (1829–1894) was a German entomologist. He wrote: *Systema Entomologiae sistens insectorum* (1876) in which he named a species from Zurich and Munich as 'Paradigma bucheckeri Landolt' according to the manuscript name furnished by Landolt. However in nomenclatorical terms he created, perhaps unintentionally, an eponym for himself.

Buchholz

Swampdamsel sp. *Leptobasis buchholzi* **Rácenis**, 1959
[JS *Chrysobasis buchholzi*]

Pond Damselfly sp. *Azuragrion buchholzi* **Pinhey**, 1971
[Orig. *Enallagma buchholzi*]

Greek Goldenring ssp. *Cordulegaster helladica buchholzi* (Lohman, 1993)
[Orig. *Sonjagaster helladica buchholzi*]

Dr Karl Friedrich Buchholz (1911–1967) worked at the Forschungsinstitut Koenig, Bonn and was better known as a herpetologist, particularly of Greek species. He was Curator of Herpetology at Museum Koenig at the time of his death. He was a colleague and friend of Rácenis and first examined the collections described by Pinhey. Buchholz described (1950–1959) ten new odonate species and subspecies, most of them from Africa.

Buck

Turquoise Flatwing *Griseargiolestes bucki* **Theischinger**, 1998

Dr Klaus Buck (1923–2006) of Wilster, Germany, was a prolific photographer of Australian dragonflies. Many of his photographs have been used in field guides and other papers.

Buckley

Blue-tipped Helicopter *Platystigma buckleyi* **McLachlan**, 1881
[Orig. *Mecistogaster buckleyi*]

Clarence Buckley (c.1801–fl.1880) was a collector of natural history specimens in Ecuador and Bolivia. He remained in South America (until c.1880), but details of his death appear to be unrecorded. Buckley collected for a number of artists and ornithologists, including Gould, running a team of collectors. The British Museum houses a large fish collection made by him, including several eponymous species. There are also over 80 species of birds collected by him in BMNH and six birds, four amphibians and a reptile are named after him. In one description of a bird in 1887, he is described as 'the late Mr Buckley'.

Buddha

Reedling sp. *Indolestes buddha* **Fraser**, 1922
[JS *Indolestes indicus* Fraser, 1922]

Gautama (Siddhartha) Buddha (c.563BC–c.483BC) was a sage from the ancient Shakya republic on whose teaching Buddhism was founded. The name Buddha translates as 'the enlightened one' and is used as a term for the first person to be 'awakened' in an era.

Buden

Fineliner sp. *Teinobasis budeni* **Paulson**, 2003

Dr Donald William Buden (b.1943) is an American zoologist who is Professor at the Division of Natural Sciences and Mathematics, College of Micronesia. He has collected widely (including the type in the Caroline Islands, especially in Micronesia). He has written many papers (1970s onward) and co-written a number with D R Paulson such as: *The Odonata of Pohnpei, Eastern Caroline Islands, Micronesia* (2006) and *Odonata of Yap, Western Caroline Islands, Micronesia* (2007). The dedication reads: *"...I take pleasure in naming this species after its discoverer, who backpacked across the island numerous times, often under adverse weather conditions, to survey its Odonata fauna."*

Buenafe

Sprite sp. *Pseudagrion buenafei* **Müller**, 1996

Alex Buenafe is an insect collector and nature guide from Negros, the Philippines. He was one of Roland Müller's regular helpers during his many expeditions to various islands of the Philippines in search of dragonflies and other insects (1985–1997) during which he collected the holotype. This species was found in Mindanao.

Bühr

Leaftail sp. *Oligoaeschna buehri* **Förster**, 1903
[Orig. *Jagoria buehri*]

Heinrich Bühr (b.1868) was a notary from Baden-Württemberg, Germany. The etymology reads: *"To Notar H. Bühr (Brettan), who's eagerness to collect brought numerous samples to my collection."* The single male was collected in Brunei.

Bum Hill

Bumhill Wisp *Agriocnemis bumhilli* Kipping, Martens & Suhling, 2012

This is a toponym; Bum Hill is a community campsite on the banks of the Kwando River, Namibia.

Burbach

Shadowdamsel sp. *Drepanosticta burbachi* Dow, 2013

Klaus Burbach (b.1963) is a German Landscaper, conservationist and entomologist. He studied at the University of Applied Sciences in Freising in late 1980s. He is currently living and working in Bavaria for the Nature Conservation Authority of Lower Bavaria as well as in freelance conservation work. He is well-known as the leading Odonata researcher in Bavaria. He co-authored: *Libellen in Bayern* (1995). The species was collected in Sarawak.

Burgeon

Giant Skimmer *Hadrothemis burgeoni* **Schoudeten**, 1934
[JS *Orthetrum austeni* (Kirby, 1900)]

Louis Jules Léon Burgeon (1884–1949) was a Belgian engineer with a passion for entomology, particularly beetles, who became Director, Zoology Department, the Royal Museum for Central Africa, Tervuren. He actively collected fauna (insects, amphibians and reptiles) and flora there (from 1917) and in East Africa (1930s). He co-wrote: *Les insects du Congo Belge* (1950). He described 126 new beetles and also found one female specimen of the Odonata species in Kindu, Belgian Congo. A skink is also named after him.

Burgos

Leaftail sp. *Phyllogomphoides burgosi* **Brooks**, 1989

Mario Burgos is a rancher in Costa Rica. The etymology reads: *"...Named in honour of Sr Mario Burgos who very kindly allowed his ranch to be purchased by Guanacastle National Park".*

Burmeister

Keyhole Glider ssp. *Tramea basilaris burmeisteri* **WF Kirby**, 1889
[A. Wheeling Glider, Orig. *Tramea burmeisteri*]

Carl Hermann Konrad Burmeister (1807–1892) was a German ornithologist and entomologist. He originally trained in medicine hoping to secure a post in the Dutch East Indies, but when that failed to materialise he began teaching. He was Professor and Director of the Institute of Zoology of Martin Luther University at Halle Wittenberg, Germany (1837–1861). He was in the Prussian civil service but got his release by using the very inventive excuse that a persistent stomach complaint was caused by arsenic emissions in the museum and the drinking water in Halle, which had high natrium (sodium) sulphate content. He sent many specimens to the zoological collections at the institute. These were largely collected during his two expeditions to South America, Brazil (1850–1852) and La Plata region of Argentina (1857–1860). Subsequently, he was resident in Argentina (1861–1892) and greatly honoured there. He was founding Director of the Institute at the Museo Nacional in Buenos Aires, remaining in post until retirement (1880). His major scientific contribution was the four volumes of (the intended five-volume) *Handbuch der Entomologie* (1832–1847). In its second volume (1839) he described over 100 new dragonfly species including many of the commonest from all parts of the world. His other works include: *Reise nach Brasilien durch die Provinzen von Rio de Janeiro und Minas Geraes* (1853). While working in his laboratory he fell from a ladder and hit his head on a glass display case, resulting in his death three months later. The President of Argentina, Carlos Pellegrini, and his ministers participated in his funeral. Later a marble statue was erected in his honour in Buenos Aires. A mammal, a reptile, three birds and two amphibians are named after him as well as other insects such as a stag beetle.

Büttikofer

Sunlight Firebelly *Eleuthemis buettikoferi* **Ris**, 1910
[Orig. *Eleuthemis büttikoferi*]

Johan Büttikofer (1850–1929) was a Swiss zoologist. He made two collecting trips to Liberia during the first of which he collected the male type (1879–1882 & 1886–1887) then he returned home due to ill health. He did, however, make one more trip there (1888). He accompanied Nieuwenhuis to Borneo (1893–1894) where they explored and collected. He was the Director of the Rotterdam Zoo (1897–1924). He wrote: *Zoological researches in Liberia. A list of birds, collected by the author and Mr F.X. Stampfli during their last sojourn in Liberia* as a note for the Leyden Museum. A number of fish, three mammals, nine birds and two reptiles are also named after him.

Buwalda

Phantom sp. *Podolestes buwaldai* **Lieftinck**, 1940
Fineliner sp. *Teinobasis buwaldai* Lieftinck, 1949

Dr Pieter Buwalda (1909–1947) was a botanist who collected (1920s–1940s) in the Dutch East Indies when working for the Buitenzorg Forest Research Institute. He studied biology at the University of Groningen, eventually achieving a doctorate (1936). Being so interested in tropical trees he accepted a job as a planter in Java (1937), as there were no botanist posts. He was soon put at the disposal of the Forestry Research Station, employed to look at forestry resources. His first reports were so well received that he was awarded the exploration of the Tenimbar and Aroe Archipelagos to study and obtain samples and seeds of potentially economically useful plants. He went on to collect in the Moluccas (1938) during which he caught malaria and also suffered quinine poisoning; nevertheless, he went on to explore in central Sumatra, then east Java, south Sumatra and southeast Borneo (1941) among others. During WW2, he was a Lieutenant in the Buitenzorg Reserve and fought against the Japanese, was taken prisoner and suffered severe hardship and ill health as a POW. He was invalided to the Netherlands (1945) and soon returned to Java but shortly after died of diseases related to his captivity. He collected the holotype, a single male specimen of *Podolestes,* in Sumatra (1939) and he collected a series of *Teinobasis* in the Aru Islands (1938).

Buxton

Bluetail sp. *Ischnura buxtoni* **Fraser**, 1927

Professor Patrick Alfred Buxton CMG FRS (1892–1955). Trinity College, Cambridge awarded a first for Part One of his bachelor's degree (1914). He enlisted that year as a private in the medical corps but after a few months was sent back to England to finish his medical studies and he gained his MB in zoology. He finished his medical training at St George's Hospital (1917). He immediately took up a commission in the Royal Army Medical Corps and was posted to northern Persia (Iraq) (1917–1919) where he collected in his spare time. He spent the next two years in a Cambridge laboratory. He then took a medical post in Palestine (1921–1923). He visited Samoa with G H E Hopkins (1924–1925) having been given a temporary appointment at the London School of Hygiene and Tropical Medicine to organize an expedition to investigate filariasis. (He was appointed Head of the Department of Medical Entomology on his return in 1926.) Although resident during that time they did make a side trip to New Hebrides and another to Tonga. In their spare time they collected insects and other arthropods, accumulating 20,000 specimens that they took back to England. The results were published by BMNH as: *Insects of Samoa* (1935), which Buxton edited and otherwise contributed to. Other works of his included an address to the Royal Entomological Society on: *Tsetse and Climate* (1955). Oddly it was J S Armstrong (q.v.) who collected this species, not Buxton who was credited in its name. However, Buxton was among the many who regularly provided odonates from Samoa.

Bwamba

Bwamba Horntail *Libyogomphus bwambae* **Pinhey**, 1961
[Orig. *Onychogomphus bwambae*]

This is a toponym. The Bwamba Forest was part of the Bwindi National Park, Uganda (now Semliki National Park).

Byers

Duckweed Firetail *Telebasis byersi* **Westfall**, 1957

Dr Charles Francis Byers (1902–1981) was an American entomologist at Florida University. He worked on material at Cornell such as James Needham's collection. He wrote many papers, such as: *Notes on Some American Dragonfly Nymphs (Odonata, Anisoptera)* (1927) and *Some notes on the Odonata fauna of Mountain Lake, Virginia* (1951) and much longer works such as: *A contribution to the knowledge of Florida Odonata* (1930). He described four new odonate species and M J Westfall wrote a biography of him (1982).

C

Cacus

Genus *Caconeura* **Kirby**, 1890
Genus *Cacoides* Cowley, 1934

Cacus is a giant in Roman mythology. Cowley's name replaced Selys' (1854) genus name *Cacus*, which he gave to a South American gomphid species. The replacement was needed, since the name Cacus was already used for a beetle genus.

Caesar

Tiger sp. *Gomphidia caeserea* **Lieftinck**, 1929

No etymology was given but undoubtedly this refers to the title 'caesar' (plural 'caesares'), cf. species names such as *Anax imperator* (Leach, 1815).

Caesar

Mediterranean Bluet *Coenagrion caerulescens caesarum* **Schmidt**, 1959
[JS *Coenagrion caerulescens* (Fonscolombe, 1838)]

Named after two people whose forenames are Cesare. (See **Cesare**, **Conci** and **Nielsen**).

Callirrhoe

Skimmer sp. *Diplacina callirrhoe* **Lieftinck**, 1953

In Greek mythology Callirrhoe was an Oceanid, a daughter of Oceanus and Tethys and the mother of Geryon one of the Oceanids. The name appears in various other myths relating to other people.

Callisto

Yellow-veined Widow *Palpopleura callista* Grünberg, 1902
[JS *Palpopleura jucunda* (Rambur, 1842)]
Sylph sp. *Gynothemis calliste* **Ris**, 1913
Cruiser sp. *Macromia callisto* **Laidlaw**, 1922

In Greek mythology Callisto was a nymph whom Zeus loved; she was transformed into a bear and set among the stars.

Calvert

The Genus *Calvertagrion* St. Quentin, 1960

Boreal Bluet *Enallagma calverti* Morse, 1895
[JS *Enallagma boreale*]

Leaftail sp. *Phyllogomphoides calverti* **Kirby**, 1897
[Orig. *Cyclophylla calverti*]
Striped Saddlebags *Tramea calverti* Muttkowski, 1910
Pond Damselfly sp. *Agrion calverti* Perkins, 1910
[JS *Megalagrion hawaiiense* (McLachlan, 1883)]
Threadtail sp. *Protoneura calverti* **Williamson**, 1915

Calvert's Emerald *Somatochlora calverti* Williamson & Gloyd, 1933
Skimmer sp. *Ypirangathemis calverti* **Santos**, 1945
Skimmer sp. *Misagria calverti* **Geijskes**, 1951
Darner sp. *Neuraeschna calverti* **Kimmins**, 1951
Skimmer sp. *Oligoclada calverti* Santos, 1951
Dancer sp. *Argia calverti* **Garrison** & von **Ellenrieder**, 2017

Forest Skimmer ssp. *Cratilla lineata calverti* **Förster**, 1903
[Orig. *Cratilla calverti*]
Mantled Baskettail *Epitheca semiaquea calverti* Muttkowski, 1915
[JS *Epitheca semiaquea* (Burmeister, 1839)]

Dr Philip Powell Calvert (1871–1961) was an American entomologist, a specialist on the neotropical odonate fauna and one of the true giants of odonatology. The University of Philadelphia awarded his certificate in biology (1892) and his PhD (1895). He undertook postdoctoral research at the Universities of Berlin and Jena (1895–1896) and was also a teacher at the University of Philadelphia as assistant instructor (1892–1897), instructor (1897–1907), assistant professor (1907–1912) and full professor (1912–1939). He was associated with the American Entomological Society for many years serving on its council, then as Vice-President (1894–1898) and President (1900–1915). He was Associate Editor (1893–1910) and Editor (1911–1943) of 'Entomological News'. He took a sabbatical year with his wife (1901) Amelia Smith Calvert (q.v.) in Costa Rica (1909–1910), where he conducted extensive field work on Odonata, his favoured research topic. They wrote (his wife was the first author) an extensive book: *A year of Costa Rican natural history* (1917). He published over 300 papers and notes on Odonata (1889–1961). His major works include: *Catalogue of the Odonata (dragonflies) of the vicinity of Philadelphia, with an introduction to the study of this group of insects* (1893), the *Odonata* parts in *Biologia Centrali Americana* (1901-1908), *Contributions to a knowledge of the Odonata of the Neotropical region, exclusive of Mexico and Central America* (1909) and *The neotropical species of the "subgenus Aeschna" sensu Selysii 1883 (Odonata)* (1956). He described c.260 new species and 18 new genera of Odonata. He was known as an outstanding teacher and was *"beloved by all who knew him"*.

Calypso

Duskhawker sp. *Gynacantha calypso* **Ris**, 1915

Calypso is one of the sea nymphs in Greek mythology. She seduced Odysseus and kept him for years away from his wife, Penelope, until Athena intervened; eventually Calypso had to let him go and even helped him to build his boat. She has both negative and positive connotations in Greek mythology: as a concealer and seductress, Calypso is a negative symbol, but as a rescuer she is a positive one. She is always compared with Penelope and thus ended up being a force for diversion and distraction.

Cambridge

Dasher sp. *Micrathyria cambridgei* **Kirby**, 1897

Reverend Octavius Pickard-Cambridge (1828–1917) was an English clergyman and zoologist. He studied theology at the University of Durham and was ordained (1858) and succeeded his father as Vicar of Bloxworth (1868). His zoological focus was on spiders but he also wrote on Lepidoptera. He assisted another arachnologist, John Blackwell (1861–1864), with his work on *British and Irish Spiders*. He published himself on spiders (1859–1917) such as: *Biologia Centrali-Americanii* (1883–1902) as well as: *The Spiders of Dorset* and others. He became a world authority on spiders, describing a considerable number of new species including the Sydney Funnel-web *Atrax robustus*. He accompanied E E Austen on his journey in Lower Amazonas in Brazil (1896) during which this species was discovered.

Camilla

Halloween Pennant *Libellula camilla* **Rambur**, 1842
[JS *Celithemis eponina* (Drury, 1993)]

No etymology is given, but the name may refer to Camilla in Roman mythology, since Rambur named another species as Lucilla. Camilla was the daughter of King Metabus and Casmilla. Driven from his throne, Metabus was chased into the wilderness by armed Volsci, his infant daughter in his hands. The river Amasenus blocked his path, and, fearing for the child's welfare, Metabus bound her to a spear. He promised Diana that Camilla would be her servant, a warrior virgin. He then safely threw her to the other side, and swam across to retrieve her. The baby Camilla was suckled by a mare, and once her *"...first firm steps had [been] taken, the small palms were armed with a keen javelin; her sire a bow and quiver from her shoulder slung."* She was raised in her childhood to be a huntress and kept the companionship of her father and the shepherds in the hills and woods.

Cammaerts

Cammaerts's Glyphtail *Isommas robinsoni* Cammaerts, 1987
Cammaerts's Hooktail *Paragomphus cammaertsi* **Dijkstra** & Papazian, 2015

Dr Roger Jean Françoise Cammaerts (b.1947) is an entomologist and zoologist at the Brussels Free University. He is a specialist of African Gomphidae and has described 14 new such species and subspecies (1967–2004). His major Odonata paper is: *Taxonomic studies on African Gomphidae (Odonata, Anisoptera). 2. A revision of the genus Neurogomphus Karsch, with the description of some larvae* (2004). One of his particular interests is animal communication. He often co-authors papers with Marie-Claire Cammaerts. The authors named this species for him as the person *"...who made great contributions to gomphid taxonomy and first recognised this species."*

Campion

River Junglewatcher *Neodythemis campioni* **Ris**, 1915
[Orig. *Allorrhizucha campioni*]
Sylvan sp. *Coeliccia campioni* **Laidlaw**, 1918
Threadtail sp. *Elattoneura campioni* **Fraser**, 1922

[Orig. *Disparoneura campioni*]
Skydragon sp. *Chlorogomphus campioni* Fraser, 1924
[Orig. *Orogomphus campioni*]
Pond Damselfly sp. *Nesobasis campioni* **Tillyard**, 1924
Tigertail sp. *Synthemis campioni* **Lieftinck**, 1971

Herbert Campion (1869–1924) was a British odonatologist, who was at the time of his death a Temporary Assistant in the Entomological Department at the BMNH (1921–1924). He was a frail child and never attended school but, according to 'Entomological News': *"…overcame the difficulties and handicaps incident to acquiring and education by his good brain, patience and great and constant love for his studies, coupled with the assistance and fellow-likings of his brother Frederick William Campion, throughout his life."* He and his brother collected Odonata in Epping Forest and gave an annual paper in 'The Entomologist' every year for seven years (1903–1909). When they moved house they continued with their reports as 'Notes on the Dragonfly Seasons' (1910–1913). Subsequently they wrote two papers: *Larval Water Mites as Dragonfly parasites* (1909) and *Prey of some Dragonflies* (1914). He became a shorthand clerk but joined (1911–1921) the newly established Imperial Bureau of Entomology then transferred to the British Museum. At the museum, he studied more exotic Odonata and wrote some papers on the Odonata of Tunisia, West Africa, New Guinea, Australia, New Caledonia and Macedonia, describing 24 new species, most of them from Asia and Africa. He published notes on: *The Fabricus Types of Odonata in the British Museum* (1917). Lieftinck honoured him *"…for his valuable help in collecting together a complete record of the Odonata of these Islands* [Fiji].*"*

Campos

Leaftail sp. *Phyllogomphoides camposi* **Calvert**, 1909
[Orig. *Gomphoides camposi*]

Professor Dr Francisco Campos Ribadeneira (1879–1962) was an Ecuadorian zoologist and medical entomologist. He taught biology at the Colegio Vicente Rocafuerte and was a medical zoology professor at the University of Guayaquil despite never having graduated himself, although he was awarded an honorary doctorate (1930). He collected insects and created the first Ecuadorian entomological collection. He wrote a systematic catalogue of the *Odonata of Ecuador* (1922, 1925) and also wrote: *Contribución al estudio de los insectos del Callejón Interandino* (1926). He contributed the type specimen to the author. Other insects are named after him including a disease carrying sandfly. He died after a long battle with stomach cancer.

Caner

Gold-studded Cruiser *Phyllomacromia caneri* Gauthier, 1987
[Orig. *Macromia caneri*]

René Caner (1956–1984). The etymology explains that the species was *"…named in the memory of the late René Caner, a prematurely deceased very good friend of the author."*

Canning

Stripe-headed Threadtail *Caconeura canningi* **Fraser**, 1919
[JS *Prodasineura sita* (Kirby, 1893)]

Charles John Canning (1st Earl Canning also Viscount Canning) (1812–1862) was Governor General of India (1856–1862). The etymologies of the Odonata are not listed in Fraser's descriptive paper of the threadtail, nor in many other of his other papers at that time. The single specimen of *C. canningi* was said to have been collected in Coonoor in Nilgiri Hills, but it later transpired that the specimen had been collected by Fraser himself, and originated from Ceylon. However, given that Major Fraser would have known of the Governor General, and that other taxa including a bird (Andaman Crake *Rallina canningi* Blyth, 1863) and a moth (*Samia canningi* Hutton, 1860) were named after him, it might be who he intended. Once removed is the town in West Bengal that was named after the Governor. This was a port through which many natural history specimens were exported and some have mistakenly been thought to have come from there. It is a shame that Fraser did not enlighten future generations.

Cantrall

Pond Damselfly sp. *Mesoleptobasis cantralli* **Santos**, 1961

Professor Irving James Cantrall (1909–1997) held the Chair of Zoology at the University of Michigan and was an internationally recognised authority on Orthoptera. The University of Michigan awarded both his bachelor's degree (1935) and doctorate (1940). During the following two years he served as an aquatic biologist with the Tennessee Valley Authority then in the US Army Airforce (1943–1946). He became an Assistant Professor of Biology at the University of Florida (1946–1949) and then (1949) Assistant Professor of Zoology at University of Michigan and Curator of the Edwin S. George Reserve. In 1956 he was appointed Curator of Insects at the University of Michigan Museum of Zoology before retiring from the faculty (1978). (In the original etymology, his forename is misspelt 'Irwing').

Capixaba

Pond Damselfly sp. *Leptagrion capixabae* **Santos**, 1965
Amberwing sp. *Perithemis capixaba* **Costa**, De Souza & **J Muzón**, 2006

Capixaba is the demonym used for residents of a municipality located in the Brazilian state of Espirito Santo where the species is found.

Capra

Banded Demoiselle ssp. *Calopteryx splendens caprai* **Conci**, 1956

Dr Felice Capra (1896–1991) was an Italian entomologist. He graduated from Turin University and became the Curator of the Genoa Natural History Museum (1924–1957), which he left finding it difficult to get on with the new director. Besides Odonata he worked on Orthoptera, Coleoptera and on cave fauna in general. His friends were often treated to his interesting war stories, as he had participated in both world wars, and of the great men of entomology that he had met. He was still walking the woods near his home, collecting insects, and remaining fit until very late in life.

Carajás

The Skimmer genus *Carajathemis* **Machado**, 2012

This is a toponym; The Serra dos Carajás are a chain of mountains in the Amazon.

Caribbe

Caribbean Darner *Triacanthagyna caribbea* **EB Williamson**, 1923
Pond Damselfly sp. *Enacantha caribbea* **Donnelly** & Alayo, 1966

These species are named for the locality, the Caribbean.

Carl Cook

Yaqui Dancer *Argia carlcooki* Daigle, 1995
Siphontail sp. *Neurogomphus carlcooki* **Cammaerts**, 2004

Carl Cook (b.1925) is an American odonatologist (Also see **Cook**).

Carlos Chagas

Emerald sp. *Neocordulia carlochagasi* **Santos**, 1967

Professor Carlos Chagas Filho (1910–2000) was a Brazilian biologist and physician in neuroscience. He was Director of the Institute of Biophysics, Federal University of Rio de Janeiro. In the etymology, the author thanks the Institute of Biophysics for allowing them to undertake a survey of the fauna on their site.

Carlota

Speckled Dasher sp. *Micrathyria carlota* **Needham**, 1942
[JS *Micrathyria dentiens* Calvert, 1909]

Carlota is a feminine forename, but the author gave no etymology, so it is not possible to know for sure if this was named after a particular person. However, the lack of a Latin declension suggests it is not an eponym.

Carlotta

Sedge Sprite *Nehalennia carlotta* Butler, 1914
[JS *Nehalennia irene* Hagen, 1861]

The author (Hortense Butler, later Heywood) gave no etymology so I cannot say who this particular Carlotta was. However, she did adopt (1931) a child whom they called Julie Charlotte, so she may have been named after a relative and the damselfly could be named after her. Interestingly the type specimens were collected by J G Needham (q.v.) so it is possible that his *carlota* (above) and her *carlotta* are one and the same.

Carmela

Limniad sp. *Sangabasis carmelae* Villanueva & **Dow**, 2014

Dr Carmela Pedroso Española (b.1975) is a conservation biologist working as an assistant professor at the University of the Philippines Diliman (2006–present, 2017). Her interests include zoology, ecology, and ethnobiology. Her BS in biology was awarded by the University of the Philippines, Visayas (1998) and her MS by the University of the Philippines Los Banos, Laguna (2006). She undertook postgraduate research at Manchester Metropolitan University which awarded her PhD (2013). Before her present post, she has been a research assistant (2001), field researcher (2001–2002), and research assistant at the Philippine Federation for Environmental Concern (2007–2008) as well as gaining much voluntary experience. She is co-founder of the Wild Bird Club of the Philippines and was a board member of the Wildlife Conservation Society of the Philippines. She was honoured in appreciation of her valuable advice on fieldwork in Luzon, which she gave the first author. Among her written papers is: *Are populations of large-bodied avian frugivores on Luzon, Philippines, facing imminent collapse?* (2013). She discovered, and co-authored the description of a new bird, *Gallirallus calayanensus* (2004) and has two other species named after her: *Hoya carmelae* (2010) and *Trachyaretaon carmelae* (2005).

Carmelita

Pondhawk sp. *Erythemis carmelita* **Williamson**, 1923

Myrtle Carmelita Carriker née Flye (1893–1960) was the wife of Melbourne Armstrong Carriker Jr. (q.v.). She was the daughter of an American engineer who became a coffee planter at Santa Marta, Colombia, which is where she met her husband while on a collecting expedition (1911) and where they collected the type (1917). She returned to the US (1927) and was later divorced (1941). An amphibian and a bird are named after her. Williamson wrote about honouring her "...*whose courage and industry as a member of Mr. Carriker's expeditions to the American tropics merit recognition.*"

Carmichael

Shadowdamsel sp. *Drepanosticta carmichaeli* **Laidlaw**, 1915
[Orig. *Protosticta carmichaeli*]

Thomas David Gibson-Carmichael, 1st Baron Carmichael GCSI, GCIE, KCMG, DL (Lord Carmichael of Skirling) (1859–1926) was a Scottish liberal politician and colonial administrator. At school in Hampshire, his devotion to entomology and scientific discovery received every encouragement. He entered St John's College, Cambridge (1877) (BA 1881, MA 1884); his second-class in history reflected parental direction, not natural bent. He was a member of parliament (1895–1900) before various crown appointments, including Governor of Victoria (1908–1911) and then Governor of Madras (1911–1912) and Governor of Bengal (1912–1917).

Carol

Shadowdamsel sp. *Protosticta caroli* **Van Tol**, 2008

Carolus Linnaeus (See **Linnaeus**)

Caroline

Skimmer sp. *Celebophlebia carolinae* **Van Tol**, 1987
Shadowdamsel sp. *Protosticta rozendalorum* Van Tol, 2000

Caroline Rozendaal née Kortekaas is a Dutch zoologist and the widow of F G Rozendaal (q.v.). She has frequently been described as having been crucial to his success by supporting his work in the field. She collected in North Moluccas, Indonesia (1985) while undertaking an ornithological survey, as she did when undertaking a similar survey in Sangihe Island near Sulawesi where she collected the type the same year. Van Tol wrote: "*This beautiful species is called after Caroline Rozendaal, who makes so many efforts with her husband (Frank G Rozendaal) to investigate the fauna of Southeast Asia*". A mammal and a bird are also named after her. (Also see **Rozendaal**)

Carolus

Dancer sp. *Argia carolus* **Garrison** & von **Ellenrieder**, 2017

Carlos Esquivel Herrera (b.1956) is a Costa Rican biologist. (See **Esquivel**)

Carolus

Rifle Snaketail *Ophiogomphus carolus* **Needham**, 1897

The author gave no etymology, so it is not possible to know for sure if this was named after a particular person. However, while the lack of a Latin declension suggests it is not an eponym, many species are named after the father of binomials Carl von Linné, otherwise known as Carolus Linnaeus (See **Carol** & **Linnaeus**).

Carové

Bush Giant Dragonfly *Uropetala carovei* **White**, 1843
[Orig. *Petalura carovei*, A. Kapokapowai]

Friedrich Wilhelm Carové (1789–1852) was a German philosopher and religious writer against the Roman Catholic faith. His PhD was awarded by Heidelberg University and he was a Professor at Breslau for a short while. He was also briefly a member of the provisional German parliament (1848). He wrote the children's book: *Story without an end* and among his religious tracts is: *Ueber das sogenannte germanische und sogenannte christliche Staatsprincip* (1843). The etymology makes no mention of him, but the identity is confirmed by Rowe's *The Dragonflies of New Zealand* (1987).

Carpenter

Little Longleg *Nilogomphus carpenteri* **Fraser**, 1928
[JS *Notogomphus dorsalis* (Selys, 1858)]
Black-splashed Elf *Tetrathemis carpenteri* Fraser, 1941
[Syn. of *Tetrathemis polleni* (Selys, 1877)]
Pearly Flasher *Oxythemis carpenteri* Fraser, 1944
[JS *Aethiothemis solitaria* Ris, 1908]

Professor Dr Geoffrey Douglas Hale Carpenter (1882–1953) was a British entomologist and physician. He graduated from St Catherine's College, Oxford (1904) then studied medicine at St Georges Hospital, London, graduating MB ChB (1908). He was a physician at the London School of Hygiene and Tropical Medicine gaining his MD in 1913. Having joined the Colonial Medical Service (1910) he worked on tse-tse flies and sleeping sickness in Uganda and Tanganyika. He was called to service in the Army Medical Corps (WW1) and was stationed in East Africa at a fort in Kakindu. He had plenty of spare time during which he started to study butterflies. He said of this: *"The hosts of butterflies at Kakindu passed beyond anything I had ever seen; some days are quite unforgettable"*. He also spent time stationed in Tanganyika (1916–1918). It was here he began to study insect mimicry, culminating in the co-written book: *Mimicry* (1933). He made a large collection of Odonata in Uganda (1927–1928) including the type series of some of these species. He became Hope Professor of Zoology at Oxford University (1933–1948). (Also see **Hale**)

Carpentier

Bluetail sp. *Ischnura carpentieri* **Fraser**, 1946

Dr Fritz Carpentier (1890–1978) was an eminent Belgian entomologist, a pioneer of the morphology of insects. He was a curator in the Liége University Museum (1922–1958) and in charge of their zoology degree course as Professor of Invertebrate Morphology and Systematics. He collected three specimens of the above species in Annam (Vietnam) (October 1920), which he sent with other material for Fraser to study.

Carvalho

Firetail sp. *Telebasis carvalhoi* **Garrison**, 2009

Professor Alcimar do Lago Carvalho (b.1960) is a Brazilian entomologist, particularly interested in aquatic insects. He is currently (2017) Professor in the Entomology Department at the National Museum, Federal University of Rio de Janeiro, Brazil, the university that awarded his bachelor's degree (1984) and masters (1985). His PhD was awarded by the University of São Paulo. His main foci in entomology are the taxonomy, biology and phylogeny of aquatic insects, with emphasis on Odonata Anisoptera of South America. He is a friend of the author as well as a fellow odonatologist. He recognized the species as new and kindly provided Garrison the opportunity to describe it.

Cassandra

Flatwing sp. *Philogenia cassandra* **Hagen**, 1862

In Greek mythology Cassandra was the daughter of King Priam and Queen Hecuba of Troy. Her beauty caused Apollo to grant her the gift of prophecy.

Cassiopeia

Leaftail sp. *Phyllogomphoides cassiopeia* **Belle**, 1975
[Orig. *Gomphoides cassiopeia*]

In Greek mythology Cassiopeia was the mother of Andromeda. The author compared his species with the closely related *Gomphoides* [presently *Phyllogomphoides*] *andromeda*.

Castellani

Mercury Bluet *Coenagrion castellani* **Roberts**, 1948
[JS *Coenagrion mercuriale* (Charpentier, 1840)]

Omero Castellani (1903–1974) was an Italian entomologist, primary school teacher, painter and poet. He founded (1945) the Associazione Romana do Entomologia. The type was collected in Italy (1946). Part of his collection is at the Museo di Zoologia, Universita di Roma.

Castor

The Genus *Castroaeschna* **Calvert**, 1952

Darner sp. *Castroaeschna castor* **Brauer**, 1865
[Orig. *Aeschna castor*]
Shortwing sp. *Perissolestes castor* **Kennedy**, 1937

[Orig. *Perilestes castor*]
Pincertail sp. *Onychogomphus castor* **Lieftinck**, 1941

In Greek and Roman mythology Castor and Pollux were twin brothers, together known as the Dioscuri. In the same paper, Kennedy named two new damselfly species of the same genus from Ecuador as *castor* and *pollux*. Lieftinck did the same for two new pincertail species from Peninsular Malaysia and Sumatra.

Catharina

Sylph sp. *Macrothenis catharina* **Karsch**, 1890
[JS *Macrothemis nubecula* (Rambur, 1842)]

Toponym named after the Santa Catarina State in Brazil, where the species was found.

Cecilia

Green Snaketail *Ophiogomphus cecilia* Fourcroy, 1785
[Orig. *Libellula cecilia*]

Banded Damoiselle ssp. *Calopteryx intermedia cecilia* **Bartenev**, 1912
[JS *Calopteryx splendens intermedia* Selys, 1887]

In Roman antiquity Cecilia indicated a relationship to the noble family of Caecelian, bishop of Carthage in the early 300s. Antoine Francis Fourcroy (1785) introduced formal binomial scientific names for insects, which Etienne-Louis Geoffroy had described in his *Histoire abregee des insectes qui se trouvent aux environs de Paris* (1762) with a French name only. Geoffroy furnished all 16 of the dragonfly (*Libellula*) species recorded by him with female French names. Fourcroy introduced 7 new binomial names on these species. Only 'La cécile' – *cecilia* – is at present recognized as a valid species name. Also some other of Fourcroy's names may have been the first ones introduced to the species in question. However, these cases have remained unsolved. Bartenev did not provide any etymology.

Celaeno

Sylph sp. *Macrothemis celeno* **Selys**, 1857
[Orig. *Libellula celeno*]
Cruiser sp. *Macromia celaeno* **Lieftinck**, 1955

In Greek mythology Celaeno (Celeno) is one of the Harpies and also one of the Pleiades and one of the Danaids among others.

Celia

Phantomhawker sp. *Planaeschna celia* **KDP Wilson** & **Reels**, 2001

Mrs Celia Stuart Wilson née Compton (b.1927) is the first author's mother. Originally from Dunbar, East Lothian, Scotland she is now residing in Brighton, England.

Celio (Lencioni)

Threadtail sp. *Idioneura celioi* **Lencioni**, 2009

Celio Lencioni (1926–2011) was the uncle of the author.

Célio Valle

Firetail sp. *Telebasis celiovallei* **Machado**, 2010

Professor Célio Murilo de Carvalho Valle (b.1933) is a Brazilian biologist, zoologist and environmentalist at the Federal University of Minas Gerais. He is a good friend and colleague of Machado and collected this species at Carajás (1978).

Cepheus

Leaftail sp. *Phyllogomphoides cepheus* **Belle**, 1980

In Greek mythology Cepheus was the husband of Cassiopeia and father of Andromeda. All three of them are in names of the present genus *Phyllogomphoides*.

Cesare (Conci & Nielsen)

Mediterranean Bluet *Coenagrion coerulescens caesarum* **Schmidt**, 1959
[JS *Coenagrion caerulescens* (Fonscolombe, 1838)]

Professor Cesare Conci (1920–2011) and Dr Cesare Nielsen (1898–1984). Schmidt named this subspecies after two Italian odonatologists (caesarum meaning 'all the caesars'). Conci was Professor of Zoology at the University of Genoa. Besides dragonflies he studied Mallophaga and biology of cave species. Nielsen was a dentist and physician in Bologna. He published about 20 papers on Italian and African dragonflies in which four new species were named including the Balkan Emerald *Somatochlora meridionalis*. Jointly Conci and Nielsen published the book: *Odonata* in the *Fauna d'Italia* series (1956). (See also: **Caesar**, **Conci** and **Nielsen**)

Ceyx

Pond Damselfly sp. *Palaiargia ceyx* **Lieftinck**, 1949.

In Greek mythology Ceyx is the son of Eosphorus and the King of Thessaly who married Alcyone. They were very happy together and according to Pseudo-Apollodorus's account, often called each other 'Zeus' and 'Hera'. This angered Zeus, so while Ceyx was at sea, the god threw a thunderbolt at his ship. Ceyx appeared to Alcyone as an apparition to tell her of his fate, and she threw herself into the sea in her grief. Out of compassion, the gods changed them both into kingfishers.

Chacón

Meadowhawk sp. *Sympetrum chaconi* **De Marmels**, 1994

Anibal Chacón Hernández (b.1943) worked as a technical assistant at the Entomology Department, MIZA (Museo del Instituto de Zoologia Agricola), Maracay, Venezuela and is described as a tireless collector of insects including the type in venezuela (1994). The etymology reads: *"The new species is named after Mr. Anibal Chacón, my invaluable companion on many collecting trips, and technical assistant at the Section of Entomology, of the MIZA [Museo del Instituto de Zoologia Agricola, Maracay, Venezuela]. He is the collector of this and many other fine Venezuelan dragonflies.* In the etymology of a Hymenoptera named after him the author says: *Mr. Anibal Chacon, who during many years has shared with*

Venezuelan entomologists his passion for entomology and insects collecting in Venezuela". He also has at least one coleopteran named after him.

Chalciope

Cruiser sp. *Macromia chalciope* **Lieftinck**, 1952

In Greek mythology Chalciope was the daughter of King Aeetes of Colchis, sister of Medea and wife of Phrixus. She remained loyal to her father even though he killed her husband.

Champion

Flatwing sp. *Philogenia championi* **Calvert**, 1901

George Charles Champion (1851–1927) was an entomologist who specialized in Coleoptera. He was taken on as a collector by Frederick DuCane Godman and Osbert Salvin and went to Guatemala (1879). He spent four years collecting in Central America, particularly Panama where the type was taken and had a collection containing 15,000 insect species when he returned to England. He stayed in Godman and Salvin's employ and saw through the printing process their 52-volume work: *Biologia Centrali-Americana*; he wrote a number of its sections. Calvert wrote: *"Dedicated to Mr G C Champion, whose field labours have furnished so much material..."* He collected a male specimen (holotype) of this new species in Panama.

Chanca

Clubtail sp. *Gomphus chancae* **Bartenev**, 1956

[JS *Shaogomphus schmidti* Asahina, 1956)

This is a toponym; the name Chanca refers to the type locality Khanka Lake in Russian Far East at the Chinese border.

Chandrabal

Lesser Blue Skimmer *Orthetrum chandrabali* Mehrotra, 1959

[JS *Orthetrum triangulare* Selys, 1878]

Professor Chandrabal was formerly the Reader in Zoology at Banaras Hindu University.

Chang

Lyretail sp. *Stylogomphus changi* **Asahina**, 1968

Bao-Sin Chang (1932–1992) was a well-known amateur entomologist in Taiwan who specialised in Lepidoptera, especially moths. He taught at high schools and acquired his knowledge of moths from Japanese enthusiasts when he was young. He found the male type of *Stylogomphus changi* in Taiwan (1965) and sent the specimen to a famous Japanese butterfly scholar. Dr Shirozu, who passed it to Asahina (1968). He is also commemorated in the names of other insects.

Chantaburi

Featherleg sp. *Copera chantaburii* **Asahina**, 1984

This is a toponym; Chanthaburi is a province in southeastern Thailand. The author collected the species there in September 1980.

Chao

Fiery Coraltail *Ceriagrion chaoi* **Schmidt**, 1964
Duskhawker sp. *Cephalaeschna chaoi* **Asahina**, 1982
Clubtail sp. *Davidius chaoi* Cao & **Zheng**, 1988
Longleg sp. *Anisogomphus chaoi* **Liu**, 1991
Pincertail sp. *Nihonogomphus chaoi* **Zhou** & Wu, 1992
Spineleg sp. *Merogomphus chaoi* Yang & **Davies**, 1993
Griptail sp. *Phaenandrogomphus chaoi* Zhu & Liang, 1994
[JS *Phaenandrogomphus tonkinicus* (Fraser, 1926)]
Tigerhawker sp. *Polycanthagyna chaoi* Yang & Li, 1994
Goldenring sp. *Anotogaster chaoi* Zhou, 1998
Grabtail sp. *Lamelligomphus chaoi* **Zhu**, 1999
Jewel sp. *Aristocypha chaoi* **Wilson**, 2004
[Orig. *Rhinocypha chaoi*]
Oread sp. *Calicnemia chaoi* Wilson, 2004
Grappletail sp. *Heliogomphus chaoi* **Karube**, 2004
Coralleg sp. *Rhipidolestes chaoi* Wilson, 2004

Dr Hsiu-Fu Chao (Xiufu Zhao) (1917–2001) was a renowned Chinese entomologist, doyen of Chinese odonatology who even as a child showed great interest in observing nature. He studied at Fuzhou Scientific and Technology Middle School, where the well-known biologist Zhongzliang Tang gave lectures on biology during Zhao's time there. Guided by Tang, he began studying biology. At Yenjing University, he studied under the guidance of the celebrated biologist Jingfu Hit. He was soon studying fruit tree pests, then parasitic wasps and before long making original discoveries. Seeing that he was interested in zootaxonomy, Professor Wu sent him a copy of Needham's *Manual of the Dragonflies of China*, to introduce him to zootaxonomy. As a postgraduate he was engaged by Qilu University, Qingdao, Shandong province, where he investigated vermin prevention. He became a teacher at Peiying Girls Middle School, Quanzhou, Fujian province (1941) then at Xieho University, Shaowu County (1942), where Professor T-C Maa (q.v.) encouraged his interest in insects and where he also studied and collected a large number of fleas, following the occurrence of bubonic plague in the region. He often went to the Wuyi Mountain area and caught large numbers of parasitic wasps and dragonflies for study. After the Spring Festival (1948) he accompanied his father-in-law to Anhai County by sea. Unfortunately, the ship overturned and a large number of passengers were killed. Zhao, an experienced swimmer, managed to swim away, saving his father-in-law and also dragging along with him a chest of specimens wrapped in a tarpaulin. It was these specimens that helped him complete his dissertation for Doctorate in Massachusetts, USA (1951). At Fujian Agricultural University, he looked at insect borne plant disease prevention and pest control. He again visited the USA (1979) to exchange ideas on the prevention and control of plant pests and parasitic wasps. From a proposal made by him, the Natural Environmental Protection Area of Wuyi Mountain was established (1979). Here over a million specimens were collected from 39 orders, 356 families and 4,697 species, many new to science. Supported by the Provincial Government, he started to publish the periodical, 'Wuyi Science Journal', for which he acted as Editor. His works include over fifty odonatological monographs and papers, including a series of papers on

Chinese Gomphidae (1953–1955), a revised and updated version of which was published as a book: *The gomphid dragonflies of China (Odonata: Gomphidae)* (1990). He established a new gomphid subfamily, named eight new genera and 64 new species of Odonata. Even during hospitalisation, he continued to work on dragonflies and his last three publications were completed in hospital (1999). He also wrote popular science literature.

Chapada

Dancer sp. *Argia chapadae* **Calvert**, 1909

This is a toponym; standing for the state of Chapada in Brazil, which was the type locality.

Chapin

Western Snorkeltail *Mastigogomphus chapini* **Klots**, 1944
[Orig. *Oxygomphus chapini*]

Dr James Paul Chapin (1889–1964) was an American ornithologist. He was joint leader of the Lang-Chapin Expedition, which made the first comprehensive biological survey of the Belgian Congo (1909–1915). He was Ornithology Curator for the AMNH and President of the Explorers' Club (1949–1950). He wrote: *Birds of the Belgian Congo* (1932), which largely earned him the award of the Daniel Giraud Elliot Gold Medal that year. He collected the holotype male of this species on the River Congo (October 1930). He died of a heart attack. Fourteen birds, two mammals, three amphibians and four reptiles are named after him.

Chararum (Xaraés)

Wedgetail sp. *Acanthagrion chararum* **Calvert**, 1909

This is a toponym; this Brazilian species was found in Cuyaba, Brazil. It was named after the nearby Charaés (Xaraés) Marshes, which had got its name from the extinct indigenous Indian tribe Xaraés once living in the region. Presently this wetland area is known as the Pantanal.

Charca

Rubyspot sp. *Hetaerina charca* **Calvert**, 1909

This species was found near Coroico in Bolivia. It was named after the Charca people, who lived in the same area prior the arrival of the Spaniards.

Charpentier

Common Bluet *Agrion charpentieri* **Selys**, 1840
[JS *Enallagma cyathigerum* (Charpentier, 1840]

Blue-eyed Goldenring ssp. *Cordulegaster insignis charpentieri* Kolenati, 1846
[Orig. *Aeschna charpentieri*]

Toussaint de Charpentier (1779–1847) was a German geologist and entomologist. He studied geology and mining at the Technische Universität Bergakademie Freiberg then the University of Leipzig. He went to Prussia (1802) to work for the Silesia mining authority in Breslau and shortly afterward took over a mining department in Schweidnitz before returning (1811) to Breslau. He was promoted and went to Dortmund (1828) and further promoted (1830) then transferred again to the Silesia authority (1836). He later (1839) transferred to Brieg for the rest of his life. He was a leading amateur entomologist and he named 46 new Odonata species from Europe, of which 18 remain valid. He also introduced several genus names including *Ischnura, Enallagma, Erythromma* and *Pyrrhosoma*. He wrote a great deal on mining and also on his hobby of entomology such as: *Horae Entomologicae* (1825). His major work on Odonata was: *Libellulinae Europaeae descriptae ac depictae* (1840) with magnificent colour plates.

Chasen

Oread sp. *Calicnemia chaseni* **Laidlaw**, 1928
[Orig. *Calicnemis chaseni*]

Frederick Nutter Chasen (1896–1942) was an English zoologist. He became Assistant Curator of the Raffles Museum (1921), then Director (1932–1942). He was a well-known authority on Malaysian birds and mammals and co-authored many scientific publications on these topics. Chasen perished at sea when fleeing Singapore during WW2. A reptile and six birds are named after him. He collected the holotype male in Pahang (June 1922).

Cheesman

Reedling sp. *Indolestes cheesmanae* **Kimmins**, 1936
[Orig. *Austrolestes cheesmanae*]
Bluetail sp. *Ischnura cheesmani* **Fraser**, 1942
[JS *Ischnura taitensis* Selys, 1876]

Miss Lucy Evelyn Cheesman (1881–1969) wanted to train as a veterinary surgeon, but in her time the restrictions on the education of women precluded it. Instead she became an entomologist, explorer and traveller, making a number of expeditions to the Galapagos Islands, the Marquesas, New Guinea, the New Hebrides and the Solomon Islands (1924–1936). She worked for many years as a volunteer at the British Natural History Museum and was the first female curator at London Zoo. She wrote many books including: *Everyday Doings of Insects* (1924), *Insect Behaviour* (1933) and *Time well spent. Scientific explorations in the Pacific* (1960). A reptile and four amphibians are also named after her as are a number of other insects including the beetle *Costomedes cheesmanae*. Fraser mistakenly gave the masculine genitive ending to the Forktail species. (Also see **Evelyn**)

Cheng

Emerald Spreadwing sp. *Megalestes chengi* **Chao**, 1947

Professor Tso-hsin Cheng (1906–1998) was a distinguished Chinese zoologist and ornithologist. He was educated at the Fujian Christian University in Fuzhou City, from where he received the degree of BSc (1926). He went to the USA and received his doctorate from the University of Michigan (1930), then returned to his old university in Fuzhou as Professor and Director of the Department of Biology. He was one of the founders (1934) of the China Zoological Society. He transferred to Peking (Beijing) (1950) to take charge as Curator of Bird Specimens in the Institute of Zoology, Academia Sinica, and (1951) he founded the Peking Natural History Museum. For over 60 years he undertook fieldwork and research in ornithology and conservation, publishing more than 10 million words in

30 books, 20 monographs, 150 scientific papers, and 260 popular articles. A bird is also named after him.

Chiang Chin-Li

Paddletail sp. *Sarasaeschna chiangchinlii* Chen & Yeh, 2014

Chin-Li Chiang (b.1954) is a friend of the authors and one of those who provided them with useful information. The type series was collected from Daxi, Taoyuan County in northern Taiwan (2010 & 2013). The etymology reads: *"The new species is dedicated to Mr. Chin-Li Chiang for his continuous effort and hard work in searching for many rare dragonfly species of Taiwan and also discovery of this new species".*

Chichibu

Clubtail sp. *Gomphus chichibui* **Fraser**, 1936
[JS *Trigomphus interruptus* (Selys, 1854)]

Chichibu (Chichibuno-Miya-Yasuhito-Shino) (1902–1953) was a member of the Japanese Imperial family, the younger brother of the Emperor. Fraser often named species for high-ranking individuals.

Chico Mendes

Wedgetail sp. *Acanthagrion chicomendesi* **Machado**, 2012
[JS *Acanthagrion apicale* Selys, 1876]

Chico Mendes (née Francisco Alves Mendes Filho) (1944–1988) was a Brazilian rubber tapper (from the age of 9) in Acre State, trade union leader, environmentalist and rainforest activist who was murdered by a rancher while defending the Amazon forest. He was born at a time when schools were prohibited on rubber plantations (for fear that the peasants might learn to read and do arithmetic!) and he only learned to read when he was 18. The Chico Mendes Institute for Conservation of Biodiversity is named in his honour.

Chilton

Mountain Giant Dragonfly *Uropetala chiltoni* **Tillyard** 1930

Dr Charles Chilton (1860–1929) was a British born biologist who emigrated as a child (1861) to New Zealand. He graduated (1880) then gained an MA in zoology (1881) from Canterbury College. He later took a BSc at Otago University (1887). He became Rector of Port Chalmers District High School (1888). In his spare time, he studied marine zoology in Otago Harbour, and (1893) he gained a doctorate in science, being the first to do so in New Zealand. He decided to switch to medicine and went to Edinburgh (1895) where he was awarded his MB and CM, becoming a surgeon there. He studied eye diseases in Heidelberg then returned to New Zealand (1900) where he practised as an ophthalmic surgeon. He was offered the post of Professor of Biology at Canterbury College where he taught (1903–1928) and from 1929 became Professor Emeritus, during which time a research station was founded. He wrote over 50 scientific papers. Tillyard wrote: *"I wish to dedicate this new species... ...to Dr Chilton as a memorial of the excellent work which he has done, and is doing, in connection with the Cass Biological Station."*

Chimantá

Pond Damselfly sp. *Aeolagrion chimantai* **De Marmels**, 1988

This is a toponym; the Chimantá Massif in Bolívar state, Venezuela, which was the type locality.

Chingola

Forest Jewel ssp. *Platycypha lacustris chingolae* **Pinhey**, 1962

This is a toponym; Chingola is a locality in Zambia.

Chini

Sergeant sp. *Chalybeothemis chini* **Dow**, Choong & **Orr**, 2007

This is a toponym; Sungei Chini was the type locality, a river connecting the Chini Lake to the Pahang River in central Pahang, Malaysia.

Chirihuana

Firetail sp. *Acanthagrion chirihuanum* Calvert, 1909;

[JS *Telebasis obsoleta* (Selys, 1876)]

The specific name is derived from that of an Indian tribe of the neighbourhood of the type locality Cuyabá, Brazil.

Chirripa

Bannerwing Sp. *Cora chirripa* **Calvert**, 1897

Calvert wrote: "*The specific name proposed is slightly altered from that of an Indian tribe*" referring to the Chirripo people after whom an amphibian is also named.

Chittaranjan

Velveteen sp. *Schmidtiphaea chittaranjani* Lahiri, 2003
[Orig. *Bayadera chittaranjani*]

Chittaranjan Lahiri was the father of the author Ashok Ranjan Lahiri (d.2012).

Chloe

Yellow-striped Flutterer ssp. *Rhyothemis phyllis chloe* **Kirby**, 1894

Chloe, from the Greek χλόη (*khlóē*) means 'young green shoot' and is one of the many names of the Greek goddess Demeter. Kirby selected many of his names from folklore and mythology.

Chloris

Eastern Amberwing *Libellula chlora* **Rambur**, 1842
[JS *Perithemis tenera* (Say, 1840)]
Slender Redskimmer *Rhodopygia chloris* **Ris**, 1911
[JS *Rhodopygia hollandi* Calvert, 1907

In Greek mythology Chloris was a nymph associated with spring, flowers and growth; the name is derived from *Khloris* Χλωρίς, from *khloros* χλωρός, meaning 'greenish-yellow' or 'pale green'.

Choi Fong

Hong Kong Tusktail *Fukienogomphus choifongae* **KDP Wilson** & Tam, 2006

Choi Fong was the grandmother of the discoverer and collector of the type, Miss Joyce Wong Kin, who wanted the dragonfly named in memory of her. Miss Wong was part of the author's dragonfly working group of the Hong Kong government's Agriculture, Fisheries and Conservation department (AFCD) (2004).

Choma

Masai Sprite *Pseudagrion chomae* **Pinhey**, 1964
[JS *Pseudagrion massaicum* Sjöstedt, 1909]

This is a toponym; Choma is a locality in Zambia.

Chongwe

Slate Sprite *Pseudagrion chongwe* **Pinhey**, 1961
[JS *Pseudagrion salisburyense* Ris, 1921]

This is a toponym; Chongwe is the type locality in Northern Rhodesia (now Zambia).

Chou

Grabtail sp. *Lamelligomphus choui* **H-F Chao** & **Liu**, 1989
Goldenring sp. *Neallogaster choui* Yang & **Li**, 1994

Professor Io Chou (sometimes Yao Zhou) (1912–2008) was a Professor at Northwestern Agricultural University, and known as the godfather of Chinese entomology. He was also Director of the Entomological Museum, National Model Worker, Academician of the International Scientific Academy of San Marino Republic and chief adviser of Wild Protection Commission in five provinces of northwest China (Shaanxi, Gansu, Ningxia, Qinghai and Xinjiang). He devoted his life to agricultural entomology, insect morphology, insect taxonomy and the history of entomology. He edited a major two-volume work on Butterflies: *Monographia Rhopalocerorum Sinensium* (2001) and wrote: *History of Chinese Entomology* (1980) as well as several other books and over two hundred scientific papers. He also discovered three hundred and thirty-seven new species and twenty-six new genera.

Chris Müller

Threadtail sp. *Nososticta chrismulleri* **Theischinger** & Richards, 2016

Chris James Müller (b.1974) is an entomologist and geologist and a research associate at the Australian Museum (Sydney). His BSc was awarded by Macquarie University (1998) followed by a first-class degree from the University of Tasmania (1999). Macquarie also awarded his PhD (2011). His career has been with mining companies in Indonesia (1999–2000), Ghana (2001), Mongolia & China (2002–2005) and with three different companies in Papua New Guinea (2005–2007, 2007–2011 & 2011–current). He has professional experience in Thailand and has conducted a number of Lepidoptera surveys for non-government organisations (NGOs), industry, academic institutions and government. He has observed more than 90% of Papua New Guinea's 950+ known butterfly species, and discovered and described more than 30 new species in the region, working closely with associates at the Natural History Museum (London), Museum Naturalis (Leiden, the Netherlands) and Australian Museum. He has published more than 50 entomological and geological papers such as: *The life history of Sabera fuliginosa fuliginosa (Miskin) (Lepidoptera:Hesperiidae) and additional hostplants for the other members of the genus in northern Queensland* (1999) and *A stunning new species of Jamides Hübner, 1819 (Lepidoptera, Lycaenidae), with notes on sympatric congeners from the Bismarck Archipelago, Papua New Guinea* (2016). He took part in a 'rapid biodiversity survey' of Papua New Guinea's Manus and Mussau Islands (October 2014) with both authors and others. The etymology reads:"…*The species is dedicated to Chris Muller who collected a number of interesting odonates for the second author, and provided much appreciated camaraderie in remote bush camps throughout Papua New Guinea. In appreciation of adventures shared*". Chris recently told me: "*I have to say that I am becoming more and more interested in the Odonates, to the point that they are competing with the Lepidoptera in fascination for me.*"

Christine (Legrand)

Western Horntail *Libyogomphus christinae* **Legrand**, 1992
[Orig. *Tragogomphus christinae*]

Christine Legrand (b.1974) is the author's second daughter. (Also see **Legrand**)

Christine (Theischinger)

Milky Flatwing *Austroargiolestes christine* **Theischinger** & **O'Farrell**, 1986
S-Spot Darner *Austroaeschna christine* Theischinger, 1993
[A. Mountain Darner]

Mrs Christine Theischinger (née Pingitzer) (b.1946) is the wife of the senior author, Dr Günther Theischinger (q.v.).

Chu

Cruiser sp. *Macromia chui* **Asahina**, 1968
Common Spineleg *Merogomphus chui* Asahina, 1968
[JS *Merogomphus pavici* Martin, 1904]

Chu Yao-Yi (b.1932) is Professor Emeritus, Department and Graduate Institute of Entomology, National Taiwan University. He made a study of the commercial processing of insects (1989–1991) as Taiwan was pre-eminent in such trade. He was co-author of: *Ecological encyclopedia of Taiwanese butterflies* (1987). He collected the type specimens of both of these species.

Chun-Liu

Lyretail sp. *Stylogomphus chunliuae* **Chao**, 1954

Chun-liu Xu was the wife of the author Dr Hsiu-Fu Chao (q.v.).

Cigana

Threadtail sp. *Phasmoneura ciganae* Santos, 1968
[JS *Forcepsioneura sancta* (Hagen in Seys,1860)]

Cigana in Portuguese means a Gypsy woman.

Circe

Golden-winged Flasher *Aethiothemis circe* **Ris**, 1910
[Orig. *Apatelia circe*]

In Greek mythology Circe is a minor goddess of magic.

Claasen

Shadowdamsel sp. *Drepanosticta claaseni* **Lieftinck**, 1938

Lieutenant J M van Ravenswaay Claasen was a patrol officer stationed in the South Sepik division in Indonesian New Guinea. He collected in New Guinea (1937 & 1938) including the holotype (1937).

Clara

Skimmer sp. *Nannodiplax clara* **Needham**, 1930

Clara is a feminine forename, but the author gave no etymology, so it is not possible to know for sure if this was named after a particular person. However, the lack of a Latin ending suggests it is not an eponym, as does Needham's frequent use of apparent forenames.

Clarilli

Damselfly sp. *Palaiargia clarillii* **Orr**, **Kalkman** & Richards, 2014

This is a joint dedication to Clara (b.1991) and Klaus-Peter ('Wassili') Seiler (b.1957) (Hence Clara & Wassili). Clara Seiler is the daughter of Klaus-Peter, and he is the husband of Mechtild Seiler, who is the sister of Martin Schorr (q.v.). They are supporters of the International Dragonfly Fund.

Clausen

Wisp sp. *Agriocnemis clauseni* **Fraser**, 1922

Curtis Paul Clausen (1893-1976) was an American scientist, explorer, author, sports enthusiast, and bon vivant. The University of California at Berkeley awarded his entomology BSc (1914) and his MSc (1920). He was an Assistant in Entomology under H J Quayle at the University's Citrus Experiment Station (1914-1915) then became Assistant to the Superintendent of the State Insectary (1915-1916). He explored in Japan, China and the Philippines (1916-1918) looking for a natural enemy of citrus scale insects to control pests in California. After military service as a second lieutenant he joined the US Department of Agriculture (1920). He spent thirteen years researching parasites and predators in the Far East. He later explored extensively in Mexico, Central and South America and Europe. He was then stationed in Washington (1931). Here he wrote: *Entomophagous Insects* (1940) a landmark in biological control of pests. He was in charge of the Division of Foreign Parasite Introductions (1934-1951) and

(1942–1951) also headed the Division of Control Investigations. He then retired. However, he soon became Professor of Biological Control and Chairman of the state-wide Department of Biological Control of the University of California (1952–1959) and then Professor Emeritus. Thomas Bainbrigge Fletcher (1878–1950) collected (1918–1919) the species in Shillong, Assam. There is no etymology so this remains speculation.

Clausnitzer, H-J

Gabon Hooktail *Paragompus clausnitzerorum* **Dijkstra**, **Mézière** & Papazian, 2015

Hans-Joachim Clausnitzer (b.1942) is the father of Dr Viola Clausnitzer (b.1970) (below). He has spent decades on his work on African dragonflies as a passionate naturalist. Together with his wife Christa Clausnitzer he has travelled widely in Kenya, Uganda, Rwanda and Ghana, photographing and recording dragonflies on their travels. Both are retired now and spend most of the German winter in Uganda, adapting a migratory behaviour. The etymology reads: "*Named in honour of Mr Hans-Joachim Clausnitzer and his daughter Dr Viola Clausnitzer, chair of the IUCN Dragonfly Specialist Group, who together made many contributions to the knowledge of African Odonata.*"

Clausnitzer, V

Gabon Hooktail *Paragompus clausnitzerorum* **Dijkstra**, Mézière & Papazian, 2015

Dr Viola Clausnitzer (b.1970) is the daughter of Hans-Joachim Clausnitzer (b.1942) (above). She is a German zoologist, ecologist, entomologist and odonatologist currently (2011–2017) affiliated to the Senckenberg Museum of Natural History, Görlitz, Germany. The title of her master's thesis at the University of Marburg (1995) was: *Behavioral ecology of Notiothemis robertsi Fraser, 1944 (Odonata: Libellulidae)* and her doctoral thesis at the same university was: *Rodents of Mt. Elgon (Uganda). Ecology, biogeography and the significance of fire* (2000). After which she held various posts, such as coordination of the bachelor curriculum 'Ecosystem Management', University of Goettingen, before joining the staff at Seckenburg. She is Chair of the SSC / IUCN Dragonfly Specialist Groups, holds various positions and commitments within the IUCN (International Union for Conservation of Nature) and is Trustee for Conservation of the Worldwide Dragonfly Association. She has been honoured for her outstanding achievements in the field of conservation and led the BIOTA project on diversity and ecology of East African Odonata. She has co-authored a number of publications with Dijkstra (q.v.) such as: *The dragonflies and damselflies of eastern Africa: handbook for all Odonata from Sudan to Zimbabwe* (2014), and has authored or co-authored more than 75 other papers such as: *Focus on African freshwaters: hotspots of dragonfly diversity and conservation concern* (2012). She is involved in global assessments of dragonflies and has written several publications on African dragonflies describing or co-describing five new species. She has travelled widely in Africa, while exploring remote areas, but also did a lot of training and publicity work. The etymology reads: "*Named in honour of Mr Hans-Joachim Clausnitzer and his daughter Dr Viola Clausnitzer, chair of the IUCN Dragonfly Specialist Group, who together made many contributions to the knowledge of African Odonata.*"

Claussen

Dancer sp. *Argia claussenii* **Selys**, 1865

Peter Claussen (1804–1855) was a Danish natural history collector, particularly of botanical specimens and fossils. Because of a fraud, he emigrated to Brazil and, having enlisted as a common soldier, he entered Rio de Janeiro with Don Pedro's army, but then lived as a peddler. He was a spy during Argentina's Cisplatine War with Brazil (1825–1828). He became a merchant in Minas Gerais and a farm owner at Curvelo. He undertook a long Brazilian journey (1833–1835) during which he met the naturalist P W Lund and German botanist Ludwig Riedel. Lund's exploration of the caves on Claussen's property led to the discovery of many fossils sparking Claussen's interest in natural history over and above any purely commercial interest. He later accompanied Castlenau (q.v.) on his great South American expedition (1843). When he returned to Europe he developed mental ill health and died in an institution in London. Neither the etymology, nor the collector's name is given in the paper on this Brazilian insect.

Clavijo

Flatwing sp. *Dimeragrion clavijoi* **De Marmels**, 1999

Dr José 'Pepe' Alejandro Clavijo Albertos (b.1953) is a Venezuelan agronomist engineer, environmentalist and entomologist who is now Director MIZA (Museo del Instituto de Zoologia Agricola) at the University Venezuela and was in charge of the entomology collection there. A graduate of the University of Venezuela, his PhD was awarded by McGill University, Montreal, Canada. His greatest influence was Dr. Francisco Fernandez Yépez who advised him to take an interest in insects (1967). He has authored more than thirty scientific papers and chapters in books, etc. He collected the type in Venezuela (1995). The original description says: "*I dedicate this species to my colleague and friend, Dr. José A. Clavijo A., who has never failed to bring back from his field trips some of the most interesting Odonata, including the present one.*"

Cleis

The Genus *Cleis* **Selys**, 1853
[presently *Umma* Kirby, 1890]

Bombardier *Lyriothemis cleis* **FM Brauer**, 1868

Cleis was the daughter of the Greek lyric poet Sappho (Cleis was also her mother's name). Selys named three African Demoiselle genera *Sapho*, *Phaon* (with whom Sappho was in love) and *Cleis* in the same publication. Unfortunately, *Cleis* was a preoccupied name and (1890) it was replaced with *Umma*.

Clelia

Red Skimmer ssp. *Orthetrum pruinosum clelia* **Selys**, 1878
[Orig. *Libella clelia*]

Clelia is a feminine given name derived from the Latin 'Cloelia' who, in Roman legend, was a maiden who was given to an Etruscan invader as a hostage, but managed to escape by swimming across the Tiber. Selys named many damselfly species with female names originating from Roman or Greek mythology.

Clément

Eastern Junglewatcher *Hylaeothemis clementia* **Ris**, 1909

Armand Lucien Clément (1848–1920) was an artist, natural historian, teacher, bee-keeper and entomologist. He became President of the Zoological Society of France (1919). He wrote and illustrated a number of books from a museum guide to a book on human anatomy including: *Destruction des Insectes et autres Animaux nuisibles* (1920) and many papers on apiculture, butterflies and ants. The type was in Selys' collection and he had noted that he received the specimen from Borneo from Clément, although it may have been via him as there is no evidence of Clément visiting Borneo.

Clementia

Shadowdamsel sp. *Palaemnema clementia* **Selys**, 1886

In Roman mythology Clementia was the goddess of forgiveness and mercy.

Clendon

Sanddragon sp. *Progomphus clendoni* **Calvert**, 1905

(See **McClendon**)

Cleopatra

Dragonlet sp. *Erythrodiplax cleopatra* **Ris**, 1911
[Orig. *Erythrodiplax connata cleopatra*]

Cleopatra V!! Philopator (69BC–30BC) was the last pharaoh of ancient Egypt. She was a member of the Ptolemaic dynasty, a family of Greek origin that ruled Egypt after Alexander the Great's death. She represented herself as a reincarnation of the goddess Isis. Ris used Selys' manuscript name for this species.

Clio

Cruiser sp. *Macromia clio* **Ris**, 1916

In Greek mythology Clio was the muse of history, a daughter of Zeus and Mnemosyne. She had one son, Hyacinth. Also, known as the Proclaimer, she is often pictured holding a set of tablets or parchment scroll.

Clymene

Sable Cruiser *Macromia clymene* **Ris**, 1921
[JS *Phyllomacromia monocerus* (Förster, 1906]
Skimmer sp. *Diplacina clymene* **Lieftinck**, 1963

In Greek mythology Clymene may refer to several characters such as an Oceanid who was mother of Atlas and Prometheus among others or variously an Amazon, a Nereid, a nymph and many others.

Coby

Coby's Longleg *Notogomphus cobyae* **Dijkstra**, 2015

Jacoba 'Coby' Dijkstra-Stutvoet (1945–1998) was the mother of Dr Klaas-Douwe Benediktus Dijkstra (b.1975). The species was discovered shortly after the tenth anniversary of her death.

Cogman

Masai Sprite *Pseudagrion massaicum cogmani* KH Barnard, 1937

[JS *Pseudagrion massaicum* Sjöstedt, 1909]

This is a toponym; the place known as Kogman's Kloof, Montagu, South Africa is the type locality.

Colenso

Ringtail sp. *Austrolestes colensonis* **White**, 1846

[Orig. *Agrion colensonis*]

Rev. William Colenso (1811–1899) was a naturalist, ethnologist, philologist and missionary in New Zealand. He was apprenticed to a printer (1826) and took a job in New Zealand (1834) working for the Church Missionary Society printing copies of Christian texts in the Maori language. Other works followed, including a Maori text of the Treaty of Waitangi (1840). (At the signing of the Treaty, his cautious representations to Lieutenant Governor William Hobson that many Maori were unaware of the meaning of the treaty were brusquely set aside.) His observations recorded at the time were published as: *The authentic and genuine history of the signing of the Treaty of Waitangi* (1890), the most reliable contemporary European account of the event. His enthusiasm for natural history was boosted by the brief visit of Charles Darwin on the *Beagle* (1835). Although primarily a collector, he did receive some systematic training (1838) from the New South Wales government botanist. The missionary aspect of his job led him on extended journeys around the North Island of New Zealand, collecting all the way until 1852, yet he still managed to become an ordained minister (1844), as well as to marry and father children. His inflexible, overbearing and humourless nature led to friction with some of the foremost Maori leaders as well as a falling out with the missionary society. What is more, he was suspended as a deacon (1852) when he fathered a child by a Maori girl, thus becoming a figure of ridicule among the Maori community whom he had enjoined against sin. He took up politics (1858) and was elected to the General Assembly (1861). Although he remained on the provincial council until its abolition, Colenso increasingly turned to writing and botany. He published a large number of scientific papers and was commissioned by the General Assembly to produce a Maori dictionary (1865), but only a section was published. He also wrote a number of historical pamphlets describing his inland explorations. In his final years, Colenso was regarded as something of a character, a man who had outlived his adversaries thus receiving a tolerance denied him in his more active years. His suspension as deacon was revoked (1894) and he was readmitted to the Anglican clergy. A bird is also named after him.

Collart

Graceful Jewel *Libellago collarti* **Navas**, 1929

[JS *Stenocypha gracilis* (Karsch, 1899)]

Albert Désiré Clément Hubert Collart (1899–1993) was a Belgian entomologist at the Royal Institute of Natural Sciences of Belgium with a passion for bookplates. He was collecting actively in the late 1920s through 1930s and many of his collections are held at the Essig Museum of Entomology. He went to the Congo (1923) as a health worker with little spare time, but he still managed to collect insects, birds and local artefacts, such as the Jewel type (May 1926). He returned home for good (1930) because of ill health and spent the rest of his career at the Royal Museum. He was research associate from (1932–1935), assistant naturalist intern (1935–1936), assistant naturalist (1936–1941), assistant curator (1941–1950), curator (1950–1952), laboratory director (1952–1964) and scientific collaborator (1965). He published papers in the 'Bulletin of the Society of Entomology in Belgium'.

Comanche

Comanche Skimmer *Libellula comanche* **Calvert**, 1907

The Comanche are a people, a Plains Indian tribe whose historic territory, known as Comancheria, consisted of present day eastern New Mexico, southern Colorado, northeastern Arizona, southern Kansas, all of Oklahoma and most of northwest Texas. A semi-autonomous nation within the USA, they have their own Department of Higher Education, primarily awarding scholarships and financial aid for members' college educations. Additionally, they operate the Comanche Nation College in Lawton, Oklahoma. They own ten tribal smoke shops and four casinos.

Common

Sprite sp. *Pseudagrion commoniae* **Förster**, 1902

Elise (Gottliebin) Förster (née Common) (1880–fl.1950) was the wife of the author Johann Friedrich Nepomuk Förster (1865–1918) (See **Förster**).

Conci

Mediterranean Bluet *Coenagrion caerulescens caesarum* **Schmidt**, 1959
[JS *Coenagrion caerulescens* (Fonscolombe, 1838)]

Professor Cesare Conci (1920–1983) was an Italian odonatologist who was Professor of Zoology in University of Genoa. Besides dragonflies, he studied Mallophaga and biology of cave species. Jointly, Conci and Cesare Nielsen (q.v.) published the book: *Odonata* in the *Fauna d'Italia* series (1956). (See also **Caesar**, **Cesare** & **Nielsen**, joint dedication)

Cook

Flatwing sp. *Heteragrion cooki* Daigle & **Tennessen**, 2000

Carl Cook (b.1925) is an American odonatologist elected Honorary Member of the Dragonfly Society of the Americas (2005) of which he was one of the founders and first President. He began collecting very young when he found a specimen of *Ophiogomphus asperses*, a rarity in Kentucky (1940) that started sixty years of fascination with gomphids worldwide, of which he has gathered a large collection. He has described seven new Odonata. His many scientific papers include: *Notes on the genus Somatochlora collected in Kentucky and Tennessee* (1947) and *Stylogomphus sigmastylus sp. Nov., a new North American dragonfly previously confused with S. ablistylus* (2004). (Also see **Carl Cook**)

Cooloola

Wallum Vicetail *Hemigomphus cooloola* **Watson**, 1991

This is a toponym; Cooloola National Park, in southern Queensland, Australia was the type locality.

Coomalie

Coomalie Pin *Eurysticta coomalie* **Watson**, 1991

This is a toponym; Coomalie is in Australia and is the locality where the species was first recognised.

Cooman

Demoiselle sp. *Atrocalopteryx coomani* **Fraser**, 1935
[Orig. *Agrion coomani*]

Révérend-Père Albert Joseph Marie de Cooman (1880–1967) was a Belgian Jesuit priest who collected insects in Vietnam throughout much of the first half of the twentieth century. He entered a seminary (1898–1902) and almost immediately afterwards he left for Vietnam where he moved around at first for some years before settling in Tonkin where he spent the next thirty-nine years. He retired to France (1951) where he stayed for the remainder of his life. He published many papers and described new species as well as working on the taxonomy of several insect groups. One such was published in the 'Proceedings of the Royal Entomological Society of London': *Description D'un Paromalus (Col. Histeridae)* (1936). Many insects collected by him were sent to MNHN and BMNH. He took part in the entomology taxonomy congress (1962) in France. The author does not provide any etymology for the name, not even the collector's name. However, since the species was found in Tonkin (Vietnam), the dedication undoubtedly refers to R.P. Albert de Cooman, who is known to have collected insects in Hoa Binh (Tonkin).

Coomans

Slendertail sp. *Leptogomphus coomansi* **Laidlaw**, 1936
Threadtail sp. *Elattoneura coomansi* **Lieftinck**, 1937
Sprite sp. *Pseudagrion coomansi* Lieftinck, 1937
Sylvan sp. *Coeliccia coomansi* Lieftinck, 1940
Flatwing sp. *Podolestes coomansi* Lieftinck, 1940
Shadowdamsel sp. *Protosticta coomansi* **Van Tol**, 2000

Dr Louis Coomans de Ruiter (1898–1972) was a Dutch ornithologist, botanist, collector and entomologist who was an administrator in the East Indies (Indonesia) (1921–1936, 1938–1942 & 1948–1952). While in Sulawasi (1938–1942) he was very interested in collecting Odonata and discovered a number of new species, which were lodged with RMNH where the shadowdamsel, which he had collected, was re-examined. He also collected in Borneo and elsewhere. A bird is also named after him.

Cora

The Genus *Cora* **Selys**, 1853

Cora's Pennant *Macrodiplax cora* **Brauer**, 1867
[A. Wandering Pennant, Coastal Glider Orig. *Diplax cora*]
Satinwing sp. *Euphaea cora* **Ris**, 1930

Selys named his new genus after Kore (the maiden, Cora in Latin), which is an alternate name for the Greek goddess Persephone. Ris used the name *cora* since the blue colour pattern of the thorax of his new damselfly resembled some *Cora* bannerwing species.

Corbet, P S

Sheartail sp. *Microgomphus corbeti* **Pinhey**, 1961
[Orig. *Microgomphus schoutedeni corbeti*]
Wisp sp. *Agriocnemis corbeti* **Kumar** & Prasad, 1978
Duskhawker sp. *Gynacantha corbeti* Lempert, 1999
Darner sp. *Castoraeschna corbeti* **Carvalho**, Pinto & Ferreira, 2009
Pond Damselfly sp. *Cyanallagma corbeti* **JM Costa**, **Santos** & I. de Souza, 2009
Ebony Gem *Libellago corbeti* van der Poorten, 2009
Damselfly sp. *Risiocnemis corbeti* Villanueva, 2009
Firetail sp. *Telebasis corbeti* **Garrison**, 2009
Pond Damselfly sp. *Tukanobasis corbeti* **Machado**, 2009

Professor Philip Steven Corbet (1929–2008) was a British entomologist whose primary focus was aquatic insects in general and Odonata in particular and who was considered one of the world authorities. Born in Kuala Lumpur, Malaysia, he was taken to England soon after (1931). His mother took him to New Zealand during WW2 where he attended school. He went to Reading University, where he gained his BSc (1950), then to Gonville and Caius Cambridge for his PhD (1953) with a thesis on the Emperor dragonfly. He was also interested in biological controls of pests. He was in Uganda as entomologist at the East African High Commission (1954–1957) and at the East African Virus Research Institute, Entebbe (1957–1962). He was then in Canada at the Department of Agricultural Research (1962–1967) and as Director of their Research Institute at Belleville, Ontario (1967–1971), then Professor of the Biology Department of the University of Waterloo, Ontario (1971–1974). He moved to New Zealand as Professor and Director of the Joint Centre for Environmental Science at University of Canterbury (1974–1978) and Chair in the Department of Zoology (1978–1980). He returned to the UK to the Chair of Biological Science at University of Dundee (1980–1990) retiring as Professor Emeritus. He wrote: *Dragonflies* (1960), *A Biology of Dragonflies* (1962) and *Dragonflies: behaviour and ecology of Odonata* (1999), which is the primary source of information on most aspects of dragonfly biology, and co-wrote *Dragonflies* (2008) which appeared posthumously. In his time, he was almost universally recognised as the doyen of odonatology and he influenced a whole generation of younger workers who have done much to popularise this insect group. He died after suffering a heart attack in his local village shop. (See **Philip**)

Corbet, S A

Bold Leaftail *Phyllogomphus corbetae* **Vick**, 1999
[Syn. of *Phyllogomphus selysi* Schouteden, 1933]

Dr Sarah Alexandra Corbet (b.1940), the sister of Dr Philip S Corbet (above), was born in New Zealand. Philip recruited her as his assistant when she was 10 years old, and later supervised her A Level project. He invited her to Entebbe before she started university, which is when she determined to become an entomologist. She was a Lecturer in both the Department of Applied Biology and later in the Department of Zoology at the University of Cambridge until she retired (1999). She was also Director of the International Bee Research Association. She had published on this species earlier under a different name. She has written many papers and a number of books including being a series Editor of the Naturalists' Handbooks, including co-authorship of several such as: *Solitary Wasps* (1983), *Bumblebees* (1987) and *Insects, Plants and Microclimate* (1991).

Corduli Santos

The Emerald genus *Cordulisantosia* Fleck & **JM Costa**, 2007

(See **Santos**)

Cornelia

Demoiselle sp. *Calopteryx cornelia* **Selys**, 1853
Orange Amberwing *Perithemis cornelia* **Ris**, 1910

Cornelia is a name from Ancient Rome – most famously the wife of Julius Caesar. The genus *Perithemis* includes many other names from ancient history or mythology.

Cornelia (Lieftinck)

Wiretail sp. *Selysioneura cornelia* **Lieftinck**, 1953
Spectre sp. *Petaliaeschna corneliae* **Asahina**, 1982

Cornelia Maria (Corry) Lieftinck née van Veen (1916–1980) was the wife of M A Lieftinck (q.v.). Asahina named the hawker species after Lieftinck's wife. Lieftinck did not provide an etymology, but likely the species name refers to his wife too. However, in the same paper he named three other *Selysioneura* species with names from antiquity (*thalia, aglaia, venilia*) so that species might belong above.

Cornelis (van Steenis)

Jewel ssp. *Heliocypha fenestrata cornelii* **Lieftinck**, 1947

Cornelis Gijsbert Gerrit Jan 'Kees' van Steenis (1901–1986) was a Dutch botanist. The University of Utrecht awarded his Master's degree (1925) and PhD (1927). He then worked as a botanist at a herbarium in Java (1927–1946). He was active in the Netherlands (1946–1949) where he was engaged in the organization of *Flora Malesiana*, a description of the flora of the Malay Archipelago and New Guinea. He spent the rest of his life (1950–1986) as Director of the Flora Malesiana Foundation but was also Professor of Tropical Botany for the Royal Tropical Institute in Amsterdam (1951) and for the University of Leiden (1953) and Professor and Director at the National Herbarium of the Netherlands there (1962–1972). He provided Lieftinck with several interesting Odonata from Sumatra, Bali and elsewhere. He collected this subspecies from Bali (1936) and also collected in Australia and New Zealand.

Coropina

Dasher Sp. *Micrathyria coropinae* **Geijskes**, 1963

This is a toponym; Coropina Creek, Surinam was the type locality.

Corycia

Cruiser sp. *Macromia corycia* **Laidlaw**, 1922

In Greek mythology Corycia was a naiad. She lived on Mount Parnassus in Phocis and her father was the local river-god, Kephisos. She bore Lycoreus with Apollo.

Coryndon

Orange-bellied Flasher *Aethiothemis coryndoni* **Fraser**, 1953
[Orig. *Lokia coryndoni*]

Robert Thorne Coryndon (1870–1925) was, among other things, Governor of Uganda (1981–1922) and Kenya (1922–1925). The Coryndon Museum was opened in 1930 and was named in his honour, but has since been renamed and is now the Nairobi National Museum. Fraser published papers in the Museum journal and the skimmer is a Uganda endemic.

Costa, A

Faded Pincertail *Onychogomphus costae* **Selys**, 1885

Achille Costa (1823–1898) was an Italian entomologist who became Director of the Zoological Museum of Naples, founding their entomological collections and describing many new species. He made the greatest contribution to the entomological exploration of southern Italy and the Italian islands in the second half of the 19th century. The type was collected in Sardinia.

Costa, J P M

(See **Janira Martins** Costa)

Costa Lima

Pond Damselfly sp. *Tuberculobasis costalimai* **Santos**, 1957
[Orig. *Leptobasis costa-limai*]

Ângelo Moreira da Costa Lima (1887–1964) was the foremost Brazilian entomologist of his time, considered the father of Brazilian entomology. He qualified as a physician (1909) and became a surgeon but soon switched to zoology. He worked at the Universidade Federal Rural do Rio de Janeiro. Among other works he wrote: *Terceiro catálogo dos insetos que vivem nas plantas do Brasil. Seus parasitos e predadores* (1922) and the 11-volume: *Insetos do Brasil* (1938–1960). More than 200 species are named after him.

Cottarelli

Cottarelli's Longleg *Notogomphus cottarellii* Consiglio, 1978

Professor Vezio Cottarelli (b.1937) is an Italian zoologist (a graduate of the University of Rome) who was (until 2009) Professor of Animal Biology and Zoology at the Dipartimento di Scienze Ambientali, Università della Tuscia. Currently he is an honorary researcher there. He has collected in a number of countries including Ethiopia (1973), where he collected the type, and the Philippines (2005). He co-wrote, among at least 83 papers: *Zoological Researches in Ethiopia, 1: Some results of the first and the second Italian zoological mission to Ethiopia, sponsored by the National Academy of Lincei (1973 and 1975)* (1978).

Couturier

Greater Shadowcruiser ssp. *Idomacromia proavita couturieri* **Legrand**, 1985

Guy Couturier (b.1939) is a friend of the author; he collected the female holotype of this species in Ivory Coast (1980). He is a French researcher (Natural History Museum in Paris) who has worked in Ivory Coast. He has published also on Diptera and Hemiptera, such as: *Les Diptères Doiichepodidae de Côte d'Ivoire : Description de trois nouvelles espèces* (1978) He has also longer works, such as the book: *Recherche et Amenagement en Milieu Forestier Tropical Humide: Le Projet Tai De Cote-D'ivoire* (1984).

Cristina

Threadtail sp. *Neonevra cristina* **Rácenis**, 1955
[JS *Neoneura bilinearis* Selys, 1860]

Cristina Rácenis (b.1953) was described as 'the dear daughter' of the author.

Croizat

Flatwing sp. *Teinopodagrion croizati* **De Marmels**, 2002
Flatwing sp. *Heteropodagrion croizati* Peréz-Gutiérrez & Montes-Fontalvo, 2011

León Camille Marius Croizat (1894–1982) was an Italian botanist, bio-geographer, and evolutionist. He served in the Italian army (1914–1919), after which he returned to Università degli Studi di Torino and graduated in law (1920). He hated the Fascists and emigrated to the US (1923), where he took any job going. He painted in watercolours with modest financial success until the Wall Street Crash (1929) killed his market. He tried his luck in Paris, without success, so returned to New York. Merrill, then Director of the Arnold Arboretum, Harvard, hired Croizat (1937) as a technical assistant. His drawings were said to be unbelievably accurate. He was sacked over a paper he had written (1946), which was published through a reputable rival outlet. He moved to Caracas, Venezuela, and held a number of academic positions there (1947–1952). He was the botanist on the Franco-Venezuelan expedition to the sources of the Orinoco (1950–1951). He resigned his academic positions (1953) to work full time on biological problems. He and his wife became (1976) the first Directors of Jardin Botanico Xerofito, Coro, which they had founded (1970). It is now named after him, as are a reptile and four birds. Peréz-Gutierrez & Montes-Fontalvo wrote: "*The species is named in honor of León Croizat, the father of panbiogeography.*"

Cruz

Forktail sp. *Ischnura cruzi* **De Marmels**, 1987

Fernando Cruz (b.1961) is a collector of insects. He sent the specimens of this new species to the author having collected them in Bogota, Colombia with J Cortés (1986).

Cupid

Threadtail sp. *Protoneura cupida* **Calvert**, 1903

In Roman mythology Cupid (or Cupido) is the god of desire, erotic love, attraction and affection; the counterpart of Eros in Greek myth. He is often portrayed as the son of Venus, goddess of love.

Curtis

Orange-Spotted Emerald *Oxygastra curtisii* **Dale**, 1834
[Orig. *Cordulia curtisii*]

Dr John Curtis (1791–1862) was a British entomologist and illustrator. He learned to engrave in his father's workshop. When his father died, his mother encouraged his study of natural history with a local naturalist, Richard Walker (1791–1870). At 16 he was apprenticed to a lawyer but devoted his spare time to studying and drawing insects and he began to make a living selling the insects he collected. Many of his illustrations were published in: *An Introduction to Entomology* (1815–1826). He moved to London, met Sir Joseph Banks who introduced him to Leach at the BMNH who in turn introduced him to Dale, who became his patron. His greatest work was: *British Entomology* (1824–1839). He slowly went blind (from 1840) and received a civil list pension. The original illustrations were put up for sale after his death, but bought by Walter Rothschild who donated them to the BMNH.

James Charles Dale discovered this species at Barley Heath near Bournemouth (June 1820), but he did not describe it until fourteen years later. The species was first thought to be endemic to Britain, but it was soon shown to occur in France, Spain and Portugal. The species later became extinct in Britain (the last record was 1963).

Cuyabá

Wedgetail sp. *Acanthagrion cuyabae* **Calvert**, 1909

This is a toponym; Cuyabá, Brazil was the type locality.

Cyclops

Northern Evening Darner *Telephlebia cyclops* **Tillyard**, 1916
[Orig. *Telephlebia godeffroyi cyclops*]

In Greek and later Roman mythology, Cyclops had a single eye in the middle of his forehead. Although Tillyard did not provide an etymology, this can be concluded from the description. One of the main distinguishing features of his subspecies was the large black rounded blotch on the frons. *NB*: There are four Odonata species and subspecies called '*cyclopica*', the name referring to the Cyclops Mountains in New Guinea.

Cycnos

Southern Skimmer *Libellula cycnos* **Selys**, 1848
[JS *Orthetrum brunneum* (Fonscolombe, 1837)]

'Cycnos' is the French version of 'Cycnus' or 'Cygnus' who in Greek mythology was a character that was transformed into a swan (there are a number of different versions).

Cydippe

Sylph sp. *Macrothemis cydippe* **Calvert**, 1898
[JS *Macrothemis hemichlora* (Burmeister, 1839)]
Cruiser sp. *Macromia cydippe* **Laidlaw**, 1922

In Greek mythology Cydippe was a priestess of Hera and the mother of Cleobis and Biton. She was on her way to a festival in the goddess' honour but the oxen, which were to pull her cart, were overdue and her sons pulled the cart the entire way (c. 8km). Cydippe was impressed with their devotion to her and asked Hera to give her children the best gift a god could give a person. Hera had the two brothers drop dead instantaneously as the best thing she could give them was for them to die at their moment of highest devotion. So be careful what you wish for!

Cynthia

Sylph sp. *Macrothemis cynthia* **Ris**, 1913

In Greek mythology Cynthia was an epithet of the goddess of the moon, Artemis. It was sometimes used as she was supposedly born on Mount Cynthus.

Cynthia (Longfield)

Blade-tipped Duskhawker *Heliaeschna cynthiae* **Fraser**, 1939
Grenadier ssp. *Agrionoptera insignis cynthiae* **Lieftinck**, 1942

Cynthia Longfield FRES (1896–1991). (See **Longfield**)

Cyrano

The Jewel Genus *Cyrano* **Needham** & Gyger, 1939

Cyrano de Bergerac was the eponymous hero in Edmond Rostand's play (1897). The character Cyrano has an extremely large nose. All jewel (Chlorocyphidae) species are characterized by a conspicuous protruding rhinarium.

Cyrene

Skimmer sp. *Diplacina cyrene* **Lieftinck**, 1953
Tigertail sp. *Synthemis cyrene* Lieftinck, 1953

In Greek mythology Cyrene was the daughter of Hyseus, King of the Lapiths. She was a fierce huntress, sometimes called the 'girl lionkiller'. When a lion attacked her father's sheep, Cyrene wrestled with the lion. Apollo, who was present, immediately fell in love with her and kidnapped her. He took her to North Africa and founded a city in her name.

D

D'Abreu

Wisp sp. *Agriocnemis dabreui* **Fraser**, 1919

Edward Alwyn D'Abreu FZS (d.1939) was a zoologist who was Assistant Curator (c.1915–c.1917) and then Curator (c.1919–1939) at the Central Museum, Nagpur. Before this he had been a teacher at Victoria School, Kurseong (c.1910–c.1915). He was a Fellow of the Zoological Society Central Museum. He wrote: *The Central Provinces Butterfly List* (1931) and a guide to Nagpur Museum (1933) as well as many papers in, e.g. 'Bombay Natural History Society Journal', particularly on birds such as: *On the distribution of Curlews and Godwits in the Central Provinces* (1934). Fraser gives no explanation of the name of the wisp but it seems clear that this is who he had in mind. He collected the holotype female of this wisp in Central provinces of India.

Daecke

Attenuated Bluet *Enallagma daeckii* **Calvert**, 1903
[Orig. *Telagrion daeckii*]

Victor Arthur Erich Daecke (1863–1918) was a German born American (1881) amateur entomologist. He was an artist and, at one time, worked as such for the US Printing Company and was an illustrator with the Philadelphia Press. He was on the advisory committee (1910) on entomology of the Philadelphia Academy of Natural Sciences having been a long-time associate (1900). After a long illness (1907–1908) he became an assistant in the Pennsylvania State Department of Zoology (1908–1918). He took the holotype when collecting in southern New Jersey (1900–1902).

Dagny

Sylvan sp. *Idiocnemis dagnyae* **Lieftinck**, 1958

Mrs Dagny Bergman (1890–1972) was a teacher specialising in natural history. She accompanied her husband Dr Sten Bergman, a Swedish zoologist, in his expeditions to Kamchatka (1920–1923) and New Guinea (1956–1959). She wrote a book: *Vildmarksår* (1948) about their stay in New Guinea. She collected a few specimens of this new species during the New Guinea trip (May 1957). Lieftinck wrote: "*I have much pleasure in naming this handsome little species after Mrs. Dagny Bergman, who made valuable collections of dragonflies in West and South New Guinea.*"

Dahl, J

Orange-nosed Jewel *Chlorocypha dahli* **Fraser**, 1956

Jörgen Dahl (c.1925–1998) was a Danish zoologist on a Danish expedition to Cameroon (1949-1950), which collected the type. He wrote: *Results from the Danish Expedition to the French Cameroons 1949–50. XXI. Coleoptera: Lagriidae, Sandalidae, Brenthidae and Scarabaeoidea* (1957).

Dahl, K F T

Grenadier sp. *Nesoxenia dahli* **Ris**, 1898
[JS *Nesoxenia mysis* Selys, 1878]

Professor Dr Karl Friedrich Theodor Dahl (1856–1929) was a German zoologist, in particular an arachnologist. He attended the universities of Leipzig, Freiburg and Kiel, achieving the status of 'Privatdozent', a postdoctoral equivalent to the US Associate Professor. He travelled to the Baltic States and the Bismarck Archipelago, New Guinea, where he collected, including the type (1896–1897). He became Curator of Arachnids at the Berlin Natural History Museum (1898) staying on until retirement. He also had interests in biogeography and animal behaviour. He wrote: *Das leben der vögel auf den Bismarckiseln* (1899). He married Maria Dahl (1872–1972), a co-worker at the Zoological Institute of Kiel who also published several works on spiders and finished editing his book on Melanesia. Two birds are also named after him.

Daigle

(See **Jerrell** James Daigle)

Daimoj

Cruiser sp. *Macromia daimoji* Okumura, 1949

This is the Japanese word for 'large'. The etymology refers to another species name in Japanese that has a shape on the body looking like the Japanese character 'Dai' which can mean 'number'. Quite how this relates to this species is not clear.

Dale

Eastern Pygmyfly *Nannophya dalei* **Tillyard**, 1908
[Orig. *Nannodythemis dalei*]

James Charles Dale (1791–1872) was a wealthy British naturalist particularly interested in entomology. Cambridge University awarded his MA (1818). When not managing his estate, or acting as a magistrate he engaged in fieldwork, often journeying up to forty miles a day. His collection of 33 cabinets of insects, which they received (1906) is now in the Hope Department of Entomology of the University of Oxford, and includes his four cabinets of Odonata. He wrote 83 notes and articles, the first of which was on Lepidoptera; others include: *A List of the more rare of the Species of Insects found on Parley Heath, on the Borders of Hampshire* (1834). He was a friend of, and regular correspondent with, Edmond de Selys Longchamps who had labelled the specimen in recognition of Dale before Tillyard described it from that specimen.

Dalloni

Orange-winged Dropwing ssp. *Trithemis kirbyi dallonia* **Navás**, 1936
[JS *Trithemis kirbyi* Selys, 1891]

Marius-Gustave Dalloni (1880–1959) was Professor of Geology at the University of Algiers (1925–1950) and a noted socialist militant who was Deputy Mayor of Algiers (1946) and a member of the Commission on the Constitution. He made a collecting expedition to the Spanish Pyrenees (1910) to survey the geology of the area. He published a great number of papers and other works, notably: *Mission au Tibesti zoologie: Étude préliminaire de la faune du Tibesti* (1935) (Tibesti, northern Chad, is where the type specimen was collected). This is the record of an expedition he led to the area (1930–1931) principally to collect mineral specimens. A mammal collected on the same expedition is also named after him.

Danaë

Black Darter *Sympetrum danae* Sulzer, 1776
[A. Black Meadowhawk]

In Greek mythology, Danaë was a daughter of the king of Argos, King Acrisius and his Queen Eurydice. After she was locked in a tower or cave due to a prophecy that her son would kill Acrisius, Zeus came to her in the form of golden rain and she became pregnant with his child. Perseus was this offspring. She was sometimes credited with founding the city of Ardea in Latium.

Daniel

Leaftail sp. *Phyllogomphoides danieli* **González-Soriano** & Novelo-Gutiérrez, 1990

Daniel Novelo-Galicia (b.1985) is the son of the second author Rodolfo Novelo-Gutiérrez (b.1955).

Daniell

Large Red Damsel *Moroagrion danielli* **Needham** & Gyger, 1939
[JS *Pyrrhosoma nymphula* (Sulzer, 1776)]

The authors do not give any etymology, nor is there any other clue in the publication that would reveal after whom this species was named. The specimen is now thought not to represent a new genus and species from the Philippines at all and was probably collected in Spain and represents the common European species *Pyrrhosoma nymphula*.

Daphne

Skimmer sp. *Huonia daphne* **Lieftinck**, 1953

Daphne is a naiad in Greek mythology; in the same paper the author named several other species after characters from Greek myth.

Dardano

Pond Damselfly sp. *Leptagrion dardanoi* **Santos**, 1968

Professor Arturo Dárdano de Andrade-Lima (1919–1981) was a Brazilian botanist. The Department of Agronomic Engineering from Escola Superior de Agricultura de Pernambuco awarded his bachelor's degree (1943). He was Associate Professor in the Department of Biology at Pernambuco Federal Rural University (1958–1981). He also studied at Kew in England (1956–1957 & 1975) and Munich (1975). He was a researcher at the Institute of Agronomic Research until retirement (1974–1978). Most of his papers were on botanical subjects.

Darwall

Darwall's Hooktail *Paragomphus darwalli* **Dijkstra, Mézière** & Papazian, 2015

Dr William Robert Thomas Darwall (b.1961) is (2017) Head of the IUCN Freshwater Biodiversity Unit (since 2002). The University of St Andrews awarded his BSc, the University of Utah his MSc and Hull University his PhD (2004). His past experience includes research and conservation projects in Malawi, Tanzania and Ireland, and he has worked in commercial aquaculture in Scotland. The etymology says that *"…Darwall and his team were instrumental in synthesising the threat status of Africa's aquatic nature. Such facilitators are the unsung heroes of conservation efforts worldwide."*

Darwin

Darter sp. *Sympetrum darwinianum* **Selys**, 1883
Glider sp. *Tramea darwini* **Kirby**, 1889
[JS *Tramea cophusa* Hagen, 1867]

Charles Robert Darwin (1809–1882) was the prime advocate, together with Wallace, of natural selection as the driver of speciation. To quote from his seminal work, *On the Origin of Species by Means of Natural Selection* (1859): '*I have called this principle, by which each slight variation, if useful, is preserved, by the term Natural Selection.*' He was naturalist on 'HMS Beagle' on her scientific circumnavigation (1831–1836). In South America, he found fossils of extinct animals that were similar to extant species. On the Galápagos Islands, he noticed many variations among plants and animals of the same general type as those in South America. Darwin collected specimens for further study everywhere he went. On his return to London he conducted thorough research of his notes and specimens. Out of this study grew several related theories; evolution did occur; evolutionary change was gradual, taking thousands or even millions of years; the primary mechanism of evolution was 'Natural Selection'; and the millions of species alive today arose from a single original life form through a branching process called 'specialisation'. However, Darwin held back on publication for many years not wanting to offend Christians, especially his wife. He is remembered in the names of numerous other taxa including twenty-three birds, three amphibians, four mammals and nine reptiles. The glider was collected on the Galapagos Islands.

Dasha

Sylvan sp. *Coeliccia poungyi dasha* **Kosterin**, 2016

A hypocoristic form of the female name Darya, named after the author's friend who preferred to remain anonymous.

Dashidordzhi

(See **Doshidordzi**)

Dau

Bombardier sp. *Lyriothemis daui* **Krüger**, 1902
[Junior JS *Lyriothemis magnificata* Selys, 1878]

Hans Dau was a friend of the author. In honouring him, Krüger describes him as one of his two faithful 'Wandergefahrten', which we interpret as 'travelling companions'. (Also see **Lau, F**)

Davenport

Reedling sp. *Indolestes davenporti* **Fraser**, 1930
[Orig. *Ceylonestes davenporti*]
Shadowdamsel sp. *Protosticta davenporti* Fraser, 1931

Mr Davenport was the owner of Toni Mudis estate, where Fraser stayed during his visit to Annaimallai and Mudis Hills in South India (1929). During his stay, other new dragonfly species were found.

David

Clubtail genus *Davidius* **Selys**, 1878

Clubtail sp. *Davidius davidii* Selys, 1878
Levant Clubtail *Gomphus davidi* Selys, 1887

Father Jean Pierre Armand David (1826–1900) was a French Lazarist priest (better known simply as Pere David) as well as a fine zoologist. He taught in Savona, and Marchese Doria and C L M d'Albertis were among his pupils. He was a missionary to China and was the first westerner to observe many animals, including birds, as well as the Giant Panda and the deer *Elaphurus davidianus* famously named for him. He arrived in China in 1862 and started collecting in 1863. The French naturalist Alphonse Milne-Edwards classified many of his specimens. He co-wrote: *Les Oiseaux de Chine* (1877). He collected several specimens of Odonata in central China and Tibet, including the type series of *D. davidii*. However, *G. davidi* was collected in Syria. Pere David collected thousands of specimens and had many plants named after him, as well as a mammal, an amphibian, four reptiles and fifteen birds.

David

Clubtail genus *Davidioides* **Fraser**, 1924

No etymology given.

Davies

Emerald sp. *Somatochlora daviesi* **Lieftinck**, 1977
Davies' Metalwing *Neurobasis daviesi* **Hämäläinen**, 1993
Gem sp. *Libellago daviesi* **van Tol**, 2007
Skydragon sp. *Chlorogomphus daviesi* **Karube**, 2001

Professor Dr David Allen Lewis Davies (1923–2003) was a professional research biochemist and immunologist and a world leader in transplantation antigen studies (1960s). He volunteered (aged 19) for military service and joined (1943) the Royal Tank Regiment and commanded his tank company at Juno Beach on D-day and was later promoted to Captain (1945). After WW2, he worked in military intelligence and was invited to advise the German Defence on military law at the Nuremberg Trials. He enrolled at Cambridge University (1946), which awarded his BSc (1949) and MSc (1953). Later he received a PhD from the University of London and DScs from both London and Cambridge. He worked as a biochemist in several institutes (1949–1988) and became Professor of Immunology at Queen Elisabeth's College, University of London (1975–1980). His interest in natural history dated back to his youth in Wales. In the 1950s he developed vacuum drying methods for preserving insect and plant samples, one of his papers on this subject being published in *Nature* (1956). His interest in dragonflies became a consuming passion, especially in the last 20 years of his life. His aim was to obtain representatives of as many genera as possible for his synoptic dragonfly collection so he travelled widely to all corners of the globe in search of rarities. Indefatigable in the field, he rediscovered *Hemiphlebia mirabilis* in Victoria (1985) and, at the age of 69, he climbed high in the Himalayas to find the enigmatic *Epiophlebia laidlawi* abundant at an altitude of 3500m. Besides collecting for himself he was eager to exchange specimens with other odonatologists, gradually amassing a large collection, presently housed in the Cambridge University Museum of Zoology. He published 40 articles on dragonflies (1981–2003), including papers on taxonomy and, with his secretary Pamela Tobin, a checklist of the dragonfly species of the world (1984–1985), the first of its kind (since 1890). He was author or co-author of 16 new dragonfly species or subspecies, most of them from New Caledonia and Yunnan, where he also collected. Allen Davies was an original thinker, who often held controversial views and hypotheses. He drove a red Aston Martin, often at extremely high speeds, along the narrow byways around Cambridge. He was a great raconteur, who entertained and encouraged younger colleagues in many ways. In September 2003, a memorial service was held at his Cambridge college, Gonville and Caius, that was attended by many noted odonatologists. (Also see **Allen (Davies)** & **Pamela**)

Davina

Cascader ssp. *Zygonyx iris davina* Fraser, 1926

Davina is a feminine forename, but not a specified one. Many other *Zygonyx* taxa have been named after forenames. The author gave no etymology, so it is not possible to know for sure if this was named after a particular person. However, the lack of a Latin ending suggests it is not an eponym, as does Fraser's frequent use of apparent forenames. On the other hand, he often did not fully identify through etymologies the people he honoured.

Davis

Sandhill Bluet *Enallagma davisi* **Westfall**, 1943

Edward Mott Davis (1888–1943) was a friend of the describer and Director (1934) of Thomas R. Baker Museum of Natural Science at Rollins College, Winter Park, Florida. He graduated from Harvard (1909) and attended Massachusetts Agricultural College (1910) before operating a fruit farm for many years (1910–1934) during which he continued his interest in ornithology and entomology.

Dayak

Skydragon sp. *Chlorogomphus dyak* **Laidlaw**, 1911
[Orig. *Orogomphus dyak*].
Shadowdamsel sp. *Telosticta dayak* **Dow** & **Reels**, 2012

Dayak is the collective term used for all the indigenous peoples of Borneo where the type was discovered.

De Beaufort

Fineliner sp. *Teinobasis debeauforti* **Lieftinck**, 1938

Professor Dr Lieven Ferdinand de Beaufort (1879–1968) was a Dutch zoologist. As a student, he participated (1902–1903) in the first scientific expedition to New Guinea, headed by the geographer Professor A Wichman. There he, with Hendrikus Albertus Lorenz, collected the type series of this new dragonfly species (July 1903). Universiteit van Amsterdam awarded his doctorate (1908). He undertook a second voyage to the Dutch East Indies (1909–1910). He regularly published (1904–1921) about birds collected in New Guinea by him, his friend, (later Professor) Cosquino de Bussy, and others. In his early years, he was an assistant of Max Weber, who he succeeded as Director of the Zoological Museum of the University of Amsterdam (1922–1949) and was Extraordinary Professor in Zoogeography (1929–1949). He was one of the founders (1901) of the *Nederlandsche Ornithologische Vereeniging* (the Dutch Ornithological Society), becoming Secretary (1911–1924) and its Chairman (1924–1956). He was also a member of the editorial board of the society's journal 'Ardea' (1924–1956) and he was zoology, ethnography and anthropology editor of 'Nova Guinea, n.s' during the time that Lieftinck was contributing entomological papers. He edited the 11-volume: *The Fishes of the Indo-Australian Archipelago* (1911–1962). A mammal, a reptile and two birds are also named after him.

De Beaux

Fineliner sp. *Teinobasis debeauxi* **Lieftinck**, 1938

Professor Dr Oscar De Beaux (1879–1955) was an Italian mammalogist, palaeontologist and conservationist. He was a scientific assistant at the Carl Hagenbeck Zoo in Hamburg (1911–1913). He was Professor of Zoology at Genoa University and became Director of Museo Civico di Storia Naturale di Genova (1934–1947). He wrote: *Biological Ethics: An Attempt to Arouse a Naturalistic Conscience* (1930). De Beaux had permitted Lieftinck to study the type series of this species collected by Dr Lamberto Loria (January 1894). A mammal is also named after him.

De Bellard

Spreadwing sp. *Lestes debellardi* **De Marmels**, 1992

Eugenio de Bellard Pietri (1927–2000) was a speleologist and conservationist. He studied at the University of Central Venezuela but transferred to Bogata University in Colombia because of political unrest. There was so much political turmoil there that he eventually went to study law at Salamanca University in Spain. He undertook his doctorate in law at Merida Los Andes University back in Venezuela. He organized the expedition to the southern part of Venezuela where the species was found. He worked for Shell de Venezuela, becoming Secretary to the Board of Directors (1969–1974) and again when it was taken into public ownership (1979–1981) and then for Corpoven (1985–1987) after which he retired. In private life, he devoted his time to cave exploration, discovering many new caverns, and organized three expeditions into the Amazonian rainforest, which collected vast amounts of natural history specimens within and around the caves during one of which the damselfly was found. He published several scientific works, including: *History of Speleology in Venezuela 1678 to 1950*.

Deam

Dancer Sp. *Argia deami* **Calvert**, 1902

Charles Clemon Deam (1865–1953) of Bluffton, Indiana, was a pharmacist, businessman, surveyor, botanist, conservationist and forester; and the first State Forester of Indiana (1909–1913 & 1917–1928) (sometimes referred to as 'good ol' Charlie Deam'). He attended DePauw University for two years but had to give up because of the cost of tuition. He part owned a pharmacy for all his working life but his passion was botany. He took up botany as an interest on the long walks he took to relieve stress. His time in forestry is notable for his move to re-establish forest on land cleared for farming, particularly to halt soil erosion. He collected many Odonata specimens in Mexico including the type, which he lent to Calvert to study through their mutual friend Bruce Williamson (q.v.). He also collected many plants in Mexico and Guatemala as well as the US over a number of years and many plants are named after him. He wrote four books about plants in Indiana: *Trees of Indiana* (1911), *Shrubs of Indiana* (1924), *Grasses of Indiana* (1929) & *Flora of Indiana* (1940). (Also see **Stella**)

Debra

Shadowdamsel sp. *Sinosticta debra* **Wilson** & Xu, 2007

Debra Lynn Wilson (1952–2005) was the senior author's late sister.

Decken

Powder-striped Sprite *Agrion deckeni* **Gerstäcker**, 1869

[JS *Pseudagrion kersteni* (Gerstäcker, 1869)]

Baron Carl Claus von der Decken (1833–1865) was a German explorer who died in Somalia. He explored in East Africa and was the first European to try to climb Mount Kilimanjaro. Decken explored the region of Lake Nyasa on his first expedition (1860). Together with a geologist, he visited the Kilimanjaro massif (1861). He ascended Kilimanjaro (1862) to 13,780 feet, seeing its permanent snow-cap; he also established its height as about 20,000 feet, and mapped the area. One expedition took him to Madagascar, the Comoro Islands and the Mascarene Islands (1863). Then, in Somalia (1865) he sailed the River Giuba where his ship, the *Welf*, foundered in the rapids above Bardera. During the trip, Dr Otto Kersten collected the sprite. Somalis killed von der Decken and three other Europeans. After his death, his letters were edited and published (1869) in book form under the title: *Reisen in ost-Afrika*. At least one fish, a bird and two mammals are named after him.

De Fonseka

Bombardier sp. *Lyriothemis defonsekai* Van der Poorten, 2009

[A Vermilion Forester]

Terence de Fonseka (1919–2000) was an odonatologist. He won an entrance scholarship to University College, Colombo and obtained a degree in zoology at the University of London. He entered the Ceylon Civil Service (1945–1970) and worked there in different governmental departments. Then he moved with his family to England, where he worked for the Department of Health until retirement (1984). He became interested in dragonflies and their conservation only a few years before leaving Sri Lanka. He wrote the book: *The dragonflies of Sri Lanka* (2000), which was published shortly after his death. In the foreword, he explained that the main reason for this book was to inspire a greater public interest in dragonflies in Sri Lanka, especially among the younger people.

Degors

Riverhawker sp. *Tetracanthagyna degorsi* **Martin**, 1895

Alfred Degors (d.1921) was a French insect collector and entomologist from Le Blanc (Indre) who was particularly interested in Coleoptera. A beetle genus and some beetle species also bear his name. Although no etymology is given the hawker is most likely named after this man.

Dejoux

Stonewash Dropwing *Trithemis dejouxi* **Pinhey**, 1978
[Orig. *Trithemis donaldsoni dejouxi*]

Dr Claude Dejoux (b.1939) is a French hydrobiologist and limnologist who has studied African and neotropical freshwater sites including a major study of Lake Titicaca during his work for ORSTOM. His BA was awarded by the University of Toulouse (1962) and his doctorate by the University of Paris (1974). His first post was as a researcher for ORSTOM Paris (1963–1965) and later he directed the research centre at N'Jamena, Chad (1972–1973), then worked in Martinique (1983) and the Ivory Coast (1985) and was Director of ORSTOM in Bolivia (1985–1989), including its Laboratoire d'Hydrobiologie. He was again in West Africa (1990). He was also a consultant for WHO (1973–1991). He was then in Mexico City (1992–1995) and afterwards an international consultant in hydrobiology. Since retirement from IRD as Director of Research he has continued to work, including a trip to Gansu province, China to look at the ecotoxicology of wetlands there (2007). Among many other books, reports and scientific papers (1975–1998) he wrote: *Synecologie des Chironomides du lac Tchad: Dipteres, Nematoceres* (1977), *La pollution des eaux continentales africaines* (1988) and *El Lago Titicaca* (1991). In private life he loves scuba diving, tennis and wildlife photography.

Delecolle

Skimmer sp. *Celebothemis delecollei* **Ris**, 1912

No etymology is given. However, it is known that a Mr Delecolle, based in Brussels, collected for Ris in Celebes at the appropriate time. I have not found any further biographical details.

Delessert

Skimmer sp. *Libellula delesserti* **Selys**, 1878
[JS *Orthetrum triangularis* (Selys, 1878)]

Adolphe François Delessert (1809–1869) was a French naturalist who collected in India (1834–1839) sponsored by his wealthy uncle, Baron Benjamin Delessert. His travels took him to Mauritius, Reunion Island, Penang, Pondicherry, Malay Peninsula, Singapore, Java, and Madras. He met Jerdon (1839) in the Nilgiri hills and gave him a new babbler, which Jerdon described as *Crateropus delesserti* and in the same vein Selys wrote: "...*a été prise en Cochinchine dans les Monts Neelgheries par M. Delessert*". Two birds are also named after him. He wrote: *Souvenirs d'un Voyage dans l'Inde de 1834 à execute 1839* (1843) illustrated by, among others, the German botanical artist Heyland (1791–1866).

Delia

Threadtail sp. *Disparoneura delia* **Karsch**, 1891
[JS *Prodasineura verticalis* (Selys, 1860)]
Sylph sp. *Macrothemis delia* **Ris**, 1913

No etymology given so I assume that both these species are named Delia, an epithet of the Greek moon goddess, Artemis rather than a real person.

De Marmels

Pond Damselfly sp. *Mesamphiagrion demarmelsi* **Cruz**, 1986
[Orig. *Cyanallagma demarmelsi*]
Blue-eyed Darner sp. *Rhionaeschna demarmelsi* von Ellenrieder, 2003
Threadtail sp. *Epipleoneura demarmelsi* von Ellenrieder & **Garrison**, 2008
Pond Damselfly sp. *Tepuibasis demarmelsi* **Machado** & **Lencioni**, 2011
[Orig. *Austrotepuibasis demarmelsi*]

Dr Jürg De Marmels (b.1950) is a Swiss-born Venezuelan entomologist. His first degree in zoology was awarded by Zürich University (1978), as was his doctorate (2001). He was Professor of Zoology (1983–86) and of Entomology (1987–2011) for undergraduates at the Institute of Agricultural Zoology, part of the Central University of Venezuela. He was also Professor of Insect Taxonomy for graduates (1981–2016), beyond his official retirement (2011). He is still an active member of the staff of the Museum of the Institute of Agricultural Zoology 'Francisco Fernández Yépez' (MIZA). He has published around one hundred scientific papers, including some on Lepidoptera, and among his Odonata works is: *Revision of Megapodagrion Selys, 1886* (2001). He has described nearly 90 new South American odonate species and erected a number of genera including *Chalcothore, Teinopodagrion, Lamproneura, Bromeliagrion, Tepuibasis* and *Andaeschna*. His research focus is taxonomy and biogeography of insects.

Demoulin

Clubtail sp. *Idiogomphoides demoulini* St. Quentin, 1967
[Orig. *Gomphoides demoulini*]

Dr Georges O J Demoulin (1919–1994) was a Belgian entomologist who specialised in mayflies, Ephemeroptera. His original intention was to become a pharmacist but he abandoned this (1937) to train as a naturalist at the Royal Museum of Natural History of Belgium. After spending a year in Africa at the Garamba National Park, he took up a post back at the museum in charge of the water-associated insect collection. He wrote numerous papers, especially on mayflies, such as: *Ephemeroptera des faunes éthiopiennes et malgache* (1970). About 20 species and one genus of insects are named after him.

Dening

Dening's Sprite *Pseudagrion deningi* **Pinhey**, 1961
[A Dark Sprite]

Richard Cranmer 'Tim' Dening (1920–2005) was a British entomologist. He joined the Indian Army and served there and in Burma until the end of WW2. After the war, having returned to Cambridge to complete his interrupted history degree, he joined the Colonial Administrative Service in Northern Rhodesia (now Zambia) and was a District Commissioner in Mwinilunga, Mumbwa and Samfya. He remained in Zambia after independence, working as an agricultural economist and later worked as a consultant in the same field, all over the world. His primary interest was Lepidoptera but he occasionally collected other insects. His other interests included travel, human health, nutrition and squash, which he played into his eighties and even after a heart attack. His travel was legendary including driving from Sri Lanka to Delhi and then all the way to West Sussex, England (1979). He collected the type in the Bangweulu Swamps in Zambia (October 1959). He donated his collection of Lepidoptera and Odonata collected in forty-two countries to Glasgow Museum.

Denise

Carnarvon Tigertail *Eusynthemis deniseae* **Theischinger**, 1977

Denise Jones née Theischinger (b.1969) is the author's daughter.

De Ville

Demoiselle sp. *Mnesarete devillei* **Selys**, 1880
[Orig. *Lais devillei*]

Émile De Ville (c.1836–1880) was the Belgian consul in Quito, Ecuador. He collected the first specimens of the new species and many other Odonata for de Selys Longchamps whom he visited (9 March 1877). Later he became consul in Zanzibar. He should not be confused with the French naturalist Emile Deville (1824–1853).

Diamang

Flasher sp. *Aethiothemis diamangae* **Longfield**, 1959

This is not the name of a person but a diamond mine! Longfield wrote of the skimmer: "*I have named it after the diamond mine in Dundo.*" The new species was found in Dundo, in Northeast Angola. The 'Diamang – Companhia de Diamentes de Angola' had supported the local fieldwork carried out by Museo do Dundo.

Diana (Carle)

Pudu Redspot *Ophiopetalia diana* Carle, 1996
[JS *Phyllopetalia pudu* Dunkle, 1985]

Diana Carle was the daughter of the author Frank Louis Carle the American entomologist who is Curator of Rutgers Entomological Museum and Director of The New Jersey Aquatic Insect Survey

Diana

Relict Dragonfly sp. *Epiophlebia diana* Carle, 2012

Diana was the goddess of the hunt, the moon and birthing in Roman mythology. The etymology states that the name is '...*an allusion to the highland huntress and protector of humankind, who if despoiled, will presage the destruction of mankind by their own handiwork, just as Actaeon was devoured by his own hunting dogs*'.

Dickson

Rock Hooktail *Paragomphus dicksoni* **Pinhey**, 1969
[JS *Paragomphus cognatus* (Rambur, 1842)]

Charles Gordon Campbell Dickson (1907–1991) was a South African lepidopterist. He discovered no fewer than thirty-eight new lepidopteran species in the mountains and hills of the Western Cape, thirteen of which are named after him. He revised some genera and described 102 new butterfly species and subspecies. He collected the holotype of the hooktail in South Africa (December 1968). Among other works (1940–1991) he co-wrote: *Life Histories of the South African Lycaenid Butterflies* (1971), which resulted in the University of Cape Town awarding him an honorary MSc, and later *Butterflies of the Table Mountain Range* (1980). Apart from butterflies his other great passion was for vintage cars and steam locomotives; he drove around in a vintage Riley.

Dierl

Sylvan sp. *Coeliccia dierli* St. Quentin, 1970

Dr Wolfgang Dierl (1935–1996) was an entomologist whose main focus was Lepidoptera. His doctorate was awarded in 1962. He took part in the Himalayan Research Scheme that was the first scientific-ecological project in northeastern Nepal to span all the earth sciences, botany, zoology, anthropology and human sciences, which consisted of thirty-eight scientists including Count Douglas Bigot de St Quentin (1899–1982). The scheme was funded by the German government and took place over several years (1960–1965). Both Dierl and St Quentin contributed to the various reports of the expeditions known as *Nepal Khumbu Himal*. Dierl took part in two extensive collecting trips to Nepal (1963–1973) when working at the Zoologische Staatssammlung Munchen (1966–1996) becoming Curator of Lepidoptera (1975–1996). During that time, he moved the collection into a new building and completely re-organised it. He wrote, among many other (73) works: *Insekten* (1991).

Dingavan

Pincertail sp. *Onychogomphus dingavani* **Fraser**, 1924

Captain G Dingavan collected the holotype male in Burma (October 1922) and also the holotype of *Paragomphus risi* (1924). There was a George Stanley Dingavan (d.1940) who was an assistant surgeon serving in the Highland Light Infantry (1911) in the subcontinent that I believe to be the same person who collected for Fraser.

Distant

Highland Dropwing *Stoechia distanti* **Kirby**, 1898
[JS *Trithemis dorsalis* (Rambur, 1842)]

William Lucas Distant (1845–1922) was a British entomologist who spent much of his working life in a tannery. His father was a whaler captain and took him on a trip to the Malay Peninsular (1867), which triggered his love of natural history and resulted in him writing a description of the butterflies there, *Rhopalocera Malayana* (1882–1886). He made two extended trips to Transvaal. Following the first he wrote: *A Naturalist in the Transvaal* (1892) and a second visit, of some four years, gave him time to amass a large collection of insects (including the type specimens of the dropwing), many of which were described in *Insecta Transvaaliensia* (1900–1911). He became editor of 'The Zoologist' (1897) and was employed (1899–1920) at the BMNH cataloguing the insect collection and describing many new species he found there. He devoted much of his time to the Rhynchota. He published many other contributions to works on various fauna. When he retired from the museum they purchased his collection of c.50,000 specimens. He died of cancer. A reptile is also named after him.

Ditzler

Darner sp. *Triacanthagyna ditzleri* **Williamson**, 1923

William Howard Ditzler (1880–1966) collected the type of this species and many other Odonata during the University of Michigan Venezuelan Expedition (1920), which included E B Williamson, Jesse H Williamson and H B Baker. He also participated, along with his wife Ivy (q.v.) and daughter Laura (q.v.) in another collecting trip, organized by Williamson (1931), from Indiana to Georgia during which all three members of the family collected enthusiastically.

Divina

Rubyspot sp. *Hetaerina divina* **Hagen**, 1854
[JS *Hetaerina auripennis* (Burmeister, 1839)]

Divina is simply Latin meaning divine or heavenly.

Do

Featherleg sp. *Matticnemis doi* **Hämäläinen**, 2012
[Orig. *Platycnemis doi*]

Captain Do Manh Cuong (b.1979) is a Vietnamese entomologist with a focus on Coleoptera and Odonata. The University of Hanoi awarded his MSc (2003). He worked (2004–2016) at the Entomology and Zoology Department, Military Institute of Hygiene and Epidemiology, Hanoi, Vietnam. He has described seven new species of dragonflies from Vietnam and seven new Asian species of beetle. He has written or co-written more than thirty articles as well as the book: *Checklist of dragonflies from Vietnam* (2007).

Dobson

Lesser Duskhawker *Gynacantha dobsoni* **Fraser**, 1951
Tropical Wisp *Agriocnemis dobsoni* Fraser, 1954
Pilbara Tiger *Ictinogomphus dobsoni* **Watson**, 1969
[Orig. *Indictinogomphus australis dobsoni*]

Roderick Dobson (1907–1979) was an English naturalist resident in Jersey, Channel Islands (1935–1948 & 1958–1979). He visited Australia (1948–1958 & 1967–1968) where he undertook extensive fieldwork on dragonflies and other aquatic insects, especially in New South Wales and Queensland. He collected the type specimens of the two species named by Fraser. He wrote: *Birds of the Channel Islands* (1952) as well as a number of articles in British Birds. (Also see **Roderick**)

Dodd

Northern River Hunter *Austrogomphus doddi* **Tillyard**, 1909

Frederick Parkhurst Dodd (1861–1937) was an Australian entomologist and amateur collector of butterflies and beetles. He was originally employed (1884) in a bank in Townsville before taking up entomology full time (1898). He moved from Victoria to Brisbane, Townsville and then Kuranda (1904), where he became known as the 'Butterfly man of Kuranda'. He collected in Queensland (1906) including the type and on expeditions to Darwin (1908–1909) and to Heberton (1910–1911). He also collected in New Guinea (1917).

Dohrn

Threadtail sp. *Alloneura dohrni* **Krüger**, 1898
[JS *Prodasineura collaris* Selys, 1860]
Duskhawker sp. *Gynacantha dohrni* Krüger, 1899
Shadowdancer sp. *Idionyx dohrni* Krüger, 1899
[JS *Idionyx yolanda* Selys, 1871]
Potbelly sp. *Risiophlebia dohrni* Krüger, 1902
[Orig. *Nannophlebia dohrni*]

Dr Heinrich Wolfgang Ludwig Dohrn (1838–1913) was a German entomologist and politician. His family came from Pommern, part of which is now in Poland. He studied entomology at Szczecin (Stettin) where he graduated (1858). He travelled extensively in West Africa (1864–1866) during which he collected on the island of Príncipe (1865). He collected insects in Sumatra (1893–1897). Leopold Krüger worked on his Sumatran dragonfly collections and published four papers about them (1898–1902). His father, Dr Karl Augustus Dohrn, was also an amateur entomologist and his younger brother was Felix Anton Dohrn (1840–1909), the founder of the Stazione Zoologica di Messina 'Anton Dohrn'. Felix Anton's godfather was Felix Mendelssohn, the composer. Two birds are also named after him.

Donald (Fraser)

Evehawker sp. *Anaciaeschna donaldi* **Fraser**, 1922
Shadow-emerald sp. *Macromidia donaldi* Fraser, 1924
[Orig. *Indomacromia donaldi*]

Donald Fraser was the author's son although Fraser does not give any etymology. From the descriptions, it is clear that these species were not named after any collector; Fraser himself collected the latter species. He also named a species after his wife (m.1905) (Ethel Grace Varall) without giving any etymology. (See **Ethel**)

Donald (Kimmins)

Skimmer sp. *Lanthanusa donaldi* **Lieftinck**, 1955

Douglas Eric Kimmins (1905–1985) of the BNHM was the intended object of the honour of the name. However, Lieftinck made a mistake and used the name Donald when he meant to use Douglas. He later pointed out his mistake but was unable to change the name as it had been already published… and under accepted rules of taxonomic naming no later change could be justified. As he said: *"…questions like this would seem to fall under the jurisdiction of metonymy and do not affect nomenclature – factum fieri infectum non potest! – I am pleased all the same to correct the mistake."* (See **Kimmins**)

Donaldson

Donaldson's Dropwing *Trithemis donaldsoni* **Calvert**, 1899
[Orig. *Pseudomacromia donaldsoni*, A. Denim Dropwing, Twig Dropwing]

Dr Arthur Donaldson-Smith (1864–1939) was a physician, traveller and big-game hunter of American birth who spent much time in East Africa. He visited Lake Rudolph (Lake Turkana) (1895 & 1899). He was in Ethiopia (1896) and may have been present at the Ethiopian victory over the Italians at the Battle of Adwa. He wrote: *Through Unknown African Countries* (1897). He was elected a Fellow of the Royal Geographical Society. He collected the dropwing at Stony Brook, a tributary of the Erer River in eastern Ethiopia (August 1894). A mammal and seven birds are also named after him.

Donnelly

Donnelly's Knobtail *Epigomphus donnellyi* **González** & **Cook**, 1988
Pinchtail *Odontogomphus donnellyi* **Watson**, 1991
Threadtail sp. *Drepanoneura donnellyi* **von Ellenrieder** & **Garrison**, 2008
Dancer sp. *Argia donnellyi* Garrison & von Ellenrieder, 2015

Bannerwing ssp. *Cora chirripa donnellyi* **Bick** & **Bick**, 1990
Emerald ssp. *Hemicordulia tenera donnellyi* **Kosterin**, **Karube** & Futahashi, 2015

Dr Thomas Wallace Donnelly (b.1932) is an American geologist and odonatologist. His geology BSc was awarded by Cornell University (1954), his MSc by the California Institute of Technology (1956) and his PhD (1959) by Princeton University. He became first Assistant Professor of Geology at Rice University in Houston (1959) and was Professor of Geology at State University of New York at Binghamton until retirement (1966–1996). His geological research started with structural geology, his doctoral thesis being on the Geology of the Caribbean Islands St. Thomas and St. John. Later he focussed on deep sea sediments and participated in two cruises on the 'Glomar Challenger' within the Deep-Sea Drilling project. His dragonfly interest started at the age of fifteen when he spent a summer washing dishes at an Audubon camp in Maine, where one of the instructors was the odonatologist Donald Borror (q.v.) and at high school in Washington, D.C., he found that the curators at the Smithsonian were enormously welcoming and helpful to a teenage youth with a real interest in odonates. He described (1961) two new species including *Protoneura ailsa*, which he named after his wife (m.1956) Ailsa (q.v.). So far, he has described 56 new species or subspecies, many of which he collected in the field. He has travelled all over the world, often several times a year, in search of odonates, visiting all continents, including Antarctica and is mostly accompanied by his wife and sometimes some of his children. The couple's latest trip (November 2016) was to Paraguay, which is still poorly known for

Odonata, where the 83-year-old veteran discovered several new or potentially new species. The trip report (2017) ended with the words: *"Will we go back? You bet!"*. Nick Donnelly has undoubtedly seen more Odonata species in the wild than any other odonatologist. He has published many papers on Odonata, including revisions of the Fijian genera *Melanesobasis* (1984) and *Nesobasis* (1990) and *The Odonata of the central Panama and their position in the Neotropical odonate fauna, with a checklist, and descriptions of new species* (1992). Donnelly himself considers his three papers *Distribution of North American Odonata. Part I, Part II* and Part III (2004), as his greatest contribution to odonatology. These papers provide detailed distribution maps for all species. (Also see **Ailsa**)

Donner

Pacific Clubtail *Gomphus donneri* **CH Kennedy**, 1917

[JS *Gomphus kurilis* Hagen, 1857]

This is a toponym; Donner Lake is in Nevada County, California, which was the type locality.

Donovan

Keeled Skimmer *Libellula donovani* **Leach**, 1815

[JS *Orthetrum coerulescens* (Fabricius, 1798)]

Edward Donovan (1768–1837) was an Anglo-Irish natural history illustrator and zoologist and author of several books. In his 16-volume: *Natural History of British Insects* (1792–1813). He named and illustrated several new species of British dragonflies, of which the name *Cordulegaster boltonii* (Golden-ringed Dragonfly) has remained valid. His beautiful book: *An Epitome of the Natural History of the Insects of China* (1798) includes two colour plates of Chinese dragonflies. He also described a species of shark. Donovan suffered financially from the activities of unscrupulous book dealers and publishers, and died in poverty.

Dora

Threadtail sp. *Nososticta dora* **Kovács** & **Theischinger**, 2016

Dora Kovács (b.1993) is the daughter of the author Tibor Kovács (b.1965).

Dorothea

Lilysquatter sp. *Paracercion dorothea* **Fraser**, 1924

[Orig. *Coenagrion dorothea*]

Spreadwing sp. *Lestes dorothea* Fraser, 1924

Fraser did not give etymology for the name Dorothea in any of these species. Possibly Dorothea was one of his relatives. He also named a species after his wife (Ethel Grace Varall 1881–1960) without giving any etymology (see **Ethel**).

Dorothea (O'Donel)

Sylvan sp. *Coeliccia dorothea* Fraser, 1933

Grenadier ssp. *Agrionoptera insignis dorothea* **Fraser**, 1927

Dorothea O'Donel was the wife of H V O'Donel (q.v.), who collected the type series in Duars, Bengal. (Also see **O'Donel**)

Dorothea (Williamson)

Demoiselle sp. *Psolodesmus dorothea* **Williamson**, 1904
[JS *Psolodesmus mandarinus* McLachlan, 1870]

Dorothea 'Dora' Williamson née Kellerman (1849–1928) was the mother of the author who named this Taiwanese demoiselle species after her.

Doshidordzi

Common Darter *Sympetrum striolatum doschidorzii* **Belyshev** & Doshidorzhi, 1958
[JS *Sympetrum striolatum* (Charpentier, 1840)]

Anudarin Doshidorzi (sometimes Doshidordzhi or Dashidorzi) (1910–1976) was a Mongolian ichthyologist and entomologist who was the first Mongolian to be Professor of Zoology at the State University in Ulan Bator. Belyshev wrote the paper and included Doshidordzhi as a co-author; Doshidorzhi did not create an eponym for himself. The description appears in an article they wrote: *On Odonata fauna of Mongolia* (1958) in *Zoologicheskii Zhurnal*. He co-authored other articles such as: *Mongolian Grayling Thymallus brevirostris Kessler from the Dzabakhan River Basin* (1972). The type was taken in Mongolia (1954). At least one beetle and at least one fish are also named after him.

Doubleday

Atlantic Bluet *Enallagma doubledayi* **Selys**, 1850
[Orig. *Agrion doubledayi*]

Edward Doubleday (1811–1849) was an English ornithologist and entomologist whose main interest was Lepidoptera. His best-known work was the co-written: *The Genera of Diurnal Lepidoptera: comprising their generic characters, a notice of their habits and transformations, and a catalogue of the species of each genus.* When visiting America (1830s) he wrote a series of articles for the 'Entomological Magazine' in London under the running title: *Communications on the Natural History of North America.* He collected there with Foster. He was later given a post at the BMNH where he was employed until his death.

Doula

Cherry-eye Sprite *Pseudagrion pseudomassaicum doulae* **Pinhey**, 1961
[JS *Pseudagrion sublacteum* (Karsch, 1893)]

This is a toponym; Doula in Cameroon was the type locality.

Dow

Dryad sp. *Pericnemis dowi* **Orr** & **Hämäläinen**, 2013

Dr Rory Alisdair Dow (b.1965) is Honorary Research Associate at the Naturalis Biodiversity Center, Leiden, the Netherlands. He was originally a mathematician, with a PhD in applied mathematics from the University of London. His research interests are in the faunistics and taxonomy of Asian Odonata. He has extensive experience of working in southeast Asia, especially in Malaysia. To quote from this species' etymology "*...named in honour of Dr Rory Dow who first recognized the species as being distinct from P. triangularis and who has made numerous and important recent contributions to the systematics and faunistics* *of Bornean Odonata.*" He is first or sole author on over 50 papers and has described 56 new species of Odonata; recent works include: *A review of the genus Bornargiolestes Kimmins, 1936 (Odonata: Zygoptera) with a description of two new species from Sarawak, Malaysia* (2014), *Revision of the genus Devadatta Kirby, 1890 in Borneo based on molecular and morphological methods, with descriptions of four new species (Odonata: Zygoptera: Devadattidae)* (2015) and *A new Bornean species of Drepanosticta allied to D. actaeon Laidlaw, with notes on related species (Odonata: Zygoptera: Platystictidae)* (2017).

Drescher

Grappletail sp. *Heliogomphus drescheri* **Lieftinck**, 1929

Friedrich Carl Drescher (1874–1957) was an entomologist who was Curator of Entomology at the Museum Zoologicum Borgoriense (1943–1957). After schooling he was employed by a trading company first in Manchester (1894) then Batavia (Djakarta) (1896). He worked for various companies until accepting a partnership (1909) in a mercantile company with premises all over the Malay Archipelago, retiring (1920) as a 'sleeping partner'. He spent his retirement pursuing his interest in entomology. In the slump of the 1930s his finances dwindled (1932) but his long-time friend and fellow naturalist, Dr Edward Jacobson, supported him both as a friend and materially. He returned briefly on business to the Netherlands (1938) but he quickly returned to Java. During the Japanese occupation, he was going to be interned but was taken, with his huge collection of beetles, to the museum at Bogor and allowed to work there. After the war his position was confirmed and he was custodian until his death at 83. His passion was Coleoptera but he was also interested in Odonata and Hemiptera. He collected in Java including the type (1928). Lieftinck wrote his obituary in 'Treubia' (1958).

Drummond

Grabtail sp. *Lamelligomphus drummondi* **Fraser**, 1924
[JS *Onychogomphus risi* (Fraser, 1922)]

Captain Drummond collected the holotype female specimen at Loimwe in Southern Shan States of Burma (1923).

Duges

Blue-eyed Darner sp. *Rhionaeschna dugesi* **Calvert**, 1905
[Orig. *Aeshna dugesi*]

Professor Alfredo Auguste Delsescautz Duges (1826–1910) was a French-born Mexican physician and naturalist. He was the son of zoologist Antoine Louis Duges (1797–1838). He is largely remembered for his extensive studies in Mexican herpetology. After studying medicine at the University of Paris, he emigrated to Mexico and settled in Guanajuato. With his brother, entomologist Eugenio Duges (1826–1895), he organized frequent field trips in order to collect specimens, including the type. He published numerous scientific papers on herpetology, botany, and entomology, et al. At Guanajuato, he became the director of the local museum, later named the *Museo Alfredo Dugès* in his honour.

Duivenbode

Lieutenant sp. *Brachydiplax duivenbodei* **Brauer**, 1866
[Orig. *Perithemis duivenbodei*]

Maarten Dirk van Renesse van Duivenbode (1804–1878) was a merchant and property owner. Although not specified by Brauer, this dragonfly species from New Guinea was undoubtedly named after him. Based on the island of Ternate in the Moluccas he served the Dutch Trade Company (Nederlandsche Handelsmaatschappij). He was merchant, captain, commander and honorary major. He was knighted in the order of the Dutch Lion and the Leopold's order. Duivenbode helped Alfred Russel Wallace to rent a house in Ternate (1858) that Wallace retained for three years. In his book *The Malaya Archipelago* Wallace wrote about him: '...*of an ancient Dutch family, but who was educated in England, and speaks our language perfectly. He was a very rich man, owned half the town, possessed many ships, and above a hundred slaves. He was moreover, well educated, and fond of literature and science—a phenomenon in these regions. He was generally known as the King of Ternate, from his large property and great influence with the native Rajahs and their subjects.*' Duivenbode's son Constantijn Willem Rudolf van Renesse van Duivenbode (b.1858) went on with the family business. The Duivenbodes were heavily involved in the feather trade as well as generally collecting natural history specimens for sale. Interestingly, the name (sometimes spelt 'Duyvenbode') means 'pigeon-post messenger' and was an honorific conferred when the family used carrier pigeons to keep in touch with William of Orange during the siege of Leiden. Eight birds are also named after the father or son.

Dumont

Flatwing sp. *Sinocnemis dumonti* **KDP Wilson** & **Zhou**, 2000

Professor Dr Henri Jean François Dumont (b.1942) is a Belgian entomologist and limnologist. His first degree, Diploma of Licenciate in biology, was awarded (1964) by the University of Ghent as was his PhD in zoology (1968) and his DSci in limnology. He then became assistant professor there rising to senior assistant professor (1969), associate professor (1975) and senior associate professor (1980). He became Professor of Ecology (1987) and Director of the Laboratory of Animal Ecology, Biogeography and Conservation (1987–1994). He is now Professor Emeritus of Limnology at the Department of Biology, State University of Ghent, Belgium and also Professor at the Institute of Hydrobiology, Jinan University, China. He was Editor-in-Chief of 'Hydrobiologia' (1980–2003). He was one of the founders of the Societas Internationalis Odonatologica (SIO) in Gent (1971) and later (1998) became the first editor of 'Pantala', the international journal of odonatology. During his career, he has made more than fifty field trips including extensive expeditions to the Middle East, Mount Everest, the Sahara and other African countries and the Far East, during all of which he collected dragonflies. He has written (1960–2014) at least 435 scientific papers, reports and books, mostly in the fields of zooplankton ecology and biodiversity, ancient lakes ecology, invasive species and molecular phylogenetics of Odonata, Cladocera and Copepoda. His books include: *Odonata of the Levant* (1991). He has described seven new odonate species and subspecies and named four new genera including *Atrocalopteryx* and *Paracercion*. The cladoceran crustacean family *Dumontiidae* is named after him, as well as the chydorid genus *Dumontiellus*, and several species in blue-green algae, rotifers, syncarids, copepods and cladocerans.

Dunkle

Threadtail sp. *Protoneura dunklei* Daigle, 1990
Dasher sp. *Micrathyria dunklei* **Westfall**, 1992

Firetail sp. *Telebasis dunklei* **GH Bick** & **JC Bick**, 1995
Pond Damselfly sp. *Mesamphiagrion dunklei* von **Ellenrieder** & **Garrison**, 2008
Pond Damselfly sp. *Denticulobasis dunklei* **Machado**, 2009

Sidney Warren Dunkle (b.1940) is a biologist, educator, ecological consultant, entomologist and odonatologist currently retired and living in Tucson, Arizona. His BSc was awarded by Baldwin-Wallace College (1962), his MSc by the University of Wyoming (1965) (during which time he was also a ranger/naturalist at Grand Teton NP) and his PhD by the University of Florida (1980). He was Assistant Professor at Cuyahoga Community College, Cleveland (1966–1970) then Instructor at Orange Coast Community College, California (1970–1971) and Assistant Professor at Santa Fe Community College, Florida (1972-1976). He then took a post as Instructor at Fresno (California) City College (1977) and was also Ecological Consultant at Gainesville (1980–1991). He became Interim Professor at the University of Florida, Gainesville (1985–1988) and Manager of the International Odonata Research Institute (IORI) (1985–1991). He has also undertaken many collecting trips to South America, especially to Ecuador and Peru. The threadtail type was taken in the Dominican Republic (1989). Altogether he has described fifteen new species (1982–1995). He wrote, among other papers and books: *Dragonflies of Florida, Bermuda, and the Bahamas* (1989), *Damselflies of Florida, Bermuda, and the Bahamas* (1990) and *Dragonflies through binoculars: a field guide to dragonflies of North America* (2000). He has been a member of the Dragonfly Society of America for many years and was secretary (from 1987); he is also a member of the Societas Internationalis Odonatologica. He has described 15 new species of Odonata. In private life he enjoys canoeing, camping, hiking, photography and square dancing.

Dupuy

Gambia Riverjack *Mesocnemis dupuyi* **Legrand**, 1982

André-Roger Dupuy (1935–2012) was an officer in the Foreign Legion, a hunter, an explorer, researcher, naturalist and conservationist. He was a much decorated (Legion of Honour, etc.) commando then foreign legionnaire who, when returning to civilian life, explored in North Africa and the Sahara. He was also at a research station in Algeria where he was able to indulge his love of birds and wildlife in general. After returning to Paris he was recruited to administer the Niokolo Koba NP in Senegal and, at the time the type was taken, he had become Directeur des parcs nationaux du Senegal about which he wrote: *Les gardiens de la vie sauvage: Avec ceux qui se battent pour les elephants* (1984) as well as other books such as *Soldat des Bêtes* (1991).

Dyer

Dusky Lilysquatter ssp. *Paracercion calamorum dyeri* **Fraser**, 1919
[Orig. *Argiocnemis dyeri*]

George Henry Thiselton-Dyer (1879–1944) was a mechanical engineer, botanist and collector who was the son of an eminent Kew Director and botanist, Sir William Turner Thiselton-Dyer (1843–1928). His maternal grandfather, Sir Joseph Dalton Hooker, was also Director at Kew (1865–1885). He collected plants in Australia (1903–1905) including the doubly eponymic *Thiseltonia dyeri*. Fraser collected one of the two type specimens (April 1919) in Thiselton-Dyer's compound at Poona. He became an engineering inspector in the Ministry of Health (c.1938).

E

Earnshaw

Demoiselle sp. *Mnais earnshawi* **EB Williamson**, 1904
[JS *Mnais andersoni* McLachlan, 1873]
Hooktail sp. *Acrogomphus earnshawi* **Fraser**, 1924
[Orig. *Onychogomphus earnshawi*]

Ralph Allen Earnshaw (1876–1940) was a veteran of the Spanish-American war who was awarded the Congressional Medal for saving passengers from the burning liner 'Hardinge'. He operated a timber business in Burma (1897–1940). He was also a renowned big game hunter. He took the types of these species while collecting (1899–1911) for Williamson in the Karenni and Toungu districts of Burma. Williamson had also provided a thorough description of the *Acrogomphus* (1907), but did not give a name for it since the specimen was badly broken. Fraser named this species after its collector, referring to Williamson's earlier description and René Martin's opinion that it was definitely a new species. Earnshaw died at the American Baptist Mission House in Peku Karen Compound, Toungoo and was buried in the Cantonment Cemetery, Toungoo.

Ebner

Wisp sp. *Agriocnemis ebneri* **Ris**, 1924
[JS *Agriocnemis zerafica* Le Roi, 1915]

Bluetail ssp. *Ischnura elegans ebneri* **Schmidt**, 1938
[JS *Ischnura elegans* (Vander Linden, 1820)]

Dr Richard Ebner (1885–1961) was an Austrian schoolteacher and entomologist, a specialist in Orthoptera. He travelled widely in Europe, North Africa, and Central Asia. He collected the type specimens of these two species, the wisp in southern Sudan (1914) and the Blue-tailed Damselfly subspecies in Syria (1928). He bequeathed his huge collection of Orthoptera to Naturhistorisches Museum Wien. Two reptiles are also named after him.

Echeverri

Knobtail sp. *Epigomphus echeverrii* Brooks, 1989

Sr Gustavo Echeverri is a Costa Rican landowner who, according to Brooks "…*kindly donated his ranches to the Guanacaste National Park*." He also managed the ranches of some other landowners and he helped in the process of some of their land being dedicated to the Park.

Eda

Emerald sp. *Hemicordulia edai* **Karube** & Katatani, 2012

Professor Dr Shigeo Eda (b.1932) is a Japanese amateur odonatologist and Honorary President of the Japanese Society of Odonatology (2011). He was Professor of Dentistry at Matsumoto Dental University in Nagano (1974–2000). He has published hundreds of articles and notes on Japanese dragonflies (since 1949) and his books include: *A guide to the collection and observation of dragonflies* (1976). He was the editor of 'Tombo – Acta Odonatologica Japonica' (1998–2009). (Also see **Masako**)

Edmond

Shadowdamsel sp. *Palaemnema edmondi* **Calvert**, 1931

Baron Edmond de Selys Longchamps created the genus name. Calvert had visited Selys in Liége twice (1895 & 1896); perhaps knowing him personally is the reason for the posthumous dedication using just his first name. (See **Longchamp** & **Selys**)

Edmund

Edmund's Snaketail *Ophiogomphus edmundo* **Needham**, 1951

Edmund Needham is the grandson of the author, James George Needham (1868–1956). (See **Needham**)

Eduard

Emerald sp. *Hemicordulia eduardi* **Lieftinck**, 1953

Professor Dr Eduard Handschin (Handschin-Hofstetter) (1894–1962) was a Swiss entomologist, palaeontologist and zoologist. He was Professor of Zoology and the Director of the Naturhistorisches Museum at Basel. He published at least 130 scientific papers. Perhaps his best-known book is *Practical introduction to the morphology of insects: a book for teachers, students and entomophiles* which went through 16 editions (1928). The etymology reads: *Dedicated to my esteemed friend Professor Dr. Eduard Handschin, the well-known entomologist and Director of the Naturhistorisches Museum at Basel".*

Eduardo

Blue-eyed Darner sp. *Rhionaeschna eduardoi* **Machado**, 1984
[Orig. *Aeshna eduardoi*]

Eduardo Ribeiro Machado (b.1972) is the son of the author Angelo Barbosa Monteiro Machado (b.1934). The dedication says: *"This beautiful species is dedicated to my companion of odonatological excursions, my son, Eduardo".*

Edwards

Blue Basker *Urothemis edwardsii* **Selys**, 1849
[Orig. *Libellula edwardsii*]

Henri Milne-Edwards (1800–1885) was born in Belgium, the 27th son of a reproductively-prolific Englishman, and went on to become a renowned French naturalist. He became Professor of Hygiene and Natural History at the Collège Central des Arts et Manufactures (1832). He succeeded (1841) Audouin (q.v.) as the chair of entomology at the Museum d'Histoire Naturelle. He wrote works on crustaceans, molluscs and corals and his name is remembered in the scientific names of many marine organisms. He sent (1845) Selys Algerian Odonata specimens for study, which had been collected by Hippolyte Lucas.

Egler

Wedgetail sp. *Acanthagrion egleri* **Santos**, 1961

Dr Walter Alberto Egler (1924–1961) was a German-Brazilian botanist who became Director of Museu Goeldi, Sao Paulo, Brazil. He died when his boat was swept over a waterfall. He had helped Santos in arranging field trips.

Eisen

Baja Bluet *Enallagma eiseni* **Calvert**, 1895

Dr Gustav Eisen (1847–1940) was a Swedish born American marine invertebrates and earthworm specialist, particularly of the eponymous genus *Eisenia*. He was a graduate (1873) of Uppsala University. He became a member (1874) and later life member (1883) of the California Academy of Science. He went on to become (1893) the Curator of Archaeology, Ethnology, and Lower Animals at the Academy, later changed to Curator of Marine Invertebrates. He was appointed (1938) as an honorary member, which is considered the highest honour from the Academy. He was a polymath with diverse interests, including art and art history, archaeology, anthropology, agronomy, horticulture, history of science, geography, cartography, cytology, and protozoology, as well as marine invertebrate zoology. He also studied malaria-vector mosquitoes, founded a vineyard in Fresno, introduced avocados and Smyrna figs (he wrote a book on the history of figs) to California, campaigned to save the giant sequoias, and wrote a multivolume book about the Holy Grail! He made a large collection of dragonflies from the Cape region of Baja California (1893–1894) that included the type. Mount Eisen, in the Sierra Nevada in California, was also named after him (1941), as were numerous worms, some algae and plants. His ashes were later interred at Redwood Meadow near the foot of the eponymous peak.

Eisentraut

Cameroon Yellowwing *Allocnemis eisentrauti* **Pinhey**, 1974
[Orig. *Chlorocnemis eisentrauti*]

Professor Dr Martin Bruno Eisentraut (1902–1994) was a German zoologist and collector. He was on the staff of the Berlin Zoological Museum, working on bat migration and the physiology of hibernation. His first overseas trip was to West Africa (1938). He left Berlin to become Curator of Mammals at the Stuttgart Museum (1950–1957), then became Director of the Alexander Koenig Museum in Bonn, where he lived for the rest of his life. He made six trips to Bioko and Cameroon (1954–1973) during one of which (1967) he collected the type. Eisentraut wrote many scientific papers and three books, including: *Notes on the Birds of Fernando Pó Island, Spanish Equatorial Africa* (1968). He also published a slim volume of poems. He facilitated Pinhey's study of Hartwig's Odonata collection made in West Cameroon and Fernando Po on one of the expeditions directed by Eisentraut. Two birds, four mammals and a reptile are also named after him.

Ekari

Pond Damselfly sp. *Papuagrion ekari* **Lieftinck**, 1949

The Ekari are a people in Papua New Guinea who live in the surroundings of Lake Paniai, where Hilbrand Boschma found the species during the Le Roux Expedition to Netherlands New Guinea (1939).

Elgner

Short-Tailed Duskdarter *Zyxomma elgneri* **Ris**, 1913

Hermann Elgner (c.1856–1913) was a German lepidopterist based in Cape York and the Torres Strait Islands, northeast Australia. He collected insects almost daily for over a decade (1900–1913), particularly intensely during 1906–1910, until his health declined (1911). He also collected on islands forming part of what was then German New Guinea. His collection is now in several Australian Museums.

Elias

Pond Damselfly sp. *Inpabasis eliasi* **Santos**, 1961
[JS *Inpabasis rosea* (Selys, 1877)]

Claudionor Elias was a Brazilian from Espirito Santo state who, like his father Pinto, was a full-time insect collector who collected the types with Santos in Amazonas (Brazil) (1959).

Elisabeta

Flatwing sp. *Philogenia elisabeta* **Calvert**, 1924

The etymology reads: "*...Its species name is suggested by that of the companion of Gretchen at the well in Goethe's Faust.*"

Elisabeth (Djakonova)

Bog Hawker ssp. *Aeshna subarctica elisabethae* Djakonov, 1922
[Orig. *Aeshna elisabethae*]

Elizaveta N Djakonova (Dyakonova) née Savelyeava was the wife of the author, Alexander Mikhaylovich Djakonov (Dyakonov) (1886–1956), who named the hawker, which he had found in Russian Karelia, after his wife. They lived in Saint Petersburg.

Elisabeth (Keiser)

Madagascar Cascader *Zygonyx elisabethae* **Lieftinck**, 1963
Elisabeth's Islander *Nesolestes elisabethae* Lieftinck, 1965

Elisabeth 'Lili' Keiser née Jenny was the wife of the Swiss entomologist Dr Fred Keiser (q.v.), a known dipterologist and friend of the author. The Madagascar species were dedicated to "*Mrs Elisabeth Keiser in recognition of her constant and invaluable assistance in the field during various expeditions.*" She had participated in an expedition (1957–1958) to Madagascar with her husband when the type was taken. (Also see **Kaiser**).

Elisabeth (Ris)

Greek Red Damsel *Pyrrhosoma elisabethae* **Schmidt**, 1948

Miss Elisabeth Ris (1872–1959) from Goldbach-Küstnacht, Kanton Zürich, Switzerland was the sister of the famous odonatologist Friedrich Ris. She was honoured '…*in appreciation of her great efforts in relieving the post-war suffering in Germany*'. (Also see **Ris**)

Elise

Calico Pennant *Celithemis elisa* **Hagen**, 1861
[Orig. *Diplax elisa*]

Johanna Maria Elise Hagen née Gerhards (1832–1917) was married (1851) to the author, Hermann August Hagen (1817–1893).

Eliseva

Lucifer Jewel *Platycypha eliseva* **Dijkstra**, 2008

Ellis Bettina Grootveld (b.1976) was the author's companion for many years. Klaas-Douwe Benediktus Dijkstra (b.1975) says in his etymology for this species: *The name Elisabeth, borne by Linnaeus's wife and some other family members, can be linked to his fascination with the abundance of creation. Biodiversity is still the muse of systematists. Elisabeth is also the root of both given names of Ellis Bettina Grootveld, my companion and inspiration for fourteen years. The name 'eliseva' (noun in apposition) is derived from the original Hebrew name Elisheba or Elisheva, in reference to her beloved Levant.* Sara Elisabeth von Linné née Moraea (1716–1806) was the wife of Carl von Linné, father of taxonomic nomenclature.

Elke

Azure Flatwing *Austroargiolestes elke* **Theischinger** & **O'Farrell**, 1986

Elke Müller (b.1942) is the wife of one of the collectors of this species, Leonard Müller who took the type in Queensland (1982). (See also **Leonard** & **Muller**))

Ellen

Spangled Shadow-emerald *Macromidia ellenae* **Wilson**, 1996

Eleanor Jane Wilson (b.1985) is the daughter of the author Keith Duncan Peter Wilson (b.1953) (q.v.). She was born in Maidstone, Kent, UK, educated in Hong Kong and studied psychology and criminology at Brighton University, East Sussex, UK, where she now lives and currently works in publishing as a commissioning and freelance editor. Interestingly, the eponymous species was used on a Hong Kong postage stamp (2000).

Ellenbeck

Ethiopian Dropwing *Trithemis ellenbeckii* **Förster**, 1906

Dr Hans Ellenbeck was a German physician. He collected material for the Berlin Museum while participating in the East Africa expedition (1899–1901) that was led by Baron Carlo von Erlanger (q.v.) and which visited Somalia, Abyssinia (Ethiopia) and Sudan. Three birds are also named after Ellenbeck.

Elliot (Pinhey)

Plump Flasher *Aethiothemis ellioti* **Lieftinck**, 1969
[Orig. *Lokia ellioti*]
Pinhey's Horntail *Tragogomphus ellioti* **Legrand**, 2002

Dr Elliot Charles Gordon Pinhey (1910–1990) (see **Pinhey**).

Elliot (Scott)

Elliot's Hawker *Zosteraeschna ellioti* **Kirby**, 1896
[Orig. *Aeshna ellioti*, A. Highland Hawker]

Professor Capt. George Francis Scott-Elliot FRSE FRGS (1862–1934) was an Indian born Scottish soldier, author and explorer who collected the holotype at Mt Ruwenzori. He was educated at Cambridge (BA 1882) and Edinburgh (MSc) and travelled widely in South Africa, Madagascar, Mauritius, Sierra Leone, Egypt, Tripoli (Libya), Uganda and Tanganyika (Tanzania). He was an officer in the King's Own Scottish Borderers and saw action in Egypt (awarded the Order of the Nile) and WW1. He was a lecturer in botany at the Royal Technical College, Glasgow and Professor of Botany at the Glasgow Veterinary College. He wrote a number of books including: The *Flora of Dumfriesshire* and *Naturalist in Mid-Africa*.

Ellison

Cruiser sp. *Macromia ellisoni* **Fraser**, 1924

Bernard C Ellison was a curator of the Bombay Natural History Society. Fraser did not provide an etymology for this name. However, it seems virtually certain that Fraser was honouring this man who was also naturalist to the shoots. He wrote a book on field sports, *Game Preservation and Game Experiments in India* in the 'Journal of the Bombay Natural History Society' (1928) and another (*H.R.H. The Prince of Wales's sport in India*, 1925) chronicling the visit of the then Prince of Wales to India and his hunt, which slaughtered numerous rhinoceroses and tiger in Nepal. Ellison and his staff were in charge of skinning the kills ready for taxidermy. He several times mentions Fraser in his account referring to their correspondence at the time when the species was named.

Elopura

Flatwing sp. *Rhinagrion elopurae* **McLachlan**, 1886
[Orig. *Amphilestes elopurae*]

This is a toponym; Elopura is a locality in Sabah northeast Borneo where the species is found.

Elouard

Legrand's Glyphtail *Isomma elouardi* **Legrand**, 2003

Jean-Marc Elouard is a French hydrobiologist and medical entomologist who was Director of Research at the IRD (e.g. ORSTOM) (1972–2005). He trained at Orsay University, Lyon University and the Pasteur Institute of Paris, which awarded his PhD. He spent much of his career in West Africa in the programme against onchocerciasis, and then to Madagascar through a programme on the insects of rivers and he has written many articles in the series 'Aquatic biodiversity of Madagascar'. His special interest is in Simuliidae (Diptera Brachycera of onchocerciasis human and animal vectors) as well as African-Madagascan mayflies. Now retired, his concerns are a database on the flora and fauna of Lauret, writing texts on significant insects in the region, as well as writing a conference slideshow on global warming.

Elsie

Darner sp. *Rhionaeschna elsia* (Calvert, 1952)
[Orig. *Aeshna elsia*]

Elsie Rosner née Lincoln (1913--2002) was an American entomologist and the wife of the psychoanalyst Albert Aaron Rosner. The University of Pennsylvania awarded her BA (1933) and MSc in entomology. She was a journalist for 'Time' magazine (1939), a medical reporter for 'MD' (1957–1978) and news editor (1969–1978) and then worked for 'Medical World News' (1979–1984) and 'Physicians' Weekly' (1984–2001). She also worked with the author and illustrated scientific papers, including his for the American Entomological society.

Elton

Skimmer sp. *Phyllothemis eltoni* **Fraser**, 1935

John Richard Elton-Bott (1881–1942). According to Fraser, Elton-Bott took two males of this new species at King Islands in Mergui, Lower Burma. (See **Bott**)

Elwes

Clearwing sp. *Notholestes elwesi* **McLachlan**, 1887
[JS *Caliphaea confusa* Hagen, 1859]

Henry John Elwes FRS (1846–1922) was a wealthy English collector and illustrator, mainly of plants, but he also collected birds, butterflies and moths. After five years in the Guards he travelled very widely over much of Europe, Asia Minor, India, Tibet, Mexico, North America, Chile, Russia, Siberia, Formosa, China and Japan. Elwes discovered many plants, some named after him. His home was Colesbourne Park, where he lived for 47 years (1875–1922). He created a collection of about 140 different varieties of snowdrops, including one named after him. His great-grandson and his wife now live in the house whose gardens are open to the public. He was interested in natural history in general as evidenced by his: *Memoirs of Travel, Sport and Natural History*. He also wrote an article on: *The Geographic Distribution of Asiatic Birds* (1873). Most of his writing was on birds, but he also wrote the 7-volume: *The Trees of Great Britain and Ireland* (1906–1913). He wrote a number of other monographs, such as one on lilies (1880). He collected the damselfly holotype in Darjeeling, India. Four birds are also named after him.

Emilia

Gabon Horntail *Libyogomphus emiliae* **Legrand**, 1992
[Orig. *Onychogomphus emiliae*]

Emilia Legrand (b.1978) is the youngest daughter of the author, Jean Legrand (b.1944), who also named species after his two other daughters.

Emily

Skimmer sp. *Diastatops emilia* **Montgomery**, 1940

Emily Joan Alward nee Montgomery (1935–2011) was the daughter of the author, she became a college librarian. She graduated, aged 19, from Purdue where her father was a professor, and gained her masters at Indiana University. (Also see **Esther** (**Montgomery**) & **Montgomery**)

Emma

Emma's Dancer *Argia emma* **Kennedy**, 1915

Emma Torinda Kennedy (b.1850) was the mother of the author Clarence Hamilton Kennedy (1879–1952) (q.v.).

Emmel

Clubtail sp. *Idiogomphoides emmeli* **Belle**, 1995

Professor Dr Thomas Chadbourne Emmel (b.1941) is Director of the McGuire Center for Lepidoptera and Biodiversity, as well as Professor Emeritus of Zoology and Entomology, and Curator of Natural Sciences with the Florida Museum of Natural History. Stanford University awarded his PhD (1967). He was Director of the University of Florida Boender Endangered Species Laboratory (1995) and the Department of Zoology's Division of Lepidoptera Research (1980–2003). He has an intense lifelong research interest in ecological and conservation issues. He has written more than 400 publications, including 35 books, particularly on butterflies and tree snails, such as: *Florida's Fabulous Butterflies & Moths* (1997), and on general biology topics. He has worked intensively (since 1984) on the endangered Schaus Swallowtail butterfly in the Florida Keys and has directed an extensive captive propagation and reintroduction effort. His research on the effects of mosquito control pesticides on non-target wildlife and humans living in south Florida have led to better control measures for the use of pesticides and enhanced wildlife survival. Emmel collected the holotype of this Clubtail in Rhondonia, Brazil (November 1990).

Endicott

Clubtail sp. *Stylurus endicotti* **Needham**, 1930
[Orig. *Gomphus endicotti*]

James Gareth Endicott (1898–1993) was a Canadian missionary and socialist born in Szechuan, China. After service in WW1, he attended University of Toronto's Victoria College, which awarded his master's degree. He was then ordained and returned to China as a missionary, staying for two decades. He became Professor of English and Ethics at West China Union University. He also became social advisor to Chiang Kai-shek and to US military intelligence (1944–1945) as a liaison between the American military and the Chinese Communist forces fighting against the Japanese. He later became a supporter of the Communist party and the church gave him an ultimatum to either modify his public statements or quit, so he resigned his ministry. He returned to Canada (1947). As western countries were backing Chiang, Endicott advised the Canadian government that the Kuomintang regime's fall was imminent and then went public with his predictions and his denunciation of the Kuomintang as corrupt. His comments were denounced as traitorous by the media and he was labelled the most reviled Canadian of the year! He became a senior figure in the peace movement (1949), serving as President of the International Institute for Peace (1957–1971). In 1982, the United Church extended a formal apology to Endicott for having denounced him three decades earlier. He collected the holotype female in Chungking, China.

Eos

Damselfly sp. *Palaiargia eos* **Lieftinck**, 1938

Eos is a goddess in Greek mythology.

Eponina

Halloween Pennant *Celithemis eponina* **Drury**, 1773
[Orig. *Libellula eponina*]

Eponina was a faithful wife in a Roman tale.

Erato

Cruiser sp. *Macromia erato* **Lieftinck**, 1950

Erato is a character from Greek mythology.

Eric (Laidlaw)

Sylvan sp. *Coeliccia erici* **Laidlaw**, 1917

Eric Laidlaw was the son of the author, and the species name was also dedicated to another relative, his nephew, who died (1915) during WW1. Laidlaw wrote: "*I have named this species after a soldier who fell in 1915 and also after my little son his cousin.*" Laidlaw also named dragonfly species after his two daughters Mary and Ruby. (Also see **Louisa**, **Mary (Laidlaw)**, **Ruby** & **Laidlaw**)

Eric (Novelo-Galicia)

Ringtail sp. *Erpetogomphus erici* Novelo-Gutiérrez, 1999

Eric Novelo-Galicia (b.1992) is the son of the author Rodolfo Novelo-Gutiérrez (b. 1955).

Erica

Clubtail sp. *Agriogomphus ericae* **Belle**, 1966
[Orig. *Ischnogomphus ericae*]

Erica Belle (b.1956) is the daughter of the author Dr Jean Belle (1920–2001) (q.v.). She was the Dutch Women's Chess champion (1975, 1980 & 1981).

Erico

Emerald sp. *Hemicordulia erico* **Asahina**, 1940

Ms Eri Abe née Esaki (b.1934) lives in Tokorozawa city in Saitama prefecture. In Japan, "-co" (or "-ko") is often added to the end of the female first name. Although Asahina did not provide any etymology, it is known that Erico was named after Eri, the youngest of the three daughters of Dr Teizo Esaki, whose Odonata material from Micronesia collected during1936–1939, which included the type (1937), Asahina had studied. Also, the other daughters and Esaki himself were credited with eponyms in the same paper (see **Esaki, Haluco** and **Lulico**).

Erigone

Skimmer sp. *Diplacina erigone* **Lieftinck**, 1953

Erigone is a character in Greek mythology, either the daughter of Icarius or the daughter of Aegisthus.

Erinys

Flatwing sp. *Allopodagrion erinys* **Ris**, 1913
[Orig. *Megapodagrion erinys*]

Erinys were characters in Greek mythology equivalent to the furies of Roman myth.

Erlanger

Barbet Percher *Philomonon erlangeri* **Förster**, 1906
[JS *Diplacodes luminans* (Karsch, 1893)]
Denim Dropwing *Trithemis etlangeri* Förster, 1903
[JS *Trithemis donaldsoni* (Calvert, 1899)]

Baron Carl (Carlo) Viktor Heinrich von Erlanger (1872–1904) was a German collector from Ingelheim in the Rhineland. He travelled in Tunisia (1893 & 1897) and wrote two expedition reports. He visited Abyssinia (Ethiopia) and Somaliland (1899–1901) during which these two species were found, accompanied for part of the time by Neumann. He named c.40 new avian taxa as well as having almost as many (37) named after him. He died in a car accident. Two mammals, a reptile and an amphibian are also named after him.

Ernst Mayr

Damselfly sp. *Palaiargia ernstmayri* **Lieftinck**, 1972

Dr Ernst Walter Mayr (1904–2005) was a German ornithologist and zoologist who began serious bird studies in the South Pacific. He is best known as an eminent writer on evolution; it even being said: '*he is, without a doubt the most influential evolution theoretician of the twentieth century*'. As a ten-year-old boy he could recognise all the local birds on sight and by song. He led ornithological expeditions to New Guinea (1928–1931); an experience that he said "…*fulfilled the greatest ambition of my youth.*" He collected 7,000 skins in 2½ years. He later joined an expedition to the Solomon Islands, before returning to his academic career at the Berlin Museum. Mayr was employed by the Department of Ornithology of the American Museum of Natural History (1931), at first as a visiting curator, to catalogue the Whitney Expedition's collection of South Sea birds in which Mayr had participated. During his first year there, he wrote 12 papers, describing 12 new species and 68 new subspecies. He went on to become the Alexander Agassiz Emeritus Professor of Zoology at Harvard University. He was the originator of the 'founder effect' idea of speciation and was a leading proponent of the Biological Species Concept. His many important books include: *Systemics and the origin of species* (1942) and *Animal species and evolution* (1963). A mammal and twenty-eight birds are also named after him. He found this new damselfly species at Arfack mountains in Vogelkop, New Guinea (April 1928).

Esaki

Skimmer sp. *Pacificothemis esakii* **Asahina**, 1940

Professor Dr Teizo Esaki (1899–1957) was a Japanese entomologist. He was chairman of the Entomological Laboratory of the Faculty of Agriculture and was appointed as Dean of the Faculty of Agriculture (1948), while concurrently holding the laboratory chairmanship at Kyushu Imperial University (then Kyushu University). A born linguist, he spoke English, German, French, Italian and Hungarian as well as Esperanto. He was an influential and erudite teacher with a wide knowledge of entomology, zoology and the history of biology, although his specialty was taxonomy of Hemiptera and he particularly loved butterflies. He collected in Micronesia (1936–1939) including this type. Most of the specimens he collected were deposited in the Kyusyu University (Japan), where Asahina studied the Odonata (1940). He was one of the International Commissioners on Zoological Nomenclature. He published a great many papers on zoography, entomology and history as well as about many insect taxa. (Also see **Erico, Haluco, Lulico**)

Escherich

Rusty Featherleg *Platycnemis escherichi* **Schmidt**, 1951
[JS *Copera rufipes* (Selys, 1886)]

Georg Escherich (1870–1941) was a German forester and politician. He became head of the anti-communist 'home guard' in Bavaria (1919) and he organised his supporters (1920) into the 'Orgesch', an anti-semitic group. He collected the specimen in Cameroon (August 1913).

Esmeralda

Skimmer sp. *Dasythemis esmeralda* **Ris**, 1910

This is a toponym; Esmeralda is a locality in Ecuador.

Esquivel

Dancer sp. *Argia carolus* **Garrison** & von **Ellenrieder**, 2017

Carlos Esquivel Herrera (b.1956) is a Costa Rican biologist. After graduating from the University of Costa Rica (UNA) he began teaching at UNA (1980). He also undertook postgraduate studies at University of Kansas (1982–1985), which awarded his masters in entomology (1985). Back at UNA he taught ecology, zoology, entomology and scientific communication, all the while continuing his research. He was an intern at the International Odonata Research Institute attached to the University of Florida (1989). He wrote: *Dragonflies of Mesoamerica and the Caribbean* (2006). His research focus is the study of the order Ephemeroptera.

Estes

Estes's Sprite *Pseudagrion estesi* **Pinhey**, 1971

Dr Richard Despard Estes (b.1928), known as the 'Guru of Gnu', is a mammal ecologist particularly interested in wildebeast and it has been suggested that Estes is responsible for most of the world's knowledge of wildebeest behaviour. His doctoral thesis in the 1960s was on the Wildebeest of the Ngorngoro Crater. He has written two guides for travellers to Africa: *The Behaviour Guide to African Mammals* (1995) (considered the standard reference of its kind) and *The Safari Companion* (1999). He collected this new species in Angola (1970).

Esther (Mongomery)

Spiketail sp. *Diastatops estherae* **Montgomery**, 1940

Esther Montgomery née Barrett (1902–1988) was the wife of the author Basil Elwood Montgomery (1899–1983) (q.v.). (Also see **Emily (Montgomery)** & **Montgomery**)

Esther (Tristan)

Esther's Threadtail *Neoneura esthera* **EB Williamson**, 1917

Senora Esther de Tristán née Castro-Meléndez (1884–1947) was the wife of Professor José Fidel Tristán (1874–1932) (q.v.) and a fellow scientist who was a teacher of natural science. Dr P P Calvert (q.v.) suggested the name as Professor Tristán had participated in the collection of the first known set of specimens from Guanacaste, Costa Rica (1910).

Etcheverry

Dragonfly sp. *Gomphomacromia etcheverryi* **Fraser**, 1957

[JS *Gomphomacromia paradoxa* Brauer, 1864)]

Professor Dr María Edith Etcheverry Campaña (1928–2012) was a Chilean entomologist and science historian. She studied at the University of Chile in Santiago de Chile (1945–1949) graduating with a degree in biology and chemistry and later was awarded her MSc in entomology from the University of Wisconsin (1959). Her PhD was awarded by the University of Chile (1962). She held teaching positions there in the Faculty of Philosophy and Education and was curator of the entomological collections, researcher and then Professor of Biology and Director of the Centre for Entomological Studies. During this time, she wrote more than forty papers and longer works including several with José Herrera (q.v.). Her interest in Diptera and Lepidoptera, together with her great scientific capacity, led her to participate in many events and make numerous collecting expeditions to Uruguay, Argentina, Bolivia, Peru, Ecuador, Costa Rica and the United States (Washington, Madison, Chicago, Kansas, Miami, Tallahassee, Tampa, St. Louis, Austin and Pittsburg). She is also commemorated in a butterfly genus.

Ethel (Fraser)

Velvetwing sp. *Dysphaea ethela* **Fraser**, 1924

Ethel Grace Fraser née Varrall (1881–1960) was the wife (m.1905) of the author Lieutenant-Colonel Dr Frederic Charles Fraser (1880–1963). (Also see **Varrall**)

Ethel (Merriman)

Threadtail sp. *Neoneura ethela* **EB Williamson**, 1917

Ethel Merriman née Williamson (b.1881) was the wife of J E Merriman, and the sister of the author Edward Bruce Williamson (1878–1933).

Ethel

Hispaniolan Malachite *Phylolestes ethelae* Christiansen, 1948

No etymology was given but presumably a relative of the author Kenneth Allen Christiansen (b.1924).

Eudoxia

Skimmer sp. *Atoconeura eudoxia* **Kirby**, 1909
[Orig. *Accaphila eudoxia*]

Eudoxia is a name from antiquity without any special connotation.

Eugenia

Forceptail sp. *Gomphoides eugeniae* **Navás**, 1927
[JS *Phyllocycla argentina* (Hagen, 1878)]

Senora Eugenia was the wife of the collector P Mühn, a Jesuit who collected in Argentina at the beginning of the twentieth century, including the type (1927).

Eungella

Eungella Darner *Austroaeschna eungella* **Theischinger,** 1993

This is a toponym; Eungella is in northeastern Queensland, Australia.

Euphrosyne

Cruiser sp. *Macromia euphrosyne* **Lieftinck**, 1952

Euphrosyne was a character from Greek mythology, one of the three 'charities'.

Eurydice

Bombardier sp. *Lyriothemis eurydice* **Ris**, 1909

Eurydice was a character from Greek mythology, one of the daughters of Apollo (and also an oak nymph). She married Orpheus who tried to bring her back from the dead with his enchanting music.

Eurynome

Cruiser sp. *Macromia eurynome* **Lieftinck**, 1942

Eurynome was a character from Greek mythology; there are around ten different Eurynomes.

Eusebia

Cascader sp. *Zygonyx eusebia* **Ris**, 1912
[Orig. *Pseudomacromia eusebia*]

Eusebia (d.360) was the second wife of the Roman emperor Constantius II.

Eustáquio

Shortwing sp. *Perilestes eustaquioi* **Machado**, 2015

José Eustáquio Santos Júnior, of the Universidade Federal de Minas Gerais, Belo Horizonte, Brazil is a 'dear friend' of the author who provided him with Odonata material.

Euterpe

Cruiser sp. *Macromia euterpe* **Laidlaw**, 1915

Euterpe was a character from Greek mythology, one of the muses, the daughters of Mnemosyne, fathered by Zeus.

Evans

Evans' (Desert) Bluetail *Ischnura evansi* **Morton**, 1919

Captain William Edgar Evans RAMC (1882–1963) was a Scottish botanist. Edinburgh University awarded his BSc (1906). He became an Assistant in Mycology at Heriot-Watt College and was then in the Royal Army Medical Corps (1916–1919) serving in the Middle East. He collected a large number (300+) of odonate specimens in Mesopotamia (1917–1918) and sent them home to his father, William Evans, who passed them to Morton to study. During these years, he collected plants from South Africa (1917), Iraq (1918) and Persia (1919). He then became an assistant in charge of herbs at Edinburgh Botanical Gardens (1919–1944).

Evelyn

Tigertail sp. *Palaeosynthemis evelynae* **Lieftinck**, 1953
[Orig. *Synthemis evelynae*]
Threadtail sp. *Nososticta evelynae* Lieftinck, 1960
[Orig. *Notoneura evelynae*]

Lucy Evelyn Cheesman (1881–1969) was an entomologist and explorer. (See **Cheesman**)

Ezoin

Bluetail sp. *Ischnura ezoin* **Asahina**, 1952
[Orig. *Boninagrion ezoin*]

Eizo Asahina (1914–2013) was the younger brother of the author Dr Syoziro Asahina (1913–2010) (q.v.). He was Honorary Professor at Hokkaido University and a well-known cryobiologist, writing widely on the subject.

F

Faasen

Tropical King Skimmer sp. *Orthemis faaseni* von **Ellenrieder**, 2012

Tim Johannes Adrianus Faasen (b.1978) is a Dutch entomologist working for an ecological consultancy (2001). He graduated in biology at University of Utrecht. He has made three Odonata study and collecting trips to Tamshiyacu – Tahuayo reserve in Amazonian Peru (2009, 2010 & 2015). Here he collected the eponymous skimmer and 146 other Odonata, 13 still undescribed. He also took many DNA samples and photographs of all collected species. An ongoing project of his, besides describing some of the collected Odonata, is photographing Dutch insects, of which he has accumulated macro-photographs of several thousand species. Many of his pictures are used in publications. Furthermore, he has published contributions to longer works and several scientific papers including: *Phoenicagrion trilobum, a new species of damselfly from Peru (Odonata: Coenagrionidae)* (2014).

Faivre

Banded Demoiselle form *Calopteryx splendens f. faivrei* **Lacroix**, 1915

Paul Faivre was a friend of the author Joseph-Louis Lacroix (1879–1939). This taxon has often been erroneously ranked as a subspecies. Since it was named for the andromorph female form of this Demoiselle species, present in some French populations, the name is not available for the zoological nomenclature.

Fassl

Bannerwing sp. *Euthore fassli* **Ris**, 1914
Sanddragon sp. *Progomphus fassli* **Belle**, 1973

Helicopter ssp. *Microstigma maculatum fassli* **Schmidt**, 1958
[JS *Microstigma maculatum* Hagen, 1860)]

Anton Heinrich Hermann Fassl (1876–1922) was a German commercial collector, particularly of Lepidoptera and Coleoptera. He maintained a dealership in Berlin and Bohemia. He travelled in Colombia (1907–1909), Central Peru (1913–1914) and Brazil (1920, 1921 & 1922), where he died – probably at the upper Madre de Dios River. Among others he supplied Hartert (q.v.) & Jordan. He wrote a number of papers including: *Einige kritische Bemerkungen zu J. Röbers "Mimikry und verwandte Erscheinungen bei Schmetterlingen"* (1922).

Fea

Clubtail sp. *Platygomphus feae* **Selys**, 1891

Leonardo Fea (1852–1903) was an Italian explorer, painter and naturalist. He was Assistant at the Natural History Museum in Genoa and liked exploring far off, little-known countries. He spent four years (1885–1888) collecting many insects and birds in Burma (Myanmar) where the type was taken. He also visited islands in the Gulf of Guinea, and the Cape Verde Islands (1898). He had intended to take a long trip to Malaysia but ill-health forced him to change his plans. Two mammals, four amphibians, five reptiles and six birds are also named after him.

Feather

Striped Siphontail *Neurogomphus featheri* **Pinhey**, 1967

Walter Feather (1872–1933) was a British entomologist who was elected a Fellow of the Entomological Society of London (1907). He collected Heterocera in British East Africa and Somaliland (1908–1912) whilst employed by the British Somaliland Fibre and Development Company. He collected the holotype male in Kenya (April 1931). He was a founder member of the Cross Hill Naturalists Society. Several moths are also named after him.

Feliculo

Limniad sp. *Sangabasis feliculoi* Villanueva & **Dow**, 2014

Feliculo Gemino Torayno (b.1920) is the maternal grandfather of the first author Reagan Joseph Torayno Villanueva (b.1981).

Félix Orion (Fleck)

Genus *Orionothemis* Fleck, **Hamada** & **Carvalho**, 2009
Skimmer sp. *Orionothemis felixorioni* Fleck, Hamada & Carvalho, 2009

Félix Orion Fleck (b.2006) is the son of the first author, Günther Fleck, formerly of the French National Institute for Agricultural Research.

Fenichel

Fiery Skimmer *Orthetrum fenicheli* **Förster**, 1898
[JS *Orthetrum villosovittatum* (Brauer, 1868)]

Sámuel Fenichel (1868–1893) was a Hungarian archaeologist, entomologist and ethnologist at the Hungarian National Museum (1888). He collected in New Guinea (1891–1893) including the type. Over 25,000 zoological specimens were sent back to the museum. While there he contracted either yellow fever or malaria and later died from what was described as kidney or gall-bladder disease. He wrote a number of archaeological and ethnographic articles. A bird is also named after him.

Fernand

Skydragon ssp. *Chlorogomphus preciosus fernandi* **Asahina**, 1986

Fernand Schmid (1924–1998). (See also **Schmid**)

Fernandez

Pond Damselfly sp. *Metaleptobasis fernandezi* **Rácenis**, 1955

[JS *Metaleptobasis diceras* (Selys, 1877)]
Pond Damselfly sp. *Bromeliagrion fernandezianum* Rácenis, 1958
[Orig. *Leptagrion fernandezianum*]
Threadtail sp. *Epipleoneura fernandezi* Rácenis, 1960

Dr Francisco José Fernandez Yepez (1923–1986) was a Venezuelan agronomist and entomologist and the founder of the insect collection of the Faculty of Agronomy of the Universitad Central de Venezuela. The Lyceum Fermín Toro, Caracas awarded his bachelor's degree (1941) and Agronomía de la Universidad Central de Venezuela awarded his further degree (1945). His first post was as an assistant at the Entomology Department of Experimental Agriculture and Animal Husbandry Institute. He then (1945) took a job as assistant entomologist at the Ministry of Agriculture, but a few months later went to the USA where he continued graduate studies in entomology at Cornell University, New York (1945–1947). Upon his return, he was appointed Professor of Entomology at the College of Agriculture and Animal Husbandry, Caracas (1947–1949). He became Professor of Economic Entomology and Head of Department of Entomology, Faculty of Agriculture, Central University of Venezuela and various other posts (1949–1965) and received his PhD (1950). He also taught Entomology at the Faculty of Agronomy of the University of East Jusepín (1966). Throughout his career, he collected insects, especially Lepidoptera. He published many papers (1943–1982). He is also commemorated in the name of an earwig. His brother Agustín Antonio Fernandez Yepez, a Venezuelan ichthyologist (1916–1977), described 42 fish species and wrote over 100 works on fish systematics and biology.

Fernando (Netto dos Reys)

Pond Damselfly sp. *Oxyagrion fernandoi* **JM Costa**, 1988
Spreadwing sp. *Lestes fernandoi* JM Costa, De Souza & **Muzón**, 2006

Luiz Fernando Netto dos Reys of Maranhão State collected the type specimens in Brazil (1983 & 2005).

Feuerborn

Slim sp. *Aciagrion feuerborni* **Schmidt**, 1934

Professor Dr Heinrich Jakob Feuerborn (1883–1979) was a zoologist who became head of the Zoological Institute of the University of Munich (1936–1945) during Frederick Seidel's absence. As a member of the Nazi party he had to withdraw from his position at the institute at the end of the war. He took part in the (1928–1929) Deutschen Limnologischen expedition to Sunda, principally Sumatra (where the type was taken), Java & Bali (Indonesia), studying the life and phenomena of fresh water, especially lakes and ponds. He collected widely and Schmidt studied the Odonata he found. Among many other papers, he wrote: *Die Larven der Psychodiden oder Schmetterlingsmücken* (1923).

Ficke

Black Proto *Protolestes fickei* **Förster**, 1899

Dr Hugo Ficke (1840–1912) was Director of the Museum of Natural History and Ethnology, Freiburg and a city councillor, a leading mason and a philanthropist. He died on a trip to India with his wife, where he was collecting mainly ethnographic artefacts.

Field

Northern Pondsitter *Austrosticta fieldi* **Tillyard**, 1908

James F Field (d.1926) of Tennant's Creek, Northern Territory, Australia made a collection of some 300 odonata specimens (1905–1906). He collected fauna of different sorts for zoologists including Orthoptera and eggs for oologists and one letter even mentions fossil kangaroo footprints he could dig up.

Finisterre

Threadtail sp. *Nososticta finisterrae* **Förster**, 1897

[Orig. *Caconeura finisterrae*]

This is a toponym; the Finisterre Range are mountains in Papua New Guinea where the species is found.

Finsch

Lieutenant sp. *Nannodiplax finschi* Karsch, 1889

[JS *Brachydiplax denticauda* (Brauer, 1867)

Friedrich Hermann Otto Finsch (1839–1917) was a German ethnographer, naturalist and traveller. He visited the Balkans, North America, Lapland, Turkestan and northwest China with Alfred Brehm and also the South Seas, spending nearly a year (1879–1880) on the Marshall Islands. Bismarck appointed him Imperial Commissioner for the German Colony of 'KaiserWilhelmLand' (1884) in New Guinea where the type was taken. He founded the town of Finschhafen there (1885), which remained the seat of German administration (1885–1918). He was the director of a number of museums at various times, including Bremen, where he succeeded Hartlaub as Curator (1884), and Braunschweig. Among other works he co-wrote: (with Hartlaub) *Die Vögel OstAfrikas*. A mammal, a reptile and twenty-five birds are also named after him.

Fiorentin

Emerald sp. *Neocordulia fiorentini* **JM Costa** & **Machado**, 2007

Professor Gelson Luiz Fiorentin (b.1960) is a professor at the Universidade do Vale do Rio dos Sinos Unissino, which awarded his first degree in science and biology (1984) with honours (1985). The Pontifical Catholic University of Rio Grande do Sul – PUC-RS awarded his Masters in Biosciences (zoology; 1989). He is currently Professor of Unisinos and Professor at the Lutheran University of Brazil. His research focus is on systematics of aquatic insects. He has written more than 115 scientific papers, book chapters and reports such as: *Identification keys to families of aquatic beetles occurring in Rio Grande do Sul, Brazil* (2006). He collected (1994) and reared the larvae of the Emerald species whose adults were described.

Fischer

Rainbow Bluet *Enallagma fischeri* **Kellicott**, 1895

[JS *Enallagma antennatum* (Say, 1840)]

Dr Paul Fischer (b.1868) was the State Veterinarian of Ohio (1902) where he collected the type. He was still in practice when 90 years old! He had been Professor of Pathology at Kansas State College and at Utah State Agriculture College and also a health officer in Florida.

Fisher

Fisher's Sprite *Pseudagrion fisheri* **Pinhey**, 1961
[A Dark-tailed Sprite]

Alfred Charles Fisher owned Hillwood Farm at Ikelenge, Mwinilunga, Zambia, where the holotype was collected near the Sakeshi stream (1960), and he also gave assistance to the author, who collected the type. (The farm is now part of the Nchila Reserve.)

Fitzgerald

Lesser Peppertail *Nesciothemis fitzgeraldi* **Longfield**, 1955
Powdered Junglewatcher *Neodythemis fitzgeraldi* **Pinhey**, 1961

Leslie Desmond Edward Foster Vesey-Fitzgerald MBE (1910–1974) was an Irish-born entomologist, ornithologist, conservationist, and plant collector. He undertook research work on biological controls of insect pests on sugar cane in Brazil, British Guiana and the British West Indies (1933–1938) and then the biological control of insect pests on coconut palms in the Seychelles, Madagascar and coastal East Africa (1936–1939). He was then Entomologist at the Rubber Research Institute in Malaya (1939–1941) before military service (1941–1942). After this he was Entomologist at the Middle East Anti Locust Units in Sudan, Saudi Arabia and Oman (1952–1947). He was Senior Assistant Game Warden in Kenya (1947–1949) then became the resident botanist as Senior Scientific Officer at the IRLCS (International Red Locust Control Service) Zambia (then Northern Rhodesia) (1949–1964). After this he became an ecologist and conservationist in the National Parks of Tanzania and went on ornithological, entomological, and botanical surveys to many places in Africa, the Middle East and West Indies. He collected the Odonata holotypes in Zambia (1949 & 1957). He wrote several accounts of the grasslands of eastern central Africa: *Central African Grasslands* (1963) as well as some articles on conservation such as: *The black lechwe and modern methods of wild life preservation* (1956). His plant collections grace a number of museums and he was honoured in the name of a reptile.

FitzSimons

Fitzsimons' Jewel *Platycypha fitzsimonsi* **Pinhey**, 1950
[P. Boulder Jewel]

Dr Vivian Frederick Maynard FitzSimons (1901–1975) was a zoologist and herpetologist. Rhodes University awarded his BSc (1921) and MSc (1923). He became Senior Assistant in Zoology at Transvaal Museum (1924). He spent six months in Botswana on the Vernay-Lang Kalahari Expedition (1930). He primarily studied reptiles, but also collected some insects for the Transvaal Museum where he was Director (1946-1966). Eight reptiles are also named after him.

Flavia

Emerald sp. *Lauromacromia flaviae* **Machado**, 2002
Wedgetail sp. *Acanthagrion flaviae* Machado, 2012

Professor Dr Flávia Ribeiro Machado (b.1967) is the author's daughter. She graduated in medicine from the Federal University of Minas Gerais (1991), then did her residency at a medical clinic in Santa Casa de Belo Horizonte and also a residency in infectious diseases at the University of São Paulo. She studied for her MSc in infectious diseases and tropical medicine at the Federal University of Minas Gerais (1996) and her PhD in infectious and parasitic diseases, University of São Paulo (2000). She is a specialist in intensive care medicine as recognised by the Brazilian Association of Intensive Medicine (1997). She is currently an associate professor and Chief of the Division of Critical Care, Department of Anaesthesiology, Intensive Care and Pain of the Federal University of São Paulo. Her area of expertise is critical care medicine, with emphasis on sepsis; she is also Vice President of the Latin American Institute of Sepsis (ILAS) and Chief Editor of the 'Journal of Critical Care Medicine', the official journal of the Brazilian Association of Intensive Medicine and the Portuguese Society of Critical Care. She is also a member of the Executive Committee and the Scientific Committee of the Brazilian Research in Intensive Care Network (BRICNet). She collected the holotype of *Acantagrion flaviae* during a medical visit to the Maribo Indian village at the Curuca River in the Amazon.

Fletcher, J

Ebony Boghunter *Williamsonia fletcheri* **EB Williamson**, 1923

James Fletcher (1852–1908) was an English-born Canadian botanist, entomologist and writer. He started working life as a clerk at the Bank of British North America in London but transferred to Montreal (1874) and Ottawa (1875). He worked as an assistant in the Library of Parliament (1876) through which he discovered his interest in botany and entomology. He established a national reporting system to help identify and control the spread of insects and weeds harmful to agriculture and became (1887) the first Dominion Entomologist and Botanist attached to the Central Experimental Farm (where a garden bears his name). He was a founder of the American Association of Economic Entomologists, from which the Entomological Society of America grew. He also initiated the Canadian National Collection (CNC) of Insects, Arachnids and Nematodes.

Fletcher, TB

Threadtail sp. *Disparoneura fletcheri* **Fraser**, 1919
[JS *Elattoneura atkinsoni* Selys, 1886]
Tiger sp. *Gomphidia fletcheri* Fraser, 1923
Spectre sp. *Petaliaeschna fletcheri* Fraser, 1927

Thomas Bainbrigge Fletcher (1878–1950) was an English entomologist. He was a naval paymaster (1910) and was later appointed Imperial Entomologist in India, succeeding Harold Maxwell-Lefroy. He was interested in Indian entomology, especially that of economic importance and also worked extensively on the *microlepidoptera*. His major publications include: *Some South Indian Insects* (1914) and *A List of Generic Names used for Microlepidoptera* (1929). He collected insects and listed publications on Indian entomology and wrote a catalogue of Indian insects. He donated the bulk of Rodborough Common in Gloucestershire to the National Trust (1937). (Also see **Bainbridge**)

Flint

Sanddragon sp. *Progomphus flinti* **Belle**, 1975
Flint's Cruiser *Macromia flinti* **Lieftinck**, 1977
Spectre sp. *Periaeschna flinti* **Asahina**, 1978
Shortwing sp. *Perissolestes flinti* **De Marmels**, 1989
Knobtail sp. *Epigomphus flinti* **Donnelly**, 1989
Threadtail sp. *Drepanoneura flinti* von **Ellenrieder** & **Garrison**, 2008

Dr Oliver Simeon Flint Jr. (b.1931) is an American entomologist who is Curator Emeritus of Neuropteroid Orders at the Smithsonian Institution. The University of Massachusetts awarded his BS (1953) and MS (1955) and Cornell his PhD (1960). He was a research fellow at Cornell (1957–1959). He became Associate Curator, Department of Entomology, Smithsonian Institution (1961–1965), then Supervisor and Curator, Division of Neuropteroids (1965–1995) since when he has been Curator Emeritus (1996–present). He has visited South America frequently and took part (1988) in the expedition to the Manu Biosphere Reserve (Peru). He worked on Trichoptera and Odonata with Louton and Garrison, being author of an amazing 7% of all known Trichoptera. He has written 226 papers and books (1956–2014), the majority on the systematics of Neotropical Trichoptera or caddisflies, such as: *The Life History and Biology of the Genus Frenesia (Trichoptera: Limnephilidae)* (1956) to *Studies on the caddisfly (Trichoptera) fauna of Nevada* (2014). He collected large numbers of dragonflies, caddisflies and other aquatic insects throughout Central and South America and the West Indies.

Fonscolombe

The genus *Fonscolombia* **Selys**, 1883
[Homonym of *Fonscolombia* Lichtenstein, 1877 in Hemiptera]

Red-Veined Darter *Sympetrum fonscolombii* Selys, 1840
[A Nomad, Orig. *Libellula fonscolombii*]
Mercury Bluet *Agrion fonscolombii* **Rambur**, 1842
[JS *Coenagrion mercuriale* Charpentier, 1840]

Étienne Laurent Joseph Hippolyte Boyer de Fonscolombe (1772–1853) (See **Boyer**)

Forel

Pond Damselfly sp. *Metaleptobasis foreli* **Ris**, 1918

Auguste-Henri Forel (1848–1931), like Ris, was a Swiss entomologist but was primarily a myrmecologist, neuro-anatomist and psychiatrist notable for his investigations into the structures of both human and ant brains. He became Professor of Psychiatry at the University of Zurich (1879). He was a pioneer in neuron theory, sexology and psychology as well as drawing many parallels between ant and human behaviour. He was a noted socialist, eugenicist and racist but later turned to the Bahá'í faith. He wrote a book on the ants of Switzerland: *Les fourmis de la Suisse* (1874), and named his home, 'La Fourmilière'. His magnum opus was the five volume: *Le Monde Social des Fourmis* (1923), which he wrote with his left hand after suffering a stroke (1912).

Forrest

Longleg sp. *Anisogomphus forresti* **Morton**, 1928
[Orig. *Temnogomphus forresti*]

George Forrest (1873–1932) was a Scottish botanist who explored in Yunnan, China's most bio-diverse province. He started work in a chemist's shop, where he learned about the medicinal qualities of plants and how to dry and preserve them. He went to Australia at the height of the (1891) gold rush and spent 10 years panning for gold before returning to Britain (1902). There he was employed as a clerk at the Royal Botanic Gardens in Edinburgh (1903). He went to Talifu, China (1904) and then (1905) set out for northwest Yunnan, close to the border with Tibet. At Tzekou his party was based at a French mission and, after a collecting foray, Forrest and his team of 17 collectors returned to the mission to find that the locals had turned on the foreigners. Of his group, only Forrest escaped with his life. Continuing to collect despite this tragedy and despite being stricken with malaria, he and his new team of collectors amassed a very considerable weight of specimens that he took back to Britain (1906). During the rest of his life, he made six further expeditions to Yunnan, Sichuan Province, Tibet, and Upper Burma. He discovered more than 1,200 plant species that were new to science as well as many birds, mammals and other fauna. He died from a heart attack in China while still collecting. Three mammals and seven birds are also named after him.

Förster

Shadowdamsel sp. *Protosticta foersteri* **Laidlaw**, 1902
Trifid Duskhawker *Platacantha försteri* **Martin**, 1909
[JS *Agyrtacantha dirupta* (Karsch, 1889)]

Johann Friedrich Nepomuk Förster (1865–1918) was a German entomologist, zoologist, botanist, teacher and collector. The year before he started university he began his odonatological studies writing to Selys because he could not buy his books. Selys was very helpful, endowing him with many reprints and specimens. They became close collaborators for the last few years of Selys' life. He studied natural science at Heidelberg University (1886–1889). He began teaching soon after graduation at Schopfheim (1889–1898) then Mannheim (1898–1899) and at Bretten (1899) and married that year. At the outbreak of WW1, he was (against his will) sent to teach in a school at Oberkirch. He wrote around fifty papers (thirty-four on odonata) sometimes with others, the following with Laidlaw: *Report on dragonflies. II. The legions Platycnemis and Protoneura* (1907) and later named two species in his honour. He named thirty-one genera and one subgenus and described c.160 species and subspecies, although nearly seventy were synonyms. On his death in hospital after a short severe illness, his wife was forced to sell his botanical and entomological collections to various museums (Lepidoptera to Tring, Coleoptera to BMNH, botany to Munich and his Odonata to UMMZ). Three birds are also named after him. (Also see **Common** & **Reinhold**)

Foster

Dancer sp. *Argia fosteri* **Calvert**, 1909
[JS *Argia albistigma* Hagen, 1865]

William T Foster (1867–1915) was a field naturalist and collector. He was a seaman who became a shopkeeper at Sapucay, Paraguay (1894–1915) and regularly supplied bird and mammal specimens to the British Museum. He collected the type in Paraguay (1899).

Fountaine
Bluetail sp. *Ischnura fountaineae* **Morton**, 1905
[Orig. *Ischnura fountainei*]

Margaret Elizabeth Fountaine (1862–1940) was an English lepidopterist and diarist and an accomplished illustrator. The eldest of seven children, she inherited wealth enabling her to be independent. She travelled extensively over five decades in sixty countries throughout Europe, southern Africa, India, Tibet, North America, Australia and the Caribbean. She published many papers and some longer works. About 22,000 specimens that she collected or had collected are housed in the Norwich Castle Museum and her four sketch books of butterflies and their life cycles are at the BMNH. Her diaries were unsealed (1978) in accordance with her will, 100 years after she began writing them in 12 leather-bound volumes amounting to 3203 pages. They are a mix of Victorian reserve and startling candour. She loved travel from an early age, but a meeting for a few days with H J Elwes (1895) sparked her interest in entomology after which she started her collection with a trip to Sicily. In Europe, she often travelled by bicycle. She met Khalil Neimy (1901) in Damascus and employed him as a 'dragoman' (guide and translator) and they became companions, despite his married status, for twenty-seven years until his death (1928). She took the *Ischnura* type in Algeria (1904). She died of a heart attack while collecting on the slopes of Mount St Benedict in Trinidad. In a note with her diaries she wrote: "*Before presenting this – the Story of my Life – to those, whoever they may be, one hundred years from the date on which it was first commenced to be written, i.e. April 15: 1878, I feel it incumbent upon me to offer some sort of apology for much that is recorded therein, especially during the first few years, when (I was barely 16 at the time it was begun) I naturally passed through a rather profitless and foolish period of life, such as was and no doubt is still, prevalent amongst very young girls, though perhaps more so then – a hundred years ago, when the education of women was so shamelessly neglected, leaving the uninitiated female to commence life with all the yearnings of nature quite unexplained to her, and the follies and foibles of youth only too ready to enter the hitherto unoccupied and possibly imaginative brain.*"

Foyle
Bluetail sp. *Ischnura foylei* **Kosterin**, 2015

Christopher Foyle (b.1943) is a British businessman, philanthropist and writer. He took over, and continues to chair, Foyles his eponymous family bookshop (1999) when it was on its last legs and, together with the management that he put in place, turned around and greatly expanded it. He founded (1977) and grew, for 30 years, the 'Air Foyle Group' of specialist passenger and cargo airlines. He has chaired a number of national and international aviation trade organisations and charities. He has written four books such as: *Pioneers to Partners – a history of British Aircraft from 1945* (2009) and contributed to two others. He is financing two fauna and flora research projects in the rain forests of Indonesia. He is founding the Foyle Research Institute to examine and investigate scientifically anomalous phenomena. Kosterin honoured him with this Sumatran species for "...*his support for research and conservation of Indonesian rainforests and their biodiversity in both Kalimantan and Sumatra.*"

Francesca

Duskhawker sp. *Gynacantha francesca* **Martin**, 1909
[Orig. *Subaeschna francesca*]

Francesca is a feminine forename, but not a specified one. The author gave no etymology, so it is not possible to know for sure if this was named after a particular person. However, the lack of a Latin ending suggests it is not an eponym.

Franklin

Delicate Emerald *Somatochlora franklini* **Selys**, 1878
[Orig. *Epitheca franklini*]

Sir John Franklin (1786–1847) was an officer in the Royal Navy and well-known explorer searching for the 'Northwest Passage'. The youngest of 12 boys, Franklin joined the navy in his youth and spent the rest of his life in its service. His first Arctic voyage (1818) was commanding a vessel trying to reach the North Pole. He attempted to find the Northwest Passage (1819): the sea route across the Arctic to the Pacific Ocean. His overland expedition returned empty-handed after two years amid rumours of starvation, murder and cannibalism. He was appointed as Governor of Van Diemen's Land (Tasmania) (1837). Franklin was involved in several more voyages to the north, before disappearing (1845) in another attempt to cross the Arctic by sea. A search was undertaken under the command of Sir Clements Robert Markham. Log books were later found confirming that he had died along with most of his officers and crew – in his case probably from lead poisoning, ironically leaching from the solder used to seal canned provisions then considered the height of technology. A mammal and two birds are also named after him.

There appears to be *sans description* of this species in de Selys Longchamps' (1878) article in the 'Bulletin Academie Royale Belgique'. Richardson had participated in the famous Coppermine Expedition led by Franklin to an area including Fort Résolution on the south shore of Great Slave Lake in the Northwest Territories of Canada (1819–1822), during which the Emerald appears to have been collected (this was the only time that they were at the type locality during the time of year that the species flies). De Selys Longchamps also had a manuscript name '*Cordulia richardsoni*' for a specimen that Richardson had collected at Fort Simpson, along the Mackenzie River, west from Fort Résolution. However, before it was formally described it was recognised to be the same as Hagen's '*Cordulia*' septentrionalis.

Fraser

Satinwing sp. *Euphaea fraseri* **Laidlaw**, 1920
[Orig. *Pseudophaea fraseri*, A. Malabar Torrent Dart]
Hooktail sp. *Acrogomphus fraseri* Laidlaw, 1925
Sylvan sp. *Coeliccia fraseri* Laidlaw, 1932
[A. Fraser's Forest Damselfly/Shadowdamsel]

Skydragon sp. *Chlorogomphus fraseri* St. Quentin, 1936
Shadowdamsel sp. *Protosticta fraseri* **Kennedy**, 1936
[JS *Chlorocypha pyriformosa* Fraser, 1947]
Shadowdamsel sp. *Drepanosticta fraseri* Lieftinck, 1955
[JS *Ceylonosticta submontana* Fraser, 1933]
Northern Riverking *Zygonoides fraseri* **Pinhey**, 1956
[Orig. *Olpogastra fraseri*]
Western Tigertail *Ictinogomphus fraseri* **Kimmins**, 1958
Forest Flashwing *Phaon fraseri* Pinhey, 1962
[JS *Phaon camerunensis* Sjöstedt, 1900]
Common Blue Skimmer *Orthetrum fraseri* Sahni, 1965
[JS *Orthetrum glaucum* (Brauer, 1865)]
Treefall Elf *Tetrathemis fraseri* **Legrand**, 1977
Tropical Pinfly *Neosticta fraseri* **JAL Watson**, 1991

Sprite ssp. *Pseudagrion pruinosum fraseri* Schmidt, 1934
River Jewel *Libellago dispar fraseri* Schmidt, 1951
[JS *Chlorocypha pyriformosa* Fraser, 1947]

Lieutenant-Colonel Dr Frederic Charles Fraser (1880–1963) was an English entomologist specializing in Odonata. He studied medicine at Guy's Medical School (1903), being awarded his MD from Brussels (1904). After two years in general practice in London he joined the Indian Medical Service. He saw active military service during WW1 in Mesopotamia where he was in charge of hospitals, including hospital ships. After the war, he held various posts in India including being superintendent at two prisons and Professor of Surgery at Vizagapatam Medical College and of Obstetrics and Gynaecology at Madras Medical school. While in India, he actively collected dragonflies, especially in the Nilgiri Hills. He published (1918–1934) a series in forty-two parts entitled *Indian dragonflies* in the 'Journal of Bombay Natural History Society'. A revised version of this work, *Odonata*, was published in three volumes (1933–1936) in the series *The Fauna of British India, including Ceylon and Burma*. These volumes are still used as a standard handbook of the Indian Odonata. He was offered (1933) a major post at Madras Medical School, but resigned from service instead to devote his time to the study of dragonflies. He returned to England and settled in Bournemouth, but often spending time working in the BMNH where his collection is housed. In the 1940s and 1950s he turned his interest to African dragonflies, which he received from various sources. Fraser published over 300 papers on Odonata, in which he named c.80 new genera and described c.560 new species and subspecies. However, about one third of his new species, and many of his genera, are synonyms. Other major publications include: *A reclassification of the Order Odonata* (1957) and *A handbook of the dragonflies of Australasia* (1960). His library is now at the Manchester Museum. Fraser's contribution to odonatology was very substantial and he was also a good all-round naturalist and observer of dragonfly behaviour and life history. Some contemporaries found him difficult to work with. Cynthia Longfield (q.v.) expressed her feelings in a letter (March 1963) to Douglas Kimmins (q.v.) as follows: *"I must say the old man had a wonderful innings in the study of Odonata. And my goodness, he turned out plenty of specimens. What a pity he was such a jealous worker and so terribly dogmatic… this just spoilt such a fine career. It was all so very unnecessary in an amateur."*

Freddie Mercury

Flatwing sp. *Heteragrion freddiemercuryi* **Lencioni,** 2013

Freddie Mercury was the artistic name of Farrokh Bulsara (1946–1991), who the author described as a superb and gifted musician and songwriter whose wonderful voice and talent still entertain millions of people around the world. Four new species were described in tribute to the 40th anniversary of the rock band 'Queen' – one for each of its original members.

Frederico

Pond Damselfly sp. *Angelagrion fredericoi* **Lencioni**, 2008

Frederico Augusto Rodrigues Lencioni (b.2007) is the son of the author Frederico Augusto de Atayde Lencioni (b.1970). Interestingly, the author's father, Professor Frederico Lencioni Neto, a Brazilian biologist, who is especially interested in ornithology and entomology, was also involved in the collecting of the type material. The etymology puts it thus: "*I name this species fredericoi (noun in genitive case) after my beloved son Frederico Augusto Rodrigues Lencioni*". In the introduction of the paper, he wrote: "*In 2005 my father… …collected in São José dos Campos, São Paulo a male of a second species which belongs to the same genus, and in more recent collecting trips to these localities we collected additional material of both species which I describe here as new*". So, in honouring his son, he in a way also honoured his father. The latter has contributed to a number of works including: *Biomes of Brazil: An Illustrated Natural History* (2005). (Also see **Lencioni**)

Frey

Siberian Bluet *Agrion freyi* Bilek, 1954
[JS *Coenagrion hylas* Trybom, 1889]

Georg Frey (1902–1976) was a wealthy businessman who was able to create (1950) his own Coleoptera museum, the Museum G Frey, recognized as the world's largest and most extensive private collection of beetles. He also began the journal 'Entomologische Arbeiten aus dem Museum G Frey' that was published in 56 volumes (until 1987). He spent five decades collecting and amassing c.150,000 species of beetles, 20,000 of which are types! Some of the museum's three million specimens were collected during 36 expeditions to all parts of the world, with a number of them led by Frey himself and he employed others to collect for him. Scarab beetles in particular, especially the *Melolonthinae* and *Rutelinae*, are well represented in the museum's collections, which resided in exquisite collections facilities at the Zoologische Staatssammlung in Munich for several years after Frey's death. After a somewhat acrimonious battle over possession of the collection, it was moved to the Naturhistorisches Museum Basel in Switzerland (1996). The type was collected in Germany (1953). (Also see **Georg Frey**)

Fruhstorfer

Fruhstorfer's Junglewatcher *Hylaeothemis fruhstorferi* **Karsch,** 1889

[Orig. *Tetrathemis fruhstorferi*]
Threadtail sp. *Alloneura fruhstorferi* **Krüger**, 1898
[JS *Nososticta insignis* Selys, 1886]
Clubtail sp. *Davidius fruhstorferi* **R Martin**, 1904
Clubtail sp. *Nepogomphus fruhstorferi* **Lieftinck**, 1934
[Orig. *Onychogomphus fruhstorferi*]

Hans Fruhstorfer (1866–1922) was a German explorer, insect trader and entomologist specialising in Lepidoptera. He both collected and described many new species. He spent two years (1888–1890) in Brazil becoming successful enough to set up as a professional collector. He was then in Ceylon and followed that with a three-year sojourn in Java and Sumatra (1890–1893) and time in Celebes, Bali and Lombok (1895–1896). He took another three-year trip (1899–1902) to USA, Japan, China, Indo-China and many pacific islands, returning via India. He settled in Geneva and wrote a number of monographs based on his own extensive collection, although these have been criticised as sometimes rushed and inaccurate. Instead of further travels he employed Werner (New Guinea) and Sauter (Taiwan) to collect for him. Much of his collection is housed by the natural History Museum in Berlin as well as at BMNH in London and MNHN Paris. He died after an unsuccessful operation for cancer. His written work included a monograph on Pieridae in Seitz's: *The Macrolepidoptera of the world* (1906). A reptile is also named after him.

Fulla

Genus *Fylla* **Kirby**, 1889
[JS *Nannophya* Rambur, 1842]

In Nordic mythology Fulla was a goddess.

Fülleborn

Fuelleborn's Bottletail *Zygonoides fuelleborni* Grünberg, 1902
[Orig. *Olpogastra fülleborni* A. Southern Riverking, Robust Bottletail]

Dr Friedrich Georg Hans Heinrich Fülleborn (1866–1933) was a German physician. He served as a military doctor with the German Army in East Africa (1896–1910). He made a collection of Odonata in Nyasaland (Malawi), including the type, which Grünberg studied. He was an expert on tropical diseases and parasitology and became Director of the Hamburg Institute for Marine and Tropical Diseases (1930). Ten birds, an amphibian and a reptile are also named after him.

Funck

Dancer sp. *Argia funcki* **Selys**, 1854
[Orig. *Hyponeura funcki*]

Nicholas Funck (1817–1896) was a botanist from Luxembourg who travelled in the neotropics in the nineteenth century. He was in Brazil (1835) and Mexico and Guatemala (1837–1840) collecting with Linden and Auguste. He was also in Venezuela (1845). He became Professor of Natural History at the Athénée de Luxembourg (1948) and later (1857) co-directed the Brussels Zoo. He was also Director of Cologne Zoo (1870–1986) and then returned home where he lived out his life.

Fylgia

Genus *Fylgia* **Kirby**, 1889

In Nordic mythology, the Fylgia is a mythical beast, a shadow soul and guardian spirit.

Fylla

(See **Fulla**)

G

Gabriele

Pond Damselfly sp. *Metaleptobasis gabrielae* von **Ellenrieder**, 2013

Gabriele Margarete Elizabeth von Ellenrieder (née Müller) (b.1945) is the mother of the author Natalia von Ellenrieder. She studied Romance languages and literature (French & Spanish) and graduated (1970) at Georg-August-Universität Göttingen. She married Guillermo von Ellenrieder (q.v.) and emigrated to Argentina, where she worked (1978–2009) as a German language teacher at the Universidad Nacional de Salta (also see **Natalia** & **Guillermo**).

Gaiani

Pond Damselfly sp. *Mesamphiagrion gaianii* **De Marmels**, 1997
[Orig. *Cyanallagma gaianii*]

Marco Antonio Gaiani Pacheco (b.1967) is an amateur entomologist who left university before graduating to start a family. Over a number of years, he combined his two passions of entomology and computer science to lead to a new line of research. He achieves this through online documentation of biodiversity filed in the Museum of the Institute of Agricultural Zoology 'Francisco Fernandez Yépez' (MIZA) through a web portal of that institution. He is Curator of Homoptera at Estado Aragua Museo del Instituto de Zoología Agrícola, Facultad de Agronomía, Universidad Central de Venezuela at Maracay. He was also Associate Editor of the 'Boletin de Entomologia Venezolana'. He also has an interest in amphibians. His research foci include systematics, biodiversity and taxonomy. He collected the type in Venezuela (1993).

Gaida

Threadtail sp. *Neoneura gaida* **Rácenis**, 1953
[JS *Neoneura bilinearis* Selys, 1860]

Gaida Rácenis née Antens (1925–2006) was the wife of the author Dr Janis Rácenis (1915–1980). They married (1947) in Germany where Rácenis was doing his PhD and left for Venezuela the following year.

Gaige

Red-Mantled Skimmer *Libellula gaigei* Gloyd, 1938

Professor Frederick McMahon Gaige (1890–1976) was an American entomologist, herpetologist and botanist who was Director of the Zoological Museum, University of Michigan. His wife, Helen Beulah Thompson Gaige, was also a well-known herpetologist – and in their honour the American Society of Ichthyologists and Herpetologists makes an annual award to a graduate student of herpetology. He also has a bird named after him. (Also see **Helenga**)

Gallard

Rubyspot sp. *Hetaerina gallardi* Machet, 1989

Jean-Yves Gallard (b.1943) is a French entomologist in French Guiana (the holotype location). After studying at the Beaux-Arts d'Angers, Rouen and Paris, he became a teacher of visual arts in Cayenne where he developed a passion for the flora and fauna of Guyana. He is primarily interested in Lepidoptera, especially *Riodinidae* about which most of his scientific papers are written such as: *Description de nouveaux Riodinidae provenant de Guyane Française* (1992), co-written with Christian Brévignon (q.v.), with whom he has collected widely. He collected the types of this *Hetaerina* (1988). At least one moth is also named after him (*Detritivora gallardi*) and he has described more than 30 Lepidoptera species.

Gambles

Gamble's Flatwing *Neurolestes nigeriensis* Gambles, 1970
Gamble's Relic *Pentaphlebia gamblesi* Parr, 1977
Pepperpants *Oxythemis gamblesi* **Longfield**, 1959
[JS *Oxythemis phoenicosceles* **Ris**, 1909]
Problematic Flasher *Aethiothemis gamblesi* **Lieftinck**, 1969
[Orig. *Lokia gamblesi*]
Great Sprite *Pseudagrion gamblesi* **Pinhey**, 1978
Lyre-tipped Cruiser *Phyllomacromia gamblesi* **Lindley**, 1980
[Orig. *Macromia gamblesi*]
Western Fingertail *Gomphidia gamblesi* Gauthier, 1987
Western Hoetail *Diastatomma gamblesi* **Legrand**, 1992

Robert Moylan Gambles (1910–1990) was a British veterinarian and odonatologist. He was educated at Westminster School, Trinity College, Cambridge and the Royal Veterinary College, Edinburgh. Apart from a two-year spell in Palestine, he was Veterinary Officer in Cyprus (1935–1948) where he met Cynthia Longfield, then served in Nigeria (1949–1962), becoming Principal of the Veterinary School at Vom before returning to England (1962). He was Research Officer with the Ministry of Agriculture, Fisheries and Food (1962–1970). Having been a classical scholar, he maintained his interest in Greek and Latin literature throughout his life and was interested in music, theology and English literature too. His wife Margaret was keen on botany and their trips to Greece and the eastern Mediterranean continued until just before her death. His main interests were Odonata, and the helminthological and entomological aspects of parasitology. He became the undisputed world authority on the Odonata of West Africa and described a dozen new species. He was a Fellow of the Royal Entomological Society of London and Honorary Associate of The Natural History Museum, London which has all his collected papers and reports. Mike Parr and Roger P Lindley (q.v.) wrote a biography and bibliography of him in 'Odonatologica' (1980) in celebration of his seventieth birthday. He died following an unsuccessful operation.

Ganesh

Clubtail sp. *Gomphidia ganeshi* Chhotani, Lahiri & Mitra, 1983

Ganesh is an Indian Hindu deity; the Remover of Obstacles.

Ganesh

Common Blue Skimmer *Orthetrum ganeshii* Mehrotra, 1961

[JS *Orthetrum glaucum* (Brauer, 1865)]

Ganesh P Dube was Principal of Ranchi College, India.

Gangi

Common Blue Skimmer *Orthetrum gangi* Sahni, 1964

[JS *Othetrum glaucum* (Brauer, 1865)]

Gangi is another name for the revered Hindu goddess Durga.

Garbe

Sylph sp. *Ophippus garbei* **Navás**, 1916

[JS *Macrothemis heteronycha* Calvert, 1909]

Pond Damsel sp. *Leptagrion garbei* **Santos**, 1961

Father Ernst (Ernesto) Wilhelm Garbe (1853–1925) was a Brazilian zoologist collecting in Brazil at the end of the 19th century, where the type was taken (1908). He worked at the Paulista Museum of the University of São Paulo with Hermann von Ihering (q.v.) and was one of many collaborators of Ihering on his Revista do Museu Paulista, as was Longinos Navás. An amphibian and a bird are also named after him. Santos examined much of his collected material in the early 20th century.

Garcia, JL

Pond Damselfly sp. *Tepuibasis garciana* **De Marmels**, 2007

Dr José Luis Garcia Rodrigues (b.1948) is a hymenopterologist, who collected several specimens of another new species (*Tepuibasis nigra*) described in the same paper. (The species was named jointly in honour of both he and R Garcia below).

Garcia, R

Pond Damselfly sp. *Tepuibasis garciana* **De Marmels**, 2007

Rafael Garcia Pena (b.1955) from Caracas (now Hamburg, Germany) is the author's former graduate student who collected this and other highly interesting species during short visits to some most inaccessible mountaintops in Pantepui, Venezuela (1991). (The species was named jointly in honour of both he and J L Garcia above).

Gardner

Skimmer sp. *Hylaeothemis gardeneri* **Fraser**, 1927

Fraser did not give an etymology. Some sources suggest it might be honouring James Clark Molesworth Gardner (1894–1970).

Garlepp

Helicopter sp. *Mecistogaster garleppi* **Förster**, 1903
[JS *Mecistogaster buckleyi* McLachlan, 1881]
Firetail sp. *Telebasis garleppi* **Ris**, 1918

Otto Garlepp (1864–1959) and his brother Gustav Garlepp (1862–1907) were German collectors in Latin America (1883–1897). Otto was a teacher until he joined his brother (1893). Gustav landed in Pará, Brazil, and crossed the continent to Peru, where he stayed for four years. After a short break in Germany he went back, this time to Bolivia. When he returned to Germany (1892) he had amassed 1,530 bird specimens. He returned to Bolivia (1893) together with his wife and his brother. He visited Germany for the last time (1900), presenting 3,000 specimens of 600 species of birds at the annual meeting of the German Ornithologists' Society, Leipzig. He settled in Paraguay (1901) but was murdered during a collecting expedition. His collection is now in the Senckenberg Museum, Frankfurt. A mammal and five birds are also named after him. Otto collected in Bolivia (1893–1897, 1902–1903), in Peru (1898–1899, 1902–1903), in Columbia (1910–1911), in Panama, Costa Rica and Paraguay (1902), in Chile (1903) and in Argentina (1897–1901, 1903–1910 & 1912–1913). Gustav was a bank employee who first went to South America to collect insects for the Natural History Museum in Dresden, but also collected birds for Berlepsch. Together they also collected insects, particularly butterflies. After WW1 Otto returned to Germany to work in the insurance business. Both species were named after Otto, who had collected the specimens in Peru (1901) and Costa Rica (1913).

Garrison

Genus *Garrisonia* Penalva & **JM Costa**, 2007

Dancer sp. *Argia garrisoni* Daigle, 1991
Firetail sp. *Telebasis garrisoni* **GH Bick** & **JC Bick**, 1995
Threadtail sp. *Forcepsioneura garrisoni* **Lencioni**, 1999
Skimmer sp. *Andinagrion garrisoni* **von Ellenrieder** & **Muzón**, 2006
Skimmer sp. *Oligoclada garrisoni* **De Marmels**, 2008
Pond Damselfly sp. *Denticulobasis garrisoni* **Machado**, 2009
Tropical King Skimmer sp. *Orthemis garrisoni* von Ellenrieder, 2012

Queensland Swiftwing *Lathrocordulia garrisoni* **Theischinger** & **Watson**, 1991

Dr Rosser William Garrison (b.1948) is an American entomologist who specialised in Odonata. He attended the University of California, Berkeley, where he was awarded his MA (1974) and doctorate (1979) with a thesis on the damselfly genus *Enallagma*. He worked there as a teaching assistant, associate, and research supervisor (1976–1981). He later became Senior Entomologist of the County of Los Angeles (1984–2004) and was also an Instructor at the Rio Hondo Community College (1999–2004). Then he became Senior Insect Biosystematicist at the Plant Pest Diagnostics Laboratory, California Department of Food & Agriculture (2004–2017) in Sacramento. He has been interested in the systematics of the neotropical odonate fauna from 1982, having that year worked as the invertebrate ecologist at the Center for Energy and Environmental Research (CEER) at the El Verde Field Station in Luquillo Forest in eastern Puerto Rico. He has written 89 publications pertaining to Odonata, including three books: *Dragonfly genera of the New World: An illustrated and annotated key to the Anisoptera* (2006), *Damselfly genera of the New World: An illustrated and annotated key to the Zygoptera* (2010) and, as co-author, *Dragonflies of the Yuangas: A field guide to the species from Argentina* (2007). He has written or co-written taxonomic revisions of over 20 genera of New World Odonata, such as *Enallagma* (1984), *Hetaerina* (1990), *Erpetogomphus* (1994), *Mnesarete* (2006) & *Telebasis* (2009). His present taxonomic work, jointly with Natalia von Ellenrieder (q.v.), focuses on the speciose genus *Argia* on which he has published several revisions (1994, 2015, 2017). His publications include descriptions of 74 new species and six new genera of Odonata. In the Americas, he has done entomological field work in USA, Mexico, Costa Rica, Panama, Guyana, French Guiana, Venezuela, Ecuador, Peru, Brazil, Argentina and Chile, and is widely recognized as the current leading expert of the taxonomy of New World Odonata. He has also collected dragonflies in Australia, Papua New Guinea, China, Thailand, Namibia and Iran. In the etymology of *Oligoclada garrisoni* De Marmels wrote: "*I dedicate this species to Rosser W. Garrison, not only for his outstanding contributions to odonatology of the Americas, but also in recognation of his unselfish readiness to collaborate whenever asked for.*" (Also see **Rosser**).

Gaucha

Threadtail sp. *Peristicta gauchae* **Santos**, 1968
Emerald sp. *Neocordulia gaucha* **Costa** & **Machado**, 2007

Gaucha is the feminine of gaucho, a name given to people born in the State of Rio Grande do Sul, Brazil.

Gaumer

Dancer sp. *Argia gaumeri* **Calvert**, 1907

Dr George Franklin Gaumer (1850–1929) collected in the Americas. Kansas University awarded his first degree (1876) and masters (1893). He collected (mostly birds) in Cuba (1878), Yucatan, Mexico (1878–1881) and southwest US (1881–1884). He was Professor of Natural Sciences at the University of North Mexico after which he moved permanently to Yucatan, practising medicine. His later collections, with his two sons, were mostly botanical, but also contained insects including Odonata which Calvert studied. He wrote: *Monografía de los mamíferos de Yucatán* (1917). Many plants, three birds and one mammal are also named after him.

Gautama

Midget sp. *Indagrion gautama* **Fraser,** 1922
[Syn. of *Mortonagrion aborense* (Laidlaw, 1914]

No etymology was given, but named after Gautama Buddha; the ascetic and sage after whose teachings Buddhism was founded. Fraser named another species *Indolestes buddha* (see **Buddha**).

Gay

Pond Damselfly sp. *Antiagrion gayi* **Selys**, 1876
[Orig. *Erythromma gayi*]

Claude Gay (originally Claudio Gay Mouret) (1800–1873) was a French botanist. He began to study medicine in Paris but quickly abandoned this in favour of natural history. He went to Chile to teach physics and natural history in a college in Santiago (1828), but was accepted by the Chilean government to undertake research for them (1830–1932). He then returned to France, donating collections he had made in Chile to the MNHN, of which he was thereafter a corresponding member. It was from Gay's collection in the MNHN that Selys was later sent the damselfly to study. He returned to Chile to explore there (1834–1840) and briefly (1839) in Peru and was later in the USA (1859–1860) studying American mining techniques. He wrote many works including his 24-volume magnum opus: *Historia Fisica y Politica de Chile* published by the government of Chile (1843–1851), which also granted him citizenship (1841). Among his other works is the interesting: *Origine de la pomme de terre* (1851). He returned to Chile and settled (1863). An amphibian and three birds are also named after him as is 'Cordillera de Claudio Gay', a region of the Atacama in Chile.

Geijskes

Sanddragon sp. *Progomphus geijskesi* **JG Needham**, 1943
Redskimmer sp. *Rhodopygia geijskesi* **Belle**, 1964
Shadowdamsel sp. *Protosticta geijskesi* **Van Tol**, 2000
Pond Damselfly sp. *Tuberculobasis geijskesi* **Machado**, 2009

Dr Dirk Cornelis Geijskes (1907–1985) was a Dutch entomologist and ethnologist. He studied biology at the University of Leiden and qualified as a teacher in biology (1931). The University of Basel, Switzerland awarded his doctorate (1935) for a study of the Röserenbach in the Swiss Jura. He worked in Suriname (1938–1965), first studying tropical pests at the Agricultural Experimental Station in Paramaribo. He became (1954) Government Biologist at the Ministry of Interior Affairs and (1956) Director of the Surinam Museum. After returning to the Netherlands he worked (1967–1972) as a curator of Odonata and neuropteroid orders at the Leiden Museum. He wrote over 160 publications, including 38 on Odonata, including the book, co-written with Jan van Tol (q.v.): *De libellen van Nederland (Odonata)* (1983). He described twelve new Odonata species and a new genus, *Lauromacromia*, from Surinam (1931–1984). His collection of more than 13,000 specimens is preserved in the Leiden Museum. About twenty-five different plant and animal taxa, including a fish and a frog, are also named after him.

Gené

Common Hooktail *Paragomphus genei* **Selys**, 1841
[Orig. *Gomphus genei*]
Island Bluetail Damselfly *Ischnura genei* **Rambur**, 1842
Spotted Darter *Libellula genei* Rambur, 1842
[JS *Sympetrum depressiusculum* (Selys, 1841)]

Carlo Giuseppe Gené (1800–1847) was an Italian author and naturalist. He studied at the University of Paria and later (1828) became assistant lecturer in natural history there. He collected in Hungary (1829) and made four trips to Sardinia (1833–1838), mostly collecting insects. On Bonelli's death (1830), Gené succeeded him as Professor of Zoology and Director of the Royal Zoological Museum in Turin, and De Fillipi, in turn, succeeded him. An amphibian and two birds are named after him.

Georg Frey

Turkish Red Damsel *Ceriagrion georgifreyi* **Schmidt**, 1953

Georg Frey (1902–1976) supplied the damsel to Schmidt. (See **Frey**)

Gerda

Brook Damselfly sp. *Salomoncnemis gerdae* **Lieftinck**, 1987

Gerda Stanny Slooff-de Vries, with her husband Dr Rudolf Slooff (b.1934) the medical entomologist, collected the type in forest at 1200ft on Guadalcanal Island, the Solomon Islands (1985).

Gerhard

Dancer sp. *Argia gerhardi* **Calvert**, 1909

William Josiah Gerhard (1873–1958) was an American naturalist and collector. He travelled to Bolivia and Peru (1898) where, funded and commissioned by A G Weeks, a known North American butterfly collector, he collected extensive material. Calvert described the majority of the dragonflies that Gerhard collected.

Gerstäcker

Cruiser sp. *Macromia gerstaeckeri* **Krüger**, 1899

Upland Sprite *Pseudagrion gerstaeckeri* **Karsch**, 1899

[JS *Pseudagrion spernatum* Selys, 1881]

Carl Edward Adolf Gerstäcker (1828–1895) was a German zoologist and entomologist. He studied medicine and natural sciences in Berlin. He became Curator of the Humbolt University Zoological Museum (1857–1876). He was also Professor of Zoology at the University of Berlin (1874–1876) then (1876) Professor of Zoology at the University of Greifswald and later (1894) lecturer at the Berlin Agricultural Institute. He wrote: *Monographie der Endomychiden* (1858) and co-wrote: *Handbuch der Zoologie* (1863–1875). He wrote two papers on African Odonata and described seven new species of which three are currently valid.

Gestro

Little Blue Marsh Hawk *Brachydiplax gestroi* **Selys**, 1891

[JS *Brachydiplax sobrina* (Rambur, 1842) A Sombre Lieutenant]

Clubtail sp. *Leptogomphus gestroi* Selys, 1891

Rainbow Yellowwing *Chlorocnemis gestroi* **R Martin**, 1908

[JS *Allocnemis nigripes* Selys, 1886]

Raffaello Gestro (1845–1936) was an Italian entomologist, one of the most important Italian coleopterists. He was Deputy Director and then Director of the Museo Civico di Storia Naturale, Genova, and President of the Societa Entomologica Italiana. He published 147 papers on Coleoptera taxonomy and described 936 new insect species, many based on specimens from other Italian collectors, including Fea, D'Albertis, and Beccari. Many other insects are named after him as well as a bird

Ghesquiere

Black Siphontail *Karschiogomphus ghesquierei* **Schouteden**, 1934
[JS *Neurogomphus martininus* (**Lacroix**, 1921)]
Congo Red Jewel *Chlorocypha ghesquierei* **Fraser**, 1959

Jean Hector Paul Auguste Ghesquière (1888–1982) was a botanist who collected for the Belgian National Herbarium. He was in Angola, Belgian Congo (DRC), Niger, Sao Tome and Principe, and Uganda (1918–1938) and collected the types in the Congo. An amphibian is also named after him.

Gideon

Clubtail *Stylyrus gideon* **Needham**, 1941
[Orig. *Gomphurus gideon*]

No etymology was given. However, the name may refer to Gideon, a character from the Bible. The name means 'Destroyer' or 'Mighty warrior'. This Chinese species may have been so named because it has greatly expanded apical abdominal segments which resemble a mace.

Gillies

Amani Flatwing *Amanipodagrion gilliesi* **Pinhey**, 1962

Dr Michael (Mick) Thomas Gillies (1920–1999) was a British medical entomologist. He studied medicine (1939) and became an Army medical officer in the Far East (Pakistan, India, Burma, Thailand and Hong Kong), and wherever he went he collected mayflies. When he had to abandon ship in the Malacca Straits he managed to save his collection, which turned out to contain several species new to science. After discharge, he took a post as Medical Officer to the British Embassy in Moscow, where he learnt Russian and spent time collecting insect larvae in local streams. When he returned to London he gave up medicine in favour of entomology and took a post (1951–1963) at the East African High Commission Medical Research Institute in Amani, Tanganyika (Tanzania), working on malarial vectors. He collected two specimens of the flatwing in the East Usambara Mountains (1959). He was later at the BMNH (1963–1965) revising their African mosquito collection, during which time he completed two books including: *Anophelinae of Africa South of the Sahara* (1968). He took a post as head of the newly formed Mosquito Biology Unit, at Sussex University (1965–1981) until retirement, although he carried on researching at Sussex for at least another two years. In retirement he travelled, collecting and studying mayflies in Africa, Asia and South America. He wrote many papers on mayflies and mosquitoes, but also on other insects as well as longer works culminating in his autobiography which he finished a few days before he died: *Mayfly on the Stream of Time* (2000).

Girard

Candy Threadtail *Elattoneura girardi* **Legrand**, 1980
Cruiser sp. *Phyllomacromia girardi* Legrand, 1991
[Orig. *Macromia girardi*]

Dr Claude Girard (b.1941) is a French entomologist working at MNHM, Paris; he is a specialist of termitophilous Tenebrionids. He collected the type specimens of *Elattoneura* in Ivory Coast (May 1968). The Coleoptera genus *Claudegirardius* is also named after him as is at least one species.

Godeffroy

Eastern Evening Darner *Telephlebia godeffroyi* **Selys**, 1883

Johann Caesar Godeffroy (1813–1885) was a member of a German trading house, which imported copra from the Pacific, and was interested in natural history. His family was French but moved to Germany to avoid religious persecution. His fleet of ships traded largely in the Pacific where he used them as floating collection bases, with paid collectors on board to search for zoological specimens as well as trade commercially. He sent his ships to the Pacific (1845), establishing c.45 trading posts and buying land and property, laying the foundations of German colonial power. He used his natural history collection to found a museum in Hamburg (1860), naming it after himself. He sent collectors to the Pacific and employed well-known naturalists in Hamburg, such as Otto Finsch and Gustav Hartlaub. Eventually he neglected commerce and the company was bankrupted (1879). A reptile and two birds are also named after him.

Godiard

Western Elf *Tetrathemis godiardi* **Lacroix**, 1921

Max Godiard sent the type to Lacroix from Koforidua, Ghana.

Godman

Apache Spiketail ssp. *Cordulegaster diadema godmani* **McLachlan**, 1878

Doctor Frederick du Cane Godman (1834–1919) was a British naturalist, ornithologist, lepidopterist and entomologist. He went to Eton but due to poor health was mostly educated at home. He attended Trinity College Cambridge (1853) where he met Osbert Salvin; they became life-long friends and learned taxidermy together. Godman qualified as a lawyer, but was wealthy and did not need to work, so devoted his life to ornithology, particularly of central America. His first overseas trips were with his tutor around the Black Sea, Turkey and southern Spain (1852), and with his brother to Norway (1857). With his friend Salvin, he visited Jamaica, Guatemala and Belize (1861) but had to return to the UK with a fever. However, he and Salvin planned (1876) and compiled the massive *Biologia Centrali Americana* (1888-1904). Among his other written works was *Natural History of the Azores* (1870). They presented their joint collection to the British Museum (Natural History) periodically (1885–1870). He also visited Russia, the Azores, Madeira, the Canary Islands, Egypt, South Africa, India and Sri Lanka the latter two with Elwes (q.v.) whose sister Edith he married (1872–1875). After she died he re-married (1891) Alice (1868–1944), the daughter of Percy Chaplin. The Godman-Salvin Medal, a prestigious award of the British Ornithologists' Union, is named after them as they were both among the founders of the BOU. He moved to Mexico (1885) for the warmth. He also collected early pottery and his collection of over 600 pieces is in the British Museum. Five reptiles, three mammals, four birds and an amphibian are also named after him.

Goliath

Emperor sp. *Anax goliath* **Selys**, 1872
[Syn. of *Anax tristis* Hagen, 1867]
Emperor sp. *Anax goliathus* **Fraser**, 1922
[Syn. of *Anax guttatus* (Burmeister, 1839)]

Goliath of Gath is a biblical character decribed as a giant Philistine warrior who, in one version, is defeated by David the future King of the Israelites. These large sized aeshnids were named after the biblical warrior giant because of their relative size.

Gomes

Dragonlet sp. *Erythrodiplax gomesi* **Santos**, 1946

Dr Alcides Lourenço Gomes (1916–1991) was a technician at the Experimental Biology and Pisciculture Station at Pirassununga, São Paulo, Brazil. Santos himself worked there (1938–1944), which is where his interest in dragonflies started. Gomes wrote some papers, mainly on fish, such as: *A small collection of fishes from Rio Grande do Sul, Brazil* (1947).

González

Tamaulipan Clubtail *Gomphus gonzalezi* **Dunkle**, 1992

Dr Enrique González Soriano (b.1951) is a Mexican odontologist. The Faculty of Sciences, National Autonomous University of Mexico, awarded his BSc (1977) and MSc (1982). He is a Professor at the Departmento de Zoologia, Instituto de Biologia, Universidad Nacional Autónoma Mexico, Distrito Federal, Mexico, and the Curator of the National Odonata Collection and has conducted entomological research throughout his career (1979–present). He has written around seventy scientific papers, book chapters and books. He has described nearly 20 new Odonata species. His publications include: *A synopsis of the genus Amphipteryx Selys 1853 (Odonata: Amphipterygidae)* (2010), *A biodiversity hotspot for odonates in Mexico: the Huasteca Potosina, San Luis Potosí* (2011) and *Biodiversidad de Odonata en México* (2014). His research focus is the taxonomy and reproductive biology of neotropical Odonata. He has edited three books. The eponymous species was named after him because he discovered a small isolated population in northeastern Mexico in the state of San Luis Potosi and part of the type series was collected by him.

Gordon

Western Red Hunter *Austroepigomphus gordoni* **Watson**, 1962
[Orig. *Austrogomphus gordoni*]

Stewart Gordon is a member of a family long associated with Millstream and Kangiangi Stations, Pilbara, Western Australia.

Gould

Southern Vicetail *Hemigomphus gouldii* **Selys**, 1854
[Orig. *Austrogomphus gouldii*]

John Gould (1804–1881) was the son of a gardener at Windsor Castle who became an illustrious British ornithologist, artist and taxidermist. Gould was born in Dorset, England,

and became acknowledged around the world as 'The Bird Man'. He was employed as a taxidermist by the newly formed Zoological Society of London and travelled widely in Europe, Asia, and Australia. He was arguably the greatest and certainly the most prolific publisher and original author of ornithological works in the world. In excess of 46 volumes of reference work were produced by him in colour (1830–1881). He published 41 works on birds, with 2,999 remarkably accurate illustrations by a team of artists, including his wife. His first book, on Himalayan birds, was based on skins shipped to London, but later he travelled to see birds in their natural habitats. Gould and his wife, Elizabeth, arrived on board 'Parsee' in Australia (1838) to spend 19 months studying and recording the natural history of the continent. By the time they left, Gould had not only recorded most of Australia's known birds, and collected information on nearly 200 new species, but he had also gathered data for a major contribution to the study of Australian mammals. His best-known works include: *The Birds of Europe*, *The Birds of Great Britain*, *The Birds of New Guinea* and *The Birds of Asia*. He also wrote monographs on the Odontophorinae, Trochilidae and Pittidae. Gould was commercially minded and pandered to Victorian England's fascination with the exotic, especially hummingbirds, with which he is particularly associated. His superb paintings and prints of these and other birds were greatly sought after, so much so that he had trouble keeping up with the demand. The large corpus of unpublished and unfinished work, which he left at his death, supports this. Five mammals, two reptiles, an amphibian and no less than forty-six birds are also named after him.

Graells

Spanish Bluetail *Ischnura graellsii* **Rambur**, 1842
[Orig. *Agrion graellsii*]

Professor Dr Mariano de la Paz Graells y de la Aguera (1809–1898) of Madrid was a botanist, entomologist and malacologist. He qualified in medicine and natural sciences at the University of Barcelona, where he was firstly an associate and later full Professor of Physics and Chemistry. There was an epidemic in Barcelona (1835) which he was prominent in combating. He moved to Madrid (1837) as Professor of Zoology at the Museum of Natural Sciences and Director of the Botanical Gardens. He joined a scientific expedition to the Pacific (1845), establishing the Spanish Natural History Society facilities to allow for the acclimatization of tropical plants. The *Phylloxera* outbreak virtually wiped out European vineyards in the latter part of the 19th century, and Graells, as a senior man in the Council of Agriculture, had to deal with its effect in Spain. He was a founding member of the Spanish Academy of Exact Sciences and was honoured in Spain and by several foreign governments. He wrote: *Subfamilia felina fauna mastodologica* (1897). A mammal and a bird are also named after him.

Graeser

Coppery Emerald *Somatochlora graeseri* **Selys**, 1887

Ludwig (Louis) Carl Friedrich Graeser (1840–1913) was a German entomologist and collector. He was assistant preparator at the Museum of Natural History, Hamburg, Germany, which houses much of his collection, the rest being held by Zoological Museum of the Zoological Institute of the Russian Academy of Sciences. He worked on biogeography, particularly of the butterfly fauna of Central Asia and the Amur and Vladivostok regions, and he collected insects, including the type, along a number of rivers in Eastern Amur. He wrote: *Beiträge zur Kenntnis der Lepidopteren-Fauna des Amurlandes* (1889). He described seven new species. Selys (q.v.) described him as: *"...an indefatigable and intelligent traveller".*

Graf

Ocellated Darner *Boyeria grafiana* **EB Williamson**, 1907

J L Graf was a Canadian naturalist collecting between the turn of the century and WW1. He collected the holotype in the Ohiopyle State Park, Pennsylvania (1906) and often collected there (1901–1906). He also collected on the Isle of Pines (1910) – the collection is held by the Carnegie Museum. Williamson wrote: *"This species is very properly named for J. L. Graf, a devoted and careful, though withal, silent student of nature, who first detected a difference in the Boyerias at Ohio Pyle. In the autumn of 1905, among a box of specimens he sent me, he indicated on the envelope of a Boyeria, 'colors peculiar' ...Hawking after sunset."*

Graffe

Yellow-veined Widow ssp. *Palpopleura jucunda graffei* **Martin**, 1912
[JS *Palpopleura jucunda* (Rambur, 1842)]

Mr Graffe was an administrator in Sikasso (Mali) at the Commission of Native Affairs, where he collected the type and sent them to the author to study.

Graham

Demoiselle sp. *Agrion grahami* **Needham**, 1930
[JS *Atrocalopteryx oberthueri* (McLachlan, 1894)]

The Revd Dr David Crockett Graham (1884–1961) was a Baptist missionary in Szechuan, China (1911–1918). He returned to the USA for postgraduate study and started to correspond with the Smithsonian with a view to collecting natural history specimens for them. He was back in Szechuan (1920–1926), during which period he collected the type (1922). The University of Chicago awarded his doctorate (1927). He was again in Szechuan (1928–1930). He became a Fellow of the Royal Geographical Society, London (1929) and during an expedition to Moupin acquired a Giant Panda skin for the Smithsonian. He taught anthropology and archaeology at West China Union University and was Curator, West China Union University Museum of Archaeology, Art and Ethnology, Chengtu, Szechuan province (1932–1948). After retiring (1948) he lived in Colorado (1949–1961). Many insects, a reptile and an amphibian are also named after him.

Grant

Socotra Bluet *Azuragrion granti* **McLachlan**, 1903
[Orig. *Ischnura granti*]

William Robert Ogilvie-Grant (1863–1924) was a Scottish ornithologist who studied zoology and anatomy at Edinburgh. He was Curator of Birds at the British Museum (Natural History) (1909–1918), having started work there as an assistant aged nineteen (1882). He undertook an expedition with H O Forbes to Socotra and Abd-el-Kuri Islands (1898–1899), during which the bluet was collected. He also succeeded Bowdler Sharpe as editor of the 'Bulletin of the BOC' (1904–1914). He enlisted with the First Battalion of the County of London Regiment at the beginning of WW1 and suffered a stroke while helping to build fortifications near London (1916). He wrote: *A Hand-book to the Game Birds* (1895) and is remembered in the names of an amphibian, a reptile, two mammals and twenty-one birds. McLachlan described the species in: *The dragon flies of Sokotra. Amphibiotica. in Forbes H.O. [ed] The natural history of Socotra and Abd-el-Kuri. Being the report upon the results of the conjoint expedition to these islands in 1898–9, by Mr. W.R. Ogilvie-Grant* (1903).

Graslin

Pronged Clubtail *Gomphus graslinii* **Rambur**, 1842

Adolphe Hercule de Graslin (1802–1882) was a French nobleman, botanist and entomologist specializing in Lepidoptera who was a founding member of the Société Entomologique de France and he published on caterpillars. He collected the holotype in Bercé, Loire, France. His collection was acquired by Charles Oberthür who at the time had the second largest entomology collection in the world!

Gravely

Shadowdamsel sp. *Protosticta gravelyi* **Laidlaw**, 1915
[A. Pied Reedtail]
Dusky Lilly-squatter *Argiocnemis gravelyi* **Fraser**, 1919
[JS *Paracercion calamorum* Ris, 1916]

Dr Frederick Henry Gravely (1885–1965) was a British arachnologist, botanist, zoologist and archaeology aficionado. He studied at the University of Manchester and went on to become Assistant Superintendent of the Indian Museum, Calcutta. Here he did pioneering classification work on the *Passalidae*. He collected the holotypes of both species in India; the threadtail (1914) and the lilly-squatter (1919). He went on to become Superintendent of the Government Museum of Madras (1920) where he revived the Bulletin of the museum, and embarked on the scientific preservation, study and interpretation of the museum's collections as well as enlarging them, especially the invertebrates. His insect and spider collection is in the Indian Museum, Calcutta. He was a prolific writer and among his many works are: *Notes on the habits of Indian insects, myriapods and arachnids* (1915) and *Notes on Hindu Images* (posthumously 1977).

Gray

Gray's Dragonfly *Procordulia grayi* **Selys**, 1871
[Orig. *Epitheca grayi*]

John Edward Gray (1800–1875) was a British ornithologist and entomologist. He started at the British Museum (1824) with a temporary appointment at 15 shillings a day, but rose to be Curator of Birds (1840–1874) and then Head of the Department of Zoology. Gray published descriptions of a large number of animal species, including many Australian reptiles and mammals and was the leading authority on many reptiles. He was also an ardent philatelist and claimed that he was the world's first stamp collector. He worked at the museum with his brother George Robert Gray (1808–1872) and together they published a *Catalogue of* the *Mammalia and Birds of New Guinea in the Collection of the British Museum* (1859). He wrote: *Gleanings from the Menagerie and Aviary at Knowsley Hall* (1846–1850), which was illustrated by Edward Lear. Gray suffered a severe stroke (1869) paralysing his right side, including his writing hand. Yet he continued to publish to the end of his life by dictating to his wife, Maria Emma, who had always worked with him as an artist and occasional co-author. Twenty-three reptiles, nine mammals, two birds (possibly many more) and an amphibian are also named after him. Selys makes no mention of Gray in his description, except in the name itself. However, they had corresponded for many years. Charles Morren wrote to Gray (1839) introducing Selys to him. According to Selys' diaries he had met Gray during his visits to London (July 1851, June 1871 & July 1873).

Green, E E

Blurry Forestdamsel *Platysticta greeni* Kirby, 1891
[JS *Platysticta maculata* Hagen, 1860]
Green's Gem *Libellago greeni* **Laidlaw**, 1924
[Orig. *Micromerus greeni*]

Edward Ernest Green (1861–1949) was a mycologist and entomologist who was born into an English family of tea and coffee planters in Ceylon (Sri Lanka). He managed his family's plantations there (1880), combining agriculture with being the Government Entomologist at the Royal Botanic Gardens, Peradeniya. He wrote the 5-volume: *The Coccidae of Ceylon* (1896–1922) and more than 200 papers, mainly on entomology. He presented many specimens to the British Museum, where two of his collections are housed today. He submitted an extensive collection of Odonata (collected 1900-1911) for Laidlaw (q.v.) to study. An amphibian is also named after him.

Green, R A G

Clasper-tailed Sprite *Pseudagrion greeni* **Pinhey**, 1961

Mr R A G Green collected extensively around the city of Ndola, Zambia (1959). He was honoured because he had *"...collected other interesting Odonata for the author"*.

Greenway

Tanzania Jewel *Platycypha greenwayi* **Pinhey**, 1950
[JS *Platycypha auripes* (Förster, 1906)]

Dr Percy James Greenway (1897–1980) was a South African botanist and collector at the East African Agricultural Research Station (1927–1950) and East African Herbarium (1950–1958). The University of Witwatersrand awarded his doctorate (1954). He collected in Kenya, Zimbabwe and Zambia. He wrote: *Dyeing and tanning plants in East Africa* (1941). A bird is also named after him as are at least two plants. He died in Nairobi.

Gregory

Goldenring sp. *Anotogaster gregoryi* **Fraser**, 1924
Demoiselle sp. *Mnais gregoryi* Fraser, 1924

Professor Dr John Walter Gregory (1864–1932) was a geologist, geographer and explorer. After his education at Stepney Grammar School, at 15 he became a clerk at wool sales in the City of London. His growing interest in the natural sciences led him to attend evening classes at the London Mechanics' Institute (Birkbeck College). He matriculated (1886) and gained his BSc (1891) and DSc (1893). Meanwhile (1887) he was appointed assistant in the geological department of the BMNH. His first journey outside Europe (1891) was to study the geological evolution of the Rocky Mountains and the Great Basin of western North America. He was seconded as naturalist to a large expedition to British East Africa (1892) and when this collapsed he set out on his own with a party of forty Africans. In five months, he completed scientific observations in fields ranging from structural geology and physical geography to anthropology, and from mountaineering and glacial geology to the malarial parasites. His major success was the study of the 'Great Rift Valley' summarized in two books: *The Great Rift Valley* (1896) and *The Rift Valleys and Geology of East Africa* (1921). Other major scientific expeditions included the first crossing of Spitsbergen (1896); in the West Indies (1899), Libya (1908), southern Angola (1912); and a 1500-mile walk with his son through Burma to southwestern China and Chinese Tibet (1922), during which the types were collected. He became Professor of Geology and Mineralogy at the University of Melbourne (1899–1904) and was also Director of the Geological Survey of Victoria (1901–1902) and Director of the British National Antarctic Expedition (1901). Later he became Professor of Geology at Glasgow University until retirement (1904–1929). In all he published twenty books and over 300 papers and received many honours and awards. He joined an expedition to Peru, aged 68, and was drowned when his canoe overturned in the Urubama River.

Greig

Variable Sprite *Pseudagrion greigi* **Pinhey**, 1961
[JS *Pseudagrion sjoestedti* Förster, 1906]

Dr Herbert Wallace Greig (1922–2002) was a physician at the Presbyterian Mission Batouri Hospital in Cameroon, which he had just opened. Pinhey made an overland expedition with Land Rover from Rhodesia through Congo to Nigeria (1958). This species was among the many novelties found during this visit. On his return Pinhey became ill due to "...*insidious germs of various kind*" and had to spend a few weeks in the hospitable care of the Grout family (q.v.) and Dr Greig at the American Presbyterian mission near Batouri in Cameroon.

Gressitt

Phantomhawker sp. *Planaeschna gressitti* **Karube**, 2002

Dr Judson Linsley 'Lin' Gressitt (1914–1982) was an American botanist and entomologist at the Bishop Museum in Hawaii. He grew up in Japan and experienced the great earthquake and fire (1923) and almost died (1925) of pneumonia and typhus. He was taken to California to recuperate, where his interest in natural history was sparked at Summer Camps that he attended with his cousin. He graduated from the American School, Tokyo (1932) and taught English for a year in a Japanese High School. At just 17 he undertook his first expedition, three months on Formosa walking several hundred miles and climbing the two highest mountains. He returned with major entomological and herpetological collections. He began to study at Stanford University but interrupted this for another long trip to Formosa, collecting plants as well as insects and herpetofauna, bringing back 50,000 specimens including several new insects. He transferred (1935) to the University of California, which awarded his BSc (1938) and MSc (1939). By the age of 21 he was already recognised as an internationally acknowledged coleopterist. He continued to collect in China and became (1939) Instructor at Lingnan University and, concurrently, Acting Director of the Lingnan Natural History Museum. At around this time (1941) he switched his focus from herpetology to entomology, realising he could not do both disciplines justice. He still collected in parts of China not occupied by Japanese forces. However, he, his wife and new baby were interred (1941–1943) then repatriated to the USA. Under a fellowship, he worked at the University of California, Berkeley (1944–1945) completing his PhD (1945). He served as a naval officer (1945–1946) in Guam, Philippines, Ryukyu Island and Japan. He returned to Lingnan University (1946) as Assistant Professor, rising to Associate Curator and Associate Professor. Under increasing pressure from the new communist government his wife and three children returned to the USA (1949) and eventually, after house arrest (1950), he followed (1951). After various posts, he was appointed (1953) Entomologist at the Bishop Museum, Hawaii and was in charge of the entomology collection there (1955–1972) when he relinquished the post to spend more time collecting, specifically in New Guinea. One of his significant achievements was using specially equipped aircraft to collect insects as high as the planes could fly; he demonstrated that the winds and jet streams over the Pacific carried an enormous number of insects. The J L Gressitt Rare Plant Sanctuary at West Maui was founded in his memory (1984). He initiated the museum's ongoing faunistic surveys of New Guinea (1984) and he was later seconded as Director of the Wau Ecology Institute in Papua New Guinea, which began (1961) as a field station of the Bishop Museum, later becoming a fully independent institution. He remained its Director until his death (he and his wife Margaret died when a commercial flight crashed in south China on April 26th 1982). He published more than 300 works (from 1934) including many large monographs. The institute posthumously published a book he had co-authored: *Handbook of Common New Guinea Beetles* (1985). A mammal is also named after him. (Also see **Linsley**)

Grey

Kaleidoscope Jewel *Chlorocypha greyi* **Pinhey**, 1961

[JS *Africocypha lacuselephantum* (Karsch, 1899)]

Terence Coffin-Grey (b.1926) is a taxidermist. He spent many years as Chief Technical Officer at the National Museum Bulawayo and was responsible for their displays. He accompanied Pinhey in an overland expedition via Land Rover from Rhodesia through the Congo to Nigeria and back to Rhodesia (1958). In his dedication Pinhey wrote that Coffin-Grey: "...*took a keen interest in helping in the collecting of dragonflies, particularly Chlorocyphids.*" Coffin-Grey told me about the species collection which is well worth sharing: "*I accompanied Elliot Pinhey to Nigeria and I was impressed by the size of the dragonflies halfway up Gorilla Mountain, which was on our way to the village of Ajassor on the Cross River. I collected an enormous species which Elliot could not reach with all the extensions to his net. I asked him to watch the beast carefully whilst I shot it down with my 12-gauge shotgun. He was aghast at the idea, but it was a last resort so I pressed on and shot with dust shot, choke barrel. Elliot, who had closed his eyes, glanced at the very high dry twig from which it had been hunting. There was no sign of it. Then I heard it fall on the leaves below. Pinhey could not believe it - no damage except for tiny holes in the wings. Our stay at the village of*

Adjassor was not very productive. The locals who Elliot sent out to collect for him brought back all sorts of mangled insects, no doubt edible but not worth papering. Elliot was struck down by a mysterious malady, which prevented him from further collecting. We were housed in the village Juju house, decorated with feather-bedecked wild pig skulls, complete with a cement Juju which had watch glass eyes. Elliot used it as a handy clotheshorse. Getting Elliot home was another story..."

Griffini

Firetail sp. *Telebasis griffinii* **Martin**, 1896
[Orig. *Erithragrion griffinii*]

Doctor Achille Griffini (1870–1932) was an Italian zoologist, ichthyologist and entomologist. The Royal University of Turin awarded his honours degree (1893). He then taught in private schools (1891–1899), but was also an assistant at the Royal University of Turin Zoological & Comparative Anatomy Museum (1895). He taught in various technical colleges: Royal College of Bologna (1899–1902), Udine (1902–1903), L'Aquila (1904–1905), Genoa (1906–1909) and again in Bologna (1910–1912). He was then at Royal High School Berchet, Milan (1912–1915) and while there took the post of Curator of Entomological Collections (1913–1915) at Milan Natural History Museum, but relinquished it when his teaching duties increased. He was in military service during WW1 (1915–1917) in the 1st Engineers Regiment and was seriously injured leading to paralysis of his right side. He resumed teaching in Bologna then moved to Bresica where he died. He was sent many specimens to describe. He wrote a book about the principal beetles of Italy and a few others from Europe: *Il Libro dei Coleotteri* (1896) and another on fish, published posthumously: *Il Libro dei Pesci* (1944) as well as more than 200 scientific papers and chapters (1896–1920). Rene Martin gave no details at all in his etymology just saying the name honoured Dr Griffini.

Grigoriev

Four-spotted Chaser *Libellula quadrimaculata grigorievi* **Schmidt**, 1961
[JS *Libellula quadrimaculata* Linnaeus, 1758, A Four-spotted Skimmer]

Boris Grigoriev (d.1913) was a Russian entomologist from Saint Petersburg, who was active around 1900. Schmidt named this subspecies from Afghanistan after Grigoriev, who had (1905) recorded a similar colour-form from Kazakhstan.

Grillot

Orange-striped Sprite *Pseudagrion grilloti* **Legrand**, 1987

Dr Jean-Pierre Grillot was Scientific Director of a Muséum National d'Histoire Naturelle project to study and preserve the biodiversity of an area of rain forest; the project was named for the area that became (1988) Dimonika National Park. His doctoral thesis was: *Les organes neurohemaux metamériques des coleoptères* (1969) and he co-wrote: *Contribution à la faune de la république populaire du Congo* (1974) and *Moustiques de l'île M'Bamou (Congo)* (1974). The only known location for the species is the Mayombe Hills in the NP, which is in western Congo.

Grinbergs

Pond Damselfly sp. *Antiagrion grinbergsi* **Jurzitza**, 1974

Professor Dr Janis M Grinbergs was (1957) founder Professor of Bacteriology at the Institute of Microbiology, at the Universidad Austral de Chile. Among his papers and books is: *Investigations about the abundance and function of symbiotic microorganisms case of wood-eating insects Chile* (1962). The 21st Chilean Congress of Microbiology (1999) was held in his honour and called "XXI Chilean Congress of Microbiology Dr. Janis M. Grinbergs". Jurzitza visited Chile (1974) at the invitation of the Universidad Austral de Chile, where he met Grinbergs, later writing: *Ein Beitrag zur Faunistik und Biologie der Odonaten von Chile*. In this paper Jurzitza thanked Professor Grinbergs and his family for their hospitality and generous help in many ways.

Grossi, J E

Threadtail sp. *Forcepsioneura grossiorum* **Machado**, 2005

José Everardo Grossi (b.1945) is an amateur Brazilian entomologist as well as a clinical pathologist (his son, Paschoal Grossi below, is an entomologist). The etymology reads: "*This species is dedicated to my friends the entomologists Everardo Grossi and Paschoal Grossi, who collected the specimens now described*". They have written a number of papers together such as: *A new species of Amblyodus Westwood, 1878 (Coleoptera, Melolonthidae, Dynastinae) from South America* (2011). Everardo, who lives on a Finca surrounded by Atlantic Forest, is particularly interested in the *Dynastinae* (a scarab beetle group) and has described a number of new species including a subspecies he named after his son *Dynastes hercules paschoali*.

Grossi, PC

Threadtail sp. *Forcepsioneura grossiorum* **Machado**, 2005

Paschoal Coelho Grossi (b.1983) is an entomologist most interested in lucanids, which he studied as a postgraduate student at the University of Parana in central Brazil where he lives (his father, above, is an amateur entomologist). They have written a number of papers together such as: *A new species of Amblyodus Westwood, 1878 (Coleoptera, Melolonthidae, Dynastinae) from South America* (2011). He is developing a key for the South America lucanid genera and revising work on the tribe Sclerostomini. The etymology reads: "*This species is dedicated to my friends the entomologists Everardo Grossi and Paschoal Grossi, who collected the specimens now described*".

Grout

Dark Dropwing *Trithemis grouti* **Pinhey**, 1961
[A Black Dropwing]

Rev Richard L Grout (1922–2013) and his wife Donna Grout née Linton (1925–2004) worked at the American Presbyterian mission near Batouri in Cameroon. He graduated in mechanical engineering from the University of Washington and spent a semester at UC Berkeley in the Marine ROTC during WW2. After the war, he married (1945) and worked as a draftsman with PACCAR. He attended Fuller Theological and San Francisco

Theological Seminaries, receiving a Master of Divinity degree, moved to Cameroun, initially as an instructor in a vocational school and later as a minister with the United Presbyterian Church. He returned to the United States (1963) and accepted a position as pastor at Cottage Lake Presbyterian Church in Woodinville, Washington (1965–1985) until retiring to Hansville, Washington. Pinhey wrote: "*This insect is named after Mr and Mrs Grout, who so kindly extended their hospitality when the author was incapacitated during the expedition.*" Pinhey made an overland expedition with Land Rover from Rhodesia through Congo to Nigeria (1958), accompanied by Terrence Coffin-Grey (q.v.). On his return he become ill due to '*...insidious germs of various kinds*' and had to spend a few weeks in the hospitable care of the Grouts and medical missionary Dr H Wallace Greig at the mission.

Grubauer

Junglehawker sp. *Indaeschna grubaueri* **Förster**, 1904
[Orig. *Amphiaeschna grubaueri*]

Albert Grubauer (1869–1960) was an ethnologist who collected in Malaya, Pahang (1901), including the type, and Indonesia (1911–1913) at the beginning of the twentieth century. Förster received, and studied his material, at UMMZ, which included Odonata collected in Malaya. He had private means to fund his travels and other pastimes such as philately. He wrote: *Unter Kopfjägern in Central-Celebes; Ethnologische Streifzüge in Südost- und Central-Celebes* (1913).

Guarani

Pond Damselfly sp. *Tuberculobasis guarani* **Machado**, 2009

The Guarani are an indigenous people who still inhabit various areas of the State of São Paulo.

Guarauno

Sylph sp. *Macrothemis guarauno* **Racenis**, 1957

The Guarauno are an indigenous people who inhabit the area in Venezuela where the species was found.

Guarella

Threadtail sp. *Peristicta guarellae* Anjos-Santos & Pessacg, 2013

Nely Coca Guarella (b.1932) is an Argentine amateur entomologist who originally studied anthropology but did not finish her degree. She worked closely with the Hymenoptera specialist Dr Luis De Santis at his laboratory (1960s) at the Museo de Las Plata, Buenos Aires.

Guerin

Yellow-Striped Hunter *Austrogomphus guerini* **Rambur**, 1842
[Orig. *Gomphus guerini*]
Satinwing sp. *Euphaea guerini* Rambur, 1842

Félix Edouard Guérin-Méneville née Guérin (1799–1874) was a French entomologist. He is most well-known for his illustrated work: *Iconographie du Règne Animal de G. Cuvier 1829-44*, a complement to Cuvier's (q.v.) and Latreille's work, which lacked illustrations; and also for introducing silkworm breeding to France. There is a Felix-Edouard Guérin-Méneville Collection of Crustacea at the Academy of Natural Sciences, Philadelphia. He was an all-rounder, having written scientific papers on plants, insects and other subjects. He was elected President of the Société entomologique de France (1846) and was also editor of *Dictionnaire Pittoresque d'Histoire Naturelle et Des Phénomènes de la Nature* (1855). A reptile and two birds are also named after him as are a number of other taxa.

Guichard

Ethiopian Sprite *Pseudagrion guichardi* **Kimmins**, 1958

Kenneth Mackinnon Guichard (1914–2002) was a British entomologist and art connoisseur. He worked on the Desert Locust Survey, Oman (1949–1950) and in the Sahara (1952). He was on Socotra Island, Yemen, as entomologist attached to the Middle East Command's expedition (1967). Many of his specimens were stored in the BMNH for c.20 years before examination. He made his living through his flair for spotting good paintings and etchings, buying cheap and selling at a profit. He collected the type in Ethiopia (1948). A reptile and a bird are also named after him.

Guillermo

Pond Damselfly sp. *Metaleptobasis guillermoi* **von Ellenrieder**, 2013

Guillermo von Ellenrieder (b.1940) is the father of the author Natalia von Ellenrieder (b.1972). He gained his PhD in physical chemistry at Universitad Nacional de la Plata and was a postdoctoral student (1967–1970) at the Max-Planck-Institute in Göttingen, Germany. There he met his future wife Gabriele Müller (q.v.). He was appointed professor at the Universidad Nacional de Salta, ultimately becoming vice-head of the 'Exact' Sciences Department there. He was also a researcher for CONICET (The National Scientific and Technological Research Council) with research on enzymes applied to industrial use. (Also see **Natalia** & **Gabriele**)

Gundlach

Eastern Pondhawk *Mesothemis gundlachii* **Scudder**, 1866
[JS *Erythemis simplicicollis* (Say, 1840), A. Common Pondhawk]

Juan Cristóbal Gundlach (christened Johannes Christoff) (1810–1896) was a German-born naturalised Cuban ornithologist and collector. He began to learn the art of dissection and taxidermy by watching his older brother who was a zoology student. An event that nearly cost him his life ironically allowed him to follow his chosen profession. During a hunting accident, he discharged a small gun so close to his nose that he lost his sense of smell. After that he could calmly dissect, macerate and clean skeletons without difficulty. He was a curator at the University of Marburg and later at the Senckenberg Museum of Frankfurt. He took part in a collecting expedition to Cuba (1830), and stayed on and collected there and in Puerto Rico. He wrote the first major work on the island's birds: *Ornitología Cubana*

(1876). He met the American explorer Charles Wright and explored the then virgin forest of what is now Alejandro de Humboldt National Park. He was zealous and single-minded, tending to keep what he collected, and describing it for science himself. Eight birds, two amphibians, two mammals and a reptile are also named after him.

Günther

Elf sp. *Risiophlebia guentheri* **Kosterin**, 2015

Dr André Günther (b.1962) is a German zoologist, ecologist and odonatologist, presently a lecturer (since 2014) at the Institute of Biosciences at Technische Universität Bergakademie Freiberg. He gained his engineering degree (1986) and later his doctorate in biology (2008) from the same institution. He has worked as a member of scientific staff at the Bergakademie Freiberg (1986–1990, 2001–2014) and was Head of the Zoological Department at the Freiberg Museum of Natural History (1990–1993). He was one of the founding members (1993) of the Freiberg Institute of Nature Conservation where he worked as scientific staff member (1993–2010) and as Head of Institute (from 1996) and also since 2014. Presently his studies focus on the reproductive behaviour of tropical Chlorocyphidae and Calopterygidae; the distribution and conservation status of dragonflies in Europe, Asia and Africa; and the spatial and temporal dynamics of biodiversity in post-mining landscapes in Germany. He has travelled extensively (from 1991) in South-east Asia and Africa studying, photographing and videoing, particularly Chlorocyphidae. He has published more than 120 papers, posters and symposium abstracts, 86 of them on Odonata. These include *Reproductive behaviour of Neurobasis kaupi (Odonata: Calopterygidae)* (2006), his thesis: *Comparative study of the reproductive behaviour of Southeast-Asian Chlorocy-phidae and Calopterygidae (Odonata: Zygoptera)* (2008), *Reproductive behaviour and the system of signalling in Neurobasis chinensis (Odonata, Calopterygidae) – a kinematic analysis* (2014) and *Signalling with clear wings during territorial behaviour and courtship of Chlorocypha cancellata (Odonata, Chlorocyphidae)* (2015). He is a co-author of the species *Africocypha varicolor* from Gabon.

Günther Peters

Skimmer sp. *Diplacina guentherpetersi* Villanueva, 2012
Phantomhawker sp. *Planaeschna guentherpetersi* **Sasamoto, Do** & Van, 2013

Professor Dr Günther Peters (b.1932) was a German zoologist. He received his diploma from the State University of Leningrad (1952) and completed his PhD (1960) and his habilitation (1972) at Humboldt University, Berlin. He was a research assistant, then custodian of their herpetological collection (1962–1984) and then director. He subsequently became Director of the Institute of Zoology and later Professor of Zoology (1975) at the University. He is one of the most important living German herpetologists and odonatologists. His key areas of research and teaching were the paleozoology of vertebrates, comparative morphology, and zoogeography, phylogeny and systematics. He was Deputy Director of the Museum of Natural History until he became Director (1990–1996). He became interested in odonata (1960s) and continued that interest, publishing well into retirement. He travelled widely in Europe, Africa, Asia and the Americas and latterly Australia (2004), collecting all the while. He has published at least 144 papers and books including thirty-two works on dragonflies, with special emphasis on *Aeshnidae*. He wrote the book: *Die Edenlibellen Europas Aeshnidae* (1987). He has named four new aeshnid genera and described two new dragonfly species, including the aeshnid *Boyeria cretensis*. The etymology reads: *"The species name is dedicated to Dr. Günther Peters, a noted specialist in Aeshnidae, in celebration of his 80th birthday."* (Also see **Peters**)

Gupta

Skimmer sp. *Orthetrum guptai* Baijal, 1955
[JS *Orthetrum cancellatum* (Linnaeus, 1758)]

Virendra Kumar Gupta (b.1932) is an entomologist who has written widely, especially on parasitic wasps such as: *Ichneumonologia Orientalis: or, A monographic study of oriental Ichneumonidae* (1972) and *A bibliography of the Ichneumonidae* with Santosh Gupta (1991). He took the lead in this group during the survey which collected the type as evidenced by: *On a collection of Ichneumonidae (Hymenoptera) (Entomological survey of the Himalayas)* (1955). The description appears in the report on the *Entomological Survey of the Himalayas. Part XIV – Notes on some insects collected by the Second entomological expedition to the northwest Himalayas (1955), with the description of three new species of Odonata* Singh, S, Baijal, H N, Gupta, V K and Mathew, K (1955) in which both the author and the honoured entomologist took part. However, only Baijal was given as the author of the new name.

Gurney

Pond Damselfly sp. *Papuagrion gurneyi* **Lieftinck**, 1949

Ashley Buell Gurney (1911–1988) was among those who collected insects when stationed in Guadalcanal as an army Captain during WW2 where he took the type (1944). He began his career working as a field agent (1935–1955) for the Bureau of Entomology and Plant Quarantine, eventually being appointed Entomologist (1955–1975) until retirement. He was a long-time member and served as Executive Secretary of the Entomological Society of America (1954). His primary research interest was the study of orthopteroids, chiefly grasshoppers and cockroaches, as evidenced by one of his early papers *Western Massachusetts Orthoptera, Part 1, Preliminary List of the Acrididae* (1935). His collected papers including correspondence with many leading entomologists are in the Smithsonian. Lieftinck described the damselfly from a specimen collected by Gurney on Bougainville Island (1944).

Guy

Swamp Tigertail *Gomphidia guyi* (**Pinhey**, 1967)
[JS *Ictinogomphus dundoensis* Pinhey, 1961]

Graham L Guy worked at and collected for the National Museum Bulawayo, Rhodesia. He regularly wrote articles for 'Rhodesiana' such as: *Notes on Some Historic baobabs* (1967) and *David Livingstone: Tourist to Rhodesia* (1969).

Gyalsey

Gyalsey Emerald Spreadwing *Megalestes gyalsey* Gyeltshen, **Kalkman** & **Orr**, 2017

HRH Crown Prince of Bhutan, The Gyalsey, Jigme Namgyel Wangchuck (b.2016) is the first child and the heir apparent of Druk Gyalpo (Dragon King) (2006, crowned 2008) Jigme Khesar Namgyel Wangchuk (b.1980) and his wife Gyaltsuen (Queen) Jetsun Pema (b.1990). In honour of the prince's birth (Gyalsey means Dragon Prince) 108,000 trees were planted in Bhutan by thousands of volunteers. The first author, Thinley Gyeltshen is Bhutanese and is at the School of Life Sciences, Sherubtse College, Bhutan. The etymology says the species is "...*named in honour of His Royal Highness Crown Prince of Bhutan, The Gyalsey, Jigme Namgyel Wangchuck, on the occasion of his first birthday.*"

H

Haarup

Blue-eyed Darner sp. *Rhionaeschna haarupi* **Ris**, 1908

[Orig. *Aeshna haarupi*]

Anders Christian Jensen-Haarup (1863–1934) was a Danish naturalist. He became a teacher (1885) but is known for his entomological collections. He went to Mendoza, Argentina (1904) where he collected and studied insects. He returned to Denmark (1905) to arrange his collection, sending many to other entomologists to identify. He returned to Argentina (1906) where he collected the type with Peter Jørgensen (q.v.). He published a series of popular accounts of his trip to Argentina describing the flora and fauna, but also wrote regular travel notes about all of his encounters. His 15 popular travel accounts were published in the Danish natural history magazine 'Flora or fauna', and parts of his diaries were edited into books: *I Sydamerika: Skildringer fra det vestlige Argentina* (1906) and *Af min sydamerikanske Dagbog (1904–1907)* (1911). He also wrote descriptions of new insect species and other papers on entomology.

Haber

Dancer sp. *Argia haberi* **Garrison** & von **Ellenrieder**, 2017

William (Bill) Allen Haber (b. 1946) is an American biologist who was at the University of Massachusetts and who is an expert on the Odonata of Costa Rica. He went to Monteverde (1973) to conduct research for his doctoral thesis from the University of Minnesota which he completed after five years (1978), but continued living in Costa Rica. He became a research associate for the Missouri Botanical Garden, identifying and collecting hundreds of specimens for their Flora of Monteverde project (1985). Among his publications is: *A new species of Erythrodiplax breeding in bromeliads in Costa Rica (Odonata: Libellulidae)* (2015). He has created an 'electronic field guide' with an inventory of more than 138 species of dragonflies and damselflies recorded in the Monteverde region of Costa Rica; not all have formal descriptions or names.

Habermeier

Migrant Hawker *Aeschna coluberculus habermayeri* Götz, 1923

[JS *Aeshna mixta* Latreille, 1805]

F Habermeier was a pharmacist and amateur entomologist from Fürth, Germany. He published two papers on local dragonfly fauna in North Bayern (1928, 1943). He also provided some Chinese Odonata specimens to Erich Schmidt for study (published in 1931), which he had received from E. Suenson (q.v.).

Haertel

Threadtail sp. *Forcepsioneura haerteli* **Machado**, 2001

Dr Luiz Arnoldo Haertel (b.1966) is a Brazilian physician. He graduated from Santa Catarina Federal University (1988) with a degree in medicine and became a cardiology specialist at São Paulo University (1992). He was Professor of Cardiology at Blumenau University (1993–2001) as well as being a cardiologist (1993) and later Head of Clinical Information Technology (2001). He also became (2009) Clinical Director at the Santa Caterina Hospital in Blumenau. Machado wrote: "*Named for my friend Dr Luiz Arnoldo Haertel who provided the facilities for collecting this interesting species close to the yard of his house in Blumenau (Santa Catarina, Brazil)."*

Hagen

The Genus *Hagenius* **Selys**, 1854

The Subgenus *Hagenoides* Carle, 1986
The Subgenus *Pseudohagenius* Carle, 1986

Hagen's Bluet *Enallagma hageni* **Walsh**, 1863
[Orig. *Agrion hageni*]
Green Hooktail *Onychogomphus hagenii* Selys, 1871
[JS *Paragomphus genei* Selys, 1841]
Black Petaltail *Tanypteryx hageni* Selys, 1879
[Orig. *Tachopteryx hageni*]
Skimmer sp. *Hypothemis hageni* **Karsch**, 1889
Hudsonian Whiteface *Leucorrhinia hageni* **Calvert**, 1890
[JS *Leucorrhinia hudsonica* (Selys, 1850)]
Thornbush Dasher *Micrathyria hagenii* **Kirby**, 1890
Painted Sprite *Pseudagrion hageni* Karsch, 1893
Jewel sp. *Rhinocypha hageni* **Krüger**, 1898

Hermann August Hagen (1817–1893) was a German physician and entomologist who emigrated to the United States (1867) and became the first professor of entomology in the country and the first great authority of the New World Odonata. Hagen started (1836) his study of medicine at Königsberg University, where his father was a professor of political science and his grandfather had been a professor. He gained his MD degree in 1840 (with a thesis on dragonflies!). He continued his medical studies in Berlin, Paris and Vienna before returning to Königsberg (1843) and starting work as a private physician, then becoming assistant surgeon in a hospital (1846–1849). Encouraged by his father, Hagen's interest in insects started early and aged thirteen he started to collect dragonflies. His first paper (1839) was a brief list of the Odonata of East Prussia. He toured major entomological collections and libraries in Norway, Sweden, Denmark and Germany with zoology professor Martin Heinrich Rathke (1839). These visits facilitated the work on his doctoral thesis *Synonymia Libellularum Europaerum*, a synonymic list of European dragonflies (1840). This catalogue brought him into contact with Edmond de Selys Longchamps (q.v.) and they started to correspond (1841) and first met in Paris (1843). Later Hagen made a one week visit (1867) to work on Selys' collection. In 1867 he was invited to move to the United States to work at the

Museum of Comparative Zoology at Harvard University in Cambridge, Massachusetts. He became Curator of the entomological collections there and was soon nominated as Professor of Entomology (1870). Hagen co-authored Selys' major works *Revue des Odonates de Europe* (1850), *Monographie des Caloptérygines* (1854) and *Monographie des Gomphines* (1857), to which Hagen also provided illustrations. Additionally, he provided a large number of new species descriptions for Selys' synopses of various odonate families (1853–1878). Hagen's best known and most valued publication is the two-volume *Bibliotheca entomologica* (1861–1862), which is an accurate and complete list of all earlier entomological literature, a total of 18,130 publications. Other major works among his c.400 publications include *Monographie der Termiten* (1855–1860), *Synopsis of the Neuroptera of North America* (1861) and *Synopsis of the Odonata of America* (1875). He described c.340 new Odonata species and named numerous genera. He made an extensive expedition (1882) to the western part of the United States (Montana, California, Washington and Oregon), where very little insect collecting had been done. He suffered paralysis (1890) and had to spend his last years in a wheelchair.

Hahnel

Sylph sp. *Macrothemis hahneli* **Ris**, 1913

Paul Hahnel (1843–1887) was a German entomologist, one of the 19th century's few 'academic' collectors, mainly in Brazil (1879–1884 & 1885–1887), but also around the higher reaches of the Amazon River in Peru and in Venezuela where the type was taken. He collected with Michael Otto in Brazil (1887) and sadly they both contracted amoebic dysentery and died in May of the same year on the Madeira River.

Hale Carpenter

Forktail sp. *Hivaagrion halecarpenteri* **Mumford**, 1942
[Orig. *Bedfordia hale-carpenteri*]
Two-banded Cruiser *Macromia halei* Fraser, 1947
[JS *Phyllomacromia contumax* Selys, 1879]

Dr Geoffrey Douglas Hale Carpenter (1882–1953). (See **Carpenter**)

Haluco

Emerald sp. *Hemicordulia haluco* **Asahina**, 1940

Ms Haru Susuki née Esaki (1929–1986) was the eldest of the three daughters of Dr Teizo Esaki, whose Odonata material from Micronesia (collected 1936–1939) Asahina had studied. The letters 'r' and 'l' are often mixed in Japanese pronunciation of English and so mis-assigned, and the suffix "-co" is often added to the end of the female first name. So, in Asahina's writing the name Haru turned to Halu and the suffix 'co' was added to make Haluco. The other daughters and Esaki himself were credited with eponyms in the same paper. (See **Esaki**, **Erico** and **Lulico**)

Hamada

Green-eyed Skimmer sp. *Aeschnosoma hamadae* Fleck & Neiss, 2012

Dr Neusa Hamada (b.1961) is a Brazilian entomologist. She graduated in biological sciences from the Universidade Estadual Paulista Júlio de Mesquita Filho (1984), and was awarded her masters in biological sciences (entomology) by the National Institute for Amazonian Research (1989) and her PhD in entomology by Clemson University, USA (1997). She is currently a senior researcher at the National Institute of Amazonian Research, developer of the State Health Secretariat of Rio Grande do Sul. Her focus in zoology is on the taxonomy of aquatic insects, environmental impact assessment, especially of the Amazon, and the popularisation of science.

Hämäläinen

Fineliner sp. *Teinobasis hamalaineni* **Müller**, 1992
Damselfly sp. *Risiocnemis hamalaineni* Villanueva, 2009
Shadowdamsel sp. *Drepanosticta hamalaineni* Villanueva, Van Der Ploeg & Van Weerd, 2011

Skydragon ssp. *Chlorogomphus nasutus hamalaineni* **Karube**, 2013

Dr Matti Kalevi Hämäläinen (b.1947) is a Finnish entomologist. He began studying biology at the University of Helsinki (1967), gained a BSc (1970) and his master's degree (major in zoology) (1972) and a Phil. Dr (1977) with a thesis on the suitability of ladybeetles in the biological control of aphids in glasshouses. He studied 'pest management and stored products entomology' at two universities in Canada and at the Canada Department of Agriculture Research Station in Winnipeg (1974–1975). He then worked (1975–1995) in the field of applied entomology, including (since 1978) part time teaching, as Adjunct Professor (Docent) of Agricultural Zoology, at the University of Helsinki. He published (1970–1980s) c.40 research or extension papers on applied entomology. He has devoted himself full-time (since 1995) to the study of the biodiversity and taxonomy of dragonflies of the Oriental region, both from his position as Docent at the University of Helsinki and (since 2010) as an Honorary Research Associate of Naturalis Biodiversity Center (Leiden, the Netherlands). An integral part of his research programme has been the collection of extensive fresh material on c.40 expeditions to several countries in South-East and East Asia (since 1982). His c.200 publications on Odonata (1967-2017) include synopses of the dragonfly fauna of the Philippines (1997) and Thailand (1999) and descriptions of 86 new species and four new genera. During the last twenty years, his studies have focussed mainly on the superfamily Calopterygoidea, the Caloptera damselflies. He has co-authored several books, most recently: *The metalwing demoiselles (Neurobasis & Matronoides) of the Eastern tropics* (2007), *A photographic guide to the dragonflies of Singapore* (2010) and *Demoiselle damselflies – winged jewels of silvery streams* (2013). He also wrote: *Catalogue of individuals commemorated in the scientific names of extant dragonflies, including lists of all available eponymous species-group and genus-group names* (2015, revised 2016) and several articles on the history of odonatology. A Hymenoptera species is also named after him. (Also see **Matti** and **Anu-Mari**)

Hamon

Blue Slim *Aciagrion hamoni* **Fraser**, 1955
[JS *Aciagrion africanum* Martin, 1908]
Citril sp. *Ceriagrion hamoni* Fraser, 1955
Hamon's Sprite *Pseudagrion hamoni* Fraser, 1955
[A Swarthy Sprite]

Dr Jacques Pierre Jean Hamon (b.1926) is a French entomologist. He was Entomologist at the Office de la Recherche Scientifique et Technique Outre-Mer (ORSTOM), Chief of Entomology Laboratory, Centre Muraz, Bobo Dioulasso, Volta Republic (c.1962) and Director, Division of Vector Biology and Control, WHO (c.1980), going on to become Assistant Director General there (1982–1986). He wrote a number of papers particularly on eradicating the insect malarial vectors, such as the co-written: *La resistance aux insecticides chez les insectes d'importance medicale* (1961). He collected in French West Africa

Handschin

Wiretail sp. *Isosticta handschini* **Lieftinck**, 1933

Professor Dr Eduard Handschin (1894–1962) was a Swiss zoologist and entomologist. He studied in Basel and went on to teach there (1921) until retirement, becoming associate professor (1927) and full professor (1941). He also worked at the Natural History Museum there (1946) eventually becoming Director (1956–1959), during which time he expanded their entomological collection. He collected in Indonesia and Northern Australia, including the type in the Northern Territory (1932). Among his longer works is: *Praktische Einführung in die Morphologie der Insekten: ein Hilfsbuch für Lehrer, Studierende und Entomophile* (1928).

Hannyngton

Clubtail sp. *Megalogomphus hannyngtoni* **Fraser**, 1923
[Orig. *Heterogomphus hannyngtoni*]

Frank Hannyngton (1874–1919) was a civil servant in India and an amateur entomologist. After studying at Trinity College, Dublin, he went to Wren's (the college Wren & Gurney) and passed the Indian Civil Service entrance exams (1897). He undertook his first post (1899) as assistant collector and magistrate in Tamil Nadu. He was later appointed Commissioner of Coorg (1912–1918) during which he began publishing on butterflies such as *Life history notes on Coorg butterflies* (1919). He was appointed Postmaster General of Bengal. Throughout his career he either collected himself or sent others to do so. He was a member (1908) of the Bombay Natural History Society, later joining their committee (1913). The Indian butterfly *Parnassius hannyngtoni* (Hannyngton's Apollo) is among the species named after him.

Hanson

Clubtail sp. *Amphigomphus hansoni* **Chao**, 1954

Dr John F Hanson (1915–2013) was an entomologist at the University of Massachusets, where H-F Chao studied for his PhD (1951). Massachusetts State College (now UMass) awarded his BS (1937), MS (1938) and PhD (1943). Having been a research fellow (1945–1947) at UMass, he became Assistant Professor of Insect Morphology (1947) and continued working there rising to full professor until retiring (1980) to become Professor Emeritus. Remarkably he had a parallel career in an engineering company and he and his wife Marie founded and operated the Ace Filament Company which for more than two decades provided engineering consulting, inventing, and manufacturing services for Indelco, Ceramic Coating Incorporated, Optical Micro Systems, Tesla Engineering and Raytheon. His passion was the study of stoneflies and he published his first paper, *Studies on* the *Plecoptera of North America I* while still a student. He made numerous field-collecting trips in eastern and western North America. On his retirement from the university, his *Plecoptera* collection, with its types and many thousands of specimens, was passed to the Museum of Natural History at the Smithsonian Institution. In retirement, he continued to work on various engineering-related projects as well as a book concerning a universal system of evolutionary progress. When still a very young man he wrote the following life objectives in his diary: "*To do everything as best I can; to teach college; to travel and collect insects; to be active in sports; to live 100 years in perfect health.*" He achieved all but the last, and only missed that by 19 months. The species patronym *Isogenoides hansoni* and the generic patronym *Hansonoperla* also honour his contributions.

Hao-Miao (Zhang)

Velvetwing sp. *Dysphaea haomiao* **Hämäläinen**, 2012

Dr Hao-Miao Zhang (b.1982) is a Chinese entomologist and odonatologist at the Kunming Institute of Zoology of the Chinese Academy of Sciences in Kunming, Yunnan. His doctorate was awarded (2012) by the South China Agricultural University, Guangzhou. He undertook postdoctoral research (2012–2016) at the Institute of Hydrobiology, Chinese Academy of Sciences, Wuhan. He has carried out extensive fieldwork in many Chinese provinces, most notably Yunnan, where he has found many new odonate species, many as yet undescribed. He has written more than thirty papers on Chinese odonata in which around thirty new species are described. He co-wrote: *A Fieldguide to the Dragonflies of Hainan* (2015) and has written the extensive two-volume: *Odonata of China* (2017). He provided the specimens and most of the photographs for the study of the eponymous species (2007). Hämäläinen wrote in his etymology: "*The new species is named after Dr Haomiao Zhang, an enthusiastic and gifted odonatologist, in recognition of his achievements in the study of Chinese dragonflies and in appreciation of his fruitful collaboration with the author and generosity in sharing specimens and information.*"

Hardy

Lesser Tasmanian Darner *Austroaeschna hardyi* **Tillyard**, 1917

George Hurlstone Hurdlestone Hardy (1882–1966) was an English-born Australian entomologist and engineer and sometime member of staff of the University of Queensland. Having trained in engineering he emigrated to Western Australia (1911) and moved to Tasmania where he became Acting Curator of the Tasmanian Museum, Hobart (1913–1917). He became a fellow in economic biology at the University of Queensland (1922–1932). His entomological work focussed on the taxonomy of brachycerous Diptera and he wrote 173 papers (1914–1966), the bulk (144) on Australian Diptera. He was a founder of both the Entomological Society of Queensland (1923), being its first treasurer (1923–1932), and the Australian Entomological Society (1964). He retired to Katoomba NSW but continued to publish right to the end of his life. In the type description of *Synthemiopsis gomphomacromioides* the author relates "*Mr G.H. Hardy, of the Tasmanian Museum, also captured a single male at Flowerdale Creek …*" (no etymology is given in the description, based on specimens collected by Tillyard, but the species is endemic to Tasmania).

Haritonov

Dwarf Darter *Sympetrum haritonovi* Borisov, 1983
Bluetail sp. *Ischnura haritonovi* **Dumont**, 1997
[JS *Ischnura aralensis* Haritonov, 1979]

Professor Dr Anatoly Yurevich Haritonov (1949–2013) was a Russian entomologist who spent his whole career at Laboratory of Insect Ecology of the Russian Academy of Sciences (Novosibirsk). He started as a junior staff researcher (1975), rising to senior staff researcher (1979) and then was elected Head of the Laboratory (1981) where he remained until his death. He was also Curator at the Siberian Zoological Museum. He attended the Biology and Geography Faculty of Chelyabinsk State Pedagogical Institute (1968–1971) and after graduating went (1972) to the Biological Institute at Novosibirsk. There his tutor, who became a friend, was B F Belyshev (q.v.). He defended his PhD thesis, *Dragonflies of Ural and Transuralia (fauna, ecology, zoogeography there,* (1975) and a further DSc thesis *Boreal odonatofauna and ecological factors of geographical distribution of dragonflies* (1991). He is considered one of the greatest and most prolific Russian odonatologists of all time. He collected across many countries of the former Soviet Union as well as the occasional foray elsewhere such as to Slovenia (1997). His main fieldwork collaborator was Dr Olga N Popova, his wife. He wrote 184 papers (1971–2012), often with both Borisov and Dumont, especially on the biology and distribution of Russian dragonflies. His very first co-authored paper was: *Fauna and biology of dragonflies in South Ural and their role as intermediate hosts of helminths* (1971) and almost his last *Biotopical groups in the populations of the dragonfly Coenagrion armatum (Charpentier, 1840)* was written with his wife. He died after a two-year battle with lung cancer.

Harkness

Harkness's Dancer *Argia harknessi* **Calvert**, 1899

Harvey Willson Harkness (1821–1901) was an American mycologist and natural historian. His original training was as a physician and he went to California in the 'gold rush' (1849). There he rose to prominence as a physician, educator, newspaper editor and real estate developer which made his fortune, allowing him to retire from medicine (1869). He joined the California Academy of Sciences in San Francisco becoming President (1887–1896). He devoted himself to the natural history of the Pacific States, publishing a number of articles on the subject. In the last 30 years of his life, he wrote a number of papers on California fungi, such as the monograph *California Hypogeous Fungi* (1899).

Harmand

Sultan sp. *Camacinia harmandi* **Martin**, 1900
[JS *Camacinia harterti* Karsch, 1890]

Dr François Jules Harmand (1845–1921) was a French Navy surgeon-naturalist, diplomat and explorer. He undertook a series of voyages to and in Indochina (1875–1877) visiting Cambodia, Annam and Laos. He was Civil Commissioner-General in Tonkin (1883), held consular posts in Thailand (1881), India (1885) and Chile (1890), and was French Ambassador to Japan (1894–1906). He collected the type in Sikkim (1890). Two birds are also named after him.

Haroldo (Travassos)

Threadtail sp. *Epipleoneura haroldoi* **Santos**, 1964

Dr Haroldo Pereira Travassos (1922–1977) was a Brazilian ichthyologist and editor of museum journals. He was a graduate of the National Veterinary School of Medicine, but made a career in the Ichthyology Department of the Brazilian National Museum (1944). He was the son of Lauro Travassos (see **Travassos**).

Harrisson

Flatwing sp. *Podolestes harrissoni* **Lieftinck**, 1953

Major Thomas 'Tom' Harnett Harrisson (1911–1976) was a British polymath, an anthropologist, ornithologist, ecologist, explorer, journalist, broadcaster, soldier, ethnologist, museum curator, mass observationist, archaeologist, film-maker, writer and conservationist. He spent much of his life in Sarawak and during WW2, after having been a radio critic (1942–1944), he was parachuted (1945) into Borneo to organise tribesmen against the Japanese. He was Curator at the Sarawak Museum (1947–1966) and undertook pioneering excavations at Niah, Sarawak, discovering a 40,000-year old skull. After leaving Sarawak he lived in USA, UK and France before being killed in a motor accident in Thailand. He wrote: *Savage Civilisation* (1937). Lieftinck wrote: "*Named in honour of Mr. Tom Harrisson, D.S.O., the able curator of the Sarawak Museum at Kuching.*" An amphibian, a bird, and a fish are also named after him.

Harte

Wedgetail sp. *Acanthagrion hartei* **Muzón** & Lozano, 2005

Miguel Harte (b.1961) is an artist who lives and works in Buenos Aires where he has had many exhibitions (early 1980s–2014). The etymology says: "*Named in honour of Miguel Harte, an Argentine plastic artist who gave the first author a new and fascinating vision of insects*".

Hartert

Sultan sp. *Camacinia harterti* **Karsch**, 1890

Ernst Johann Otto Hartert (1859–1933) was a German ornithologist and oologist. He travelled extensively, often on behalf of his employer Walter (Lord) Rothschild. He collected the type in Sumatra (1889). He was the ornithological curator of Rothschild's private museum at Tring, which later became an annexe to the British Museum (Natural History), housing all of the bird skins. He wrote: *Die Vögel der paläarktischen Fauna* and many other works often in collaboration with others (1890s–1920s). A reptile and sixty-one birds are also named after him.

Hartmann

Hartmann's Talontail *Crenigomphus hartmanni* **Förster**, 1898
[A. Clubbed Talontail, Orig. *Onychogomphus hartmanni*]

Dancing Jewel *Libellago caligata hartmanni* Förster, 1897
[JS *Platycypha caligata* Selys, 1853]

Dr Karl Eduard Robert Hartmann (1832–1893) was a German physician, anatomist, anthropologist, ethnographer and naturalist. He qualified in medicine at Humboldt-Universität, Berlin, where he later became Professor of Anatomy (1873), having previously (1865) taught comparative zoology and physiology at the agricultural academy in Proskau. He conducted anatomy studies of marine species in Sweden and Italy and made expeditions to Africa. He accompanied Baron Adalbert von Barnim on an expedition to Egypt, Sudan and Nubia (1859–1860), based around the course of the Blue Nile. Adalbert died on the trip (1860). He also collected the types in Transvaal (1896 & 1897). He wrote: *Reise in Nordost Africa* (1863). Among his other written work was *Der Gorilla* (1881) in which he posited that humans and apes have a common ancestor. A reptile is also named after him.

Hartwig

Leaftail sp. *Phyllogomphus hartwigi* **Buchholz**, 1958
[JS *Phyllogomphus coloratus* Kimmins, 1931]
Superb Dropwing *Trithemis hartwigi* **Pinhey**, 1970

Wolfgang Richard Hartwig (b.1928) is a German illustrator and collector formerly of Zoologisches Forschungsmuseum Alexander Koenig, Bonn. He illustrated the *Collins Field Guide to the Birds of West Africa* (1988) and *Birds of West Africa* (1977). His work also adorned Eisentraut's book *Im Schatten des Mongo-ma-loba* (1982). Pinhey wrote a paper (1974) to record a second collection of Odonata made by Hartwig in West Cameroon and Fernando Po, the material from which he studied (1951) through the offices of Professor Eisentraut (q.v.). He collected the leaftail type there (1958). A mammal is also named after him.

Haseman

Dancer Sp. *Argia hasemani* **Calvert**, 1909

John Diederich Haseman (1887–1969) was a zoologist and ichthyologist. He graduated from Indiana University (1905) and taught there (1905–1906). As a graduate student, he was sent to Brazil as a last-minute substitute to represent the Carnegie Museum and to collect fishes on a museum-sponsored expedition. (He had previously gone on an expedition to Cuba with Dr. Eigenmann and then under Eigenmann's guidance he went again on his own to Cuba at a later date.) Upon arriving in Brazil where he collected the dancer type (1907) in Bahia, he found the main expedition about to set out. It was decided that he should run his own solo expedition, which lasted two and a half years (1908–1911) and covered large areas of Argentina, Bolivia, Brazil, Paraguay, and Uruguay. He never went back to the university but continued to study ichthyology in the field. He wrote, among other works: *Some Factors of Geographical Distribution in South America* (1912). He said, of his travels through the wilds of South America: "*After the noises of the day the hush which comes at night-fall causes even the hardened traveler at times to shudder. No man over fifty years of age should attempt to enter this region. A hard heart and cold blood are useful to him who invades it.*" A reptile is also named after him.

Hatvan

Velveteen sp. *Bayadera hatvan* **Hämäläinen** & **Kompier**, 2015

The specific epithet hatvan (a noun in apposition) is based to the Vietnamese term 'hát văn', a contraction of 'hát chầu văn'. It refers to a traditional folk art form in northern Vietnam which combines singing and dancing. This is in accord with the meaning of the genus name

Bayadera: bayadére is the French version of the Portuguese word bailadeira, which refers to a Hindu dancing girl in Indian temples.

Hau

Emerald Spreadwing sp. *Megalestes haui* **KDP Wilson** & **Reels**, 2003
Phantomhawker sp. *Planaeschna haui* KDP Wilson & Xu, 2008

Dr Billy Chi-hang Hau (b.1967) is an ecologist and environmentalist from Hong Kong. The University of Hong Kong awarded his BSc (1991), MSc (1994) and PhD (1999). He began his career in conservation (1991–1994) at WWF focusing on the impact of urban Hong Kong on the natural environment and then worked in the Kadoorie Farm and Botanic Garden (1998–2001) on a native tree project and a South China biodiversity conservation project. He has been Senior Teaching Consultant (2009–2012), Assistant Professor (2001–2009), and Deputy Programme Director and Coordinator of the MSc in Environmental Management Programme as well as Principal Lecturer at the Kadoorie Institute School of Biological Sciences at the University of Hong Kong (2001–present). His research interests include restoration of degraded tropical ecology and environmental impact assessment. He has written around 90 scientific papers, books, chapters, reports and addresses. He collected the type specimens in China with the authors.

Hauxwell

Demoiselle sp. *Mnesarete hauxwelli* **Selys**, 1869
[Orig. *Lais hauxwelli*]

Helicopter ssp. *Mecistogaster lucretia hauxwelli* Selys, 1886

John Hauxwell (1827–1919) was a commercial collector from England. He collected from the mid-19th century over nearly 40 years in the upper Amazon basin in Peru where he took the types. He was Mayor of Pebas for almost ten years, where he stayed and collected. The majority of his odonatological material went to Selys Longchamps. Henry Walter Bates mentions him in *The Naturalist on the River Amazon* (1864). Seven birds are also named after him.

Hayashi

Sylvan sp. *Coeliccia hayashii* Phan & Kompier, 2016

Fumio Hayashi (b.1957) is a professor at the Department of Biology, Tokyo Metropolitan University. His research topics include behaviour, ecology, evolution, taxonomy and phylogeny. He is a specialist of the insect order Megaloptera and has published around 120 papers (1982–2017). His odonatological publications include: *Macro- and microscale distribution patterns of two closely related Japanese Mnais species inferred from nuclear ribosomal DNA, its sequences and morphology (Zygoptera: Calopterygidae)* (2004).

Hayward

Skimmer sp. *Oligoclada haywardi* **Fraser**, 1947

Captain Kenneth John Hayward (1891–1972) was a British engineer and entomologist. He was an electrician who worked on the Aswan Dam in Egypt (1912) before joining up in WW1 and serving in France, Greece and Cyprus, becoming captain (1919). He returned to Aswan and then to London (1922). He became an engineer for La Forestal Company in

Argentina (1923–1929). During this time, he amassed a large natural history collection, which he gave to the BMNH. He was appointed (1934) Entomologist at the Concordia Experiment Station, Entre Rios, Argentine and later (1940) at the Agricultural Experimental Station at Tucuman. Here he joined (1944) the Instituto-Fundacion Miguel Lillo of the National University of Tucuman, which conferred his honorary doctorate (1950) later being conferred the status of professor emeritus. He took the type in Argentina (1945). He collected insects and wrote nearly 300 papers, many on Argentine insects.

Hearsey

Shadowdamsel sp. *Protosticta hearseyi* **Fraser**, 1922

T N Hearsey was Extra-Assistant Conservator of Forests (1898) in South Tenasserim, Burma and later that year in Ganjam, Orissa, India. He collected in Madras (1912), Kerala (1921) and Malabar (1929) and also published at least one article in the Bombay Natural History Society journal. The description contains no etymology, but it is known that Hearsey had supplied dragonflies to Fraser, such as the holotype of *Lamelligomphus malabarensis*, collected by Hearsey at Palghat (June 1921).

Helen (David)

Little Blue Sprite *Pseudagrion helenae* **Balinsky**, 1964

Helen David née Balinsky (b.1949) is the daughter of the author, Boris Ivan Balinsky (1905–1997) (q.v.). She assisted him during his visit to Okavango swamps in Botswana, where this species was found.

Hélèn (Legrand)

Sunset Jewel *Chlorocypha helenae* **Legrand**, 1984

Hélèn Legrand is the author's wife. (Also see **Legrand**)

Helena

Skimmer sp. *Edonis helena* **Needham**, 1905
Flatwing sp. *Philogenia helena* **Hagen**, 1869
Reedling sp. *Indolestes helena* **Fraser**, 1922
[JS *Indolestes cyaneus* (Selys 1862)]
Keeled Skimmer *Orthetrum helena* **Buchholz**, 1954
[JS *Orthetrum coerulescens* (Fabricius, 1798]

Helena is a feminine forename, but not a specified one. The authors gave no etymology, so it is not possible to know for sure if the species were named after particular people. However, the lack of a Latin ending suggests it is not an eponym, as do Fraser's and Needham's frequent use of apparent forenames. On the other hand, Fraser often did not fully identify the people he honoured through etymologies.

Hélène

Leaftail sp. *Phyllogomphus helenae* **Lacroix**, 1921
[JS *Phyllogomphus aethiops* Selys, 1854]

Hélène Lacroix was the daughter of the author J-L Lacroix (q.v.).

Helenga

Duskhawker sp. *Gynacantha helenga* E B **Williamson** & **Jesse** H Williamson, 1930

Helen Beulah Thompson Gaige (1886–1976) was an American herpetologist. She studied at the University of Michigan and then became Assistant Curator (1910–1923), and later Curator (1923), of Reptiles and Amphibians at the Museum of Zoology of the University of Michigan. Her specialism was neo-tropical frogs. She was married to Professor Frederick McMahon Gaige (q.v.) who was an American entomologist, herpetologist and botanist and Director of the Museum. In their honour the American Society of Ichthyologists and Herpetologists makes an annual award to a graduate student of herpetology. She co-wrote: *The Herpetology of Michigan* (1928). She became (1937) Editor in Chief of 'Copeia' and wrote many articles therein. A number of reptiles are also named after her (also see **Gaige**).

Hempel

Wedgetail sp. *Oxyagrion hempeli* **Calvert**, 1909

Adolph Hempel (1870–1949) was an American entomologist who became a Brazilian citizen. He graduated (1892) at the University of Illinois. He went to Brazil and took a post at the Paulista Museum and that year (1900) he collected the type in Sao Paulo. He headed up the Economic Entomology department of the Biological Institute at Sao Paulo (1927) until retirement (1938). He was also associated with the Academy of Natural Sciences of Philadelphia.

Henrard

Eastern Red Threadtail *Elattoneura henrardi* **Fraser**, 1954
[JS *Elattoneura acuta* Kimmins, 1938]

Paul Henrard (1899–1952) was a Belgian mycologist. His PhD thesis was: *Polarité, hérédité et variation chez diverses souches d'Aspergillus* (1934). He collected the type in the Belgian Congo (1938).

Henry

Brook Hooktail *Paragomphus henryi* **Laidlaw**, 1928
[Orig. *Mesogomphus henryi*]

George Morrison Reid Henry (1891–1983), was an English ornithologist and entomologist. He was born in Ceylon (Sri Lanka), where his father was a tea plantation manager, and educated at home with his ten siblings. He was a good draughtsman, getting a job as such and as a laboratory assistant at the Ceylon Company of Pearl Fishers when just 16 (1907–1910). He became a draughtsman at the Colombo Museum (1910) and was then promoted (1913) to the newly created post of Assistant in Systematic Entomology where he stayed until retirement (1946). He wrote or co-wrote more than twenty articles, scientific papers and at least one longer work: *A Guide to the Birds of Ceylon* (1955). He collected the holotype at 3,000ft at Urugalla in Ceylon (April 1924).

Henshaw

Spreadwing sp. *Lestes henshawi* **Calvert**, 1907

Samuel Henshaw (1852–1941) was an American entomologist. He never went to university but was given an honorary MA (Harvard, 1903). He became a member of (1871), and worked on (1874) the insect collection at, the Boston Society of Natural History as assistant to Professor Hyatt. He was Secretary and Librarian there (1892–1901). During this time, his wife of fourteen years died (1900). Partly concurrently (1891–1898) he was part-time assistant in entomology at the Museum of Comparative Zoology, Cambridge, Massachusetts. He became full-time assistant (1898–1903), then Curator (1903–1911), going on to become Director (1912–1927), and lastly Director Emeritus. In most of his time there he edited its publications. He was very interested in the early English natural historian Gilbert White and collected everything he could about the man and his life as a naturalist, his family and the village of Selborne (1876–1931), including c.100 editions of his: *Natural History of Selborne*. In the last years of his life he was greatly affected by arthritis and became quite reclusive. He did not publish widely just a few papers on entomology and some bibliographies reflecting the fact he was an avid bibliophile.

Herbert

Dancer sp. *Argia herberti* **Calvert**, 1902

Herbert Huntingdon Smith (1851–1919). (See **Smith, H H** & also see **Thisma**)

Hermes

Shadowdamsel sp. *Drepanosticta hermes* **Van Tol**, 2005

In Greek mythology Hermes is the second youngest of the Gods; the god of transitions and boundaries. He conducted souls into paradise, was a trickster and is often portrayed with a winged cap and sandals.

Hermione

Goldenring sp. *Neallogaster hermionae* **Fraser**, 1927

[Orig. *Allogaster hermionae*]

In Greek mythology Hermione was the only child of King Menelaus of Sparta and his wife Helen of Troy.

Hermosa

Wedgetail sp. *Oxyagrion hermosae* **Leonard**, 1977

[Orig. *Acanthagrion hermosae*]

This is a toponym, the type locality, in the vicinity of Pampa Hermosa, Peru by F Woytowski (1935).

Herrera

Sanddragon sp. *Progomphus herrerae* **Needham** & **Etcheverry**, 1956

José Valentín Herrera González 'Pepe' (1913–1992) was a Chilean zoologist. He graduated with a teaching degree from the teacher training institute of the University of Chile, Santiago (1934), then enrolled for a further degree in physical education graduating again in 1936. He took a six-month teaching post in Punta Arenas teaching PE and science at a boy's school and stayed fourteen years. He joined (1946) the University of Chile teacher training institute as Professor of Zoology and moved to the Faculty of Philosophy and Education as head of the biology department (1956–1960), and then as Director of Entomology (1960–1964) and subsequently Extraordinary Professor of Entomology (1964). He was (1981) Professor of Science at the Academy of Science, becoming Professor Emeritus on retiring (1982). He came out of retirement (1984) to become Director of the Entomological Centre and then the Entomological Institute that he founded (1985). As an educator, he travelled widely and was able to collect and visit many museums in the US, Europe, UK, and the neo-tropics. His chief interest was always the butterflies of Chile. He published widely (1949–1991) such as, with Etcheverry: *Mariposas communes a Chile y Peru (Lepidoptera, Rhopalocera)* (1972).

Herve

River Cruiser *Phyllomacromia hervei* **Legrand**, 1980
[Orig. *Macromia hervei*]

Jean-Pierre Hervé is a medical entomologist at l'Office de la recherche scientifique et technique outre-mer (ORSTOM). He sent Odonata specimens from Central Africa for Legrand to study. He collected one of the paratypes in Sierra Leone (1975). Among his written work is: *Hydro-agricultural facilities and health (Senegal river valley)* (1998).

Hilary

Jewel sp. *Aristocypha hilaryae* **Fraser**, 1927
[Orig. *Rhinocypha hilaryae*]

No etymology was given. Colonel Frank Wall (q.v.) collected the damselfly (q.v.) so the Hilary in question might refer to a relative or friend of his, or someone else known to Fraser.

Hilbrand

Emerald sp. *Hemicordulia hilbrandi* **Lieftinck**, 1942

Professor Dr Hilbrand Boschma (1893–1976) was a Dutch zoologist. He studied botany and zoology at the University of Amsterdam which awarded his doctorate (1920), then went to the Dutch East Indies (1920–1922). He joined a Danish expedition to the Kai Islands (1922) where he sampled and studied corals. He returned to the Netherlands, becoming Chief Assistant at the Zoological Laboratory, State University at Leiden and began giving lectures to medical students (1925) becoming Professor of Zoology in 1931. He became Director of the Rijksmuseum van Natuurlijke Historie, Leiden until retirement (1934–1958), apart from the period (1943–1945) when he was dismissed by the German occupying forces (Lieftinck was Curator there from 1954). Boschma continued to lecture as Professor Emeritus (1958–1963) and continued to publish articles (until 1974), having published more than 250 articles, papers and contributions to longer works. He joined the Le Roux Expedition to New Guinea (1939–1940) during which the type was taken, when he was in charge of general zoological investigation. Lieftinck wrote a paper (1942) on: *The dragonflies (Odonata) of New Guinea and neighbouring-islands. VI. Results of the Third Archbold Expedition 1938–'39 and of the Le Roux Expedition 1939 to Netherlands New Guinea.*

Hildebrandt

Striped Junglewatcher *Neodythemis hildebrandti* **Karsch**, 1889

Johann Maria Hildebrandt (1847–1881) was a German botanist and explorer. He pursued mechanical engineering until the loss of an eye as a result of an explosion forced him to choose another profession. He collected and travelled in Arabia, East Africa, Madagascar, where the type was taken, and the Comoro Islands (1872–1881). He was also interested in languages and wrote: *Fragmente der Johanna –Sprache* in *Zeitschrift für Ethnologie* (1876), which deals with the vocabularies of dialects in the Johanna Islands. Most of his writings are about his travels and the peoples he encountered, their lives and rituals as well as about the natural history of those places. He died of yellow fever in Madagascar. Two birds, three reptiles, an amphibian and two mammals are also named after him.

Hildegard

Wedgetail sp. *Acanthagrion hildegarda* Gloger, 1967

Hildegard Gloger was the wife of the author Hellmut Gloger, who wrote a few papers on Odonata, such as: *Bemerkungen über die Odonaten-Fauna der Galapagos-Inseln nach der Ausbeute von Juan Foerster, 1959* (1964).

Hill

Ethiopian Bluetail *Ischnura hilli* **Pinhey**, 1964
[JS *Ischnura abyssinica* Martin, 1907]

Dr Bob G Hill was an entomologist. He was the head of Department of Arts and Sciences of the Imperial Ethiopian College of Agricultural and Mechanical Arts. He wrote articles on agricultural pests and their control in Ethiopia such as: *The Insect Pests of Vegetables in Harar Province, Ehtiopia* (1965) as well as *A guide to insect collecting* (1965). He collected the type specimens in Ethiopia (March 1962) when surveying Harar province with a local entomologist, Hadera Gebramedhin (1960–1964).

Hincks

Demoiselle sp. *Mnesarete hincksi* Fraser, 1946
[JS *Mnesarete metallica* (Selys, 1869)]

Walter Douglas Hincks (1906–1961) was an English entomologist. He trained at Leeds College in pharmacy and was employed by a large manufacturing chemist. In his spare time, he began to study insects. He left the pharmaceutical industry and became (1947) Assistant Keeper in Entomology at the University of Manchester Museum, in charge of Coleoptera. The post was renamed (1957) Keeper in Entomology, a post he occupied until his untimely death. In this time, he acquired a number of important collections. His most important work was the co-written: *Check List of British Insects* (1945).

Hine

Redskimmer sp. *Rhodopygia hinei* **Calvert**, 1907
Lavender Dancer *Argia hinei* **Kennedy**, 1918
Hine's Emerald *Somatochlora hineana* **Williamson**, 1931

Professor James Stewart Hine (1866–1930) was an American zoologist, ornithologist and entomologist. Ohio State University awarded his BSc (1893) and he spent his entire career there or at the University Museum. He was Assistant in Horticulture (1894), Assistant in Entomology (1895–1899), Assistant Professor of Zoology (1899–1902) and then Associate Professor of Entomology (1902–1925). He organised the Archaeological Museum's first division of Ohio Natural History working there whilst still teaching (1925–1927) and then full-time at the Museum (1927–1930). His particular area of interest was *Tabanidae,* but he was interested in all aspects of entomology and other areas of natural history. He collected on the US Gulf Coast (1903) and Central America, including Guatemala (1905), where he was particularly interested in water-associated insects and took the type of the skimmer. Other trips were to California, Arizona, where he collected the dancer, Mexico (1907), Alaska (1917 and 1919) and Florida and Cuba (1923). He took part in the Katmai Expeditions (1915–1916) under the auspices of the National Geographical Society as part of the University of Ohio. He also spent a month studying at the BMNH (1925). He collected the eponymous emerald in Ohio (1929). He was also a keen beekeeper and was President of the Ohio Beekeepers Association. He wrote a number of books and papers over several decades including: *Tabanidae of the western United States and Canada* (1904) and *Robberflies of the genus Erax* (1919). He died of a heart attack while preparing for Christmas. A fish is also named after him.

Hintz

Hintz's Skimmer *Orthetrum hintzi* **Schmidt,** 1951
[A. Dark-shouldered Skimmer, Slow Skimmer]

Exquisite Jewel *Chlorocypha Libellago hintzi* Grünberg, 1914
[JS *Chlorocypha hintzi* (Selys, 1879)]

Eugen Hintz (1868–1932) was a German engineer, entomologist and collector in Cameroon, where he collected the Jewel holotype (1910) as part of the Second German Central Africa Expedition (1910–1911) under the leadership of Adolf Friedrichs, Duke of Mecklenburg. He wrote the account of the expedition (1919) and also wrote at least one scientific description that same year, of a water beetle. He collected other zoological and botanical specimens, particularly beetles. He also collected in Portuguese Guinea (now Guinea-Bissau) including the skimmer type (1927). He also described a series of Australian beetles.

Hippolyte

Dasher sp. *Micrathyria hippolyte* **Ris,** 1911
Skimmer sp. *Diplacina hippolyte* **Lieftinck,** 1933

In Greek mythology, Hippolyta, Hippoliyte, or Hippolyte was the Amazonian queen who possessed a magical girdle she was given by her father, Ares the God of War.

Hirao

Coralleg sp. *Rhipidolestes hiraoi* Yamamoto, 1955

Tadayoshi Hirao collected this damselfly in Sasagamine, Ehime prefecture, Japan (August 1954). He was also collecting odonata and other insects (c.1946–c.1957) in western Japan around Fukuoka and may have been associated with Kyushu University. He took the type (1954).

Hiroaki

Threadtail sp. *Nososticta hiroakii* **Sasamoto**, 2007

Hiroaki Shibata is a lepidopterist. He graduated from Kyoto University, Zoology department, where he later worked. He wrote or co-wrote such papers as: *Off-hostoyiposition by two firitillary species (Nymphalidae, Argynninae) and its relation to egg predation* (2010). He collected this species in Biak Island (August 2002) which was named after him by his friend Akihiko Sasamoto (q.v.).

Hirose

Four-spot Midget *Mortonagrion hirosei* **Asahina**, 1972

Makoto Hirose (b.1933) is an amateur Japanese entomologist. He graduated (1956) from the Faculty of Arts and Sciences of Ibaraki University and was a teacher in elementary and junior high schools until retirement (1994).

Hiten

Skydragon sp. *Chlorogomphus hiten* **Sasamoto, Yokoi** & **Teramoto**, 2011
[Orig. *Sinorogomphus hiten*]

To quote the authors' etymology: *The species name 'hiten' is the Japanese noun, meaning the religious imaginary holy people who fly around and admire Buddha and Amida in Buddhism.*

Hodges

Hodges's Clubtail *Gomphus hodgesi* **Needham**, 1950

Robert Shattuck Hodges (1875–1964) was an American chemist, mycologist and odonatologist. He graduated from the University of Alabama, Birmingham (c.1898) and was an assistant on the State Geological Survey (1901). He collected in Alabama for the Alabama Museum of Natural History, Tuscaloosa. He often collected nymphs and reared them. He wrote several articles including: *Alabama dragonflies (Odonata)* (1937). He was also an early and enthusiastic photographer.

Hodgkin

Pilbara Dragon *Antipodogomphus hodgkini* **Watson**, 1969
[Orig. *Antipodogomphus neophytus hodgkini*]

Dr Ernest Pease Hodgkin (1906–1997) was a British born Australian entomologist at the Department of Zoology, University of Western Australia and at the Department of Conservation and Environment, Western Australia. He graduated BSc zoology & entomology from Manchester University (1930) and was appointed Government Medical Entomologist in the Federated States of Malaya. He was interned in a civilian POW camp (1942–1946), then he joined his wife and children who had been evacuated to Australia where he became a lecturer at University of Western Australia. He completed his doctorate there (1950), continuing until retirement (1973) although he continued to work as an environmental consultant. As a marine biologist, most of his publications are about the biology and conservation of estuaries. His works include: *Environmental Study of the Blackwood River Estuary* (1976) and *An Inventory of Information of the Estuaries and Coastal Lagoons of South Western Australia* (1989). Hodges provided the early material on which Watson's project of studying the Odonata of northwestern part of Western Australia was based. Watson collected the types of this new species.

Hoffmann, E A

Ringed Cascader *Zygonyx hoffmanni* Grünberg, 1903
[JS *Zygonyx torrida* Kirby, 1889]

Ernst August Hoffmann (1837–1892) was a German entomologist, the first to be employed (1869) on a full-time basis at the State Museum of Natural History, Stuttgart. He was primarily a lepidopterist and wrote: *Die Großschmetterlinge Europas* (1894) and further works on butterflies and beetles. Grünberg gave no etymology but *hoffmanni* was a manuscript name originally given by Friedrich Moritz Brauer and was adopted by Grünberg, so it is probable that the species is named after this contemporary of Brauer's (1832–1904).

Hoffmann, J

Pond Damselfly sp. *Protallagma hoffmanni* **Hunger** & **Schiel**, 2012
Threadtail sp. *Prodasineura hoffmanni* **Kosterin**, 2015

Dr Joachim Hoffmann (b.1953) of Hamburg is a German entomologist. He was employed for many years at Museo de Historia Natural, Lima, Peru where he has studied the dragonflies. His papers include: *Summary catalogue of the Odonata of Peru* (2009). He has described two new *Perithemis* species. Hunger & Schiel wrote in their etymology: "*This species is named hoffmanni after our good friend and colleague Joachim Hoffmann who worked at the Museo de Historia Natural in Lima for many years and is an expert on Odonata from Peru. We found the new species on a field trip led by him.*"

Hoffmann, W E

Hooktail sp. *Paragomphus hoffmanni* **Needham**, 1931
[Orig. *Gomphus hoffmanni*]

Professor William 'Bill' Edwin Hoffmann (1896–1989) was an American entomologist. After WW1 service, he graduated from Kansas University and was awarded his masters by the University of Minnesota and worked at both universities and also for the states. He took part in the Minnesota Pacific Expedition (1924) mostly studying marine life. He held a chair at Lingnan University, Canton, China (1924–1951) and was Curator of the Lingnan Natural History Survey and Museum. During his time there, he made many collecting expeditions into the province and later to the Philippines (1927, 1940 & 1941) and Taiwan. He twice took leaves of absence to do research in England and Europe and travelled extensively, visiting 75 countries. During the war, he was interned by the Japanese and repatriated (1943) in an exchange of prisoners. He was Associate Curator of Insects at the Smithsonian (1944–1947) then returned to China for four years (1947–1951) and later worked at the University of Kansas until retirement (1962). The hooktail is only known from the single holotype female specimen collected from Nodoa, Hainan by Hoffmann (1929), a locality where he collected insects over a number of years. Other taxa are also named after him, particularly beetles. He wrote about a variety of insects including silkworms and crop pests such as: *An Abridged Catalogue of Certain Scutelleroidea (Plataspidae, Scutelleridae and Pentatomidae) of China, Chosen, Indo-China and Taiwan* (1935) and an interesting paper for the Entomological Society of Washington *Insects as Human Food* (1948). He was editor of the Lingnan University Science Journal. A naturalist at heart, he collected and observed insects, fish, birds, reptiles and plants and he is also commemorated in the name of at least one freshwater fish.

Hoi Sen

Limniad sp. *Amphicnemis hoisen* **Dow**, Choong & Ng, 2010

Academician Professor Emeritus Dr Hoi Sen Yong (b.1939) is a Malaysian biologist and zoologist. The University of Malaya awarded his BSc degrees (1964 & 1965) and his PhD (1968) during which time he was student demonstrator and tutor, following which he continued his research until becoming Lecturer in the School of Biological Science there (1968-1974). This was followed by another period of research as a staff fellow (1974). He became Associate Professor in the Department of Genetics and Cellular Biology (1974-1986) and then took the Chair of Zoology (1986-1998). Since 'retirement' he has been Senior Research Fellow (1999), Research Associate (2000-2007) and is still Professor Emeritus (2007). Among his continuing research interests are genetics, immunology and molecular biology of parasites, vectors and hosts, systematics, genetics and evolutionary biology of mammals, amphibians, reptiles and fish. His interests also include butterflies and dragonflies. He has received numerous awards and published more than 300 articles and four books as well as editing others. Among his few papers on Odonata is *Molecular phylogeny of Orthetrum dragonflies reveals cryptic species of Orthetrum pruinosum* (2014). He says of himself: "*I think my greatest success is in pioneering the local publication of scientific works by Malaysians, and founding and editing the first-ever Malaysian full-colour quarterly 'Nature Malaysiana' for some 20 years.*" At least 12 species of various taxa are also named after him.

Holderer

Bluet sp. *Coenagrion holdereri* **Förster**, 1900
[Orig. *Agrion holdereri*]

Dr Julius Holderer (1866-1950) was a German entomologist. He collected a number of Odonata in Central Asia (1898), during his Tibet and China sojourn (1897-1899), that Förster studied and described. He later became a civil servant and district magistrate until retiring (1931). Two birds and a mammal are also named after him.

Holger Hunger

Skimmer sp. *Diplacina holgerhungeri* Villanueva, 2012

Dr Holger Hunger (b.1968) is a German biologist who is a partner and co-founder (1995) of INULA (Institut fur Naturschutz und Landsschaftsanalyse). He studied biology at the University of Freiburg and University of Michigan (1989-1996) and for his PhD under a scholarship awarded by the German Federal Environment Foundation at the University of Vechta (2001-2004). He took two multi-month internships as a biologist at the Division of Resource Management of the Grand Canyon National (1993 & 1994). He was also a founding member and member of the board of Schutzgemeinschaft Libellen in Baden-

Württemberg (2001–2012), a dragonfly conservation organisation. Among other works, he co-wrote: *Libellen in Kiesgruben & Steinbrüchen* (2011).

Holland

Slender Redskimmer *Rhodopygia hollandi* **Calvert**, 1907

Dr William Jacob Holland (1848–1932) was a Jamaican-born American Presbyterian minister, entomologist and palaeontologist. He attended Moravian College and Theological Seminary, and Amhurst College, where he was awarded his BA (1869) and later (1874) Princeton Theological Seminary leaving to become pastor of a church in Pittsburgh. He was naturalist for the US Eclipse Expedition (1887), which explored Japan. He was Chancellor of the University of Pittsburgh (1891–1901), where he taught anatomy and zoology, and Director of the Carnegie Museum until retirement (1901–1922), having been hired by his friend Andrew Carnegie. He also donated his own collection of 250,000 specimens to the museum. He appears to have been a difficult man to work with. Given to tantrums, he was somewhat of a sycophant towards his 'betters' (including Carnegie), and seemingly condescending when dealing with employees. His primary interest was in Lepidoptera and he wrote: *The Butterfly Book* (1898) and *The Moth Book* (1903). His trip to Argentina (1912) to install a replica of a *Diplodocus*, at the behest of Carnegie, is told by him in his travel book *To the River Plate and Back* (1913). He loaned the specimens to Calvert to study. A bird and a shark are also named after him.

Holthuis

Cruiser sp. *Macromia holthuisi* **Kalkman**, 2008

Dr Lipke Bijdeley Holthuis (1921–2008) was a Dutch carcinologist (the scientific study of crustaceans). He was born in East Java and took his degrees at Leiden University, culminating in a doctorate (1946). He was the most prolific carcinologist of the 20th century, publishing 620 papers totalling 12,795 pages. This steady stream of publications resulted in the description of 428 new taxa: 2 new families, 5 subfamilies, 83 genera and 338 species. Sixty-Seven taxa were named after him (1953–2009). He worked at the Rijksmuseum van Natuurlijke Historie in Leiden and was Member of International Commission of the Zoological Nomenclature (1953–1996), acting also as its President (1965–1972). When the Emperor Hirohito of Japan (a keen marine biologist himself) visited the Netherlands (1971) he wanted to meet his research colleagues Holthuis and Willem Verwoort (q.v.) at Leiden museum. Holthuis collected the holotype of this Cruiser at Biak Island (1954).

Hoogerwerf

Coraltail sp. *Ceriagrion hoogerwerfi* **Lieftinck**, 1940

Andries Hoogerwerf (1906–1977) was a Dutch athlete (he held the Dutch mile and 1500 metre records), naturalist, ornithologist and conservationist. He moved to the Dutch East Indies (1931) where he worked at the Bogor Botanical Gardens and wrote much on Indonesian fauna, particularly from Java and Sumatra (1940s–1970s). He became the nature protection officer for the colony's nature reserves (1935). His works include: *De avifauna van Tjibodas en omgeving* (Java) (1949) and *Udjung Kulon: The Land of the Last Javan Rhinoceros*

(1970). He is believed to have been the only person to photograph the now-extinct Javan Tiger (1938). He collected the holotype of the damselfly at a mountain lake in northern Sumatra, Indonesia (1937). He moved back to the Netherlands (1957) but continued to visit Indonesia. He then moved to New Guinea (1963) as a scientific officer at an experimental agricultural station there. A mammal and three birds are also named after him.

Hosana

Sylph sp. *Macrothemis hosanai* **Santos**, 1967

Dr Hosana from Planaltina, Brazil. He was the owner of the farm where the dragonfly was found.

Hose

Threadtail sp. *Prodasineura hosei* **Laidlaw**, 1913

[S. *Disparoneura hosei*]

Dr Charles Hose (1863–1929) was a naturalist who lived in Sarawak and Malaysia (1884–1907). He was 'Resident of Baram' (Borneo) (1891–1903), then transferred to Sibu (1903–1907). Hose successfully investigated the principal cause of the disease beri-beri. He was also a good cartographer who produced the first reliable map of Sarawak. He sent huge collections of zoological, botanical and ethnographic material to many museums and institutions including the British Museum and at least four British universities, one of which (Cambridge) awarded him an honorary DSc (1900). He was an extremely bulky man, which meant that when he went to visit the local tribes in their long houses, it was necessary for them to reinforce the floors. He was still remembered (1995) for his extreme size! I am indebted to a member of his family for some reminiscences of him: He successfully put a stop to the head-hunting raids among the various villages by the simple expedient of organizing a boat-race in the style of the University Boat Race over a similar distance and a similar sinusoidal course (although there were twenty-two dugouts with crews of seventy plus) (1899). This satisfied their honour, apparently. Another story concerns a journey on the Trans-Siberian Railway when the train stopped near Lake Baikal and he acquired three live Baikal Seals, which he put in the luggage rack! Not surprisingly they died. As each succumbed, he skinned them on the train, to the interest and surprise of his fellow travellers. He returned to Sarawak several times after retirement, very possibly in connection with the development of the oil fields at Miri (they are still producing). He became an expert on the production of acetone (used in the manufacture of cordite) as he ran a factory for it at Kings Lynn, Norfolk, UK (WW1). The raw materials for making acetone were maize and horse chestnuts. He wrote *Fifty Years of Romance and Research* (1927) and *The Field Book of a Jungle Wallah* (1929). Laidlaw wrote, in his description that the type was: "...*Taken by Mr Hose some years ago with other material from Baram, Borneo*". Fort Hose in Sarawak, now a museum, was named after him. Nine mammals, five birds and three amphibians are also named after him.

Houghton

Knobtail sp. *Epigomphus houghtoni* Brooks, 1989

Greg Houghton accompanied (1988) John Paul, the collector of the type in Costa Rica (1988), but was tragically killed shortly afterwards.

Howe

Pygmy Snaketail *Ophiogomphus howei* Bromley, 1924

Dr Reginald Heber Howe Jr (1875–1932) was an American collector, naturalist, lichenologist, mycologist, ornithologist and odonatologist. He was educated at Harvard and the Sorbonne. He also founded the Belmont Hill School and was its first headmaster (1923–1932). He wrote a number of books, primarily on ornithology, such as: *The Birds of Massachusetts* (1901). Among his dozen odonatological publications is: *Manual of the Odonata of New England* (1917).

Hua

Jewel sp. *Rhinocypha huai* **W-B Zhou,** & Zhou, 2006
[Orig. *Heliocypha huai*]
Threadtail sp. *Prodasineura huai* W-B Zhou, & Zhou, 2007

Professor Dr Hua Li-Zhong (b.1931) is a Chinese entomologist with a particular interest in Lepidoptera. Among his written works are: *Latin-Chinese-English names of Chinese insects* (2013) and the four-volume: *List of Chinese Insects* (Vol I 2000–Vol IV 2007), volume two of which he has dedicated to J L Gressitt (q.v.). Past works have included: *Economic insect fauna of China* (1985). He collected the holotypes in China (1981).

Huaorania

Darner sp. *Coryphaeschna huaorania* **Tennessen,** 2001

This species is dedicated to the Ecuadorian Huaorani people, who live in the Department of Orellana, Ecuador, where the holotype was collected.

Hubbell

Pond Damselfly sp. *Inpabasis hubelli* **Santos,** 1961

Professor Dr Theodore Huntington Hubbell (1897–1989) was an American zoologist and entomologist. His family lived in the Philippines (1907–1913), which is where he developed an interest in natural history in general and entomology in particular. After service in WW1 he graduated at Michigan University (1920), remaining for two years as an assistant in the Museum of Zoology. He then went to the Bussey Institution at Harvard and (1923) the University of Florida. While teaching in Florida, Hubbell was Honorary Associate Curator of Orthoptera in the Museum of Zoology and worked on his doctorate, awarded in 1934 by Michigan University. He became Curator of Insects in the Museum of Zoology and Professor of Zoology at Michigan (1946). He was Acting Director of the Museum (1955), then Director (1956–1968) until retirement. He undertook extensive research expeditions to much of Europe and Latin America, as well as many parts of the United States, and wrote many articles. The damselfly holotype was collected in 1940. Hubbell's name was misspelled in the species name, but in the dedication it was correct. Hubbell was honoured because he had facilitated Santos's work during his museum visit to Ann Arbor (1959).

Hummel

Featherleg sp. *Platycnemis hummeli* **Sjöstedt,** 1932

Pincertail sp. *Gomphus hummeli* Sjöstedt, 1932
[JS *Nihonogomphus brevipennis* (Needham, 1930)]

David Axelsson Hummel (1893–1984) was a Swedish physician, explorer and resistance leader in WW2. He trained in medicine at the Karolinska Institute in Stockholm where he graduated (1923), following which he held a variety of medical appointments. He joined the Sven Hedin expedition (1927–1934) and collected the type in China (1930). He also collected plants in Gansu, as well as zoological specimens, and drew maps. He was a district medical officer on the Norwegian border (1932–1956). During the war years, he helped both Norwegian refugees and resistance agents for which he was later honoured. He later (1953) took part in an archaeological expedition to India and was Ship's Physician on whaling expeditions in the Southern Ocean (1957–1963).

Hurley

Flutterer Sp. *Rhyothemis hurleyi* **Tillyard**, 1926

James Francis 'Frank' Hurley, OBE (1885–1962) was an Australian photographer and adventurer who participated in a number of expeditions to Antarctica and served as an official photographer with Australian forces during both world wars. He is most famed for his part in the ill-fated Imperial Trans-Antarctic Expedition led by Ernest Shackleton (1916). When the *Endurance* became trapped in the pack ice and was slowly crushed all hands had to abandon ship and camp on the sea ice way out at sea. As the ship sank Hurley dived into the freezing water to retrieve his films and plates. When it was decided to start a long trek across the sea ice, he persuaded Shackleton to let him keep 120 of the glass plates and his small pocket camera. He later photographed Shackleton leaving on the lifeboat for a whaling station on South Georgia to send rescue for the rest of the shipwrecked party. Hurley later (1917–1918) served as official photographer with Australian forces, taking the only colour plates of WW1. Tillyard wrote: "*This magnificent species, which I dedicate to Captain Frank Hurley, leader of the expedition to the Lake Murray...*" The types were collected by Allan R. McCulloch in New Guinea (1922).

Hymenae

Spot-Winged Glider *Pantala hymenaea* **Say**, 1840
[Orig. *Libellula hymenae*]

Hymenaios was the Greek god of marriage ceremonies, inspiring feasts and song.

I

Ibsen

Pond Damselfly sp. *Phoenicagrion ibseni* **Machado**, 2010

Admiral Ibsen de Gusmão Câmara (1923–2014) was a naval officer and environmentalist. He attended the Brazilian Naval Academy (1941–1944) and served in the Atlantic during WW2 and during his military career visited every continent. He was assigned to patrolling the Amazon (1967–1969) and, always interested in nature, became aware of how much deforestation was happening. He was promoted to admiral (1972) and became the Brazilian Armed Forces Deputy Chief of Staff. He retired in 1981 and was that year elected President of Brazilian Foundation for the Conservation of Nature. When (1983) Brazil began the reintroduction of international zoo-born Golden Lion Tamarins to the Poço das Antas Biological Reserve, he was the foremost adviser on that pioneering path. He was also co-founder of the Right Whale Project. Machado wrote in his etymology: *"…in honor of my good friend and environmentalist Admiral Ibsen de Gusmão Câmara, in recognition for his outstanding contribution for the protection of Brazilian biodiversity".* The announcement of his death described him as *"…one of Brazil's greatest defenders of its forests and seas".*

Icarus

Skydragon sp. *Chlorogomphus icarus* **KDP Wilson & Reels**, 2001
[JS *Chlorogomphus usudai* Ishida, 1996]

In Greek mythology Icarus was the son of master craftsman Daedalus. He attempted to escape from Crete by means of wings that his father constructed from feathers and wax. Despite being warned he flew too close to the sun, the wax melted and he fell into the sea and drowned.

Ida

Nighthawker sp. *Heliaeschna idae* **Brauer**, 1865
[Orig. *Gynacantha idae*]

Ida Laura Pfeiffer (née Reyer) (1797–1858) was an Austrian traveller and travel book author. One of the first female explorers, her popular books were translated into seven languages. She liked 'boy's activities' and preferred wearing boys' clothing and her father indulged this. She received the education usually given a boy. Her first long journey was a trip to Palestine and Egypt when she was five. The experience remained with her. Following the death of her father when she was nine, her mother – disapproving this unconventional upbringing – persuaded Ida to wear girls' clothing and to take up piano lessons. When her family grew up she began to travel. She travelled widely in Europe and the Middle East (1842–1845) then started (1846) her first trip around the world visiting many places in South America, China, India, Persia, etc. before returning home (1848) and writing: *A Woman's Journey round the*

World (1854). She write more and went on to travel (1851) to South Africa, Malaya and Indonesia, Australia and North America. She set out (1857) to explore in Madagascar but was expelled for unwittingly being involved in a plot to depose the Queen. She contracted an illness (probably malaria) and died the following year. Her son published her last book *Reise nach Madagascar* (1861) that included her biography.

Igarashi

Long Skimmer ssp. *Orthetrum trinacria igarashii* **Asahina**, 1973
[JS *Orthetrum trinacria* (Selys, 1841)]

Dr Suguru Igarashi (1924–2008) was an amateur Japanese entomologist and lepidopterist. His main interest was in the early stage morphology of butterflies. He studied at the Faculty of Engineering, the University of Tokyo, and after graduation he held many high-ranking posts in business, including being a board member of Taisei Corporation (1979–1985) and President of Shin-Etsu Handotai Co., Ltd. (1990–1996). He also became an accomplished novelist. His business activities took him all over the world and, whenever he could, he collected and studied local insects. He wrote about these studies in: *Papilionidae and their early stages* (1979) for which Kyoto University awarded him a doctorate (1983). He became chairman of the editorial committee of the Lepidopterological Society of Japan and was the first President and President Emeritus of the Butterfly Society of Japan. His other entomological works include the two volume: *The Life Histories of Asian Butterflies* (1997 & 2000). He collected the type in Iraq (1970). His collection forms the basis of 260,000 insects and over 5000 papers held at the University of Tokyo Museum and was bequeathed by his wife, Yoshiko who had shared his interest in insects and helped his collecting and studies. He died from stomach cancer.

Ihering

Dasher sp. *Micrathyria iheringi* **Santos**, 1946

Rodolpho Teodoro Gaspar Wilhelm von Ihering (1883–1939) was a zoologist and biologist sometimes called the father of fish farming in Brazil. He graduated at São Paulo University (1901) and was appointed (by his father who was Director) as a Deputy Director of the Museu Paulista. He spent a year (1911) at the Estação Biológica de Nápoles and then worked at the Muséum National d'Histoire Naturelle in Paris. When Brazil entered WW1, he resigned to set up a factory, where he worked for a decade, but continued to collect, study and research. He then went to work in the Laboratory of Parasitology, Faculty of Medicine of São Paulo University (1926–1927) but after this began in earnest as an ichthyologist, working at the Institute of Biological and Agricultural Defence in São Paulo and then (1931) researched fish farming and led (1932–1937) the Technical Committee on Fish Farming in the Northeast. He co-wrote: *Catálogos da fauna brasileira: As aves do Brazil* (1907) with his father, and alone wrote, among others: *Dictionary of the animals from Brazil* (1940). One reptile is also named after him.

Iida

Shadowdancer sp. *Idionyx iida* **Hämäläinen**, 2002

Iida-Maija Anneli Kalmanlehto (until 2002 Virtanen) (b.1984) is the daughter of a Finnish friend of the author. She changed her last name (2002) in favour of her mother's last name.

The University of Tampere awarded her MSc (2014). She currently works in the field of international cooperation.

Ijima

Whiteface ssp. *Leucorrhinia intermedia ijimai* **Asahina**, 1961

Kazuo Ijima (c.1930–2016) was an amateur Japanese entomologist. He collected the type in Hokkaido, Japan (1957) where he lived in Kushiro City. He donated the main part of his entomological collection to Kushiro municipal museum. He published several papers including the co-written: *Catalogue of caddisflies (Trichoptera) collected by Kazuo IJIMA* (2014).

Ikom

Blue Cascader *Zygonyx ikomae* **Pinhey**, 1961
[JS *Zygonyx natalensis* Martin, 1900, A. Powdered Cascader, Scuffed Cascader]

This is a toponym; it is named after the locality where the species was found in eastern Nigeria.

Imms

Little Blue *Ischnura immsi* **Laidlaw**, 1913
[Syn. of *Amphiallagma parvum* (Selys, 1876)]

Augustus Daniel Imms (1880–1949) was an English entomologist. He studied zoology at London University, which awarded his BSc (1903). After a further two years' study in Birmingham he was granted a scholarship to Christ's College Cambridge which awarded his BA (1907) the same year that Birmingham awarded his DSc. Unable to find a job as entomologist, he took the post of Professor of Biology at the University of Allahabad. He had to teach alone for four years until he had trained a lecturer and two demonstrators. He was then offered and took the post of Forest Entomologist by the Indian Government based at Dehra Dun (1911) studying pests of conifers. He collected the type (1912) and then (1913) returned to England, becoming Reader in Agricultural Entomology at Manchester University. He three times tried to enlist during WW1, but was rejected because of his poor health. He became (1918–1931) Chief Entomologist at the Experiment Station of the Ministry of Agriculture. During this time, he also wrote text books and visited the US, Canada and Hawaii. He took the post of Reader in Entomology at Cambridge until retirement (1931–1945). He wrote many papers and a number of books, the most important being his: *General Textbook of Entomology* (1925), which has had many subsequent editions.

Indraniel

Midget sp. *Mortonagrion indraniel* **Dow**, 2011

Professor Dr Indraneil Das (b.1964) is an Indian herpetologist now based at University of Malaysia, Sarawak, where he is an associate professor (1998). After early education in India, he received his doctorate in animal ecology from Oxford. He was then a postdoctoral research fellow at the University of Brunei, Darussalam, and a Fulbright Fellow at Harvard University. His research focus is on ecology, systematics and biogeography of the amphibians and reptiles of tropical Asia. He has written many books, including: *Biogeography of the Reptiles of South Asia* (1996). Two reptiles are also named after him. He was honoured as the person who first introduced the author to the UNIMAS peat swamp forest, where the type series were collected.

Inglis

Lyretail sp. *Stylogomphus inglisi* **Fraser**, 1922
Pincertail sp. *Lamellogomphus inglisi* Fraser, 1924
[JS *Onychogomphus risi* (Fraser, 1922)]

Charles McFarlane Inglis (1870–1954) was a Scottish naturalist and planter, who went to India (1888). He was Curator, Darjeeling Museum (1926–1948). He collected the clubtail holotype, a male, from the Tista River area of Darjeeling District (1920) and the pincertail (1923).

Ingrid

Sprite sp. *Pseudagrion ingrid* **Theischinger**, 2000
Grapian Darner *Austroaeschna ingrid* Theischinger, 2008

Ingrid Jones (b.1999) is the granddaughter of the author Dr G Theischinger (b.1940).

Injibandi

Pilbara Archtail *Nannophlebia injibandi* **Watson**, 1969

The people of the Injibandi tribe were the original inhabitants of Millstream.

Inkiti

Bladetail *Lindenia inkiti* Bartenev, 1929
[JS *Lindenia tetraphylla* (Vander Linden, 1825)]

This is a toponym; Inkiti Lake is in Georgia, West Caucasus, and the type locality.

Inoue, K

Slendertail sp. *Leptogomphus inouei* **Karube**, 2014

Kiyoshi Inoue (b. 1932) is a Japanese amateur odonatologist from Osaka. He graduated (1957) from the Faculty of Engineering, Osaka University and worked as an engineer in companies such as Unitika Ltd. and Itochu Corporation. He started to collect dragonflies before entering elementary school. He has published (since 1956) almost 300 papers and notes on Japanese dragonflies and a few papers on other Oriental Odonata. He is co-author of: *The dragonflies of Japan in colour* (1985; reprinted in 1992 and 2005), one of the major books on the Japanese Odonata and is co-author of two Japanese gomphid subspecies (1973) and first author of *Somatochlora taiwana* (2001). His other dragonfly books include *All about dragonflies* (1999). He has worked actively within both the local and international odonatological communities and was awarded the Membership of Honour of the Societas Internationalis Odonatologica (1989) and was similarly honoured by the Japanese Society of Odonatology (2011). He is a keen photographer.

Inoue, Y

Satinwing sp. *Euphaea inouei* **Asahina**, 1977
[Orig. *Euphaea guerini inouei*]

Yasuo Inoue is an amateur Japanese entomologist, who collected 81 species of dragonflies during his stay in southern Vietnam (1958–1962), including collecting the type in South Vietnam (1958).

Inyang

Inyanga Jewel *Platycypha inyangae* **Pinhey**, 1958

This is a toponym; Inyanga is an area in southeastern Africa, which also contains Mount Inyanga and is the area where this species is found and was first collected.

Ioganzen

Banded Demoiselle ssp. *Calopteryx splendens johanseni* **Belyshev**, 1955
[JS *Calopteryx splendens ancilla* Selys, 1853]

Professor Bodo Germanovich Ioganzen (1911–1996) was one of the initiators of the development of scientific ecology in the USSR. His textbook: *Basics of Ecology* was for a long time the only one in Russia. After graduating (1932) from Tomsk University he worked at the West Siberian Fish Breeding Station in Tomsk. He was also the Rector of Tomsk Pedagogical Institute (1964–1971). He later worked at Tomsk State University (1974–1985). He participated in (and organised) more than forty expeditions over West Siberia, published about forty books, twenty brochures, and a large number of papers. His scientific interests were in fish breeding and introduction, ecology, hydrobiology and conservation of nature. Belyshev wrote (translated): "*The form described is suggested to be named Calopteryx splendens johanseni Belyshev subsp. nova in honour of B.G. Ioganzen (Tomsk), who favoured my odonatological studies*". Since Ioganzen was Russian German, Belyshev used the Western spelling of his name – Johansen.

Irala

Dancer sp. *Argia iralai* **Calvert**, 1909

Domingo Martínez de Irala (1509–1556) was a Spanish Basque conquistador. He left for the Americas (1535) under Pedro de Mendoza and took part in the founding of Buenos Aires. He explored the Parana and Paraguay Rivers with Juan de Ayolas, when the Paragua Indians killed the latter with most of his men and Irala was elected as commander (1538). He abandoned Buenos Aires (1541) and moved the citizens to Asuncion, Paraguay, which he founded – the country where the type was collected and the city where he died. He was appointed Governor by Charles V (1552) and ruled by force, establishing public buildings and churches and subjugating the local people.

Irene

Western Spectre *Boyeria irene* **Fonscolombe**, 1838
Sedge Sprite *Nehalennia Irene* **Hagen**, 1861
Bannerwing sp. *Cora irene* **Ris**, 1918

The Greek goddess Eirene was one of the Horae.

Irene (Princess)

Threadtail sp. *Nososticta irene* **Lieftinck**, 1949

[Orig. *Notoneura irene*]

Flutterer ssp. *Rhyothemis princeps irene* Lieftinck, 1942

Princess Irene Emma Elisabeth of the Netherlands (Irene van Lippe-Biesterfeld, b.1939). Both species are from New Guinea. The holotype of *Nostosticta* was collected (March 1939) a few months before the Princess was born.

Irma

Damselfly sp. *Megalestes irma* **Fraser**, 1926

Irma Bailey (née Cozens-Hardy) (1896–1988) was the daughter of William Hepburn Cozens-Hardy, 2nd Baron Cozens-Hardy (1869–1924) and the wife (1921) of Lieutenant-Colonel Frederick 'Eric' Marshman Bailey (1882–1967). He was a British intelligence officer and a keen naturalist working in the Himalayan region. Irma followed her husband to the outposts where he was stationed and is said to have been almost certainly the first European woman to enter Bhutan. She travelled with him to Sikkim, Bhutan and Tibet. Charles McFarlane Inglis had collected the type in Sikkim and asked Fraser to name the species after the Hon. Irma Bailey, whose husband had contributed so much to the collecting of dragonflies in Sikkim and Tibet.

Isa

Cascader ssp. *Zygonyx iris isa* **Fraser**, 1926

Isa is a feminine forename, but not a specified one. The authors gave no etymology, so it is not possible to know for sure if this was named after a particular person. However, the lack of a Latin declension suggests it is not an eponym, as does Fraser's frequent use of apparent forenames and that many *Zygonyx* species have feminine names.

Isabel (Conesa García)

Mediterranean Bluet *Coenagrion caerulescens isabelae* Conesa García, 1995

[JS *Coenagrion caerulescens* (Fonscolombe, 1838)]

Isabel María Muñoz Castillo (b.1954) is the wife of the author Miguel Ángel Conesa García (b.1954). She is head of pharmacy at a Malaga hospital.

Isabel (O'Farrell)

Sydney Flatwing *Austroargiolestes isabellae* **Theischinger** & **O'Farrell**, 1986

Mrs Isabel O'Farrell née Millen was the wife of the second author Professor Anthony Frederick Louis O'Farrell (1917–1997). (Also see **O'Farrell**)

Isabela

Limniad sp. *Amphicnemis isabela* Gapud, 2006

[JS *Luzonobasis glauca* Brauer, 1868]

This is a toponym; it is named after Isabela Province, The Philippines.

Isabelle (Ysabel)

Damselfly sp. *Lieftinckia isabellae* **Lieftinck**, 1987

This is a toponym; Santa Ysabel Island is part of the Solomon Isles where the type was collected.

Ishida

Shadow-Emerald sp. *Macromidia ishidai* **Asahina**, 1964

Shozo Ishida (b.1930) is a Japanese entomologist, odonatologist and educator. He graduated from the Agriculture Department of Mie University. He is a representative director of a company in Yokkaichi, Mie. Among other works he co-wrote: *Dragonflies of the Japanese Archipelago in Colour* (2001), one of the major handbooks of Japanese Odonata. He described three new dragonfly species from the Ryukyu islands. He collected part of the type series in Ishigaki Islands (May 1963). (Also see **Shozo**)

Ishigaki

Phantomhawker sp. *Planaeschna ishigakiana* **Asahina**, 1951
Velveteen sp. *Bayadera ishigakiana* Asahina, 1964

[Orig. *Bayadera brevicauda ishigakiana*]

This is a toponym; Ishikagi Island is where these species were found.

Isidro Mora

Large Blue Sprite *Pseudagrion isidromorai* Compte Sart, 1967
[JS *Pseudagrion glaucoideum* Schmidt in Ris, 1936]

Isidro Mora is the brother of José Mora and assistant to Fernando Martorell. He collected the holotype in Spanish Guinea (August 1962). (See **José Mora** & **Martorell**)

Itatiaia

Threadtail sp. *Forcepsioneura itatiaiae* **Santos**, 1970
[Orig. *Phasmoneura itatiaiae*]

This is a toponym; Itatiaia is a National Park and Brazilian municipality of the state of Rio de Janeiro where the species was found.

Ivanov

Azure Bluet *Coenagrion puella ivanovi* **Belyshev**, 1955
[JS *Coenagrion puella* (Linnaeus, 1758)]

The etymology is not given.

Ivy

Shining Clubtail *Stylurus ivae* **Williamson**, 1932

Ivy Ditzler née Lesh (1882–1976), along with her husband (W H Ditzler) and daughter Laura (q.v.), was a member of the party that collected the holotype. The collecting trip (1931) was from Indiana to Georgia and all three members of the family collected enthusiastically.

J

Jackson (Port)

Emerald sp. *Procordulia jacksoniensis* **Rambur**, 1842

[Orig. *Cordulia jacksoniensis*]

This is a toponym; Port Jackson is a natural harbour in New South Wales, Australia that contains Sydney. Many taxa are named after the area because they originate there. However, many also were exported through there by naturalists of the time collecting in Australia, such as Guerin from whose collection Rambur obtained this holotype. Sometimes the museum labels refer not to the actual type locality, but the port of export.

Jackson, T H E

Yellow-sided Jewel *Stenocypha jacksoni* **Pinhey**, 1952
[Orig. *Chlorocypha jacksoni*]
Red-veined Dropwing *Trithemis jacksoni* Pinhey, 1970
[JS *Trithemis arteriosa* **Burmeister**, 1839]

Variable Sprite ssp. *Pseudagrion sjoestedti jacksoni* Pinhey, 1961

Thomas Herbert Elliot 'Pinkie' Jackson (1903–1968) was an English entomologist who collected in East Africa. He attended Wellington College, as his soldier father hoped he would enter the military, but instead he went on to study at Harper Adams Agricultural College. He visited Kenya (1923) but went to India to work on an indigo plantation. However, he returned to Kenya (1924) and settled there as a coffee grower buying a farm, where his father joined him. He joined the army at the outbreak of WW2 and was sent to raise a company of local soldiers which fought on the border of Ethiopia. By the end of the war he had risen to the rank of Lieutenant Colonel. Returning to coffee growing he continued with his boyhood interests in ornithology, botany and entomology and created a garden at his plantation. He is best known for his interest in African butterflies and he took part in the BMNH Ruwenzori Expedition (1935) as entomologist. He also trained local collectors and built a large collection, as well as discovering a number of new species. He donated the collection, half to BMNH and half to the National Museum of Kenya. He collected the Jewel holotype in Mafuga Rain Forest, Kigezi, Uganda (1951). He wrote or co-wrote a number of papers such as: *Notes on the Rhopalocera of the Kigezi district of Uganda with descriptions of new species and subspecies.* (1956). A 'criminal gang' murdered him at his home.

Jacob

Tawny Spreadwing *Lestes jacobi* **R Martin**, 1910
[JS *Lestes ictericus* Gerstäcker, 1869), A. Yellow Spreadwing]

The author gave no etymology so there is no way to tell if this was after a particular Jacob and if so whom.

Jacobson

Clubtail sp. *Burmagomphus jacobsoni* **Ris**, 1912
[JS *Burmagomphus inscriptus* (Hagen, 1878)]

Edward Richard Jacobson (1870–1944) was a Dutch businessman and skilled amateur naturalist. He managed a trading company in Java, but also lived for some years in Sumatra. He made extensive collections for Dutch museums, leaving his business (1910) to devote himself to natural history. His main interest was entomology and he collected the type in Java (1910). He died in an internment camp during the Japanese occupation. Two birds, two reptiles and an amphibian are also named after him.

Jagor

The genus *Jagoria* **Karsch**, 1889
[S *Oligoaeschna* Selys, 1889]

Professor Dr Andreas Fedor Jagor (1816–1900) was a German ethnographer, explorer and naturalist who travelled in Asia in the second half of the 19th century, collecting for the Berlin Museum. He spent much time in the Philippines, but was also in India (1859–1861) and the Dutch East Indies (1873–1876 & 1890–1893). He collected the type in Luzon (1889). He wrote: *Reisen in den Philippinen* (1873). He described the country thus: "*Few countries in the world are so little known and so seldom visited as the Philippines, and yet no other land is more pleasant to travel in than this richly endowed island kingdom. Hardly anywhere does the nature lover find a greater fill of boundless treasure.*" He also wrote about the Dutch East Indies and southern Malaya. His ethnographical and art collection was bequeathed to the Berlin Museum. Two mammals, two reptiles and a bird are also named after him.

Jalmos

Threadtail sp. *Peristicta jalmosi* Pessacq & **JM Costa**, 2007

Jalmos Costa (1954–2012) is the husband of the second author, Professor Dr Janira Pedreira Martins Costa (b.1941) (q.v.). (See **Janira** below)

Jan (Van Tol)

Shadowdamsel sp. *Telosticta janeus* **Dow & Orr**, 2012
Limniad sp. *Sangabasis janvantoli* Villanueva & Dow, 2014

Dr Jan van Tol. (See **Van Tol** & also **Tol**)

Jane

Jane's Meadowhawk *Sympetrum janeae* Carle, 1993
[A. Cherry-faced Meadowhawk]

Jane Carle is the wife of the author, Dr Frank Louis Carle, the *Curator of the Rutgers Entomological Museum and Director of the New Jersey Aquatic Insect Survery.*

Janet

Coralleg sp. *Rhipidolestes janetae* **KDP Wilson**, 1997

Hannah Janet Wilson née Killa (b.1957) from Barry, South Wales, is the wife (1980) of the author Keith Duncan Peter Wilson (b.1953) (q.v.). They were both living in Hong Kong at the time of the discovery of the flatwing on Lantau Island.

Janice (Costa)

Threadtail sp. *Peristicta janiceae* Pessacq & **JM Costa**, 2007

Janice Martins Costa (b.1968) is the daughter of the second author, Professor Dr Janira Martins Costa. (See **Janira** below)

Janice (Peters)

Flatwing sp. *Caledargiolestes janiceae* **Lieftinck**, 1975

Janice Goldthwaite Peters (b.1939) is an entomologist who was Curator of the W L Peters Aquatic Entomology Research Collections at Florida Agricultural and Mechanical University, Tallahassee, Florida, which is affiliated to the Florida State Collection of Arthropods. The University of Colorado awarded her BS and her MS was awarded by Florida A&M University. She also has experience as an illustrator, research associate and editor. Together with her husband Dr William Lee Peters (1939–2000) she collected the type in the forests of New Caledonia during a 3-month research trip sponsored by the National Geographic Society (1972). She co-authored a number of works with her husband and others, mostly concerned with mayfly taxonomy and the behavior of *Dolania americana* (American Sand-burrowing mayfly). She co-edited *Proceedings of the First International Conference on Ephemeroptera* (1973). She received the Lifetime Achievement Award (2008), from the International Conferences on Ephemeroptera, and was Pioneer Lecturer for the Florida Entomological Society (2009).

Janira

Forceptail sp. *Aphylla janirae* **Belle**, 1994
Threadtail sp. *Phasmoneura janirae* **Lencioni**, 1999
Threadtail sp. *Epipleoneura janirae* **Machado**, 2005
Threadtail sp. *Drepanoneura janirae* von Ellenrieder & **Garrison**, 2008

Professor Dr Janira Pedreira Martins Costa (b.1941) is a Brazilian entomologist and odonatologist. She graduated in biological sciences (zoology) in Gama Filho University (1967) and was awarded her masters (1977) and doctorate (1985) by the Federal University of Rio de Janeiro (UFRJ). She became Assistant Professor of Zoology at UFRJ (1974) and later Professor and Dean. She was Director of Museum Nacional UFRJ (1994–1998). Her main areas of interest are systematics, taxonomy, biology and ecology of Brazilian Odonata, and the systematics and ecology of aquatic insects in general. Her fieldwork has mainly been in south-eastern Brazil, but also elsewhere in Brazil and in Paraguay, Uruguay and Argentina. She has published more than 100 papers spanning five decades, including: *Revisão do gênero Oxyagrion Selys 1876 (Odonata, Coenagrionidae)* (1978) and *A especiação em Mnesarete pudica (Hagen in Selys, 1853) Cowley, 1934 com a descrição de uma nova subespécia (Odonata: Agrionidae)* (1986). She has described c. 40 new South American Odonata species and named three new genera. (Also see **Janice**)

Janny

Western Scissortail *Microgomphus jannyae* **Legrand**, 1992
Janny's Knobtail *Epigomphus jannyae* **Belle**, 1993

Professor Dr Janny Margaretha van Brink (1923-1993). (See **Van Brink**)

Jarol

Relict Damsel sp. *Amphipteryx jaroli* Jocque & Argueta, 2014

Jarol Ramón Estrada Ramos (b.1981) is a nature guide in Honduras where he collected the type (2012). The etymology states that it was named "...*after our friend and guide through the cloud forest on our first expedition in Pico Bonito National Park (2012); Jarol Estrada. Jarol collected the first specimen of this species.*"

Jeannel

Clear-winged Piedface *Thermochoria jeanneli* **Martin**, 1915
[Orig. *Passeria jeanneli*]

René Gabriel Jeannel (1879–1965) was a French entomologist and coleopterist. He collected in east Africa (1911–1912) with Charles Alluaud (q.v.) during which time the type was taken; he also co-wrote the report on the scientific results of that expedition: *Voyage de Ch. Alluaud et. R. Jeannel en African Orientale.* He also worked in Africa and specialized in Leiodidae, but wrote a lot on other Coleoptera. His most important work was on the insect fauna of caves in the French Pyrenees and the Romanian Carpathians (as a member of the Romanian Academy). He was Director of the MNHN (1945–1951). Among other works he wrote: *Faune cavernicole de la France* (1940) and *La genèse des faunes terrestres* (1942).

Jean-Yves Meyer

Bluetail sp. *Ischnura jeanyvesmeyeri* Englund & **Polhemus**, 2010

Dr Jean-Yves Meyer (b.1968) is a French ecologist. The University of Montpellier awarded his PhD (1994), and he was a postdoctoral research scholar, University of Hawaii (1997). He then became Scientific Director, Conservatoire Botanique National de Mascarin, La Réunion Island (2001) and Research Associate in Botany, National Tropical Botanical Garden of Hawaii (2008). In 2002 he took his present post as Chargé de Recherche, Délégation à la Recherche, Ministère de l'Education, de l'Enseignement Supérieur et de la Recherche Gouvernement de Polynésie Française, which funded the two collecting trips to the Austral Islands made by the senior author. He is primarily a plant ecologist, field botanist and conservation biologist, and a recognized expert on invasive alien species. He is actively involved in research studies on the native and alien flora and fauna of French Polynesia, and organizes multidisciplinary field-expeditions to remote islands. He collaborates as a consultant on numerous biodiversity and biological invasion initiatives across the Pacific (South Pacific Regional Environmental Programme, Pacific Invasives Learning Network,

Conservation International Micronesia-Polynesia Hotspot, IRD New-Caledonia, National Park of Rapa Nui, Service de l'Environnement of Wallis & Futuna). He has been a guest lecturer for the Université de Polynésie Française and for the Gump South Pacific Research Station on Moorea and he advises and mentors students working in French Polynesia, with projects in the areas of botany, plant ecology, conservation biology, terrestrial biodiversity and invasive species management. He is also an executive committee member for the Moorea Biocode project. He also has several plants named after him.

Jedda

Dusky Riverdamsel *Pseudagrion jedda* **Watson** & **Theischinger**, 1991

Jedda was a film (1955), parts of which were filmed in the type locality, Katherine Gorge, Northern Territory, Australia.

Jerrell

Spreadwing sp. *Lestes jerrelli* **Tennessen**, 1997

Jerrell James Daigle (b.1950) is treasurer and a past president of the Dragonfly Society of The Americas. He worked for the Florida Department of Environmental Regulation. The spreadwing was dedicated to him "...*in recognition of his contributions to New World Odonatology and his enthusiasm for the study of dragonflies.*" Among his publications are: *A Checklist of the Odonata from orange County, Florida* (1978) and *Florida damselflies (Zygoptera): A species key to the aquatic larval* (1991). He has described 15 new species of South American Odonata.

Jesse

Pond Clubtail sp. *Agriogomphus jessei* **Williamson**, 1918
[Orig. *Ischnogomphus jessei*]
Purple Skimmer *Libellula jesseana* Williamson, 1922
Duskhawker sp. *Gynacantha jessei* Williamson, 1923
Wedgetail sp. *Acanthagrion jessei* **Leonard**, 1977

Jesse Hunter Williamson (1884–1964) was the cousin of the author, Edward Bruce Williamson (q.v.). He accompanied his cousin on a three and half month expedition to Colombia and Panama (1916–1917) specifically to collect Odonata. He spent six months (1920) on the Cornell University Entomological Expedition to South America, collecting dragonflies in Venezuela and Peru. He had to leave the expedition ahead of schedule due to a severe case of dysentery. When he recovered, he was arrested and jailed in San Ramón, Peru on suspicion of being a Chilean spy, only being released due to the intervention of the Peruvian President (A B Leguía). Interestingly, his passport application had noted that he had no permanent address due to his work as a special agent of the US Department of Justice and US Army. In it he offered the US Secretary of State to be of 'any service to your department en route'. Later he participated in expeditions to Brazil (1922) and Mexico (1923). He worked (1940s) in the office of the Caylor-Nickel Clinic, Blufton. Three of the eponymous species were collected during these South American sojourns and *Libellula jesseana* was collected by him in Florida (1921). He was co-author with his cousin of sixteen new odonate species descriptions and of one new genus, published in four papers, including: *The genus Perilestes (Odonata)* (1924) and *Five new Mexican dragonflies* (1930), having collected many of the new species himself. (Also see **Williamson** & **Bruce**)

Jillian

Jill's Shadowcruiser *Idomacromia jillianae* **Dijkstra** & Kisakye, 2004

Jillian Dorothy Silsby (b.1925) is a British odonatologist. The etymology reads: "...*this species is named in honour of a unique lady in odonatology, Jill Silsby, who has done so much for this field of science and has produced a handbook which is of great value to all workers.*" (See **Silsby**)

Jo Allyn

Dancer sp. *Argia joallynae* **Garrison** & von **Ellenrieder**, 2015

Jo Allyn Garrison (b.1948) was the first wife of the author Dr Rosser William Garrison (b.1948). (Also see **Garrison**)

Joana

Threadtail sp. *Neoneura joana* **Williamson**, 1917

Jane Atkinson was the wife of Dr D A Atkinson of the Carnegie Museum of Natural History, Pittsburgh, USA, about whom the author wrote: "...*a companion on numerous collecting trips to whom I am indebted for many specimens.*" She collected the type in British Guiana (1912).

Joanetta

Shadowdamsel sp. *Palaemnema joanetta* **Kennedy**, 1940

Mary Janet Bur née Kennedy (1931–2015) was the daughter of the author, Dr Clarence Hamilton Kennedy (1879–1952). She became a nutritionist working as a dietary aide at Howard Young, having been awarded a home economics degree at Michigan State University. She met her husband (1956) (Donald Bur) at the University of Michigan biological station. (Also see **Kennedy**)

Joannis

Blue-eyed Darner sp. *Rhionaeschna joannisi* **Martin**, 1897
[Orig. *Aeshna joannisi*]

Abbé Joseph de Joannis (1864–1932) was a French entomologist who was twice President of la Société Entomologique de France (1908 & 1916). Apart from his religious education, he also held a PhD in physics and mathematics. (An older relative, also called Joseph, went to Egypt with Napoleon (1798–1801) and was given the job of making sure that the obelisk that Napoleon took got to Paris, where it stands in the Place de la Concorde.) He worked for a time with his brother obtaining specimens from missionaries. His brother prepared the specimens and Joseph classified and published descriptions. He also had long stays in the island of Jersey, Canterbury (England) and in Louvesc (France: Ardeche) where he collected Lepidoptera and discovered several new species. Every year he went to Leiden to study the material not available in Paris. He was asked (1902) to study the Lepidoptera of

Tonkin (Indochina), on which he published (1928–1929). He was also very knowledgeable on the fauna of the Mascarenes, on which he published many papers. He was a familiar figure in the NMNH in Paris where he studied each week and his collection is held there.

Joaquim

Leaftail sp. *Phyllogomphoides joaquini* Rodrigues, 1992

Joaquim Rodrigues Capitulo (1910–1980) was the father of the author Alberto Rodrigues Capitulo (b.1953). The type was found in Argentina much earlier (1938). Although his name ended with an 'm' the patronym uses an 'n'.

Jocaste

Helicopter sp. *Mecistogaster jocaste* **Hagen**, 1869

Jocaste is a character from Greek mythology. She was a daughter of Menoeceus and Queen consort of Thebes and, more famously, the mother of Oedipus.

Johansen

(See **Ioganzen**)

Johanson

Arctic Bluet *Coenagrion johanssoni* Wallengren, 1894

[Orig. *Agrion johanssoni*]

Carl Hans Johanson (1828–1908) was a Swedish biologist considered by his peers to be an 'outstanding' teacher. He gained his master's degree (1854) and started teaching natural history at a lyceum in Vesterås. He retired (1894) and was granted an honorary doctorate (1904). He wrote: *Odonata Sueciae: Sveriges trollsländor* (1859), a thorough handbook of the 49 dragonfly species known from Sweden at that time. It included a description of a new species, *Agrion concinnum*. His compatriot H D J Wallengren noticed that the name was preoccupied and he replaced it (1894) with *Agrion johanssoni*, accidentally making a spelling error in the species name. Johanson was interested in many insect orders and made a large collection of Swedish insects, but did not publish much. His other interests included gardening and mushrooms. He was also an accomplished pianist.

John Deacon

Flatwing sp. *Heteragrion johndeaconi* **Lencioni,** 2013

John Richard Deacon (b.1951) is a bass guitarist and songwriter, about whom the author wrote: "*…whose wonderful sound and magnificent lyrics have enchanted the world for over four decades*". Four new species were described in tribute to the 40th anniversary of the rock band Queen – one for each of its original members.

Johnsen

Clubtail sp. *Burmagomphus johnseni* **Lieftinck**, 1966

Palle Johnsen (1921–2002) worked at the Zoological Institute, Naturhistorisk Museum, Aarhus, Denmark. He collected the type in Thailand (November 1961). He wrote on

Orthoptera and also *Parasitic worms in humans* (1983). *Notes on Fishes along the River Kwae Noi in Western Thailand* (1963) is about his collecting trip as part of the Thai-Danish Prehistoric Expedition (1961–1962).

Johnson

Siberian Hawker *Aeshna johnsoni* Steinmann, 1997
[JS *Aeshna crenata* Hagen, 1856]

The late Henrik Steinmann gave no etymology. This is almost certain to remain unknown. Six new names introduced by Steinmann (1997) were unnecessary replacement names and have not been used since.

Jones

Jones' Forestwatcher *Notiothemis jonesi* **Ris**, 1921
[A. Eastern Forestwatcher, Eastern Elf]

W E 'Mamba' Jones (fl.1868–fl.1950) was an Australian businessman and naturalist who went to the Colony of Natal (now KwaZulu Natal), South Africa on hearing reports of gold strikes there, where he settled in Mfongosi (c.1886). Gold soon petered out and he ran a guest house there until his death. He was an amateur naturalist known best for his commercial snake collecting, and also collected Lepidoptera which he donated to the SAM (1903). Around 1909–1910 he appears to have been employed in the Department of Agriculture of Natal, as evidenced by the Seventh Report of the Government Entomologist for that period, in which it is mentioned that W E Jones, a member of staff, published a single article in the 'Natal Agricultural Journal'. He was the first to collect this dragonfly at Mfongosi, Zululand (1911). He collected mainly entomological specimens including ants and flies and particularly in Zululand (1911–1935). He discovered (1929) a stone-age settlement near his home and collected around 150 stone artefacts particularly of Neanderthals for BMNH and SAM, Cape Town. He also collected some fossils.

Jordans

Lesser Emperor *Anax parthenope jordansi* **Buchholtz**, 1955
[JS *Anax parthenope* Selys, 1839]

Professor Dr Adolf von Jordans (1892–1974) was a German ornithologist. He met Alexander Koenig (1912) who supported him as a patron for many years. He was involved (1921) with the construction of the Bonner Private Ornithological Museum and became Curator and later Deputy Director there. He became Director of the Koenig Zoological Museum (1947–1957) Bonn, Germany, after Koenig's death and later (1951) professor there. He wrote more than 40 papers and longer works (1913–1970). His main focus was on zoological geographical variation of birds in particular. Buchholtz dedicated the species to Jordans, being grateful that Jordans had requested that Buchholtz undertake studies of Aegean species. Shortly after retiring he had an operation on his larynx (1957), after which his health deteriorated. He was on holiday when he had emergency surgery for an intestinal problem, but died after the operation. Five birds are also named after him.

Jørgensen

Sanddragon sp. *Progomphus joergenseni* **Ris**, 1908
Dancer sp. *Argia joergenseni* Ris, 1913

Peter 'Pedro' Jørgensen (1870–1937) was a Danish naturalist and entomologist. He was educated as a teacher of English and German (1889). He suffered from tuberculosis (1892) all his adult life, which prompted his decision to leave Denmark for a drier climate. He accompanied Jensen-Haarup (q.v.) to Argentina (1906) where they collected insects which were mostly sold to the German entomologist Heinrich Friese to cover their travel expenses. For some time, he worked in the service of the Ministry of Agriculture at Buenos Aires and he remained in South America for the rest of his life, eventually settling to farm and study insects in Villa Rica, Paraguay. He continued to send plants and insects back to European and American Museums to help finance his work. He wrote several papers about his travels as well as a number revising various insect families. He was found murdered on his farm. Ris received the dragonfly specimens collected by him and Jensen-Haarup (1904–1911) for study via Esben-Petersen (q.v.).

Jorina

Cadet sp. *Pseudothemis jorina* **Förster**, 1904

Jorina is a feminine forename, but not a specified one. The authors gave no etymology, so it is not possible to know for sure if this was named after a particular person. However, the lack of a Latin ending suggests it is not an eponym, as does Förster's frequent use of apparent forenames.

José Mora

Gabon Threadtail *Elattoneura josemorai* Compte Sart, 1964

José Mora, brother of Isidro Mora (q.v.) and assistant of Fernando Martorell (q.v.), collected the type in Spanish Guinea (now Equatorial Guinea) (1960). (Also see **Isidro Mora** & **Mortarell**).

Julia

Chalk-fronted Corporal *Ladona julia* **Uhler**, 1857
[Orig. *Libellula julia*]
Julia Skimmer *Orthetrum julia* **Kirby**, 1900

Julia is a feminine forename that originally comes from the Latin meaning youthful, soft-haired, beautiful or vivacious. It is the feminine form of Julius. It is possible that Uhler's name refers to the white pruinescence associated with this dragonfly.

Juliana

Dragonlet sp. *Erythrodiplax juliana* **Ris**, 1911
[Orig. *Erythrodiplax nigricans juliana*]

Ris used the manuscript name of Selys. Its etymology is not known, but Juliana is feminized version of Julianus, which is derived from the name Julius.

Juliana

Flutterer ssp. *Rhyothemis regia juliana* **Lieftinck**, 1942

Queen Juliana of the Netherlands – Juliana Louise Emma Marie Wilhelmina (1909–2004), Queen regnant of the Netherlands (1948–1980). Lieftinck named (1942) three New Guinean *Rhyothemis* subspecies after members of the Dutch royal family without providing any etymology. Undoubtedly this was done in a surge of emotional patriotism at the time when Lieftinck and other Dutch citizens living in the Dutch East Indies felt particularly isolated and worried about the fate of their nation, which had been occupied by German troops (May 1940). (Also see **Beatrix** & **Irene**).

Julius

Anax parthenope julius **Brauer**, 1865
[Orig. *Anax julius*]

Obviously named after Gaius Julius Caesar (100BC–44BC), a Roman general, statesman and consul. (Also see **Imperator** & **Junius**)

Junghuhn

Clubtail sp. *Megalogomphus junghuhni* **Lieftinck**, 1934

Friedrich Franz Wilhelm Junghuhn (1809–1864) was a German-Dutch botanist. He studied medicine in Halle and Berlin (1827–1831), but even during that time he was publishing on botanical subjects. He suffered from depression and was once involved in a duel in which he was injured, but his opponent died. He fled and joined the Prussian army as a surgeon, but was discovered and sentenced to ten years in prison. He feigned insanity and escaped (1833) and was briefly in the French Foreign Legion, but released because of his poor health. He enlisted in the Dutch colonial army also as a physician and was posted to the East Indies (1835). Settling in Java he studied the country, its people and natural history, making many expeditions about which he wrote extensively including: *Die Topographischen und Naturwissenschaftlichen Reisen durch Java* (1845). He returned to the Netherlands in ill health (1849) where he wrote socialist and humanist books. He returned to Java (1855) where he remained until his death from liver disease. On his deathbed in his house near Lembang on the slopes of the volcano Tangkuban Perahu just north of Bandung, Java, Junghuhn asked the doctor to open the windows, so he could say goodbye to the mountains that he loved. In Lembang there is a small monument to his memory in a grassy square named after him, planted with some of his favourite trees among which is the Cinchona.

Junius

Common Green Darner *Anax junius* **Drury**, 1773
[Orig. *Libellula junia*]

Lucius Junius Brutus was the founder of the Roman Republic. He was consul (509BC).

Jurzitza

Spreadwing sp. *Lestes jurzitzai* **Muzón**, 1994
Shadowdamsel sp. *Drepanosticta jurzitzai* **Hämäläinen**, 1999

Threadtail sp. *Neoneura jurzitzai* **Garrison**, 1999
Clubtail sp. *Microgomphus jurzitzai* **Karube**, 2000

Eastern Hawk ssp. *Austrocordulia refracta jurzitzai* **Theischinger**, 1999

Professor Dr Gerhard Roman Anton Jurzitza (1929–2014) was the doyen of German odonatology. His first degree was from Karlsruhe University, which also awarded his doctorate (1959). He worked at Braunschweig Institute of Pharmacology for two years before returning to the Karlsruhe Department of Botany for the rest of his professional career. He became associate professor (1973) and full professor (1979) until retiring as professor emeritus (1992). His work generally focused on the relationships between insects and plants. He undertook numerous odonatological trips and expeditions ranging from France (1960s) through to the US (1970s & 1990s) and Japan (1980s). He made four major expeditions to Uruguay, Chile, Argentina and Brazil (between 1974 and 1988), one lasting a full year. He published two field guides: *Unsere Libellen: Die Libellen Mitteleuropas in 120 Farbfotos* (1978) and *Welche Libelle is das? Die Arten Mittel- und Südeuropas* (1988) and the botanical handbook: *Anatomie der Samenpflanzen* (1987). His numerous papers included descriptions of eight new species of Odonata. He was a keen and excellent photographer of Odonata and an outspoken conservationist. He suffered poor health for the last decade of his life. During that time, he generously settled his odonatological legacy to ensure nothing would be lost and it would continue to be studied – by transferring his Odonata collection to the Senckenberg Museum and donating his rich collection of reprints to Dr Florian Weihrauch. (Aslo see **Margarete**)

K

Kafwi

Bog Skimmer *Orthetrum kafwi* **Dijkstra**, 2015

This is a toponym; the species is named after the Kafwi River in Katanga, the source of which is the type locality.

Kahl

Shortwing sp. *Perilestes kahli* **Williamson** & Williamson, 1924

Dr Paul Hugo Isador Kahl (1859–1941) was a Swedish-born American entomologist. He graduated in Sweden, studied briefly at Kansas University and was awarded an honorary doctorate by the University of Pittsburgh (1927) and was advisory Professor of Biology there for many years. He was Curator of the Entomological Museum at Kansas University (1894–1901). He was Custodian of Entomology Collections (1902–1923) then Curator of Entomology, Carnegie Museum (1923–1940). His special interest was Diptera and he was, according to Avinoff, the Carnegie Director: *"…noted for his masterly command of entomological bibliography."* The King of Sweden knighted him (1940) for his work, broadening relations between Sweden and the US.

Kaiser

Damselfly sp. *Igneocnemis kaiseri* **Gassmann** & **Hämäläinen**, 2002
[Orig. *Risiocnemis kaiseri*]

Markus Kaiser (b.1943) is a Swiss conservationist and former archivist at the State Archive of St Gallen. He participated in Roland Müller's expedition to Samar in 1997, when the type specimen was collected. On Müller's request the species was named after Kaiser in: *"… appreciation of his contribution to the protection of natural habitat in eastern Switzerland."*

Kaize

Threadtail sp. *Nososticta kaizei* **Theischinger** & Richards, 2015

Yohannes (John) L A Kaize (b.1985) is from Papua New Guinea, Indonesia. During his time as a student of the Cenderawasih University he collected dragonflies on several entomological expeditions organised by Henk van Mastrigt (2006 at Japen island and Borme, 2008 at Biak-Supiori Island, Lelambo and Walmak and 2009 at Mioswaar Island). He also collected (2007, 2008 & 2009) in the southern lowlands of Papua, Indonesia. Most of the material collected by him has been published in 'Suara Serangga Papua' and 'Entomologische Berichten'. During the expeditions to the island of Biak-Supiori he collected the types of *Nososticta kaizei*.

Kali

Velveteen sp. *Bayadera kali* Cowley, 1936

Kali is the Hindu mother goddess, symbol of dissolution and destruction.

Kalkman

Archtail sp. *Nannophlebia kalkmani* **Theischinger** & Richards, 2011

Vincent Jeroen Kalkman (b.1974) is a Dutch entomologist at the Naturalis Biodiversity Center, Leiden, who works for the European Invertebrate Survey. Saxion University in Deventer awarded his first degree (1996) and Leiden University his PhD (2013). He worked on the dragonflies of southwest Asia, making a series of trips to Turkey (1999–2003), and later shifted his attention to New Guinea where he participated in three collecting expeditions. During a survey of Conservation International in the Muller Range in 2009 he collected the type of this eponymous archtail species. Material collected during these expeditions was used for a series of publications on the Odonata of New Guinea, many of which were made in co-operation with Albert Orr, Gunther Theischinger and Stephen Richards. He conducted fieldwork on dragonflies during trips to Australia (2003, 2009, 2011, 2012, 2014), Bhutan (2015, 2016), Brunei (2003, 2004), China (2005), India (2000), Indonesia (2006, 2008), Malaysia (2002, 2004), Mexico (2009), Namibia (2007), Papua New Guinea (2009), Philippines (2004), South Africa (2007) and Thailand (2001). He has worked extensively on odonates publishing papers on faunistics, conservation, taxonomy and biogeography. He has published many papers focused on Europe, Southwest Asia and the Australasian region. His major publications include: *Global diversity of dragonflies in freshwater* (2008), *European Red List of Dragonflies* (2010), *Generic revision of Argiolestidae (Odonata), with four new genera* (2013) and the *Field Guide to the damselflies of New Guinea* (2013), *Field Guide to the dragonflies of New Guinea* (2015). He is co-editor of *Atlas of European dragonflies and damselflies* (2015). He was a founder member of the Dutch Dragonfly Association (1997) and was chair of the IUCN Dragonfly Specialist Group (2007–2009). His publications include descriptions of 34 new dragonfly species and four new genera. He is currently (since 2015) involved in the study of the Bhutanese dragonfly fauna.

Kalliste

Yellow-spotted Emerald *Hermicordulia kalliste* **Theischinger** & **Watson**, 1991

Kalliste was the home of Dr M A Lieftinck (q.v.) and his wife Corrie, in Rhenen, Netherlands.

Kalumburu

Spot-winged Threadtail *Nososticta kalumburu* **Watson** & **Theischinger**, 1984

This is a toponym; Kalumburu is a locality in the Shire of Wyndham-East, Kimberley, Australia.

Kameliya

Bombardier sp. *Lyriothemis kameliyae* **Kompier**, 2017

Kameliya Petrova. The etymology reads: *"This species is named after Kamelia Petrova in gratitude for her assistance with my book 'Dragonflies of the Serra dos Orgaos' and unrelenting support for my continuing research on Odonata in Vietnam."* The author informed me that he had consulted her and she preferred not to share any further personal details.

Kao

Paddletail sp. *Sarasaeschna kaoi* Yeh, Lee & Wong, 2015

Yui-Ching Kao (Yao-Ting Gao) (1969–2010) was a Chinese mammalogist. The authors say in their etymology: *"The new species is named to memorize Mr. Yui-Ching Kao who passed away untimely in 2010 during field work. He was an approved and talented researcher in TFRI specialized in the ecology of aquatic plants and fishes, and he kindly encouraged and guided I L Lee during Lee's stay at his laboratory."*

Kaori

Wedgetail sp. *Acanthagrion kaori* **Machado**, 2012
[JS *Acanthagrion apicale* Selys, 1876]

Kaori is a Wai-Wai Indian who guided ABM Machado through the forest during his visit to the Wai-Wai Indian village at the Mapuera River where one paratype was collected (1982).

Karaja

Pond Damselfly sp. *Phoenicagrion karaja* **Machado**, 2010

The Karaja people live in the area where the species was first collected.

Karitiana

Pond Damselfly sp. *Tuberculobasis karitiana* **Machado**, 2009

The Karitiana people inhabit the municipality of Porto Velho, where the holotype was collected.

Karny

Threadtail sp. *Caconeura karnyi* **Laidlaw**, 1926
[JS *Prodasineura verticalis* Selys, 1860, A. Red-striped Black Bambootail]
Flatwing sp. *Argiolestes karnyi* **Fraser**, 1926
[JS *Celebargiolestes cincta* (Selys, 1886)]
Emerald sp. *Procordulia karnyi* Fraser, 1926

Dr Heinrich Hugo Karny (1886–1939) was an Austrian physician and entomologist specialising in Thysanoptera and Orthoptera. He worked as a physician for the Dutch East India Company. Among his works were: *Zur Systematik der Orthopteroiden Insekten, Thysanoptera* (1921) and *Beiträge zur Malayischen Orthopterenfauna* (1924). His collections are housed in a number of institutions including the Swedish Museum of Natural History and National University of Singapore.

Karsch

Genus *Karschia* **Förster**, 1900
[Homonym of *Karschia* Walter 1889 in Arachnida; replaced by *Plattycantha* Förster, 1908]
Genus *Karsciogomphus* **Schouteden**, 1934
[JS *Neurogomphus* Karsch, 1890]

Jewel sp. *Calophlebia karschi* **Selys**, 1896
[A. Prettywing]
Jewel sp. *Rhinocypha karschi* **Krüger**, 1898
[JS *Sundacypha petiolata* Selys, 1859]
Lancet sp. *Xiphiagrion karschi* **Ris**, 1898
[JS *Xiphiagrion cyanomelas* Selys, 1876]

Grenadier sp. *Agrionoptera karschi* Förster, 1898
[JS *Agrionoptera longitudinalis biserialis* Selys, 1879]

Ferdinand Anton Franz Karsch (or Karsch-Haack) (1853–1936) was a German zoologist, arachnologist, entomologist and anthropologist. He had a very early, systematic interest in nature and while still at school published a catalogue of spiders found in Westphalia (1870–1873) under the pseudonym Paul Grüne. He was educated at the Münster Academy for four terms then changed (1875) to Friedrich Wilhelm University in Berlin and published a revision of the gall wasps as his doctoral thesis (1877). He was given a position (1878) at the Museum für Naturkunde Berlin Zoological Museum of Berlin University, where he became curator until retirement (1899–1921). He also taught at Berlin Agriculture College (1881–1892) becoming professor. He published (1870–1900) c.270 papers on spiders and insects and on taxa the museum received from explorers and naturalists in Africa, China, Japan and Australia. Thirty-five of his papers were on Odonata, in which he described 107 new species (of which 77 remain valid) and named 36 genera (of which 17 are still in use). He was editor of 'Entomologische Nachrichten' (1884–1900) and 'Berliner Entomologische Zeitschrift' (1886-1895). Karsch unexpectedly ended his publication activity on Arthropoda (1900). Thereafter he published widely (using Karsch-Haack as a family name, Haack being his mother's maiden name) on same-sex sexual activity throughout the animal kingdom, among native peoples, and in all non-Western cultures. He has been ranked as one of the 'great authorities' in the area of sexual science. Later when living in Berlin he was open about his homosexuality. Around 30 taxa are named after him.

Karube

Spectre sp. *Boyeria karubei* **Yokoi**, 2002

Haruki Karube (b.1966) is a Japanese entomologist and Zoology Curator at the Kanagawa Prefectural Museum of Natural History. He started his research work in the 1980s and has collected dragonflies, beetles and other insects in many Asian countries and Australia. Most often he has visited Vietnam. He began (2004) conservation projects in Ogasawara islands aimed at endemic dragonflies but his work has recently expanded to other conservation projects aimed at protecting many endangered species in Japan. He is also addressing the problem of alien species such as invasive alien cray fish and water plants. He has found many new dragonfly species and published numerous papers on South-east Asian dragonflies, with special focus on the families Chlorogomphidae and Gomphidae. His taxonomical papers include: *Watanabeopetalia gen. nov., a new genus of the dragonflies (Odonata, Cordulegastridae, Chlorogomphinae)* (2002) and *Survey of the Vietnamese Chlorogomphidae (Odonata), with special reference to grouping* (2013). His publications include descriptions of c.60 new Odonata species or subspecies and seven new genera or subgenera. (Also see **Sachiyo**)

Kate

Cruiser sp. *Macromia katae* **KDP Wilson**, 1993

Kate Celia Wilson (b.1987) is the daughter of the author Keith Duncan Peter Wilson (b.1953) (q.v.).

Katambora

White-faced Citril *Ceriagrion katamborae* Pinhey, 1961

Katambora is a type of grass.

Katharina

Jewel sp. *Indocypher katharina* **Needham**, 1930
[Orig. *Rhinocypha katharina*]

Katharina is a feminine forename, but not a specified one. The author gave no etymology, so it is not possible to know for sure if this was named after a particular person. However, the lack of a Latin ending suggests it is not an eponym, as does Needham's frequent use of apparent forenames.

Kaudern

Malagasy Bluet *Azuragrion kauderni* **Sjöstedt**, 1917
[Orig. *Ischnura kauderni*]

Dr Walter Alexander Kaudern (1881–1942) was a Swedish ethnographer and zoologist. He was also well versed in botany, geology and geography. The University of Stockholm awarded his zoology PhD (1910). He was curator of the geological and mineralogy department of the Gothenburg Museum (1928–1932) and then director for the rest of his life (1932–1942). His ethnographic interest was awakened by his natural history collecting expeditions, notably to Madagascar (1906–1907), and again (1911–1912), when the damselfly was collected. He wrote many zoological papers and reports on his travels such as *På Madagaskar* (1913). His ethnography collection is held by the Ethnographical Museum, Stockholm. He also travelled on an expedition to Sulawesi (1916–1921) where he collected widely, including at least one fish later named after him, the Banggai Cardinalfish.

Kaup

Great Blue Metalwing *Neurobasis kaupi* **Brauer**, 1867

Johann Jakob Kaup (1803–1873) was a German zoologist, ornithologist and palaeontologist. He started at Göttingen University, which did not meet his expectations so he moved on to Heidelberg and, still not satisfied, moved to Leiden to spend two years at the Rijksmuseet where he obtained a post under C J Temminck, focussing on amphibians and fish. He travelled around Europe before arriving home in 1828. He became a temporary assistant (1830), rising to 'Inspector' (1837), of the Grand Duke's natural history 'cabinet' in Darmstadt, remaining in their employ until his death. Giessen University awarded him an honorary doctorate and he became a professor (1858). During this time, he led scientific missions for several months to England and France. He was a versatile, largely self-educated zoologist, publishing important works (1829 onwards) on palaeontology (*Beiträge zur näheren*

Kenntniss der urweltlichen Säugethiere,1855–1862), ichthyology, ornithology (*Classification der Säugethiere und Vögel*,1844) and entomology (*Monographie der Passaliden*, 1871). He had provided odonate material from the Dutch East Indies, collected by Hermann von Rosenberg (q.v.) and Renesent van Duivenbode (q.v.), to Friedrich Brauer for his study. Brauer described a total of eighteen new odonate species from this material, some of which are often erroneously listed as having Kaup as the author. He was a proponent of 'natural philosophy'; he believed in an innate mathematical order in nature and he attempted biological classifications based on the Quinarian system (the nonsensical idea that all taxa being divisible into five sub-groups) and declared himself against Darwin's doctrines when they were published. An amphibian and four birds are also named after him.

Kaxuriana

Threadtail sp. *Epipleoneura kaxuriana* **Machado**, 1985

Named after the indigenous people who inhabit the area in Brazil where the species type was found.

Kazuko

Sylvan sp. *Coeliccia kazukoae* **Asahina**, 1984

Mrs Kazuko Hasegawa is the wife of Dr Megumi Hasegawa (q.v.), after whom Asahina named another *Coeliccia* species in the same paper. (Also see **Megumi**)

Keiser

Keiser's Forktail *Macrogomphus keiseri* **Lieftinck**, 1955
[Orig. *Macrogomphus annulatus keiseri*]

Dr Alfred 'Fred' Jakob Keiser (Keiser-Jenny) (1895–1969) was Curator at Naturhistorisches Museum at Basel, Switzerland. His studies were disrupted by service in WW1, but he acquired his secondary school teaching certificate (1918). For many years, he and his wife (1920) travelled, collected and studied flies. He volunteered in the entomology department of the Basel Natural History Museum (1942) and finally joined them (1955). He was head of the entomological collection for a decade and was particularly interested in Diptera. He made an expedition to Sri Lanka (1953) where the forktail was collected and to Madagascar (1957 & 1958) collecting craneflies and also to Greece, Tunisia and Morocco. (Also see **Elizabeth (Keiser)**)

Kellicott

Lilypad Forktail *Ischnura kellicotti* **Williamson**, 1898

Professor Dr David Simons Kellicott (1842–1898) was an American entomologist. He graduated from Syracuse University (1869), took a second degree (1874) and was also awarded his PhD (1881). He taught at various schools (until 1888). He then became Professor of Zoology & Comparative Anatomy (1888) at Ohio State University and, when the department divided (1891), he was appointed Professor of Zoology and Entomology (1891–1898). Among his many publications was: *The Odonata of Ohio* (1899).

Kellogg

Tiger sp. *Gomphidia kelloggi* **Needham**, 1930
Shadeshifter sp. *Macromidia kelloggi* **Asahina**, 1978

Claude Rupert Kellogg (1886–1977) was a zoologist and entomologist who worked and collected in the Foochow District of China (1911–1941). Apart from teaching zoology at Anglo-American College, Foochow, and Fukien Christian University, he was a beekeeper and a missionary. He collected the holotype in China (1928).

Kemp

Sylvan sp. *Indocnemis kempi* **Laidlaw**, 1917
[JS *Indocnemis orang* Förster, 1907]

Dr Stanley Wells Kemp (1882–1945) was a zoologist and anthropologist. He joined (1903) the Fisheries Research Section, Department of Agriculture, Dublin, as Assistant Naturalist. He then joined the Indian Museum, Calcutta (1911) as Superintendent, Zoological Section. There he worked very closely with the Scottish zoologist Dr Nelson Annandale (q.v.) and in 1925 he wrote Annandale's obituary in the *Records of the Indian Museum*. He was on the Abor Punitive Expedition (1911–1912), during which government scientists made extensive natural history collections including in the Garo Hills. The type was taken in Cherrapunji, India. Kemp later joined the Colonial Office (1924) as Director of Research, Discovery Committee, and led the second Antarctic Discovery Expedition in 1924 (relating to whale fisheries). He was Director of the Plymouth Marine Laboratory (1936–1945). He lost all his personal possessions, his library and his unpublished works as the result of a German air raid (1941). Three amphibians and a bird are also named after him and his wife.

Kennedy

Kennedy's Emerald *Somatochlora kennedyi* **Walker**, 1918
Pond Damselfly sp. *Anisagrion kennedyi* **Leonard**, 1937
Wedgetail sp. *Acanthagrion kennedii* **Williamson**, 1916

Dr Clarence Hamilton Kennedy (1879–1952) was an American odonatologist who was Professor of Entomology, Ohio State University and Fellow of the Zoology Section of the Ohio Academy of Science. Indiana University awarded his first degree (1902) and his masters (1903). Stanford University further awarded a masters degree (1914) and Cornell his PhD (1919). He was an assistant at Indiana University (1902–1903) and then accepted a position as scientific illustrator on the staff of the US Bureau of Fisheries. He toured Oregon and Texas (1908) as a collector for the herbarium of Mount Holyoke College and was then (1913–1915) a scientific illustrator. He became an instructor in biology and limnology at Cornell (1915–1917) and at North Carolina University (1918). He was Instructor in Entomology at Ohio State University (1919), becoming Assistant Professor (1921), Associate (1930) and full Professor until retirement (1933–1949) then Professor Emeritus. He was also in charge of entomology at the Franz Stone Laboratory (1920–1938). He was President of the American Entomological Society (1935) and Managing Editor of its Annals. He was a specialist on the Odonata of the United States publishing 57 papers including c.2000 drawings, many of which are considered the finest of their kind. His most famous publications were: *Notes on the life history and ecology of the Dragonflies (Odonata) of Washington and Oregon* (1915) and *Notes on the life history and ecology of the dragonflies of Central California and Nevada* (1917). He described c.50 new odonate species. He also named 42 new odonate genera, many of which had been ranked as synonyms, but more recently considered valid. Towards the end of his life he tired of working on dragonflies and switched to study ants. He died as the result of a cerebral haemorrhage. (Also see **Albert, Bruce, Emma** & **Joanetta**)

Kenyah

Sylvan sp. *Coeliccia kenyah* **Dow**, 2010

Named for the Kenyah people, who inhabit the Tinjar valley in the shadow of Mt. Dulit, where the type series was collected.

Kerckhoff

Rusty-tipped Proto *Protolestes kerchoffae* **Schmidt**, 1949

Louise Eshman Kerckhoff (1859–1946) was the wife of the German-American lumber and electric power millionaire William G Kerckhoff (1856–1929). The foundation, which bears his name, was founded by his wife and supports young scientists in their research. She also endowed a student's union building in his name at UCLA – Kerckhoff Hall.

Kerr

Forktail sp. *Macrogomphus kerri* **Fraser**, 1932
Pincertail sp. *Onychogomphus kerri* Fraser, 1933
Skimmer sp. *Amphithemis kerri* Fraser, 1933

Jewel ssp. *Aristocypha iridea kerri* (Fraser, 1933)
[Orig. *Rhinocypha iridea kerri*]

Dr Arthur Francis George Kerr (1877–1942) was an Irish physician. He was an amateur botanist particularly noted for his study of the flora of Thailand and a number of plants are named after him. Trinity College Dublin awarded his bachelor's degree in botany (1897) and he went on to take his MD (1901) and gained a post as physician on a ship to Australia. He was posted to Siam (1902) as an assistant to a Dr Campbell Highet and later to the British Legation, Bangkok, and was Principal Officer of Health to the government there (1904–1914). During WW1, he served in the Royal Medical Corps in France (1915–1918) but had to withdraw through ill health. He returned to Bangkok but soon went into private practice. He was then appointed Director of the Botanical Section of the Ministry of Commerce of Siam (1920–1932). For many years (1920–1929) he took an annual botanical tour. He finally returned to England (1932) and much of his botanical collection is held at Kew. He collected the types in Siam (now Thailand) and Laos (1931–1932).

Kersten

Kersten's (Powder-faced) Sprite *Pseudagrion kersteni* Gerstäcker, 1869
[P. Powder-striped Sprite, Orig. *Agrion kersteni*]

Dr Otto Anton Rudolf Kersten (1839–1900) was a German traveller, explorer and chemist, who collected the sprite holotype in Tanzania whilst on Baron von der Decken's unsuccessful expedition to climb Mount Kilimanjaro (1859–1861). He later published his memoirs in six volumes (1869–1879).

Kerville

Kerville's Featherleg *Platycnemis kervillei* **Martin**, 1909
[A. Powdered Featherleg Orig. *Psilocnemis kervillei*]

Henri Gadeau de Kerville (1858–1940) was a noble French zoologist, entomologist, botanist and archaeologist, as well as a keen photographer. Lycée Pierre Corneille awarded his bachelor's degree (1877). He collected around his home in Normandy, discovering a species of beetle new to science when only fifteen. He later travelled in Asia Minor and collected (1912) in Syria in particular where the featherleg type was taken. Admitted to the Société des Amis des Sciences Naturelles de Rouen at just twenty-one, he went on to be its secretary then president. He created a laboratory to study the biology of caves (1910). During WW1, he was a volunteer nurse in a Rouen hospital. He was made Chevalier de la Légion d'Honneur (1933). His best known published work is: *Les Insectes phosphorescents: notes complémentaires et bibliographie générale (anatomie physiologie et biologie) : avec quatre planches chromolithographiées* (1881). Most of his collections are held in Paris.

Khalid

Hajar Wadi Damsel *Arabineura khalidi* **Schneider**, 1988
[Orig. *Elattoneura khalidi*]

This is a toponym; Wadi Bani Khalid is a wadi in the Hajar Mountains of Oman, the type locality.

Kiauta, MAJE

Dryad sp. *Pericnemis kiautarum* **Orr** & **Hämäläinen**, 2013

Maria (Marianne) Antonetta Johanna Elisabeth Kiauta, née Brink (b.1948) is the Dutch wife of Professor Dr Milan Boštjan Kiauta. She is also a linguist, poet, biologist, reflexologist and Tibetophile. She first studied biology at the University of Utrecht which awarded her BSc in 1978, then she studied oriental languages and cultures, and Tibet studies, there and at Leiden. As laboratory assistant (histology, cytology) she was associated with the State University of Utrecht (1968–1991). She served as the Assistant Editor of 'Odonatologica' (1986–2013) and was, for various lengths of time, on the editorial boards of other entomological periodicals. She is the author or joint author of over 40 odonatological publications, author of 5 papers on Tibet (mostly Tibetan linguistics), of very many haiku publications and of 4 haiku books. She has been managing (since 2008) her own sanctuary for old or disabled horses and other abandoned animals. This species is named after her and her husband (below). (See **Marianne** & also see **Kiauta, M B**)

Kiauta, M B

Shadowdamsel sp. *Protosticta kiautai* **Zhou**, 1986
Skimmer sp. *Elasmothemis kiautai* **De Marmels**, 1989
[Orig. *Dythemis kiautai*]
Damselfly sp. *Risiocnemis kiautai* **Hämäläinen**, 1991

Kiauta's Hooktail *Paragomphus kiautai* **Legrand**, 1992
Cruiser sp. *Macromia kiautai* Zhou, Wang, Shuai & **Liu**, 1994
Emerald sp. *Navicordulia kiautai* **Machado** & **Costa**, 1995
Dancer sp. *Argia kiautai* Steinmann, 1997
[Unnecessary replacement name for *Argia apicalis* Matsumura, 1913; JS *Rhipidolestes okinawanus* Asahina, 1951]
Flatwing sp. *Priscagrion kiautai* Zhou & **KDP Wilson**, 2001
Billabongfly sp. *Austroagrion kiautai* **Theischinger** & Richards, 2007
Threadtail sp. *Neoneura kiautai* Machado, 2007
Fineliner sp. *Teinobasis kiautai* Theischinger & Richards, 2007
Dryad sp. *Pericnemis kiautarum* **Orr** & Hämäläinen, 2013

Professor Dr Milan Boštjan (Bastiaan) Kiauta (b.1937) is a Dutch entomologist of Slovenian extraction, a Charter Member of the Societas Internationalis Odonatologica (1971), its Member of Honour (1981) and Past President (1988–1989). His first degree was in zoology (1959) at the University of Ljubljana. After departure to The Netherlands (1962), the State University of Utrecht awarded his doctorate (1969), his thesis being: *Studies on karyotypic evolution in Odonata*. His primary interests are the cytogenetics of Odonata and of invertebrates in high mountains and subterranean cave environments. He was a bio-speleologist at the Karst Research Institute of the Slovenian Academy of Sciences in Postojna (1961–1962), a fellow of the Institute for Protection of Monuments in Ljubljana and co-founder and first editor of its periodical 'Nature Conservation'. He worked at the Netherlands State Institute of Nature Conservation Research (RIVON) in Bilthoven and Zeist (1963-1964) and at the Institute of Genetics of State University of Utrecht (1964–2002) where he is now Professor Emeritus. He was one of the founders of International Odonata Research Institute in Gainesville, Florida, USA (1986) and served as a member of its board of directors. He has given many lectures and seminars in Switzerland, Nepal, India, Thailand, Taiwan, the Philippines, USA, Japan and elsewhere and led several national and international biological research expeditions to the Nepal Himalayas. He was made a Knight of the Order of Oranje-Nassau (2002) and a Member of the Slovenian Academy of Sciences and Arts (2007). He has written more than 350 publications (including c.200 in the field of odonatology) alone, with his wife or with other co-authors. He was Editor of 'Odonatologica' (1972–2013). His other major editorships include 'Genetica' (Netherlands Journal of Genetics, 1972–1991), 'Notulae odonatologicae' (1978–2013) and 'Opuscula zoologica fluminensia' (1984–2008). A biography and odonatological bibliography were published in the festschrift dedicated to him on his 70th birthday: B K Tyagi, [Editor], *Odonata: biology of dragonflies* (2007). The last species listed is named after him and his wife. (See **Bastiaan** and also see **Marianne & Kiauta, MAJE**)

Kimmins

Skydragon sp. *Chloropetalia kimminsi* **Fraser**, 1940
[Orig. *Chlorogomphus kimminsi*]
Chalk-marked Dragonlet *Erythrodiplax kimminsi* Borror, 1942
Tigertail sp. *Palaeosynthemis kimminsi* **Lieftinck**, 1953
[Orig. *Synthemis kimminsi*]
Kimmins' Cruiser *Phyllomacromia kimminsi* Fraser, 1954

[P. Crescent-faced Cruiser, Orig. *Macromia kimminsi*]
Kimmins' Metalwing *Neurobasis kimminsi* Lieftinck, 1955
Sylvan sp. *Idiocnemis kimminsi* Lieftinck, 1958
Damselfly sp. *Lieftinckia kimminsi* Lieftinck, 1963
Sanddragon sp. *Progomphus kimminsi* **Belle**, 1973

Douglas Eric Kimmins (1905–1985) was a British entomologist. He became personal assistant to Professor H Maxwell Lefroy at the Royal College of Science, a post he held until Lefroy's death (1923–1925). He then joined the BMNH for many years (1925–1970). At first, he was employed part-time as an 'Unofficial Scientific Worker', spending the rest of his week preparing scientific drawings for other entomologists. During WW2, he was responsible for moving the 'neuropteroid' collections to the Freshwater Biological Association in the Lake District and collected locally during his time there (1930–1943). Entomologists often applied the whimsical term 'Kimminsoidea' to the group of orders, embraced by the Linnaean name Neuroptera, which he curated. He was called up and served in the RAFVR during which time he heard that his own house in southeast London had been bombed. He was posted to Bengal (1945) where he collected as he did when transferred to Calcutta, but illness forced his repatriation and he resumed work at the museum in 1946. He joined the permanent staff (1948) as an Experimental Officer and Principal Scientific Officer (1958), remaining in the entomology department for the rest of his career (1970) continuing part time for a few more years (1973). He wrote 259 papers, describing 39 new genera and 648 new species, of which three genera and 44 species were Odonata. He published lists of type-specimens of Odonata in the British Museum (Natural History) (in four parts 1966–1970). Numerous insects are named after him including three genera. (Also see **Donald**)

Kimpa Vita

Angola Longleg *Notogomphus kimpavita* Dijkstra & **Clausnitzer**, 2015

Beatriz Kimpa Vita (1684–1706) is the patron saint of Universidade Kimpa Vita after whom it was named. She was a Kongo Empire prophet and leader of a Christian movement, Antonianism, which taught that Jesus and other early Christian figures were from the Kongo Empire (part of modern day Angola). The species was discovered in the new campus grounds.

Kimura

Sylvan sp. *Coeliccia kimurai* **Asahina**, 1990

Yunosuke Kimura collected the holotype in Thailand (1988). (See **Yunosuke**)

King

Damselfly sp. *Agrion kingii* **MacLeay**, 1827
[*nomen oblitum*, obviously an *Episynlestes* species]

Rear-Admiral Philip Parker King (1793–1856) was an Australian born British marine surveyor. He was also a collector who travelled in South America (1827–1832) as Commander of the British South American Survey. His father was Philip Gidley King, the 3rd Governor of New South Wales. The King family returned to England (1807) and Philip Parker King entered the navy as a young man. He was in command of the cutter *Mermaid*

(1818) and made a number of discoveries including the Goulburn Islands, which he named after Major Goulburn who was Colonial Secretary in New South Wales. He carried out the first survey of the Great Barrier Reef (1819). In command of *Bathurst* (1821), he carried out a second survey of the reef and the Torres Strait. He collected the type. Six reptiles and two birds are also named after him.

Kinnara

Velveteen sp. *Bayadera kinnara* **Hämäläinen**, 2013

Kinnara is a mythical, half human creature. The holotype was collected in Kachin State, Burma by a group of Japanese entomologists (2000). To quote Hämäläinen's etymology: "*The specific epithet kinnara (a noun in apposition) refers to the half-human and half-bird character from the Buddhist mythology of many Southeast Asian countries. Kinnara are renowned for their skills in dance and song. This is in accordance with the meaning of the genus name Bayadera: bayadére is the French version of the Portuguese word bailadeira, which refers to a Hindu dancing girl in Indian temples. Selys Longchamps (1853) introduced the genus-group name Bayadera for his Indian species Epallage indica.*"

Kinzelbach

Clubtail sp. *Gomphus kinzelbachi* **Schneider**, 1984

Professor Dr Ragnar Karl Konrad Kinzelbach (b.1941) is Professor Emeritus of Zoology and Ecology at the Institute of Biodiversity, University of Rostock where he was Professor of Zoology (1996–2006). He conducted research and taught at the University of Mainz and was Professor of Ecology and Zoology at the Technical University of Darmstadt (1993–1995) before joining Rostock. He was a member of the DFG Tübingen Atlas of the Middle East and has been involved in the study of fauna and zoogeography of that region throughout his career, as is Wolfgang Schneider. He has written many papers and books, especially since 'retiring'. His interests continue to be ornithology, entomology, limnic ecology, zoogeography, historical zoology and he travels for field research and collecting. He is honoured in the names of around thirty different species from a variety of taxa including gastropods, spiders, scorpions, beetles and other insects.

Kirby

Kirby's Dropwing *Trithemis kirbyi* **Selys**, 1891
[A. Orange-winged Dropwing]
Slender Duskhawker *Gynacantha kirbyi* **Krüger**, 1898
Flatwing sp. *Wahnesia kirbyi* **Förster**, 1900
Fineliner sp. *Teinobasis kirbyi* **Laidlaw**, 1902
Velveteen sp. *Bayadera kirbyi* **KDP Wilson** & **Reels**, 2001

William Forsell Kirby (1844–1912) was an English entomologist and folklorist. He was educated privately, and became interested in butterflies and moths at an early age. He published his first entomological note at the age of 12 years, published a list of British butterflies aged 14, and a book: *A manual of European butterflies* aged 18 (1862). He worked for a publishing firm for six years (1860–1866) then became Assistant Naturalist at the

Museum of the Royal Dublin Society (1867). He was transferred to BMNH until retirement (1879–1909). His main entomological work was confined to Lepidoptera, Odonata and Orthoptera. Kirby had a wide range of interests, knew many languages and fully translated Finland's national epic, the Kalevala, from Finnish into English, carefully reproducing the meter *Kalevala, The Land of Heroes* (1907) which was a major influence on the writings of J R R Tolkien, who first read it in his teens. Kirby also named two odonate genera, *Aino* and *Untamo* (now synonyms) and the valid butterfly genus *Tellervo* after characters from Kalevala. Kirby also wrote: *The new Arabian nights, select tales not included by Galland and Lane* (1882) and *The hero of Esthonia, and other studies in the romantic literature of that country* (1895). He also provided many footnotes to Sir Richard Burton's translation of the *Arabian Nights*. His numerous entomological publications include several books such as: *A synonymic catalogue of diurnal Lepidoptera* (1871, with a supplement in 1877), *European butterflies and moths* (1878–1882), *Elementary text-book of entomology* (1885), *A synonymic catalogue of Lepidoptera-Heterocera, or moths, Vol. 1 Sphinges and Bombyces* (1892), *A handbook to the Order Lepidoptera* (5 volumes 1894–1897), *A synonymic catalogue of Orthoptera* (in 3 volumes 1904,1906,1910). He wrote 27 publications on Odonata, including *A revision of the Subfamily Libellulidae, with descriptions of new genera and species* (1889) and *A synonymic catalogue of Neuroptera, Odonata or dragonflies* (1890) – the first catalogue of the Odonata of the world. In Odonata Kirby described 147 new species (of which 62 valid) and named 71 new genera (43 of them presently in use).

Kiritschenko

Clubtail sp. *Anormogomphus kiritschenkoi* **Bartenev**, 1913

Alexey Nikolaevich Kiritschenko (Aleksandr Nikolayevich Kiric(s)henko) (1884–1971) was a Russian entomologist whose prime interest was Heteroptera. He was for many years the Curator of Heteroptera at the Zoological Institute, St Petersburg. He was the author of numerous papers and also reviewed the work of other entomologists collecting in Iran. He wrote the section entitled: *Insecta Hemiptera* in the *Fauna of Russia* (1913) and *Poluzhestkrylyia (Hemiptera-Heteroptera)* (1916). He collected the type in Turkestan (1912).

Kirsch

Tiger sp. *Gomphidia kirschii* **Selys**, 1878

Theodor Franz Wilhelm Kirsch (1818–1889) was a German entomologist specialising in Coleoptera. He became Curator of Entomology at the Staatliches Museum für Tierkunde Dresden where his collection is still housed. He described a number of beetles and at least one is named after him. Among his published papers is: *Beitrag zur kenntniss der Lepidopteren-Fauna von Neu Guinea* (1877).

Kishor

Flangetail sp. *Ictinogomphus kishori* Ram, 1985

No etymology is given. I believed a strong candidate to be Dr Brij Kishore Tyagi (b.1951) who is an Indian zoologist. However, he has confirmed it was not named after him and he has set out to discover who was honoured.

Kitawaki

Clubtail sp. *Sinorogomphus kitawakii* **Karube**, 1995

[Orig. *Chlorogomphus kitawakii*]

Wakoh Kitawaki (1957–1996) was a Japanese entomologist. He collected the holotype in China (1992). His collection of odonata is housed by Osaka Museum of Natural History. He collected in Tibet.

Kitchingman

Pale Claspertail *Onychogomphus kitchingmani* **Pinhey**, 1961

R M Kitchingman collected the holotype (1957) in Northern Rhodesia (now Zambia). He lived in the Mwinilunga area and sent extensive collections to the Rhodes-Livingstone Museum. He wrote the paper: *The history of Kalene Hill* (1958) and *Some species and cultivars of Hemerocallis* (1985).

Klages

Duskhawker sp. *Gynacantha klagesi* **Williamson**, 1923

Samuel Milton Klages (1875–1957) was an American collector. He left the United States for South America (1891) and collected in a range of countries; Venezuela (1898–1913), Trinidad & Tobago (1912–1913), Venezuela again (1913–1914 & 1919), French Guiana (1917–1918) where he collected the eponymous type (1917) and in Brazil (1919 & 1920–1926). Most of these trips were undertaken for the Bird Section of the Carnegie Museum of Natural History in Pittsburgh. He continued to make trips to South America until 1932. He then lived for a while with a niece, Alva Held, a high school counsellor, her mother and his sister, Mary Amelia, in Knoxville. Apparently, he annoyed Alva when his hair and moustache dye stained her bathroom sink and he left to set up home in an apartment. He must have made bad investments as he is listed in the (1943) Pittsburgh city directory as a writer at a Perrysville Avenue address and in receipt of the state dole. He wrote an unpublished romantic novel and offered a publisher an account of his travels, but his manuscripts were all burnt after Klages died of heart disease, having already been diagnosed as senile. A number of Lepidoptera and six birds are also named after him.

Klapperich

Duskhawker sp. *Cephalaeschna klapperichi* **Schmidt**, 1961

Johann Friedrich Klapperich (1913–1987) was a German entomologist, taxidermist and collector. He worked as entomology preparator of the Koenig Museum, Bonn, one of the last to work for Alexander Koenig himself (1935–1952). He was sent on a nearly two-year long collection expedition (1937–1938) to Fukien, from where he made a large collection of insects, mammals, birds as well as bird's nests and eggs. He collected in four areas in Fukien: Shaowu, Woping, Kwangtseh and Kuatun at different elevations in the mountainous region on the border with Kiangsi. He also undertook several expeditions to Afghanistan at his own expense (1952–1953) and collected the eponymous type there (1952). He wrote about the latter expedition in: *Auf Forschungsreisen in Afghanistan* (1954). He later stayed in Ammam, Jordan from where he made excursions to Turkey, Iran, Iraq,

Syria and the Arabian Peninsula. He was awarded the Star of Jordan for his services there. He also travelled to the Dominican Republic. He retired (1975) and afterwards travelled with his wife to Sierra Leone, Taiwan, Sumatra and South America.

Klene

Bannerwing sp. *Cora klenei* **Karsch**, 1891

Heinrich Klene (1845–1922) was a German Jesuit priest and entomologist. Karsch's article only gives a description of the species, there is no mention of a collector, nor an etymology and the locality is no better delineated than 'Ecuador'. Circumstantial evidence leads to the belief that this is who was intended.

Kling

Stream Junglewatcher *Neodythemis klingi* **Karsch**, 1890
[Orig. *Allorrhizucha klingi*]

Hauptmann (Captain) Eugen (or possibly Erich) Kling (1854–1892) was a German Army officer and explorer. He joined the colonial service (1884) and studied the languages, fauna and flora and peoples of Africa. He even spent some time at an observatory so he could navigate by the stars. He also learned how to collect, prepare and store specimens. He applied to join the expeditions to Africa (1886). He travelled (1888) and collected with Oskar Büttner in Togo, where he collected the type (1888) and Congo following the German Congo Expedition (1884–1886). He explored in Dahomey (1889) under Dr Ludwig Wolf and took over leadership when Wolf died. He made a point of initiating friendly relations with the local people and respected local traditions and beliefs. He resigned his commission (1890) and was appointed to lead an expedition (1891) to explore the Togo hinterland. On returning to Berlin he died of dysentery contracted in Togo. He left travel diaries with drawings and maps of the areas he visited.

Kloss

Goldenring sp. *Anotogaster klossi* **Fraser**, 1919

Cecil Boden Kloss (1877–1949) was an ethnologist and zoologist. He was a member of the staff of the museum in Kuala Lumpur (1908) for which he travelled extensively as a collector, including spending three months in southern Vietnam (1918) where the dragonfly was collected in Annam. He started to work under Herbert Christopher Robinson who was Curator of Birds for the Federated States Museums, Malaysia (1908). He was the Director of the Raffles Museum in Singapore (1923–1932) and established its Bulletin (1928). Fifteen birds, three mammals and four reptiles are also named after him as well as many plants.

Klots

Yellow-spotted Duskhawker *Cephalaeschna klotsae* **Asahina**, 1982
[Orig. *Cephalaeschna klotsi*]

Dr Elsie Klots née Broughton (c.1902–1991) was an American entomologist and natural historian. She was assistant professor in Elmira College before becoming an instructor at Cornell (1929–1932), which awarded her PhD (1932); her thesis was: *A Venational*

Study of the Gomphinae (order Odonata). She married (1927) Alexander (Bill) Barrett Klots (1903–1989), a lepidopterologist and research associate of AMNH, and they often collected together. Among her many papers and books are: *Chinese dragonflies (Odonata) in the American Museum of Natural History (American Museum novitates)* (1947) and *The New Field Book of Freshwater Life* (1966), as well as seven books written with her husband. She described four new Odonata species. During WW2, Bill's letters to Elsie were censored; the army cut away any mention of places, dates, etc. Bill and Elsie found this most bothersome, so they soon devised an interesting way to let each other know his whereabouts. In each letter, he would include mention of several butterfly species. Elsie would take these letters to the American Museum of Natural History, where, with a little work in the collection, she could pin down his general whereabouts. (The species was originally named *klotsi* but the Latin has been corrected to reflect that the person honoured was female.)

Klug

Shortwing sp. *Perissolestes klugi* **Kennedy**, 1941

Threadtail sp. *Protoneura klugi* Cowley, 1941

Rubyspot sp. *Hetaerina klugi* Schmidt, 1942

[JS *Hetaerina laesa* Hagen, 1853]

Clubtail sp. *Zonophora calippus klugi* **Schmidt**, 1941

Guillermo Klug (1875–1945) was a German naturalist. He was in Peru (1924) and initially worked with Bassler, a geologist, who spent ten years in Northern Peru for the Standard Oil Co. He collected extensive material in Peru, including the types (1929–1938), which he mainly sent to museums or sold privately in North America and Europe.

Knopf

Pond Damselfly sp. *Metaleptobasis knopfi* **Tennessen**, 2012

Dr Kenneth William Knopf (b.1950) is an American entomologist. He was at the Department of Entomology & Nematology, University of Florida. He has written a number of papers such as *Life History and Biology of Samea Multiplicalis (Guenee) (Lepidoptera: Pyralidae)* (1974) and *A new marking technique for studying the mating behaviour of Odonata* (1974) and co-written articles such as: *Description of the Nymph of Enallagma minusculum (Odonata: Coenagrionidae)* (1975) written with Tennessen (q.v.). Knopf collected the damselfly type in Ecuador (1980). He is also the co-author with May (q.v.) of *Neocordulia campana* from Panama (1988).

Koch

Darting Cruiser *Macromia kochi* Grünberg, 1911

[JS *Phyllomacromia picta* Hagen, 1871]

Dr Heinrich Hermann Robert Koch (1843–1910) was a German physician and pioneer microbiologist. He was very gifted academically and taught himself to read and write before going to school. He studied medicine at the University of Göttingen, graduating (1866) with the highest distinctions. He was almost immediately practising as a surgeon during the Franco-Prussian War. He was an administrator and professor at Berlin University (1885–1890) as well as government advisor with the Imperial Department of Health (from 1880). He was one of the founders of the science of bacteriology and discovered the tuberculosis bacterium (and those for anthrax and cholera). He was winner of the Nobel Prize for Medicine (1905). While staying in Sese Island in Lake Victoria, Uganda conducting studies on sleeping disease (1905), Koch collected the holotype of this dragonfly species. (Also see **Troch**)

Kocher

Azure Bluet *Agrion puella kocheri* **Schmidt**, 1960
[JS *Coenagrion puella* (Linnaeus, 1758)]

Colonel Louis F J Kocher (1894–1972) was a French soldier (he was Chief of Staff, Marakesh 1946) and entomologist who started his entomological work during his military career when visiting Mediterranean countries. He resigned his commission and became (1949) the head of the Laboratoire d'Entomologie de l'Institut Scientifique Chérifien in Rabat, Morocco for two decades and Secretary General of the Society of Natural & Physical Science of Morocco. His principal interest was in Coleoptera and, in particular, Chrysomelidae. He wrote, among other papers (1949–1969): *Prospection entomologique dans la Djebel Sarro* (1949) and *Catalogue commente des Coleopteres du Maroc. Fascicule X bis Nouveaux addenda et corrigenda* (1969). He had also served in the French Foreign Legion.

Koepcke

Bannerwing sp. *Polythore koepckei* Börzsöny, 2013

Professor Dr Hans-Wilhelm Koepcke (1914–2000) was a zoologist, herpetologist and ornithologist. He studied at Kiel University, culminating in a PhD (1947). He took a post at the Javier Prado Museum of Natural History in Lima, Peru and spent much of his life studying South American, and particularly Peruvian, fauna. He was, among other things, founder of the Biological Station Panguana, which has conducted research in tropical rainforests for over four decades. The new species of damselfly was collected nearby. When he eventually returned to Hamburg he worked at the Herpetology Department and taught at the Zoological Institute and the Museum of the University of Hamburg. He published widely with the most significant work being the two-volume: *Die Lebensformen: Grundlagen zu einer universell gültigen biologischen Theorie* (1971 & 1973).

Kollmannsperger

Skimmer sp. *Orthetrum kollmannspergeri* **Buchholz**, 1959
[JS *Orthetrum brevistylum* Kirby, 1898]

Dr Franz Kollmannsperger (1907–1997) was a German journalist, writer, zoologist, ecologist, and ornithologist at the University of Saarbrücken. He studied in Freiburg, Munich and Berlin culminating in a PhD in zoology. He was a student counsellor and then followed an academic career. After WW2, he made several expeditions to Africa, including leading the international Sahara and Sudan expedition (1953–1954) and collecting in the Ennedi Mountains, French Equatorial Africa (1957). He collected the type in Chad (1957). He wrote a number of papers such as: *Das Ausmaß der Auswirkungen biologischer Gleichgewichtsstörungen nordafrikanischer Klima- und Vegetationszonen* (1956) and *Von Afrika nach Afrika* (1965). A bird and a mammal are also named after him.

Kolthoff

Hawker sp. *Aeschnophlebia kolthoffi* **Sjöstedt**, 1925
[JS *Aeschnophlebia longistigma* Selys 1883]

Kjell Gustaf Adolf Henrik Kolthoff (1871–1947) was a Swedish taxidermist, as was his father, but he was also a fine painter, restorer and writer. At university, he studied Japanese and later archaeology in Japan. He was employed to paint the backgrounds of diorama at the Malmö Natural History Museum. He was a good shot and participated in a number of collecting expeditions, notably to Greenland. He collected the type in China (1921–1922) during an expedition to collect birds. Apart from articles he also wrote and illustrated several natural history books including: *My life in nature. Memories of home area, Arctic and the Far East* (1939).

Kompier

Grisette sp. *Devadatta kompieri* Phan, **Sasamoto** & Hayashi, 2015
Clubtail sp. *Trigomphus kompieri* **Karube**, 2015
Tiger sp. *Gomphidictinus kompieri* Karube, 2016

Thomas (Tom) Maurits Franciscus Kompier (b.1967) is a Dutch water and climate specialist working at the Dutch embassy in Hanoi, Vietnam. He studied Japanese at university and went on to study archaeology at the University of Osaka. In private life, he is a keen exponent of Shotokan karate (having trained in the UK, Holland and Japan), being a Black Belt 5th Dan and an instructor of some note, but described as a modest and humble man. He runs a blog on dragonflies and damselflies of Vietnam. Among his publications is the book: *Dragonflies and Damselflies of the Serra dos Orgaos* 2015. He has described or co-described over twenty new Odonata species from Vietnam (2015–2017), such as *Bayadera hatvan*, *Anisogomphus neptunus* and *Lyriothemis kameliyae*.

Koolpinyah

Threadtail sp. *Nososticta koolpinyah* **Watson** & **Theischinger**, 1984

This is a toponym; Koolpinyah Station is in Northern Territory, Australia.

Koomina

Pilbara Emerald *Hemicordulia koomina* **Watson**, 1969

This is a toponym; Koomina Pool is near Tanberry Creek, Western Australia.

Koongarra

Citrine Threadtail *Nososticta koongarra* **Watson** & **Theischinger**, 1984

This is a toponym; Koongarra is a locality near Mount Cahill, Northern Territory, Australia.

Kosterin

Shadowdamsel sp. *Drepanosticta kosterini* **Dow**, 2017

Dr Oleg Engelsovich Kosterin (b.1963) is a Russian biologist born in Omsk. After finishing his studies at Novosibirsk State University (Novosibirsk, Russia), he became a research student (1985–1987) at the Laboratory of Evolutionary Genetics, Institute of Cytology and Genetics, Novosibirsk, Russia, then a Junior Researcher (1987–1992), Staff Researcher (1992–1998), Senior Staff Researcher (1998–2008) at the same laboratory and then Section Head (2008–2012) and Head of Laboratory of Genetics and Evolution of Legume Plants (2012 onwards). In the meantime, he defended his PhD thesis: *Inheritance and properties of a histone H1 fraction specific to young tissues of the garden pea (Pisum sativum L.)* (1995). He has also been a lecturer in general biology at the Psychological Faculty at Novosibirsk State University (2006–2011) and is (since 2009) a lecturer in genetics at the Faculty of Natural Sciences and Medical Faculty at Novosibirsk State University. He has been on many entomological expeditions across the Asian part of Russia and abroad such as: nine to Kazakhstan (1983–2015), Thailand (2005, 2006, 2009), one to Ethiopia (2012) and Indonesia (2014) and nine to Cambodia (2010–2017). He has a long-term interest in the butterflies of Siberia and Central Asia, and the Odonata of Asia in general and Cambodia in particular. He has (2017) 236 publications, including four monographs, three intra-university textbooks and many scientific papers. He has described twelve new species and five new subspecies of Odonata and nine new subspecies of Lepidoptera. In private life, he enjoys photography and ballroom dancing. Four species and two sub-species of Lepidoptera are also named after him as are three species of Diptera. He also named a dragonfly after his wife (see **Priydak**).

Koxinga

Spreadwing sp. *Orolestes koxingai* Chen, 1950
[JS *Orolestes selysi* McLachlan, 1895]
Longleg sp. *Anisogomphus koxingai* **Chao**, 1954

Zheng Chenggong (1624–1662) was a Chinese military leader who was born in Hirado, Japan to the Chinese merchant/pirate Zheng Zhilong and his Japanese wife. He died on the Island of Formosa (Taiwan). Koxinga is the customary Western spelling and means literally 'Lord of the Imperial Surname'.

Kraepelin

Black-tailed Skimmer *Orthetrum kraepelini* **Ris**, 1897
[JS *Orthetrum cancellatum* (Linnaeus, 1758)]

Professor Karl Matthias Friedrich Magnus Kraepelin (1848–1915) was a German naturalist, zoologist, entomologist, herpetologist and teacher who became Director/Curator, Naturhistorisches Museum zu Hamburg (1889–1914), which was destroyed during WW2. He studied at Göttingen and Leipzig culminating in his PhD (1873) and was Professor of Mathematics and Sciences in Leipzig and Hamburg. He undertook a five-month trip to Ceylon, India, Singapore and Java (1903–1904) and also visited the USA (1908). He worked on myriapods (centipedes etc) (1901–1916). Among other papers and books, he wrote: *Scorpiones und Pedipalpi* (1899). He was most well-known as an educator and he wrote several textbooks including: *Leitfaden für den botanischen Unterricht an mittleren und höheren Schulen*. Two reptiles are also named after him, as are several insects and arthropods etc.

Kreyenberg

Clubtail sp. *Stylurus kreyenbergi* **Ris**, 1928]
[Orig. *Gomphus kreyenbergi*]

Dr Martin Kreyenberg (1872–1914) was a physician and zoologist. He was aboard the German ship *Jaguar* (1901–1905), during when he made herpetological and ichthyological collections at various coastal locations from Hong Kong to Korea and Australia. He stayed on for a while in China but returned to Germany (1906) when he was taken ill. However, he returned to China (1908) as a physician to a railway building project which was under German management. He collected many zoological specimens which he donated to the Magdeburg Museum. After this was completed he travelled, continuing to collect (particularly fish), including the type (1912). He wanted to take an extended collecting expedition but had insufficient funds. At this point he was persuaded to invest in a large coconut plantation on an island near Manila, with two German merchants. He liked the idea of studying the fauna of a tropical island; however, it was not profitable and it was devastated by two tropical storms. He was on passage to Manila aboard the steamer *Sontua* when he suffered appendicitis and died. He was buried in Manila. A snake and several fish and other aquatic organisms are also named after him.

Kricheldorff

Demoiselle sp. *Matrona kricheldorffi* **Karsch**, 1892
[JS *Matrona basilaris* Selys, 1853]

Franz Kricheldorff (1854–1924) was a German trader and naturalist who was also a commercial collector. Three generations of his family were entomologists: his father founded the Entomological Institute in Berlin (1873) and his sons took over the business when he died. He accompanied his fellow naturalist, commercial collector and explorer Antwerp Edgar Pratt (1850–1920) in Sichuan (March 1889 to October 1890) on behalf of wealthy private collectors and others, especially James Henry Leech (1862–1900), during which he collected the eponymous type series (just two males). He settled down in Berlin as a butterfly dealer.

Kristensen

Ethiopian Skimmer *Orthetrum kristenseni* **Ris**, 1911

Gunnar Kristensen (b.1870) was an insect dealer who led a collection expedition to the Harar Mountains, Ethiopia, which collected many natural history specimens including the type (1910).

Krug

Antillean Bluet *Enallagma krugii* Kolbe, 1888
[JS *Enallagma coecum* (Hagen, 1861)]

Carl (Karl) Wilhelm Leopold Krug (1833–1898) was a German businessman who worked in Puerto Rico (1857–1876), where he became the German Vice-Consul, acquired ownership of the firm he worked for and married a wealthy landowner's daughter. It is also where he collected the type. His personal hobbies were zoology and botany. Kolbe and Gundlach (q.v.), who was Krug's guest, collected together. A reptile and a bird are also named after him.

Krüger

Tiger sp. *Gomphidia kruegeri* **Martin**, 1904
[Orig. *Gomphidia krügeri*]
Shadowdamsel sp. *Drepanosticta krugeri* **Laidlaw**, 1926

Leopold Karl Wilhelm Krüger (1861–1942) was Curator of Entomology at the Museum of Natural History in Stettin (now Szczecin, Poland). He studied at Greifswald University, graduating (1884) as a qualified teacher of botany, zoology, chemistry, mineralogy and mathematics. He took various teaching jobs (1884–1891) and then a permanent position as High School Professor teaching mathematics, chemistry and natural sciences (1891) in Stettin where the local Entomological Society published some of his articles. Around this time, Heinrich Dohrn (q.v.) sent him an insect collection to identify and describe. He became Vice President (1910) and then President (1913–1927) of the Stettiner Entomologischer Verein and while still teaching, became (1919) Curator of Natural History and later of entomology at Städtische Museum Stettin, remaining until retirement (1927). Among his publications was: *Die Odonaten von Sumatra,* a series of four papers (1898–1903). He described 42 new Odonata species, over half of them presently synonyms. He also named the genus *Pornothemis,* a curiously vulgar (for the time) name, never properly explained but apparently based on the peculiarly shaped base of abdomen with protruding male genitalia.

Kuan

Pondcruiser sp. *Epophthalmia kuani* Jiang, 1998

Professor Chung-Chich Kuan (1925–2005) taught at the Department of Entomology, National Chung Hsing University, Taichung, Taiwan.

Kubo Kaiya

Cruiser sp. *Macromia kubokaiya* **Asahina**, 1964

Mr Kaiya Kubo (b.1934) is a Japanese amateur entomologist. He graduated from Komaba High School, Tokyo (1953) and worked as a company director. He is an executive of the Lepidopterological Society of Japan. He wrote: *Butterfly Hunting in the Mountains of Central Japan* (1960). He collected one paratype of this Cruiser in Okinawa (1962).

Kuchenbeiser

Goldenring sp. *Anotogaster kuchenbeiseri* **Förster**, 1899
[Orig. *Cordulegaster kuchenbeiseri*]

Fr. Kuchenbeiser was the Head (1898) of the German enclave in Tientsin, China, which was conceded to Germany (1899). Forster dedicated the dragonfly to his friend.

Kuiper

Limniad sp. *Amphicnemis kuip*eri **Lieftinck**, 1937

Frederik Jan Kuiper (1895–1981) was a Dutch zoologist who collected (principally for Johan Gottlieb van Marle) in the Dutch East Indies (Indonesia) (1935–1937) including the type on Billiton Island (1935). A bird is also named after him.

Kükenthal

Treehugger sp. *Tyriobapta kuekenthali* **Karsch**, 1903
[Orig. *Monocoloptera kükenthali*]

Professor Dr Wilhelm 'Willy' Georg Kükenthal (1861–1922) led the Bremen Geographical Society's expedition in the yacht *Berentine* to Kong Karls Land in the Arctic (1889) and was there again later (1893–1894). They ran aground and the ship was crushed by ice. Luckily the *Cecilie Maline,* a sealing vessel, saved everyone four days after they were stranded. With support from the Senckenberg Natural History Society, he also participated in an expedition to the Moluccas and to Borneo (1894) where he collected the type. He specialised in the study of Octocorallia, a taxonomic subclass that includes sea pens, sea fans and soft corals. He also conducted embryological and comparative anatomical investigations of whales and other marine mammals. His large collection of zoological specimens is now housed at the Senckenberg Museum in Frankfurt. He is remembered in the names of about twenty other taxa including two reptiles.

Künckel

Darter sp. *Sympetrum kunckeli* **Selys**, 1884
[Orig. *Diplax kunckeli*]

Philippe Alexandre Jules Künckel d'Herculais (1843–1918) was a French entomologist. After graduating (1860), he entered École des Mines (1861) but preferred (1864) to follow less theoretical courses at Collège de France, at Muséum National d'Histoire Naturelle and at the Sorbonne. He met Émile Blanchard, becoming his pupil and private secretary. In 1866 he published his first mémoire, which was on the anatomy of Hemiptera. He entered the MNHN (1869), assisting Blanchard, replacing Alphonse Milne-Edwards. He became one of the first teachers at the l'Institut National d'Agronomie (1876) leaving to study grasshoppers in Argentina for several years (c.1885). He also studied crop pests in Algeria and Corsica. He was elected President of the Société Entomologique de France (1908 & 1909).

Kunjina

Pilbara Wisp *Agriocnemis kunjina* **Watson**, 1969

This is a toponym; Kunjina Spring, on Daniel's Well Station, is in Western Australia.

Kununurra

Kimberley Pin *Eurysticta kununurra* **Watson**, 1991

This is a toponym; Kununurra is a locality in Western Australia.

Kurahashi

Emerald sp. *Megalestes kurahashii* **Asahina**, 1985

Dr Hiromu Kurahashi (b.1941) is a Japanese medical entomologist. He has worked at the International Department of Dipterology, Higashikurume-shi, Tokyo and for the National Institute of Infectious Diseases, also in Tokyo. He has published extensively, especially on Calliphoridae and Sarcophagidae from Japan and the Oriental Region such as: *The blow flies of New Guinea, Bismarck Archipelago and Bougainville Island (Diptera: Calliphoridae)* (1987) and *Blow Flies (Insects: Diptera: Calliphoridae) of the Philippines* (2000). He collected the type in Thailand (1975).

Kurentzov

Banded Darter *Sympetrum pedemontanum kurentzovi* **Belyshev**, 1956

[JS *Sympetrum pedemontanum* (Allioni 1766)]

Professor Dr Alexey Ivanovich Kurentsov (Kurentzov) (1896–1975) was a prominent Russian entomologist and bio-geographer. He specialised in beetles of the family *Ipidae* (forestry pests) and butterflies, of which he described many new species. After graduation from Leningrad University he was relocated to Ussuriysk, and later to Vladivostok (1932), where he worked at the Institute of Zoology and Pedology of the Far Eastern Branch of the USSR, later Russian Academy of Sciences, Vladivostok. He undertook numerous expeditions to remote and uninhabited regions of the vast Far East of Russia (late 1920s onwards). He published 230 papers, including 10 monographs, such as: *Ipidae of the Far East of the USSR* (1941) and a guide book of the butterflies of the Russian Far East (1970).

Kuroiwa

Demoiselle sp. *Psolodesmus kuroiwae* Oguma, 1913

[Orig. *Psolodesmus dorothea kuroiwae*]

Hisashi Kuroiwa (1858–1930) was a Japanese naturalist and collector with a special interest in aquatic species (Japan). He wrote or co-wrote a number of books and papers such as: *A list of phanerogams collected in the southern part of Isl. Okinawa one of the Loochoo Chain* (1900).

Kusum

Kinkedwing sp. *Anisopleura kusumi* Sahni, 1965

[JS *Anisopleura lestoides* Selys, 1853]

No etymology was given so there are a number of possibilities. Kusum is a girl's name (meaning flower) and she is a leading character in a number of folktales of the area where the species was first collected, 'Bhowali', Naintal in the Kumaon foothills of the Himalayas. There is also a tree *Schleichera oleosa* (known as the Lac tree, Ceylon Oak or Macassar oiltree) which in Jim Corbett National Park (adjacent to Naintal) where it flourishes, it is known as kusum. It could even be a toponym, named for the Himalayan Mountain of Kusum Kanguru, although this is somewhat further away in Nepal and there is no obvious connection. All of the above is, of course, speculation in the absence of a proper etymology.

L

Lacroix

Ivory Pintail *Acisoma lacroixi* **Martin**, 1905
[JS *Acisoma trifidum* Kirby, 1889]

Joseph-Louis Lacroix (1879–1939) was a French dentist and amateur entomologist who collected (1913–1925) in France. He may have also collected in Guadeloupe and Peru or just acquired specimens from those countries. His collection of Orthoptera is housed in the Museum Bernard d'Agneci at Niort and the Museum of Natural History, La Rochelle, as well as Paris. Among other papers he wrote: *Contribution à l'étude des Névroptères de France (Fin)* (1914), *Deux Odonates nouvelles* (1920 & 1921) and *Faune des Planipennes de France – Coniopterygidae* (1923). He described eight new odonate species

Lafaec

Reedling sp. *Indolestes lafaeci* Seehousen, 2017

This is not an eponym but rather a Latinised version of the word for crocodile in the local Tetum language of Timor, the type locality; the word being 'lafaec'. Interestingly, because of its shape, Timor is known as the land of the sleeping crocodile, and also has a creation legend called 'lafaek diak' (the good crocodile).

Laglaize

Fineliner sp. *Teinobasis laglaizei* **Selys**, 1878
[Orig. *Telebasis laglaize*]

Léon François Laglaize (fl.1898) was a French traveller, collector and insect dealer in Gabon, Senegambia, Indonesia, New Guinea (1878–1897), and Venezuela and Paraguay (1898). He was also something of an apologist for the feather trade. The Muséum National d'Histoire Naturelle, Paris (MNHN) purchased a number of birds and other material from New Guinea from him (1878).

Laidlaw

Threadtail sp. *Prodasineura laidlawii* **Förster**, 1907
[Orig. *Disparoneura laidlawii*]
Spectre sp. *Periaeschna laidlawi* Förster, 1908
[Orig. *Caliaeschna laidlawi*]
Treehugger sp. *Tyriobapta laidlawi* **Ris**, 1919
Relict Himalayan Dragonfly *Epiophlebia laidlawi* **Tillyard**, 1921
Sprite sp. *Pseudagrion laidlawi* **Fraser**, 1922
Clubtail sp. *Burmagomphus laidlawi* Fraser, 1924

Myristica Sapphire *Calocypha laidlawi* Fraser, 1924
[Orig. *Rhinocypha laidlawi*]
Satinwing sp. *Euphaea laidlawi* **Kimmins**, 1936
[JS *Euphaea subcostalis* Selys, 1873]
Shadowdancer sp. *Idionyx laidlawi* Fraser, 1936
Fineliner sp. *Teinobasis laidlawi* Kimmins, 1936
Flatwing sp. *Burmargiolestes laidlawi* **Lieftinck**, 1960

Dr Frank Fortescue Laidlaw (1876–1963) was a British physician, botanist, odonatologist and malacologist, principally interested in malacology in later life, but earlier studied Odonata. He gained a double first in zoology at Trinity College, Cambridge (1898) then was Zoologist on the Skeat Expedition to the Malay Peninsular (1899–1900) along with Annandale (q.v.) and Evans. He was then Assistant Lecturer and Demonstrator at Owens College, Manchester (1900–1903). He took up medicine (1903) at St. Bartholomew's Hospital, London, qualifying as a surgeon (1909) and was in general practice in a village in Devon, except for WW1 service (RAMC), until retirement (1945). He occasionally acted as a locum, and F C Fraser (q.v.) helped him with a locum on one occasion. R J Tillyard (q.v.) was a close friend with whom he corresponded for many years. When he could he spent time at the BMNH generally meeting there with D E Kimmins (q.v.). He wrote many papers on molluscs (1908–1963) and odonates (1901–1955) including: *On a Collection of Dragonflies made by Members of the Skeat Expedition in the Malay Peninsula in 1899–1900* (1902), *Dragonflies (Odonata) of the Malay Peninsula with descriptions of new species* (1931), *Revision of the genus Coeliccia (orer odonata)* (1932) and *A survey of the Chlorocyphidaee (Odonata: Zygoptera)* (1950). Chlorocyphids (Jewels and Gems) were his favourite Odonata group. He wrote (1950): "*I was much attracted by their beauty at the time and have felt a strong interest, I think I may say an enthusiasm, for the family ever since*". He described c. 140 new odonate species from the oriental region and named c.15 new genera. He is also commemorated in the names of a number of snails. He married Maud Wright and they had a daughter Mary Louisa, then twins Ruby and Eric. (Also see **Louisa, Mary (Laidlaw), Ruby & Eric**)

Laird

Damselfly sp. *Lieftinckia lairdi* **Lieftinck**, 1963

Professor Dr Marshall Laird (1923–2007) was a New Zealand medical entomologist. He began his career in the Royal New Zealand Air Force as a medical entomologist at New Britain (Papua New Guinea) in the South Pacific during WW2. After the war, he returned to study for his PhD which was awarded (1948) by Victoria University of Wellington, New Zealand. He returned to the South Pacific and made aquatic entomological collections in the Solomon Islands including Guadalcanal, where the eponymous damselfly is endemic and he took the type (1953). He also collected fish from inter-tidal pools in Norfolk Island. He continued this study over four decades and as many continents with academic posts at the Department of Parasitology, University of Malaya, at the WHO in Geneva and the McGill University, Montreal, Canada. He was Professor and Head of the Department of Biology (1967–1972) and Director of the Research Unit on Vector Pathology (1972–1981), both at the University of Newfoundland. He retired to New Zealand (1983), but was an

Honorary Research Fellow at the University of Auckland. He wrote many articles including: *Mosquitos and Malaria in the Hill Country of the New Hebrides and Solomon Islands* (1955), a *Bibliography of the Natural History of Newfoundland and Labrador* (1980) and longer works such as: *Biocontrol of Medical and Veterinary Pests* (1981) and *Avian Malaria in the Asian Tropical Subregion* (1998).

Lamberto

Sanddragon sp. *Progomphus lambertoi* Novelo-Gutiérrez, 2007

Dr Luis Lamberto González-Cota (1948–2006) was a Mexican dentist and entomologist who combined his two passions, paediatric dentistry and collecting Lepidoptera and other insects. He was an entomologist at Centro de Estudios Faunísticos y Ecológicos del Sud-occidente de México. After his death, his family decided (2010) to donate his collection of c.65,000 specimens (mostly Lepidoptera but at least 1800 beetles and some arachnids) to the municipality of Uruapan, Michoacán, which could conserve it for display and study. Rudolfo Novelo-Gutiérrez wrote: *"This species is dedicated to the memory of the late Dr Lamberto González Cota, friend and colleague, for his encouragement in the study of the entomofauna of Michoacán State."*

Lambertus

Skimmer sp. *Lanthanusa lamberti* **Lieftinck**, 1942

Dr Lambertus Johannes Toxopeus (1894–1951) was a Dutch lepidopterist mainly working in Indonesia where he was born. He was educated at Amsterdam University culminating in his PhD (1930). He went to Java and taught in several secondary schools and became Lecturer in Zoology at the University of Indonesia (1946) and later professor at Bandoeng (1949). He took part in an expedition to Boeroe (1921–1922) and the Third Archbold Expedition (1938–1939), establishing collecting stations in the Snow Mountains of Papua New Guinea during when the type was taken. He wrote a number of papers including a report on the Archbold trip published in 'Treubia': *'Nederlandsch-Indisch Amerikaansche expeditie naar Nieuw-Guinea (3e Archbold expeditie naar Nieuw Guinea 1938–39). Lijst van verzamelstations'*. (1940). Lieftinck studied the insects collected at Leiden and also wrote his obituary in 'Idea', the entomological journal of Indonesia.

Lamotte

Western Double-spined Cruiser *Phyllomacromia lamottei* **Legrand**, 1993
[Orig. *Macromia lamottei*]

Dr Maxime Lamotte (1920–2007) was a pioneer French ecologist who became an Honorary Professor of Zoology at the University, Paris. He collected in West Africa (1942 & 1946) when sent to explore the Nimba Mountains (1941) in Guinea. He initiated research programmes there and in Côte d'Ivoire, where he co-founded (1962) a research station called Lamto; a combination of the first part of the names of the co-founders **LAM**otte and **TO**urnier. He also trained others to collect and conserve specimens. He became (1957) Director of the Laboratory of Zoology of the École Normale Supérieure in Paris. Lamotte is also honoured in the names of a spider genus, four mammals and three amphibians.

Landolt

Southern Migrant Hawker *Aeshna landoltii* **Buchecker**, 1876
[JS *Aeshna affinis* Vander Linden, 1829]

Elias Landolt (1821–1896) was a Swiss forestry scientist who became the Chief Forester in Zürich canton. He had extensive schooling in forestry (1842–1848) in Zürich and Württemberg. He began his career (1849) rising through the ranks right up to retirement (1893) and, while doing so, was (from 1855) Forestry Professor at the Federal Polytechnic in Zürich and also its Director (1867–1870). He published a number of papers and longer works on various aspects of forestry including: *Die forstliche Betriebslehre mit besonderer Berücksichtigung der schweizerischen Verhältnisse* (1892).

Lane

Bluewing sp. *Zenithoptera lanei* **Santos**, 1941

Dr Frederico Lane Jr (1901–1979) was a Brazilian entomologist at the Museu Paulista, São Paulo. He wrote more than 20 papers (1935–1946) such as: *Comentários sôbre o livro VII (Insetos) da Historia naturalis brasiliae de Jorge Marcgrave* (1946). At least two butterflies are also named after him.

Lankester

Flatwing sp. *Philogenia lankesteri* **Calvert**, 1924

Charles Herbert Lankester (1879–1969) (aka Don Carlos) was a British naturalist and lover of orchids who was a resident of Cachi, Costa Rica at the time the holotype was taken. When living in London he applied for a position to work as an assistant to the recently founded Sarapiquí Coffee Estates Company in Costa Rica, and was hired. He was at first interested in birds and butterflies, but the region's botanical diversity was immense and he began collecting orchids in the nearby woods; many were species unknown to science. He corresponded with Kew's orchid expert, R A Rolfe, and collected live plants and planted them on the farm he bought. He returned to England (1920) to enrol his five children in English schools then travelled to Africa (1920–1922) for the British Government researching coffee plantations in Uganda. Back in England, he found that Rolfe had died, leaving many of Lankester's orchids without identification. From Costa Rica (1922), he corresponded with Oakes Ames who discovered over 100 new species among the specimens sent to him. After his wife's death (1955) he decided to sell his farm and moved to Moravia, a suburb of San José, but retained a piece of land called 'El Silvestre', which contained his garden. As 'Jardín Botánico Lankester', it was donated to the University of Costa Rica (1973). Schaus and Barnes, who were staying with Lankester, collected the flatwing. He had also accommodated Calvert on a number of occasions.

Lansberge

Clubtail sp. *Leptogomphus lansbergei* **Selys**, 1878
Threadtail sp. *Prodasineura lansbergei* Selys, 1886
[Orig. *Alloneura lansbergei*]

Johan Wilhelm van Lansberge (1830–1905) was Governor-General of the Dutch East Indies (Indonesia) (1875–1881). He was born in Bogota, Colombia and brought up in Venezuela, where his father was the Dutch Consul General. He studied at Leiden (1848–1854) and held a number of diplomatic posts in Paris, Madrid, St Petersberg and Brussels before his appointment in the East Indies. He was also a keen general naturalist and entomologist, working on the taxonomy of Coleoptera. Among his publications are: *Révision des Onthophagus de l'Archipel Indo-Néerlandais, avec descriptions des espèces nouvelles* (1883) and *Description d'un Cérambycide de Sumatra, appartenant à un genre nouveau de la tribu des Disténides* (1886). Selys described him as an: 'entomologiste très-distingué'. He collected the clubtail in Java and the threadtail in Borneo (both between 1875 and 1881) and sent them to Selys.

Lascelles

Blue-spotted Yellowwing *Chlorocnemis lascellesi* **Pinhey**, 1961
[JS *Allocnemis wittei* (Fraser, 1955)]

Peter Lascelles helped in collecting the type series in Northern Rhodesia (now Zambia) (1960). He and his African assistant, Rafael Mpala, were with Pinhey in February and March that year visiting the Mwinilunga District and aided his collecting. Lascelles is acknowledged by Pinhey as the person whose enthusiasm was responsible for much of the available knowledge of the dragonflies of Zambia.

La Selva

Dragonlet sp. *Erythrodiplax laselva* Haber, Wagner & de la Rosa, 2015

This is more of a toponym being named for the fiftieth anniversary of La Selva Biological Station, Organisation for Tropical Studies.

Latihami

Dambo Skimmer *Orthetrum latihami* **Pinhey**, 1966

Not an eponym but a reference to a body part – the hamules.

Latreille

Yellow-winged Darter *Libellula latreillei* **Selys**, 1850
[JS *Sympetrum flaveolum* (Linnaeus, 1758)]

Professor Pierre André Latreille (1762–1833) was a French entomologist; his particular focus was arthropods. He trained as a Roman Catholic priest but during his studies became very interested in natural history and never carried out a ministry. During the French Revolution, he was imprisoned (1793) when he refused to swear allegiance to the state, narrowly avoiding the guillotine. During his imprisonment, he identified a rare beetle *Necrobia ruficollis* to the prison doctor who came upon him inspecting it. He was so impressed he managed to get Latreille released. He became a teacher and corresponded with other entomologists as well as writing papers such as *Précis des caractères génériques des insectes* (1796) published at his own expense. He was put under house arrest (1797) but this was lifted due to the pleas of Cuvier, Lacépède and Lamarck and he was appointed

(1798) to work alongside the latter at the MHNH, curating the arthropods and he succeeded him as Professor of Zoology (1829–1832). He left Paris to avoid the cholera epidemic, later returned, but died there of a bladder disease. He described (1805) the Migrant Hawker *Aeshna mixta*. A bird is also named after him as are at least seven other taxa.

Lattke

Flatwing sp. *Sciotropis lattkei* **De Marmels**, 1994

Dr John Edwin Lattke Bravo (b.1959) is an entomologist and a specialist of the taxonomy of ants. The University of Central Venezuela awarded his bachelor's degree in agronomical engineering (1987) and the University of California his PhD in entomology (1999). He was Professor of Entomology at the University of Central Venezuela until 2014 and is now visiting Professor the University of Loja, Ecuador. He found this new species in Venezuela (1993). He has written numerous papers (1994–present), mainly on ants, such as: *Ant fauna of the French and Venezuelan Islands in the Caribbean* (1994) and *Revision of the ant genus Leptogenys in the New World* (2011).

Lau, F

Bombardier sp. *Lyriothemis laui* **Krüger**, 1902
[JS *Lyriothemis magnificata* Selys, 1878]

Dr Friedrich Lau (1869–1947) was a German historian and state archivist at the Dusseldorf City Museum. He wrote a: *History of the City of Dusseldorf* (1921). In honouring him Krüger describes him as one of his two faithful 'Wandergefahrten', which I interpret as travelling companions. (Also see **Dau**)

Lau, M

Coralleg sp. *Rhipidolestes laui* **KDP Wilson** & **Reels**, 2003

Dr Michael Wai Neng Lau (b.1962) is a Chinese conservationist and has written articles with Graham Reels (q.v.). He studied zoology and botany at the Australian National University, Canberra. He worked at the Fauna Conservation Department, Kadoorie Farm and Botanic Garden, Hong Kong. He worked for WWF (1980s) at Mai Po Nature Reserve, Hong Kong and, after gaining his PhD at the University of Hong Kong, returned (2011) to work for WWF, becoming Director of Wetlands Conservation. He has contributed much to herpetological research and conservation in Southern China. He has written or co-written a number of articles and books such as: *Sustaining the Pulse: Managing for Biodiversity Conservation in South China's Forest Nature Reserves* (2012) & *Hong Kong amphibians and reptiles* (1986). He discovered the new species in Hong Kong (May 2002). A reptile is also named after him.

Laubmann

Downy Emerald *Cordulia aenea laubmanni* Götz, 1923
[JS *Cordulia aenea* (Linnaeus, 1758)]

Alfred Louis Laubmann (1886–1965) was a German veterinary surgeon and ornithologist, and professor at the Zoological Museum, Munich. Living in eastern Germany immediately after WW2, he was forced into heavy labour in a factory and not allowed to visit the museum where his own library and books were housed. Among other works he wrote: *Die Eisvögel der Insel Sumatra: Versuch einer monographischen Darstellung* (1925) and *Vögel* (1930). Eight birds are also named after him. Although no etymology was given, Götz certainly named a woodpecker subspecies after him, so this makes it almost certain this is who he had in mind.

Laura

Laura's Clubtail *Stylurus laurae* **Williamson**, 1932

Laura Christine Buis née Ditzler (1911–2008), along with her mother Ivy (q.v.) and father William Ditzler (q.v.), was a member of the party that collected the holotype in South Carolina. This was a trip organized by Williamson from Indiana to Georgia (1931). She attended DePauw University and Indiana University (1928–1933), marrying directly afterward.

Laurent

Vesper Bluet *Enallagma laurenti* **Calvert**, 1919
[JS *Enallagma vesperum* Calvert, 1919]

Philip Laurent (1858–1942) was an industrialist and amateur naturalist and a well-known collector of natural history objects including butterflies and other insects. At least one moth is also named after him. The type was taken in Florida. His collection of over 40,000 pinned insects, mostly Lepidoptera and Coleoptera from Florida and the western US states went to the Academy of Natural Science in Philadelphia.

Lawrence

Lyretail sp. *Stylogomphus lawrenceae* B **Yang** & DAL **Davies**, 1996

Patricia Lawrence (Davies) (d.1998) was the author's intimate companion at the time, although they never married. She is referred to as Davies in the etymology, which reads: "*Named in honour of Patricia Lawrence Davies, our expedition colleague.*" She did a lot of capable collecting and worked with the second author on many things. The male holotype and female allotype were collected in Jiangcheng Co., Yunnan (1993). (Also see **Davies**)

Leach

Twinspot Tigertail *Synthemis leachii* **Selys**, 1871

William Elford Leach (1791–1836) was an independently wealthy British zoologist, entomologist and marine biologist and malacologist. As a twelve-year old schoolboy in Exeter, he collected marine samples along the Devon coast. He studied medicine at St Batholomew's London, Edinburgh and St Andrews Universities graduating MD (1812), but he did not practice and instead was appointed at the British Museum Zoology Department as Assistant Librarian (1813). He began sorting the collections and was made Assistant Keeper of Zoology. Here he became a world-renowned expert on crustaceans and molluscs, and also worked on insects, birds and mammals. He was elected a Fellow of the Royal Society (1817). He suffered a nervous breakdown (1821) due to overwork and resigned from the museum in 1822. He went to Europe with his elder sister to convalesce, and travelled through France, Italy and Greece. Leach died of cholera in Italy, where he had settled with his sister. He wrote a number of books including: *The Zoological Miscellany* (1814), chapters in other works, entries for the *Encyclopaedia Britannica* including eight entries on Odonata genera (such as the new genera *Calopteryx, Gomphus* and *Anax*), and other encyclopaedias as well as descriptions, etc. He named many animal species including 27 after his friend John Cranch and nine after 'Caroline' without revealing her identity. Among other taxa, four birds, two mammals and four reptiles are also named after him.

Leakey

Mealy Dropwing *Trithemis leakeyi* **Pinhey**, 1956
[Orig. *Lokithemis leakeyi*]

Louis Seymour Bazett Leakey (1903–1972) was a British archaeologist, palaeontologist and naturalist, most famous for his work on human evolutionary development following his discoveries of early man fossils in Olduvai Gorge. He was also very influential in conservation in East Africa. He was bought up by his missionary parents among the Kikuyu people; spoke the language fluently and was initiated into the Kikuyu. He was sent to study in England aged sixteen and later entered Cambridge where he matriculated (1922). He spent a year fossil hunting in Tanganyika (1924–1925) then returned to Cambridge and took a degree in anthropology and archaeology (1926) when he left again for East Africa to study early human remains at various sites until his return to Cambridge (1929). For a number of years, he travelled back and forth until finally settling back in Kenya as an unpaid Assistant Curator at the Coryndon Museum (1941) until being appointed Curator (1945). He wrote many books over the years, mostly about early man, but also on politics and his memoires; from his doctoral thesis *The Stone Age Culture of Kenya Colony* (1931) to the posthumously published *The Southern Kikuyu before 1903* (1977). Arthritis forced him to stop excavating and he turned to lecturing in the UK and USA. Pinhey, who worked under Dr Leakey at the Coryndon Museum in Nairobi (1949–1955), wrote: "*I have pleasure naming this species after Dr. L.S.B. Leakey.*" The species was found at Lake Chila in Northern Rhodesia. He died of a heart attack while staying with Jane Goodhall in London. She had been a researcher for him, as had Dian Fossey with whom he had a brief romance.

Lebas

Spectre sp. *Periaeschna lebasi* **Navás**, 1930

Father F Lebas was a Jesuit priest at the Gayaganga Mission who collected (1929) the species in Kurseong, Sikkim.

Lefèbvre

Lefebvre's Dragonfly *Diplacodes lefebvrii* **Rambur**, 1842
[A. Black Percher, Orig. *Libellula lefebvrii*]
Pale Pincertail *Onychogomphus lefebvrii* Rambur, 1842
[Orig. *Gomphus lefebvrii*]

Alexandre Louis Lefèbvre de Cérisy (1798–1867) was a French entomologist. He was a solicitor's clerk when J B Godart (1775–1825) introduced him to entomology and he became interested in Lepidoptera. He travelled throughout Europe sometimes accompanying entomologists such as A R Percheron (1797–1869) in Provence and G Bibron (1805–1848) in Sicily. He was the collector on an expedition to Egypt (1828–1829), during which he collected the female type series of the percher. He went on to be a founding member of the Societé entomologique de France. At least two butterflies are also named after him, *Allancastria cerisyi* (Godart, 1819); *Smerinthus cerisyi* (Kirby,1837)

Legrand

Legrand's Cruiser *Phyllomacromia legrandi* Gauthier, 1987
[Orig. *Macromia legrandi*]
Robust Dropwing *Trithemis legrandi* **Dijkstra**, Kipping & **Mézière**, 2015

Dr Jean Legrand (b.1944) was Curator of the Muséum National d' Histoire Naturelle, Paris, where Gauthier was his colleague. He is a specialist in African Odonata and has written many papers including: *Nouvelles additions aux représentants afrotropicaux du genre Elatto neura; groupe acuta-vrijdaghi (Odonata; Protoneuridae)* (1980) and *Note sur les Odonates actuellement connus des Monts Nimba* (1983) that was referenced in the paper describing the cruiser. In his etymology, Gauthier writes: "*C'est un grand plaisir pour moi de dédier cette nouvelle espèce à mon collègue Jean Legrand du Muséum National d'Histoire Naturelle de Paris*". The etymology for the dropwing is even more effusive stating: "*Named in honour of Dr Jean Legrand, collector of the first material of this species, and successor to Elliot Pinhey as the leading specialist of African Odonata of his time...*" He has described 43 new odonate species, most of which are eponyms. At least one fossil odonate is also dedicated to him. (Also see **Christine, Emilia, Hélèn** & **Virginie**)

Lehmann

Bluetail sp. *Agrion lehmanni* Kolenati, 1856
[*nomen oblitum*; obviously an *Ischnura*-species]

Alexander Lehmann (1814–1842) was a Russian explorer, biologist and collector of Baltic-German descent from Dorpat (now Tartu, Estonia). He travelled to Siberia (1839–1840) and to Turkestan as leader of an expedition (1841–1842) mounted to collect for the St Petersburg Botanical Garden. He was taken sick and died on his way home. His notes and accounts were published after his death as: *Alexander Lehmann's Reise nach Buchara und Samarkand in den Jahren 1841 und 1842* (1852). A reptile and a fish are also named after him.

Lempert

Lempert's Hooktail *Paragomphus lemperti* **Dijkstra** & Papazian, 2015

Jochen Lempert (b.1958) is a German biologist and photographer. He studied biology at the University of Bonn and only turned to photography in the 1990s, but has since built an international reputation as a photographer of nature. He has, for twenty-five years, been 'engaged in an ongoing project that deals with the perception of nature and creatures within the blurry contexts of scientific research, subjective perception and man-made environments'. Among his scientific papers is: *Gynacantha corbeti spec. Nov., a new dragonfly from West Malaysia (Anisoptera: Aeschnidae)* (1999). The etymology says it honours him "*...whose 1988 study of Liberian Odonata was one of the first to address tropical odonate ecology in detail and who first recognised this species as well as Eleuthemis umbrina.*" The etymology refers to his Diplomarbeit at Rheinischen Friedrich-Wilhelms-Univerität zu Bonn.

Lencioni

Demoiselle sp. *Mnesarete lencionii* **Garrison**, 2006

Frederico Augusto de Atayde Lencioni (b.1970) is among the foremost Brazilian odonatologists. The University of Mogi das Cruzes awarded his biology degree (1996). He is an employee of the Court of Justice of the State of São Paulo (since 1993). He wrote the two-volume: *Damselflies of Brazil* (2005 & 2006) among other works. He has described fifteen new species of odonata and named two new genera. He is also a very fine illustrator and was recently invited to be associated editor of the 'Proceedings of the Biological Society of Washington'. A shy man who, when I asked if he would care to add to the entry, said that writing about himself was torture. He went on to say *"...maybe you can include that in my last paper,* Diagnoses and discussion of the group 1 and 2 Brazilian species of *Heteragrion,* with descriptions of four new species (Odonata : Megapodagrionidae) (2013). *I described the new species in honour of the 40 years of foundation of the rock band Queen (my favourite rock band (although my connection to music is just as a listener)), and I described in it four new species, one for each member of the band* (Heteragrion freddiemercuryi, H. brianmayi, H. rogertaylori and H. johndeaconi)". He was honoured at the founding meeting of the Sociedad de Odonatologia Latinamericana (SOL) by the organizers of the meeting with a plaque bearing the following words: *"The Latin American Odonatology Society thanks Frederico AA Lencioni for his enormous contribution to Latin American Odontology."* (Also see **Frederico**)

Lenko

Firetail sp. *Telebasis lenkoi* **Machado**, 2010

Dr Karol Lenko (1914–1975) was a Brazilian biologist, zoologist and entomologist who was curator of insects at the Museu de Zoologia, São Paulo until retirement (1969). He collected on a number of expeditions in various parts of Brazil. He wrote a number of entomological papers, usually with others, including descriptions of new species such as: *Uma nova espécie de Gnamptogenys de Mato Grosso (Hymenoptera, Formicidae)* (1964). Machado's etymology says that he named the species: *"...in honor of my good friend the entomologist Karol Lenko who collected the type series and has provided me with many other valuable Odonata including another species of Telebasis described herein."* A number of other taxa, such as flies and crustaceans are named after him too.

Lent

Dragonlet *Erythrodiplax lenti* **Ris**, 1919
[JS *Erythrodiplax amazonica* Sjöstedt, 1918]

Alonzo Wesley 'Lent' Williamson (1845–1914) was the father of E B Williamson (see **Williamson, EB**)

Leobopp

Archtail sp. *Nannophlebia leoboppi* **Orr** & **Kalkman**, 2015

Leopold Bopp (b. 2014) is the first grandchild of Hanns-Jürgen and Ursula Babette Roland, who had generously supported the research on New Guinean Anisoptera.

Leon

Two-banded Cruiser *Phyllomacromia leoni* **Fraser**, 1928

[JS *Phyllomacromia contumax* Selys, 1879]

The etymology is not given by Fraser, as so often was the case, and it is impossible to tell who he had in mind, if indeed it was a particular person.

Leonard (Brass)

Sylvan sp. *Idiocnemis leonardi* **Lieftinck**, 1958

Leonard John Brass (1900–1971) was an Australian and American botanist, collector and explorer. He trained at Queensland Herbarium and collected for them (1930s–1960s). He served in the Canadian Army during WW2 and afterwards became a naturalized US citizen (1947). Florida State University awarded him an honorary doctorate (1962). He participated in several international expeditions to New Guinea and the Solomon Islands during one of which, the Fifth Archbold Expedition (1956–1957), he collected the eponymous type. He was an associate curator of the Archbold Expeditions with the AMNH (1939–1966) at the Archbold Biological Station, Florida. He also collected for the Arnold Arboretum, Massachusetts and was Director of Field Operations for the Upjohn Tropical Africa Expedition (1949–1950). He also helped Archbold establish the Corkscrew Swamp Sanctuary (1955). He retired to Queensland.

Leonard (Müller)

Sydney Hawk *Austrocordulia leonardi* **Theischinger**, 1973

Leonard Müller (b.1942). The etymology says: "*I want to thank my friend Mr Leonard Müller, who spent many days with me on trips in Australia, for his valuable help*". (See **Müller, L & Elke**)

Leonard, JW

Wedgetail sp. *Acanthagrion leonardi* **Jurzitza**, 1980

[JS *Acanthagrion cuyabae* Calvert, 1909]

Dr Justin Wilkinson 'Doc' Leonard (1909–1975) was an aquatic entomologist, conservationist, and ecologist. He studied for a chemistry and zoology degree at University of Michigan, obtaining his bachelors' (1931), masters' (1932) and PhD (1937) there. He was stationed in the Solomon Islands during WW2 as a medical entomologist (1944–1946) achieving the rank of Major. He was a fisheries biologist (1934–1943 & 1951–1965) at the Institute for Fisheries Research and (1964–1975) Professor of Natural Resources and Zoology, University of Michigan. He specialised in the study of mayflies and wrote more than seventy papers and longer works (1937–1975) including such articles as: *A new Baetis from Michigan* (1950) and the book *Mayflies of Michigan Trout Streams* more than fifty of which were with his wife (1943) Fannie A Leonard (née Divelbess) (1962). He also wrote on other subjects including Odonata, such as: *A Revisionary Study of the genus Acanthagrion (Odonata: Zygoptera)* (1977), which was a posthumous publication of his PhD thesis forty years after its compilation and acceptance. This revision included descriptions of 14 new species. Earlier (1937) the author had described one new damselfly species.

Leonardo (Co)

Shadowdamsel sp. *Drepanosticta leonardi* Villanueva, Gapud & Lin, 2011

Leonardo Legaspi Co (1953–2010) was a Fillipino botanist and plant taxonomist who, during his lifetime, was considered the foremost authority in ethnobotany in the Philippines.

Leonardo (Machado)

Threadtail sp. *Neoneura leonardoi* **Machado**, 2005

Leonardo Machado Haertel (b.1999) is the grandson of the author, Professor Dr Angelo Barbosa Monteiro Machado (b.1934).

Leonardo (Zusne)

Stripe-fronted Dryad *Nephepeltia leonardina* **Rácenis**, 1953

Leonardo Zusne 'assisted the author enthusiastically' in collecting dragonflies in Venezuela (1952).

Leonora

Firetail sp. *Teinobasis leonorae* **Lietinck**, 1937
[JS *Teinobasis rajah* Laidlaw, 1912]
Damselfly sp. *Rhyacocnemis leonorae* Lieftinck, 1949
Maroon Proto *Protolestes leonorae* **Schmidt**, 1951
Flatwing sp. *Philogenia leonora* **Westfall** & Cumming, 1956
Wedgetail sp. *Acanthagrion leonora* Gloger, 1967
[JS *Acanthagron obsoletum* (Förster, 1914)]
Leonora's Dancer *Argia leonorae* **Garrison**, 1994
Clubtail sp. *Gomphidia leonorae* Mitra, 1994

Red-veined Pennant ssp. *Celithemis bertha leonora* Westfall, 1952
[JS *Celithemis bertha* Williamson, 1922]

Leonora Katherine Gloyd née Doll (1902–1993) was an American odonatologist. Kansas State Agricultural College awarded both her BSc (1924) and MSc (1925) in zoology. She was offered (1929) a job as assistant to E B Williamson (q.v.) at the Zoological Museum of the University of Michigan, Ann Arbor. Her first task was to add references on cards of Williamson's 'Specific Reference Index'. Then she began to identify specimens of the museum collection, starting with the *Enallagma* material. Her first two papers on Odonata (1932, 1933) were supervised by Williamson. After Williamson's death (1933) she inherited the huge task of monographing the speciose genus *Argia*. This work became more difficult (1936) after she and her husband (m.1925), herpetologist Howard K. Gloyd, moved to Chicago. However, she continued to work at home and occasionally visited Ann Arbor. She took (1947) a part-time job as laboratory assistant at Illinois State Natural History Division in Urbana. She was awarded a National Science Foundation grant to support her study on *Argia* (1965–1968). For this work she visited several museums in the US, but unfortunately her monograph on *Argia* never materialized. Recently Rosser Garrison (q.v.) and Natalia von Ellenrieder (q.v.) have continued her work. Her extensive manuscript notes on members of this genus, currently on loan to Garrison, has provided much help. Much of her work came from the study of the museum's collection, but she also collected widely, mostly in the US, particularly Texas. She published c.40 papers on Odonata, which include descriptions of 10 new species and one new subspecies. Her papers include: *Three new North American species of Gomphinae (Odonata)* (1936) and *The dragonfly fauna of the Big Bend Region of Trans-Pecos Texas* (1958). Bastiaan Kiauta (q.v.) wrote (1977) "*Through her meticulous and significant scientific work but also through her personal charm and good comradeship towards colleagues and young dragonfly students and admirers she gained a very unique position in the large odonatological family*."

Leontine

Dragonlet sp. *Libellula leontina* **Brauer**, 1865
[JS *Erythrodiplax connata* (Burmeister, 1839)]

Leontine Brauer née Boschetty (d.c.1879) was the wife (1856) of the author Friedrich Moritz Brauer (1832–1904) (q.v.).

Leopold

Paradise Metalwing *Neurobasis leopoldi* **Fraser**, 1932
[JS *Neurobasis asustralis* Selys, 1897]
Emerald sp. *Procordulia leopoldi* Fraser, 1932

HRH Prince Leopold (1901–1983) née Léopold Philippe Charles Albert Meinrad Hubertus Marie Miguel became Leopold III, King of the Belgians (1934) and abdicated (1951) in favour of his son, Baudouin. Inspired by A R Wallace, Prince Leopold and his first wife, Princess Astrid (1905–1935), made a voyage to the Dutch East Indian Archipelago (1928–1929), during which many zoological specimens were collected. The party also visited New Guinea and the two eponymous odonate species were found at Arfak Mountains. Fraser's descriptions (1932) appeared in a paper in the voluminous publication series entitled: *Resultats scientifiques du voyage aux Indes Orientales neerlandaises de LL. AA. RR. le Prince et la Princesse Leopold de Belgique*. Leopold III was a keen amateur entomologist and social anthropologist.

Le Roi

Bannerwing sp. *Euthore leroii* **Ris**, 1918

Dr Otto le Roi (1878–1916) was a German zoologist, particularly malacologist, ornithologist and entomologist who studied under Alexander Koenig and became his assistant (1898–1899). The University of Bonn awarded his PhD (1906). He participated in the Spitsbergen Expedition (1907–1908) and travelled in Sudan (1910 & 1913). He wrote: *Birds of Rhine Province* (1906) and he collected in Egypt (1910). Ris dedicated the bannerwing to le Roi, who had 'fallen for the fatherland' (he was killed in the Carpathian Mountains on the Eastern Front during October 1916 when serving as a lieutenant in the German army). He wrote a number of papers on German and African Odonata and described eleven new species from Africa, four of which are still valid. A number of other taxa, such as molluscs and insects are also named after him.

Leroy

Leroy's Longlegs *Notogomphus leroyi* **Schouteden**, 1934
[A Clubbed Longleg, Orig. *Podogomphus leroyi*]

Jean V Leroy was a Belgian entomologist in Congo-Kinshasa near Lake Kivu. He collected for the MRAC (Musee Royal de l'Afrique Centrale, Tervuren, Belgium) (1933–1951) and BMNH and is named in other taxa such as the wasp: *Afrevania leroyi* (Benoit, 1953). He wrote several long pamphlets, all of which were published by the Institut National pour l'étude Agronomique du Congo Belge, including: *Observations relatives* à quelques insects attaquant le caféier (1936) and one on a genus of Shield Bug – *Les Antestis spp. Au Kivu* (1942).

Leticia

Dragonlet sp. *Erythrodiplax leticia* **Machado**, 1996

Leticia Machado Haertel (b.1995) is the first grandchild of the author, Professor Dr Angelo Barbosa Monteiro Machado (b.1934). His etymology reads (in translation): *"The dragonfly is dedicated to my first granddaughter, Leticia, born a few days after the discovery of the species in Pratinha Pond in Bahia."*

Letitia

Threadtail sp. *Drepanoneura letitia* **Donnelly**, 1992
[Orig. *Epipleoneura letitia*]

Letitia 'Letty' Jane Morton née Bryce (b.c.1940) was the wife of Dr Eugene Siller Morton (b.1940) (q.v.). The type was collected in Panama (1970–1971). (See also **Morton, ES**)

Lever

Pond Damselfly sp. *Nesobasis leveri* **Kimmins**, 1943

Robert John Aylwin Wallace Lever (b.1905) was Government Entomologist on Fiji (1937). He studied at the University of Durham for his BSc, then trained in tropical entomology at Imperial College London, following which he spent a year at Imperial College of Tropical Agriculture in Trinidad. He was appointed (1930) Government Entomologist to the British Solomon Islands Protectorate during when he made entomological and anthropological collections for the BMNH. He was then posted to Fiji (1937) and Malaya. Lever collected the holotype in Fiji (January 1943). Among his published papers are: *The female Ectatomma tuberculatum (Oliv.) var Punctigerum Emery (Hymenoptera, Formicidae)* (1930), *Trinidad Chrysomelid Beetles on Leguminous Plants* (1930) and *Pests of the Coconut Palm* (1969).

Lewis

Bombardier sp. *Lyriothemis lewisii* **Selys**, 1883
[JS *Lyriothemis pachygastra* (Selys, 1878)]

George Lewis (1839–1926) was an English entomologist and Fellow of the Linnaean Society with a particular interest in Coleoptera in general and Japanese species in particular. Among other works: he wrote (1873–1905), *On the Buprestidae of Japan* (1895) and *A systematic catalogue of Histerid* (1905). The type was taken in Japan where Lewis collected several times. A number of beetles are also named after him.

Lewis

Mount Lewis Bluestreak *Lestoidea lewisiana* **Theischinger**, 1996

Toponym; Mount Lewis is an area in Queensland, Australia.

Li

Flatwing sp. *Rhipidolestes lii* **Zhou**, 2003

Professor Zi-Zhou Li was a Chinese zoologist at Dali University, Dali, Yunnan, China.

Lieftinck

The Damselfly genus *Lieftinckia* **Kimmins**, 1957

Sylvan sp. *Coeliccia lieftincki* **Laidlaw**, 1932
Grappletail sp. *Heliogomphus lieftincki* **Fraser**, 1942
Australian Tiger sp. *Ictinogomphus lieftincki* **Schmidt**, 1934
[Orig. *Ictinus lieftincki*]
Eastern Bronze Cruiser *Macromia lieftincki* Fraser, 1954
[JS *Phyllomacromia aeneothorax* Nunney, 1895]
Pincertail sp. *Nihonogomphus lieftincki* **Chao**, 1954
Red Arrow *Rhodothemis lieftincki* Fraser, 1954
Duskhawker sp. *Gynacantha lieftincki* Compte Sart, 1964
[JS *Gynacantha africana* Palisot de Beauvois, 1805]
Coraltail sp. *Ceriagrion lieftincki* **Asahina**, 1967
Leaftail sp. *Phyllogomphoides lieftincki* **Belle**, 1970
[Orig. *Negomphoides lieftincki*]
Emerald Spreadwing sp. *Megalestes lieftincki* Lahiri, 1979
Spectre sp. *Petaliaeschna lieftincki* Asahina, 1982
Kinkedwing sp. *Anisopleura lieftincki* Prasad & Ghosh, 1984
[JS *Anisopleura subplatystyla* Fraser, 1927]
Lesser Shadowcruiser *Idomacromia lieftincki* **Legrand**, 1984
Redbolt sp. *Rhodothemis mauritsi* Lohmann, 1984
Clubtail sp. *Shaogomphus lieftincki* **Chao**, 1984
Goldenring sp. *Neallogaster lieftincki* Lohmann, 1993
[JS *Coratulegaster jinensis* Zhu & Han, 1992]
Lieftinck's Sprite *Archibasis lieftincki* Conniff & Bedjanič, 2013
Fineliner sp. *Teinobasis lieftincki* **Theischinger** & **Kalkman**, 2014

Glider ssp. *Tramea liberata lieftincki* **Watson**, 1967
[Orig. *Trapezostigma liberata lieftincki*]

Dr Maurits Anne Lieftinck (1904–1985) was a Dutch entomologist and the leading authority on Sundaland and New Guinea Odonata. He studied biology at Amsterdam University (1923–1929) and was then appointed (1929) entomologist at the Zoological Museum at Buitenzorg, Java in the former Dutch East Indies, eventually becoming Head of the Museum (1939) until retiring (1954). He was a POW (1942–1945) and left to recuperate in the Netherlands (1946–1947) before returning to Java, during which time he used the opportunity to visit a number of European museums examining types of Odonata and bees. The work he carried out at Buitenzorg was internationally recognised, as was his co-founding of the Dutch Entomological Society. He was also editor of 'Treubia' (1939–1954). He was granted an honorary doctorate by the University of Basle (1950). For four months, he led the Netherlands part of the Swedish-Dutch ornithological expedition to New Guinea (1948). During his stay in Java he also made short visits to Sumatra (1935, 1940), Ceylon (1938), Malaya and Sarawak (1950) and the South Moluccan Islands (1948, 1949). He returned to the Netherlands (1954) and was appointed (1954–1969) Curator of the National Museum of Natural History in Leiden. After retirement, he settled in the countryside but still wrote and continued with study trips, visiting Taiwan (1976), western Himalaya (1978) and Sulawesi (1982). In his later years, he made short nature films. More than half of his over 320 publications were on Odonata. He also wrote several taxonomic papers on Apoidea (bees), his second specialty. His major odonatological publications include: *Odonata Neerlandica* (1925–1926), written at the early age of 21 years; the seven-part series (totalling c.900 pages): *The dragonflies (Odonata) of New Guinea and neighbouring islands* (1932–1949) and *Handlist of Malaysian Odonata: A catalogue of the dragonflies of the Malaya Peninsula, Sumatra, Java and Borneo* (1954). He described c.610 new species and 24 new genera of Odonata. Only Selys had described more new odonate species than he. (See also **Maurits**).

Lien

Paddletail sp. *Sarasaeschna lieni* Yeh & Chen, 2000
[Orig. *Oligoaeschna lieni*]

Dr Jih-Ching (Ri-ging Lian) Lien (b.1927) (aka 'Doctor Mosquito') is a Taiwanese entomologist who continued to study mosquitoes well into his mid-eighties, having been chief of the Medical Entomology Division of the Institute of Tropical Medicine of the National University of Taiwan. He developed an interest in mosquitoes while working as a typing assistant to a Japanese professor of tropical medicine at Taipei Imperial University (now National Taiwan University) during the Japanese colonial period. National Normal University awarded his BA in English after which he undertook his army national service. After being discharged from the military he returned to the Institute of Malaria and his papers and lectures on his new discoveries won him a doctorate in medicine from Nagasaki University, Japan. Sponsored by WHO and the National Service Council, he travelled to Malaysia to research further. He was honoured because of his "...*continuous guidance and encouragement rendered to the senior author (W C Yeh) during his course of studying dragonflies*". Around twenty mosquitoes are also named after him and he discovered twenty-seven new species.

Lillian

Pond Damselfly sp. *Metaleptobasis lillianae* Daigle, 2004

Lillian Agnes Russett Daigle (1912–1995) was the mother of the author Jerrell (q.v.) J Daigle. He honoured her in the etymology, saying: "*The species is named after my mother, Lillian Agnes Russett Daigle, for encouraging me to explore the wonders of the natural world.*"

Lima

Two-spined Darner sp. *Gynacantha limai* **Navás**, 1916

[JS *Gynacantha convergens* Förster, 1908]

F Lima collected the type in Brazil (1910) but the author provided no further clue to who this man was.

Lindblom

Cascader sp. *Tilithemis lindblomi* **Sjöstedt**, 1917

[JS *Zygonyx natalensis* (Martin, 1900)]

Dr Karl Gerhard Lindblom (1887–1969) was a Swedish ethnographer. The University of Uppsala awarded his PhD (1916). He worked in East Africa in the second decade of the twentieth century. He was the principal author of materials on the Akamba peoples of Kenya, where the type was taken. Among his written work are: *Outlines of a Tharaka Grammar, with a list of words and specimens of the language* (1914) and *Ethnological and anthropological studies in Sweden during the war* (1946). He was also a member of the Berlin Brandenburg Academy of Sciences (1950).

Linda

Swampdamsel sp. *Leptobasis linda* Johnson, 2016

Linda Cecilia Pritchard (b.1965) is the wife of the author James (Jim) Thomas Johnson (b.1967).

Linden

Genus *Lindenia* De Haan, 1826

Blue-eye *Erythromma lindenii* **Selys**, 1840

[Orig. *Agrion lindenii*]

Professor Pierre Léonard Vander Linden (1797–1831) was a Belgian entomologist specializing in the study of Hymenoptera and Odonata. From an early age, he collected Lepidoptera. He was awarded a scholarship by the University of Boulogne, Italy (1817). After completing his medical training in Bologna (1821), he went to Paris for further medical study. He was awarded his MD by the University of Leuven (1823). He worked (1825) as a physician in Brussels, but was soon invited to become Secretary of the Royal Academy of Science and later started to manage the Academy's Natural History Museum and its botanical garden. He was appointed (1830) Professor of Zoology there and taught natural history at the Brussels Atheneum, but soon after starting died of 'stomach encephalitis'. While in Bologna he sent insects, plants and other material to P A Latreille (q.v.) in Paris. There he also published (1820) two brief papers on Italian Odonata: *Agrionines Bononiensis descriptae* and *Aeshnae Bononiensis descriptae*, later (1825) publishing his *Monographiae Libellulinarum Europaearum specimen*, which treated 37 species and was the first synopsis of the European Odonata. Altogether he described 21 new Odonata species from Europe, 11 of which are valid. Among his other publications is the two-volume: *Observations sur les Hyménoptères d'Europe de la famille des Fouisseurs* (1827 & 1829). (See also **Vander**).

Lindgren

Shadowdamsel sp. *Protosticta lindgreni* **Fraser**, 1920
[JS *Protosticta himalaica* Laidlaw, 1917]
Hooktail sp. *Paragomphus lindgreni* Fraser, 1923
[Orig. *Onychogomphus lindgreni*]

Oscar Lindgren (1857–1946) was a tea planter and amateur naturalist. He sent the types to Fraser from the Turzum tea estate, Darjeeling, where he had collected them (1920). He was also known as the self-styled 'founder' of Makum town in Upper Assam. He wrote the autobiographical: *The Trials of a Planter* (1933).

Lindley

Dusky Threadtail *Elattoneura lindleyi* **Legrand**, 1980
[Orig. *Elattoneura acuta lindleyi* Legrand, 1980]

Dr Roger Philip Lindley (b.1953) is an English pathologist and odonatologist. He published a number of papers on African Odonata (1970–1980) and described four new species and one new genus. He collected in west central Africa, particularly Côte d'Ivoire and Central African Republic. His written papers include: *On a new genus and species of libellulid dragonfly from the Ivory Coast* (1970) and *Macromia gamblesi pec. Nov. from the Central African Republic (Anioptera: Corduliidae)* (1980).

Lindner

Pond Damselfly sp. *Homeoura lindneri* **Ris**, 1928
[Orig. *Acanthagrion lindneri*]

Professor Dr Erwin Lindner (1888–1988) was a German entomologist whose main interest was diptera. He was Head of the Department of Entomology at Stuttgart Natural History Museum (1913–1953). He edited the twelve-volume *Die Fliegen der paläarktischen Region* on the systematics and anatomy of the flies of the Palaearctic. He collected on a number of expeditions to little-studied parts of Europe, including Dalmatia, Anatolia, the Alps, Italy, Spain and parts of East Africa. Ris honoured him as he held him in 'high esteem.' A number of taxa are named after him including an amphibian.

Linnaeus

Downy Emerald *Cordulia linaenea* **Fraser**, 1956
[JS *Cordulia aenea* (Linnaeus, 1758)]
Damselfly sp. *Asthenocnemis linnaei* **Gassmann** & **Hämäläinen**, 2008
Shadowdamsel sp. *Protosticta linnaei* **van Tol**, 2008

Brown Hawker ssp. *Aeschna grandis linnaei* Ander, 1953
[JS *Aeshna grandis* (Linnaeus, 1758]

Carl von Linné (1707–1778) is now much better known by the Latinised form of his name, Carolus Linnaeus (or just Linnaeus). Late in life (1761) he was ennobled and so could call himself Carl von Linné. In the natural sciences, he was undoubtedly one of the great heavyweights of all time, ranking with Darwin and Wallace. Primarily a botanist, he nevertheless established the system of using binomial names, published in *Systema naturae* that is still in use today, albeit with modifications, for naming, ranking, and classifying all living organisms. He entered Lund University (1727) to study medicine and transferred to Uppsala University (1728). At that time, botany was studied as part of medical training. His first expedition was to Lapland (1732) and then he mounted an expedition to central Sweden (1734). He went to the Netherlands (1735) and finished his studies there as a physician before enrolling at the University at Leiden. He returned to Sweden (1738), lecturing and practising medicine in Stockholm. Appointed Professor at Uppsala (1742), he restored the university's botanical garden. He bought the manor estate of Hammarby outside Uppsala, where he built a small museum for his extensive personal collections (1758). This house and garden still exist, run by Uppsala University. His son, also Carl, succeeded to his professorship at Uppsala, but never was noteworthy as a botanist. When Carl the Younger died (1783) with no heirs, his mother and sisters sold the elder Linnaeus' library, manuscripts and natural history collections to the English natural historian, Sir James Edward Smith, who founded the Linnean Society of London to take care of them. Linnaeus named 20 species of Odonata, all placed in a single genus *Libellula*. Surprisingly few taxa have been named after Linnaeus, and even fewer which are still regarded as valid, but five birds, four reptiles, three mammals and two amphibians are among them. (Also see **Carol**)

Linsley

Reedling sp. *Indolestes linsleyi* **Lieftinck**, 1960

Dr Judson Linsley Gressitt (1914–1982). (See **Gressitt**)

Lintner

Ringed Boghunter *Williamsonia lintneri* **Hagen**, 1878
[Orig. *Cordulia lintneri*]

Dr Joseph Albert Lintner (1822–1898) was an American economic entomologist. He was in business (1837–1868), but continued to study under the Mercantile Library Association. He started collecting insects (1853) and published his first paper on entomology (1862). He was appointed as Zoological Assistant in the New York State Museum (1868), a post he held until his death. He was later put in charge of entomology there (1874). He was appointed as New York State Entomologist (1880–1898). He wrote over 900 papers and reports including 13 of the *Report on the injurious and other insects of the State of New York*. He discovered many new species, mainly Coleoptera, Lepidoptera and Diptera. His collection resides in Albany Museum of Natural History. Lintner collected the male and female type specimens that Hagen described, in New York State. He died suddenly in Rome, having been granted a six month leave of absence because of his long service.

Lisa

Damselfly sp. *Hylaeargia lisae* **Theischinger** & Richards, 2013

Lisa Capon (b.1960) is the wife of the second author, herpetologist Dr Stephen J Richards of the South Australia Museum.

Lisa

Skimmer sp. *Diplacina lisa* **Needham** & Gyger, 1941

Lisa is a feminine forename, but not a specified one. The authors gave no etymology, so it is not possible to know for sure if this was named after a particular person. However, the lack of a Latin declension suggests it is not an eponym, as does Needham's frequent use of apparent forenames.

Liu

Phantomhawker sp. *Planaeschna liui* Xu, Chen & Qiu, 2009

Professor Chang-Ming Liu is a professor at the College of Plant Protection, Fujian Agriculture and Forestry University, Fuzhou, Fujian. Among his many and wide-ranging written papers are: *The genus Lasiochalcidia from China with descriptions of two new species (Hymenoptera:Chalcididae)* (2002) and the co-authored: *Study on the insect fauna of six different habitats with Malaise traps* (2008). The etymology reads: "*Named in honour of our teacher Liu Chang-Ming*".

Liveringa

Malachite Threadtail *Nososticta liveringa* **Watson** & **Theischinger**, 1984

This is a toponym; the Lower Liveringa Pool is a locality near Camballin, Western Australia.

Lizer

Threadtail sp. *Peristicta lizeria* Navás, 1920

Carlos Alfonso Lizer y Trelles (1887–1959) was an Argentine entomologist. He graduated as an agronomist at the Faculty of Agronomy and Veterinary Medicine, University of Buenos Aires (1911) and then enrolled in the University of La Plata, which awarded his PhD (1915). He worked (1912–1942) for the Department of Agriculture, becoming Head of Agricultural Zoology. He also taught at his old universities (1915–1946). Among almost 100 other papers and books he wrote (1914–1956): *An Introduction to the History of Entomology* (1947) and *Insectos y otros enemigos de la quinta* (1941). The type was taken in Buenos Aires, Argentina (1917), where Lizer was born and lived.

Loew

Common Glider *Tramea loewii* **Brauer**, 1866
[Orig. *Tramea löwii*]

Friedrich Hermann Loew (1807–1879) was a German entomologist who specialized in Diptera. The University of Halle-Wittenberg awarded his degree in mathematics, philology and natural history and they then employed him to teach mathematics. He went on (1830) to teach in Berlin and then as private tutor to an aristocrat. He took a post as senior teacher (1834) in what is now Poznan in Poland. He took part in an expedition to the Far East (1841–1842), the results of which formed much of his later writing. He was elected to the German parliament (1848), but was disillusioned and depressed following the death of his daughter and he left politics (1850) to become director of a technical school. The following few years (1851–1854) were plagued with ill health and he retired. His pension (1868) allowed him to study Diptera full time although he did serve on the local council as well (1870–1876). He suffered a stroke (1878) and died the following spring. Over his lifetime, he delineated the taxonomically useful characters of more than 4,000 species of Diptera and amassed a huge collection now in museums in UK, Germany, US and others. He wrote a great many papers from *Dipterologische Notizen*. In: *Wiener Entomologische Monatsschrift* (1837) to *Neue nordamerikanische Dasypogonina* (1874). At least 17 insect species are named after him.

Longchamp

The genus *Longchampia* Kirby, 1890
[JS *Diastatomma* Burmeister, 1839]

Baron Michel Edmond de Selys Longchamps (1813–1900). (See **Selys** & **Edmond**)

Longfield

Darner sp. *Castoraeschna longfieldae* **Kimmins,** 1929
[Orig. *Coryphaeschna longfieldae*]
Forest Bluet *Enallagma longfieldae* **Fraser,** 1947
[Syn. of *Africallagma vaginale* (Sjöstedt, 1917)]
Hybrid Duskhawker *Heliaeschna longfieldae* **Gambles,** 1967
[JS *Heliaeschna sembe* Pinhey, 1962]
Yellow-winged Elf *Tetrathemis longfieldae* **Legrand,** 1977

Cynthia Longfield FRES (1896–1991) was an Anglo-Irish odonatologist, ornithologist and explorer, later called 'Madame Dragonfly' because of her work on odonata. During WW1, she was a driver in the Royal Army Service Corps (1914–1916) and then worked in an aircraft factory (1916–1918). After the war she travelled all over the world in a series of long trips. Her first exploration was for three months in South America (1921–1922) to the Andes and Lake Titicaca. Her second was to Egypt (1923), then she took part in the St George Expedition to the Pacific as unpaid assistant entomologist (1924–1925), accompanying Miss Evelyn Cheesman (q.v.). There followed an expedition to Brazil (Mato Grosso) and Paraguay (1927). She made a six-month journey to South-East Asia (1929–1930), visiting several countries. Next was a trip to Canada (1932) and then a six-month expedition (alone) to East Africa and Congo (1934). This was followed by another visit to East Africa (1937–1938). During all her travels, she collected dragonflies, butterflies and other insects. Her first research paper (1929) was on dragonflies of Mato Grosso, Brazil, followed by a paper (1931) on the Odonata of British Somaliland, which included the description of her first new species. Several other papers followed and she was established as an expert of the dragonflies of the British Isles and Africa. She had worked voluntarily at the BMNH (since 1927) and was made (1948) an honorary associate of the museum. Her private funds made possible an independent life and travel. During WW2, she was control officer at Brompton Fire Station. On one occasion her prompt action helped to save the Botany Department of BMNH when a bomb had set it on fire. After retirement (1957), she returned to the family estate in County Cork, Eire living there for the rest of her life. Her major publications include the books: *The Dragonflies of the British Isles* (1937) and the co-authored *Dragonflies* (1960), and the research papers: *Studies on African Odonata, with synonymy, and descriptions of new species and subspecies* (1936), *The Odonata of N. Angola Part I* (1955) and *The Odonata of N. Angola Part II* (1959). She described 38 new species and 3 genera of African Odonata. Her great-niece, Jane Hayter-Hames, wrote her biography: *Madam Dragonfly: The life and times of Cynthia Longfield* (1991). (Also see **Cynthia**)

Loogali

Clubtail sp. *Microgomphus loogali* **Fraser**, 1923
Sylvan sp. *Coeliccia loogali* **Laidlaw**, 1932

Nothing is given in the etymology and this is neither an eponym, nor a toponym. It may originate from the Burmese word 'Loogali' meaning 'lad' or 'youth'.

Lorentz

Pond Damselfly sp. *Oreagrion lorentzi* **Ris**, 1913
Threadtail sp. *Notoneura lorentzi* **Lieftinck**, 1938
[JS *Nososticta nigrifrons* (ris, 1913)]

Dr Hendrikus Albertus Lorentz (1871–1944) was a Dutch explorer who studied law and biology. He participated in the expedition (1901) of Professor Wichmann to northern (Dutch) New Guinea and he himself led expeditions (1905–1906, 1907 & 1909–1910) in southern New Guinea, leading to the discovery of Wilhelmina Peak (named after the Dutch Queen Wilhelmina) in the Snow Mountains. He drew up a map of the places they visited. Upon his return, he entered the Dutch consular services, becoming ambassador in Pretoria, South Africa (1929). Ris described the damselfly in his paper: *Die Odonata von Dr. H.A. Lorentz' Expedition nach Südwest-Neu-Guinea 1909 und einige Odonata von Waigëu* (1913). The Lorentz Range of mountains and the Lorentz River in New Guinea are also named after him as are a fish, five birds and three mammals.

Lorenz

Sprite sp. *Pseudagrion lorenzi* Gassmann, 2011

Lorenz Bier-Schorr (b.1993) is the son of Mr Martin Schorr of the International Dragonfly Fund (IDF). Martin Schorr gave financial support to the author for his work on New Guinea odonata and also the work on the description of this species. He made a special request to the author to name the species for his son's eighteenth birthday.

Lorenzoni

Goldenring sp. *Aeschna lorenzonii* Disconzi, 1865
[Syn. of *Cordulegaster boltonii* (Donovan, 1807) or *C. bidentata* Selys, 1843]

Antonio Lorenzoni (d. before 1865) was the author's late pupil from Vicenza, Italy.

Loringa

Titan sp. *Philoganga loringae* **Fraser**, 1927
Sylvan sp. *Coeliccia loringae* **Laidlaw**, 1932
[JS *Coeliccia didyma* Selys, 1863]

Nothing is given in the etymology and this may not be an eponym, nor can any reason be found to believe it is a toponym.

Lorquin

Archtail Sp. *Nannophlebia lorquinii* **Selys**, 1869
[Orig. *Neophlebia lorquinii*]
Fineliner sp. *Teinobasis lorquini* Selys, 1877
[Orig. *Telebasis lorquini*]

Pierre Joseph Michel Lorquin (1797–1873) was a French naturalist, entomologist and traveller. He was particularly interested in Coleoptera and Lepidoptera. He first collected in France (1836), then southern Spain and Algeria (1847–1848). He was in California (1849/1850) (some sources say to join the gold rush) where he stayed some years before collecting in Oregon (1858), then going to China and the Philippines (1859–1860), Dutch East Indies (Celebes, Moluccas, Aru & New Guinea,1860–1865) and Colombia, returning to California to collect again (1866–1869). Selys received the types from Lorquin, who collected them in the Moluccas. He is also commemorated in the names of at least two butterflies, including Lorquin's Admiral *Limenitis lorquini*.

Lort

Scarlet Darter *Orthetrum lorti* **Kirby**, 1896
[JS *Crocothemis erythraea* (Brullé, 1832)]

Ethelbert Edward Lort-Phillips (1857–1943) collected the types with his wife in Somaliland (February 1895). (See **Phillips, L**)

Louisa

Limniad sp. *Amphicnemis louisae* **Laidlaw**, 1913
[JS *Amphicnemis wallacii* Selys, 1863]

Laidlaw gives no etymology, but the species is obviously named after his daughter, Mary Louisa Fortescue Folkard nee Laidlaw (b.1912–fl.1994), who was also honoured in another species (Also see **Eric**, **Laidlaw** & **Mary (Laidlaw)** and **Ruby**).

Louis Sirius

Emerald sp. *Aeschnosoma louissiriusi* Fleck, 2012

Louis Sirius Fleck (b.2009) is the son of the author, Günther Fleck.

Louton

Demoiselle sp. *Mnesarete loutoni* **Garrison**, 2006
Threadtail sp. *Drepanoneura loutoni* von **Ellenrieder** & **Garrison**, 2008
Dancer sp. *Argia loutoni* Garrison & von Ellenrieder, 2015

Dr Jerry A Louton (b.1944) is an American biologist, limnologist and odonatologist. His PhD in zoology and ecology was awarded (1982) by the University of Tennessee, Knoxville, with the dissertation: *Lotic dragonfly (Anisoptera: Odonata) nymphs of the Southeastern United States: identification, distribution and historical biogeography*. He was a museum information technology specialist with the Department of Entomology at the Smithsonian NMNH having started as a computer specialist (1982–1984). He has travelled to Peru several times (periodically 1987–1989 & 1992–1997) within the BIOLAT (Biological

Diversity in Tropical Latin America) project, conducting biodiversity surveys of the Odonata based at the Pakitza Biological Station in the Manu National Park. He worked on the project and collected the dragonflies together with Oliver Flint (q.v.) and Rosser Garrison. At the same time, he also studied the odonate fauna. Among other papers and longer works, he is one of the authors (the others being von Ellenrieder and Garrison) of both *Damselfly Genera of the New World: An Illustrated and Annotated Key to the Zygoptera* (2010) and *Dragonfly Genera of the New World: An Illustrated and Annotated Key to the Anisoptera* (2006).

Lu

Claspertail sp. *Nychogomphus lui* X Zhou, **W-B Zhou** & Lu, 2005

Sheng-Xian Lu teaches at Malipo National Middle School, Wenshan, Yunnan. The species epithet refers to the third author of that name. Adding the third author and naming the species after him is, to say the least, unusual. We believe this was a courtesy for the person who collected the type and who arranged the local fieldwork for the primary author. He collected the type in Yunnan (2000).

Lucas

Threadtail sp. *Neoneura lucas* **Machado**, 2002

Lucas Machado Tomellin (b.2000) is the grandson of the author Dr Angelo Barbosa Monteiro Machado (b.1934).

Lucas

Algerian Clubtail *Gomphus lucasii* **Selys**, 1849

Pierre Hippolyte Lucas (1814–1899) was a French entomologist. He was Assistant Naturalist at the MNHN Paris. He studied (1839–1842) the fauna of Algeria as part of the Scientific Commission on the Exploration of Algeria where the Clubtail was found.

Lucia

Lucia Widow *Palpopleura lucia* Drury, 1773
[Orig. *Libellula lucia*]
Hawker sp. *Aeshna lucia* **Needham**, 1930

Lucia was a figure in classical history who ended her life as a martyr in the time of Emperor Diocletian. It is also a feminine forename, but not a specified one. The authors gave no etymologies, so it is not possible to know for sure if these species were named after a particular person. However, the lack of Latin declension suggests they are not eponyms, as do Needham's frequent use of apparent forenames and that most of Drury's dragonfly names came from Roman antiquity.

Lucia

St Lucia Basker *Urothemis luciana* **Balinsky**, 1961

Toponym; named after the type locality, St. Lucia Bay, in KwaZulu-Natal province, South Africa where the species is endemic.

Lucia (Machado)

Threadtail sp. *Forcepsioneura lucia* **Machado**, 2000

Lucia Ribeiro Machado Haertel (b.1965) is the daughter of the author Dr Angelo Barbosa Monteiro Machado (b.1934).

Lucie

Fineliner sp. *Teinobasis luciae* **Lieftinck**, 1937

Damselfly ssp. *Papuargia stueberi luciedecknerae* **Orr** & Richards, 2017

Juliana Brighantine 'Lucie' Deckner (1902–1985) was "...*the wife and constant companion of Mr. Stüber, who has given a great deal of assistance in her husband's field work*". (Also see **Stüber**)

Ludovica

Banded Demoiselle *Libellula ludovicea* Fourcroy, 1785

[JS *Calopteryx splendens* (Harris, 1780)]

Beautiful Demoiselle *Calepteryx ludoviciana* Stephens, 1835

[JS *Calopteryx virgo* (Linnaeus, 1758)]

Princess Louise Ulrika (Lovisa Ulrika) (1720–1782) was Queen consort of Sweden (1751–1771). She was the first individual to be honoured with an eponym in Odonata.

Luis

Leaftail sp. *Phyllogomphoides luisi* **González-Soriano** & Novelo-Gutiérrez, 1990

Luis Enrique González-Figueroa (b.1982) is the son of the first author Dr Enrique González Soriano (b.1951). (See **Gonzales**)

Luis Moojen

Cruiser sp. *Lauromacromia luismoojeni* **Santos**, 1967

[Orig. *Neocordulia luis-moojeni*]

Luis Edmundo Moojen was the collector (1963) of the holotype and he collected other specimens for Santos in Brazil.

Luiza

Firetail sp. *Telebasis luizae* **Lencioni**, 2010

Luiza Maria de Atayde Lencioni (b.1948) is the mother of the author Frederico Augusto de Atayde Lencioni (b.1970). (Also see **Lencioni**)

Luiz Felipe

Flatwing sp. *Heteragrion luisfelipei* **Machado**, 2006

Luiz Felipe Machado Haertel (b.2000) is the grandson of the author Dr Angelo Barbosa Monteiro Machado (b.1934). (Also see **Machado**)

Luja

Albertine Longleg *Notogomphus lujai* **Schouteden**, 1934
[Orig. *Podogomphus lujai*]

Édouard-Pierre Luja (1875–1953) was a horticulturalist, professional collector, botanist and entomologist from Luxembourg. He studied in Belgium, France and England but turned to biology and exploration. He went to the Congo and Mozambique and to South America, visiting Brazil in particular. There he collected for a number of individuals and institutions, specialising in insects and particularly ants, but all animals and plants. He discovered at least 80 new plants and 120 animals, including fish. Most species named after him bear the name *lujae* rather than *lujai*. He collected the eponymous holotype in the Congo.

Lulico

Emerald sp. *Hemicordulia lulico* **Asahina**, 1940

Ms Ruri Teshima (b.1932) is one of the three daughters of Dr Teizo Esaki (q.v.), whose Odonata material from Micronesia (collected 1936–1939) Asahina had studied. She now lives in Fukuka city. The letters 'r' and 'l' are often mixed in Japanese pronunciation, and the suffix "-co" is often added to the end of the female first name. So, in Asahina's writing the name Ruri turned to Lulico. Also, the other daughters Erico and Haluco, and Esaki himself, were credited with eponyms in the same paper. (See **Erica**, **Haluco** and **Esaki**)

Lundqvist

Reedling sp. *Indolestes lundqvisti* **Lieftinck**, 1949
[Orig. *Lestes lundqvisti*]

Eric Lundqvist (1902–1978) was a Swedish forestry officer, traveller and author, who spent nearly 30 years in the Netherland's East Indies as a government forestry officer. He worked in both Borneo and New Guinea (he collected this damselfly in New Guinea, July 1941) and Banka Island, off Sumatra (1930s). He settled in Java, having married a local girl (1940), but was imprisoned when the Japanese invaded, although he managed to escape, hiding in his wife's village. Later, when the Dutch colonial rule was challenged, he had to flee to Sweden but returned and was re-united with his wife (late 1940s). He was appointed Head of Forestry in New Guinea (1949) but left after a few years to go to Sweden with his wife. In the 1950s he became Professor of Forestry Science at Jakarta University but left again (1960) due to political unrest and they settled in Sweden for good, apart from a brief spell in India. He died as a result of Parkinson's disease. He wrote 17 books about his life and travels.

Luz Marina

Threadtail sp. *Neoneura luzmarina* **De Marmels**, 1989

Luz Marina Escobar González (b.1957) is the wife of the author Dr Jürg Carl De Marmels (b.1950). (Also see **De Marmels**)

Lydia

Common Whitetail *Plathemis Lydia* Drury, 1773
[A. Long-tailed Skimmer Orig. *Libellula lydia*]

Lydia is not for a particular person but a name from Roman antiquity.

Lyell

Swamp Bluet *Coenagrion lyelli* **Tillyard**, 1913
[Orig. *Agrion lyelli*]

George Lyell (1866–1951) was an Australian naturalist. He worked for a cheese and bacon factory rising through the ranks from junior clerk to head of the dairy machinery branch. He became a partner in a firm that made dairy machines and also entomological equipment (1890). Whilst testing equipment he caught a butterfly in a local park (1888), which stimulated his serious collecting and he joined his local field naturalists club in Victoria. Here he made friends with a number of Australian naturalist pioneers. He built up an enormous collection of butterflies and moths, at first from country areas near Melbourne and then from the Gisborne area and other states. A correspondence, begun with the National Museum of Victoria (1902), resulted in his donating his collection to the museum (1932) and working (until 1946) to amalgamate it with the museum's existing Lepidoptera holdings. He continued to add to the collection until his death when it stood at 51,216 specimens, representing 6,177 species including 534 types; it still forms the major part of the museum's Lepidoptera collection. He wrote a number of notes and papers (1890–1929) and co-wrote a book: *Butterflies of Australia* (1914). He was still collecting in the field when he was 84.

Lygaea

Dragonlet sp. *Erythrodiplax lygaea* **Ris**, 1911

Ris did not give etymology of the name, but wrote that he had replaced Selys' old 'collection name' with this new name… Lygea (var. Ligia). This is a woman's name of ancient Greek origin, which may also be the origin of *lygaea*; a type of grass.

Lynn

Columbia Clubtail *Gomphus lynnae* **Paulson**, 1983

Mary Lynn Erckmann née Ward (b.1942) collected the holotype, drew many of the figures for the description and had helped Paulson collect thousands of specimens in his first five years in Washington State (they were married at that time). She studied zoology at the University of Miami, which awarded her BS (1964). From her arrival in Seattle until she retired (1967–2013), she worked at the University of Washington Zoology Department as a technician in laboratories that studied avian physiology, bird behaviour and ecology, and plant/herbivore interactions. She has done extensive field research in Argentina, Costa Rica, Mexico, Alaska, Arizona and Washington. She remains keen to play a part in studies that promote wildlife conservation, particularly with endangered species, loves animals and is heavily involved in dog rescue operations. Interestingly, her husband Jim Erckmann is a good friend of Paulson's and did the illustrations for his *Shorebirds of the Pacific Northwest* (1998).

Lytton

Duskhawker sp. *Gynacantha lyttoni* **Fraser**, 1926
[JS *Gynacantha bayadera* Selys 1891]

Victor Alexander George Robert Bulwer-Lytton, 2nd Earl of Lytton (1876–1947) was an India-born English politician and colonial administrator. He worked at the Admiralty (1916–1920), was appointed Under-Secretary of State for India (1920–1922) and became the Governor of Bengal (1922–1927). He wrote two books, one about his grandfather and the other dealing with his Indian experiences *Pundits and Elephants* (1942).

M

Maa

Emerald Spreadwing sp. *Megalestes maai* Chen, 1947
Dragonhunter sp. *Sieboldius maai* **Chao**, 1990

Professor Dr Tsing-Chao Maa (1910–1992) was a Chinese entomologist who worked in different agricultural organizations in Lienching, Chungan and Shaowu. He was then a lecturer in the Department of Biology of Fukien Christian University in Shaowu (1939–1945). Hsiu-fu Chao (q.v.) came to teach at the same department (1942), where Maa greatly encouraged his interest in insects. Maa left mainland China and went to Taiwan (1946) before moving to the USA but returned to Taiwan after retirement, until his failing eyesight led to him giving up taxonomic work when he again went to the USA (1988). He made a collection of Odonata in Fujian Province, China (1939–1945) and at Fukien, Ta-Chu-lan (1947–1948). He worked at the Bishop Museum, Hawaii (1958–1975). He wrote a number of articles such as: *A critical review of the Chinese Machaerotidae* (1947) and books, including the monograph *A review of the Machaerotidae (Hemiptera: Cercopoidea), (Pacific insects monograph)* (1963) and as co-author of *Catalogue of Taiwanese Dragonflies* (1984). He is also commemorated in the names of a number of insects, many of which he collected.

Maack

Longleg sp. *Anisogomphus maacki* **Selys**, 1872
[Orig. *Gomphus maacki*]

Richard Otto (Karlovic) Maack (1825–1886) was an Estonian, Russian naturalist, geographer and anthropologist. After studying natural sciences in St Petersburg, he became (1852) Professor of Natural Science at the Irkutsk Gymnasium and later its Director. Following this he was Superintendent of all schools in northern Siberia (1868–1879). He explored in the Russian far-east and Siberia (1850s) and made extensive collections there, including from the Amur River basin (1855–1856) and the Ussuri River (1859), which he described later in his book: *A voyage to Amur* (1859). Other taxa are named after him including at least three shrubs and Russia's largest butterfly, the Alpine Black Swallowtail *Papilio maackii*.

Maathai

Maathai's Longleg *Notogomphus maathaiae* Clausnitzer & **Dijkstra**, 2005

Wangari Muta Maathai (1940–2011) was a Kenyan environmental and political activist, honoured in the name of this Kenyan endemic. She was educated at Benedictine College, Kansas (BSc, 1964), then the University of Pittsburgh for her MSc (1966) and took a research post at the University of Nairobi where she was later awarded her PhD (1971). She founded the Green Belt Movement (1970s) that championed women's rights, environmental conservation and the planting of trees. She was awarded the 'Right Livelihood Award'

(1986) and was the first African woman to receive the Nobel Peace Prize (2004) for her contributions to 'sustainable development, democracy and peace'. Elected to parliament, she served as an assistant minister for environment and natural resources (2003–2005). She died of complications arising from ovarian cancer.

MacCulloch

(See McCulloch)

Machado, ABM

Threadtail sp. *Epipleoneura machadoi* **Rácenis**, 1960
Pond Damselfly sp. *Inpabasis machadoi* **Santos**, 1961
Pond Damselfly sp. *Oxyagrion machadoi* **Costa**, 1978
Threadtail sp. *Psaironeura machadoi* **De Marmels**, 1989
[JS *Psaironeura bifurcata* (Sjöstedt, 1918)]
Emerald sp. *Cordulisantosia machadoi* Costa & Santos, 2000
[Orig. *Santosia machadoi*]
Bannerwing sp. *Chalcopteryx machadoi* Costa, 2005
Demoiselle sp. *Mnesarete machadoi* **Garrison**, 2006
Threadtail sp. *Epipotoneura machadoi* von Ellenrieder & **Garrison**, 2008
Emerald sp. *Neocordulia machadoi* Santos, Costa & Carriço, 2010
Shadowdamsel sp. *Drepanosticta machadoi* **Theischinger** & Richards, 2014

Clubtail ssp. *Zonophora campanulata machadoi* St. Quentin, 1973
[Orig. *Zonophora machadoi*]

Professor Dr Angelo Barbosa Monteiro Machado (b.1934) is a Brazilian physician, writer of children's books and odonatologist at the Departamento de Zoologia, Instituto de Ciências Biológicas, Universidade Federal de Minas Gerais, Brazil (where he created the laboratory of neurobiology). The same university awarded his medical degree (1958) although he never practised, instead researching and teaching neurobiology and making significant discoveries relating to the pineal gland. He gained his PhD (1963) at UFMG and was on a postdoctoral program at Northwestern University in Chicago (1965–1967). He was nominated assistant professor (1966) and later Professor of Neuro-anatomy at the Department of Morphology at the Institute of Biological Sciences UFMG (until 1987) and then became Associate Professor of Entomology in the Department of Zoology of the same Institute (1987–2004). After retirement, he became Professor Emeritus (2005). Since 1987 he has focused his research on Odonata, which had been his hobby since the age of 16, and he still continues his research, publishing actively and advising former students. He has published well over 80 papers on the taxonomy of Brazilian and other Neotropical Odonata (1953 to present). Over 20 of his papers treat the species of the family Protoneuridae. He has described c. 100 new species of Odonata and named 11 new genera. Several of his new species are named after his children and grandchildren. Having turned his former hobby on dragonflies into his work, he took up a new hobby writing books (c.40) about biology for children, including: *O menino e o trio* (The Boy and the River) (1989) and *O Velho da Montanha: uma Aventura Amazonica* (The old man from the mountain: an Amazonian adventure) (1993). In many of these books dragonflies appear as secondary characters.

As an environmentalist, he has participated in major movements for nature conservation in Brazil, being founder and President (until 2014) of the Foundation Biodiversitas and currently a member of its presidential council. The journal *Zootaxa* published a jubilee issue in his honour (2016). About 60 taxa have been named after him, including species of many insect orders, spiders and even a fungus. (See also **Angelo**)

Machado, AdeB

Machado's Skimmer *Orthetrum machadoi* **Longfield**, 1955
[A Highland Skimmer]
Forest Hooktail *Paragomphus machadoi* **Pinhey**, 1961

António de Barros Machado (1912–2002) was a Portuguese zoologist and arachnologist. The university of Porto awarded his BSc (1929), where he was appointed as an assistant in zoology (1934). He then worked at the MNCN in Madrid and studied for his PhD, but this was interrupted by the civil war and he returned to Porto (1936). He then worked at the NMHN, Paris before teaching privately and then going to Angola because he was unable to get work in a Portuguese University due to his politics. He was Director of the Dundo Laboratory and Museum of Zoology and Anthropology in Angola (1947–1974). He became President of the Portuguese Society of Ethnology (1978) and finally received a doctorate from the Abel Salazar Biomedical Institute of the University of Porto (1990) where he became Professor Emeritus.

Machado, JP

Mantled Spiderlegs *Planiplax machadoi* **Santos**, 1949

Joaquim Pereira Machado Filho (b.1916) was a Brazilian entomologist particularly interested in Lepidoptera and Dermaptera. He described a number of species in both orders and is also honoured in the name of a Lepidoptera genus.

MacIntyre

Threadtail sp. *Protoneura macintyrei* **Kennedy**, 1939

William Clarke-Macintyre (1881–1952) was a professional insect collector and dealer in Ecuador, mostly around the Valley of the Rio Pastaza and Lake Runtun, overlooking Banos, near Mt. Tungurahua. He employed local Indians to collect from the headwaters of River Napo. He collected the holotype.

Mackwood

Threadtail sp. *Caconeura mackwoodi* **Fraser**, 1919
[JS *Elattoneura centralis* (Hagen, 1860)]

Frank Mitchell Mackwood (1843–1931) was an English businessman and director of 'Mackwoods', an agricultural company in Ceylon. He was also an amateur entomologist and lepidopterist. His collections are in the National Museum, Columbo and BMNH. He married (1873) while in Australia. He died in Colombo, Sri Lanka, where he had collected widely.

Bo Beolens

Maclachlan

Forest Wisp *Agriocnemis maclachlani* **Selys**, 1877
[S. *Agriocnemis aligulae*]

(See **McLachlan**)

Maclean

(See **McLean**).

Macleay

Southern Giant Darner *Telephlebia racleayi* **Martin**, 1909
[JS *Austrophlebia costalis* Tillyard, 1907]

William Sharp Macleay (1792–1865) was a British born Australian naturalist and collector. He read classics at Trinity College Cambridge being awarded his BA (1814) and MA (1818). He was then appointed as an attaché to the British Embassy in France. There he met, probably through his famous scientist father Alexander Macleay (1767–1848), many of the leading naturalists; Cuvier, Lamarck, Latreille (q.v.), Savigny and St Hilaire. His interest in entomology was sparked and he began adding to his father's insect collection. His writing began with *Horae Entomologicae; or, Essays on the Annulose Animals* (1819–1821) and *Annulosa Javanica* (1825). When his father became Colonial Secretary in New South Wales, he tried for a posting there but was unsuccessful, being posted instead to Havana (1825) as part of the Anglo-Spanish Commission on the abolition of slavery. He became a judge in the same organisation (1830–1836). There he continued to pursue natural history in his leisure hours. He returned to England via the USA and retired (1837) from public service. He then left England (1838) for Australia, arriving in Sydney (1839), and studied the (especially marine) natural history around Port Jackson where he settled. Macleay named three new dragonfly species from the voyages of his friend Philip Parker King when the latter asked him to describe the Annulosa he had accumulated on his voyage around Australia (1818 & 1822). Macleay was a keen supporter of the Australian Museum, being a committeeman or trustee (1841–1862). A mammal and a reptile are also named after him. (Also see **Racleay**)

Macuxi

Pond Damselfly sp. *Tuberculobasis macuxi* **Machado**, 2009

Macuxi are an indiginous people that inhabit an area of northern Brazil of the same name, where the holotype was collected.

Madelena

Limniad sp. *Amphicnemis madelenae* **Laidlaw**, 1913

Laidlaw gives no hint as to why he used this name. Obviously, it is a female given name, but no particular person is identified. However, he did name a number of species after family members, although only one (*erici*) is identified in an etymology. Madelena tends to be the Portuguese version of Madeleine or Magdalene.

Madi

Northern Fingertail *Gomphidia madi* **Pinhey**, 1961
[JS *Gomphidia bredoi* (Shoutenden, 1934)]

This is a toponym; West Madi is an area of Nile Province, Uganda.

Maes

Dark-rayed Duskhawker *Gynacantha maesi* **Schouteden**, 1917
[JS *Gynacantha sextans* McLachlan, 1896]
Blackwater Cruiser *Phyllomacromia maesi* Schouteden, 1917
[Orig. *Macromia maesi*]

Dr Joseph Yvon Maes (1882–1960) was an anthropologist at the Belgium Royal Museum of Central Africa for which he wrote a guide: *Le Museedu Congo Belgea Tervuren: Guide illustree du visiteu* (1925). He was awarded his doctorate by the University of Ghent (1905). He became Curator of the Department of Ethnography at Tervuren (1909) and led an expedition to Bandundu, Congo (1913–1914), where he collected the type specimens and many anthropological artefacts. He was made Professor of the Colonial University in Antwerp (1924) and at the University of Ghent (1925). He wrote many books, mostly on anthropology and native artefacts.

Magdelena

Shortwing sp. *Perissolestes magdalenae* **EB Williamson** & JH Williamson, 1924
[Orig. *Perilestes magdalenae*]

This is a toponym; the species occurs in the Magdalena Valley in Columbia.

Magela

Kakadu Vicetail *Hemigomphus magela* **Watson**, 1991

This is a toponym; Magela Creek is a location in western Arnhem Land, Australia.

Mago

Pond Damselfly sp. *Hylaeonympha magoi* **Racénis**, 1968

Professor Francisco Mago Leccia (1931–2004) was a distinguished Venezuelan ichthyologist who specialized in electric fish of the rivers and lagoons of Venezuela and South America and a good friend of Rácenis. He attained the degrees of docent in biology and chemistry, graduating from the Instituto Pedagógico de Caracas, MSc (marine biology) from the University of Miami, Florida, USA and doctor in sciences from the University of Central Venezuela. He was a founding member of the Instituto Oceanográfico de la Universidad de Oriente in Cumaná Sucre state and of the Instituto de Zoologia Tropical (IZT), University of Central Venezuela, Caracas. He was also Professor of Animal Biology, Vertebrate Biology and Systematic Ichthyology at the Biology School of Sciences Faculty there and became director of its museum. He wrote five books, many articles in longer works and many scientific papers. Perhaps his best-known book was: *Electric Fishes of the continental waters of América* (1994) although the others were all on fish found in Venezuela. He described at least 23 species and is honoured in the names of seven others apart from the eponymous Venezuelan endemic damselfly.

Mahecha

Forktail sp. *Ischnura mahechai* **Machado**, 2012

Dr Germán Arturo Bohórquez Mahecha (b.1946) is a professor of morphology and veterinary anatomy in the Department of Morphology at the University of Minas Gerais, Brazil. He has a degree in veterinary medicine from the National University of Cordoba (1975), a master's degree from the Federal University of Minas Gerais (1984) and a PhD from the University of São Paulo (1989). His focus is on reproduction in wild birds. Among his very many published papers is: *Pampatherium Paulacoutoi, a new species of Giant Armadillo from Bahia, Brazil (Ledentata, Dasypodidae)* (1983). The author wrote in his etymology that the species is named: *"...in honor of my good friend Prof. Germán Arturo Bohorquez Mahecha who collected the type specimens in Colombia."* (1989)

Mahesh

Sylvan sp. *Calicnemis maheshi* Sahni, 1964

[JS *Coeliccia renifera* (Selys, 1886)]

No etymology was given.

Makiritare

Flatwing sp. *Heteragrion makiritare* **De Marmels**, 2004

The Makiritare (or Yekuana) are a people of the Amazon.

Malaise

Flatwing sp. *Rhipidolestes malaisei* **Lieftinck**, 1948
Spreadwing sp. *Lestes malaisei* **Schmidt**, 1964
Coraltail sp. *Ceriagrion malaisei* Schmidt, 1964

René Edmond Malaise (1892–1978) was a Swedish entomologist, art collector and explorer who is remembered for inventing the Malaise trap. He was a member of the Swedish expedition to Kamchatka (1920–1922) and returned to Sweden (1923) via Japan where he witnessed the great earthquake of that year. He returned to Kamchatka (1924) and stayed in the Soviet Union until 1930. He went on an expedition to northern Burma (Myanmar) (1933–1935) during which he collected about 100,000 insects, many previously unknown to science, including the types of these eponymous species. He supervised the entomological department of the Naturhistoriska Museet, Stockholm (1952–1958). He wrote: *Atlantis, en geologisk verklighet* (1954). He is also remembered for making a notable collection of fishes in Burma, one of which is named after him.

Malaita

Damselfly sp. *Lieftinckia malaitae* **Lieftinck**, 1987

This is a toponym; Malaita is the largest island of the Malaita Province in the Solomon Islands.

Malcolm (Donnelly)

Pond Damselfly sp. *Nesobasis malcolmi* **Donnelly**, 1990

Malcolm Donnelly (b.1965) is the son of the author Dr Thomas Wallace Donnelly (b.1932), and he collected the holotype in Viti Levu, Fiji (1980) while he was there with his parents.

Malcolm (Smith)

Lieutenant sp. *Brachydiplax malcolmi* **Fraser**, 1922
[JS *Brachydiplax chalybea* Brauer, 1868]

Malcolm Arthur Smith (1875–1958) (See **Smith, MA**)

Malkin

Malkin's Forceptail *Phyllocycla malkini* **Belle**, 1970

Boris (Borys) Malkin (1917–2009) was a Polish born American anthropologist and entomologist with a particular interest in Amerindian cultures. He emigrated to the US (1938) and studied (1941) at the University of Oregon, then joined the US Air Force and saw action in the Pacific. His free time was spent collecting spiders for the AMNH and insects for the Smithsonian. After WW2 he studied anthropology at University College London and archaeology at the LSE. The California Academy of Science funded an expedition for him (1948) to Africa. Later he devoted his time to collecting natural history specimens, as well as artefacts documenting the material culture of the indigenous peoples of South and Central America. He collected the holotype of this dragonfly species in Brazil (February 1966).

Mallory

Clubtail sp. *Davidius malloryi* **Fraser**, 1926

George Herbert Leigh Mallory (1886–1924) was a British mountaineer who took part in the first three British Everest Expeditions, disappearing with Sandy Irvine on the last when just 800 feet from the summit. His body was not found until an expedition that looked for his and Irvine's bodies (1999). Fraser wrote: "*I have named this interesting species after Mr. Mallory, who so nobly laid down his own life in the cause of science on the slopes of Mt. Everest.*" Thomas Bainbrigge Fletcher (q.v.) collected the specimen in the Khasia Hills (April 1924), a few weeks before Mallory disappeared (June 1924).

Mamfe

Cameroon Horntail *Libyogomphus mamfei* **Pinhey**, 1961
[Orig. *Tragogomphus mamfei*]

This is a toponym; Mamfe (or Mamfé) is a city and capital of the Manyu department of the Southwest Province in Cameroon. The holotype was collected there on Gorilla Mountain.

Mandraka

Mandraka Emperor *Anax mandrakae* Gauthier, 1988

This is a toponym; Mandraka is an area in Madagascar from which the dragonfly hails.

Manau

Skydragon sp. *Chlorogomphus manau* **Dow** & Ngiam, 2011

Manau anak Budi (b.1962) is a guide and collector from Borneo who is a friend and colleague of the authors. He was a member of the expedition that collected there (2011 & 2013) and where he collected the type (2011).

Mann

Darner sp. *Rhionaeschna manni* **Williamson** & Williamson, 1930
[Orig. *Aeshna manni*]

William Montana Mann (1886–1960) was an American entomologist. He attended Staunton Military Academy, Virginia (1902–1905), then worked as a rancher in Texas and New Mexico, all the while collecting entomological specimens. Washington State College (1909–1911) awarded his BA and the Bussey Institution, Harvard his ScD in Entomology (1915), where he studied under W M Wheeler. He collected overseas, including on the Stanford Expedition to Brazil (1911), Haiti (1912), Cuba and Mexico (1913) and on the Philip Expedition to the Middle East (1914), Fiji & Solomon Islands (1915–1916). He was an entomologist for the Bureau of Entomology, US Department of Agriculture (1916–1925) and made further collecting trips to Spain, Columbia, Mexico and Cuba. He married (1926) Lucile Quarry (1897–1986). He was Superintendent of NZP – National Zoological Park – (1925), becoming Director (1927) until retirement (1956), then Director Emeritus. He was an Honorary Curator at the USNM to whom he left his collection. He wrote or co-wrote at least two dozen articles and an autobiography: *Ant Hill Odyssey* (1948).

Manolis

Pond Damselfly sp. *Tepuibasis manolisi* **Machado** & **Lencioni**, 2011
[Orig. *Austrotepuibasis manolisi*]

Dr Timothy (Tim) Dean Manolis (b.1951) is an American biological consultant, artist and an illustrator generally interested in wildlife, in particular birds and insects, especially Lepidoptera and Odonata. The University of Colorado, Boulder awarded his PhD (1982). Before becoming an independent consultant, he worked for the Animal Protection Institute. Among the books he has illustrated is: *Field Guide to the Spiders of California and the Pacific Coast States* (2013) and among those he wrote and illustrated is: *Dragonflies and Damselflies of California* (2003). The etymology reads: *"Named manolisi (noun in the genitive case) in honour of Dr Timothy D. Manolis, a dragonfly and bird lover, who collected the holotype and allowed us to describe it."*

Mantiqueira

Flatwing sp. *Heteragrion mantiqueirae* **Machado**, 2006

This is a toponym; the Serra da Mantiqueira is the area in Brazil where the specimens were collected.

Marcela

Sanddragon sp. *Progomphus marcelae* Novelo-Gutiérrez, 2007

Marcela Novelo-Gutiérrez (b.1956) is the wife of the author Rofolfo Novelo-Gutiérrez (b.1955).

Marcella

Hyacinth Glider *Miathyria marcella* **Selys**, 1857
[Orig. *Libellula marcella*]

Marcella is a feminine forename, but not a specified one. The authors gave no etymology, so it is not possible to know for sure if this was named after a particular person. However, the lack of a Latin ending suggests it is not an eponym, as does Selys' frequent use of apparent forenames.

Marchal

Helicopter sp. *Mecistogaster marchali* **Rambur**, 1842
[JS *Mecistogaster l. lucretia* Drury, 1773]
Darner sp. *Rhionaeschna marchali* Rambur, 1842
[Orig. *Aeschna marchali*]
Mauritius Dropwing *Thalassothemis marchali* Rambur, 1842
[Orig. *Libellula marchali*]

M Marchal (fl.1840–d.c.1846) was a resident of Isle de France (Mauritius) and amassed a collection of natural history specimens, much of which was sold by his widow.

Marcia

Skimmer sp. *Libellula marcia* Drury, 1773
[Syn. of *Rhyothemis variegata* (Linnaeus, 1763)]

Marcia Aurelia Ceionia Demetrias, better known as Marcia, was the mistress and one of the assasins of 2nd century AD Roman Emperor Commodus.

Margaret (Stevenson)

Texas Emerald *Somatochlora margarita* **Donnelly**, 1962

Margaret Stevenson was the collecting companion of the describer Thomas Wallace Donnelly (b.1932) (q.v.) and Ailsa Donnelly (q.v.) in Texas (1961). His etymology reads: *"Miss Margaret Stevenson, a delightful companion of my wife and myself on all of our collecting trips during the spring of 1961, and a very great help to us during the bizarre manoeuvres, which were required in order to net this most elusive and beautiful insect."* (Also see **Donnelly** and **Ailsa**)

Margaret (Westfall)

Mountain River Cruiser *Macromia margarita* **Westfall**, 1947

Margaret Lucille Westfall neé Shepherd (1922–2000) was the wife of the author Dr Minter Jackson Westfall Jr (1916–2003). She graduated at Cornell University (1946), where she had met Minter and married him (1945). She accompanied her husband on numerous collecting trips and to international symposia of Odonatology. (Also see **Westfall**)

Margarete

Darner sp. *Castoraeschna margarethae* **Jurzitza**, 1979

Margarete Jurzitza née Dembicky (1924–2012) was the wife (1954) of the author Dr Gerhard Roman Anton Jurzitza (1929–2014). (Also see **Jurzitza**)

Margarita (Alexander)

Clubtail sp. *Fukienogomphus margarita* **Chao**, 1954
[JS *Fukienogomphus promineus* Chao, 1954]

Mabel Marguerite Alexander (neé Miller) (1894–1979) was the wife of Dr Charles Paul Alexander (1889–1981) (q.v.) who was Chao's teacher in Massachusetts (1948–1951). Chao gave no etymology, but it is almost certainly named after her. (Also see **Alexander**)

Margarita (Compte Sart)

Leaftail sp. *Phyllogomphus margaritae* **Compte Sart**, 1967
[JS *Phyllogomphus coloratus* Kimmins, 1931]

Margarita Sart Alemany (1913–1979) was the mother of the author Arturo Compte Sart (b.1933). (Also see **Raymond (Compte Sart)**).

Margarita (De Selys)

Demoiselle sp. *Echo margarita* **Selys**, 1853
Flatwing sp. *Philogenia margarita* Selys, 1862

Valentine Emilie Marguerite de Selys Longchamps (1848–1852) was the daughter of the author Baron Michel Edmond de Selys Longchamps (1813–1900). She died in his arms when just four years old, having contracted meningitis. Selys wrote in his diary: *"…I passed the night from yesterday to today close to Marguerite… Then, at twenty to four, the poor child, for whom I have cried so much, rendered her soul to God in my arms. As she died I felt her heartbeat fail in my hands while I flooded her cheeks with my tears, having kissed her half a minute prior to the fatal separation, and later I kissed three times her poor little still warm hands… …So much despair, so many tears."* (Also see **Selys**)

Maria

Cuban Blue Threadtail *Neoneura maria* **Scudder**, 1866
[Orig. *Agrion maria*]

Maria is a feminine forename, but not a specified one. The authors gave no etymology, so it is not possible to know for sure if this was named after a particular person. However, the lack of a Latin declension suggests it is not an eponym.

Maria (Belle)

Skimmer sp. *Libellula mariae* **Garrison**, 1992

Maria Belle née Offers (1918–1986) was the wife of Dr Jean Belle (q.v.). The species was named at the request (April 1990) of this Dutch odonatologist for his *"late beloved wife Maria…."* (Also see **Belle**)

Maria (de Selys)

Yellow-patched Lieutenant *Brachydiplax maria* **Selys**, 1878

[JS *Brachydiplax chalybea* Brauer, 1868]

Marie Denise de Selys Longchamps née Gandolphe (1777–1857) was the mother of the author Baron Michel Edmond de Selys Longchamps (1813–1900). (Also see **Selys**)

Maria (Laidlaw)

Pondhawk sp. *Epithemis mariae* **Laidlaw**, 1915

[Orig. *Amphithemis mariae*]

(See **Mary Laidlaw**)

Maria (van Steenish)

Jewel sp. *Heliiocypha mariae* **Lieftinck**, 1930

[Orig. *Rhinocypha mariae*]

Maria Johanna van Steenish-Kruseman (1904–1999), with her husband Dr van Steenish, collected the type series in Sumatra (October 1929).

Maria (Walsh-Heldt)

Limniad sp. *Amphicnemis mariae* **Lieftinck**, 1940

Slendertail sp. *Leptogomphus mariae* Lieftinck, 1948

Maria Ernestine Walsh-Held (1881–1973) was a Swiss botanist and entomologist who after marrying in Australia (1911) settled in Soekaboemi, West Java. She collected insects as a hobby (1911), but when widowed (1913), collected for a living and even employed others to collect for her. She collected insects, plants and fossils in Timor (1928–1929), southern Sumatra and in Borneo (1937), where she collected the holotypes of these species. She retired to Switzerland (1958). Several plants are also named after her, such as *Vernonia walshae*.

Mariana (Machado)

Demoiselle sp. *Mnesarete mariana* **Machad**o, 1996

Mariana Fonseca Machado (b.1996) is the granddaughter of the author Dr Angelo Barbosa Monteiro Machado (b.1934) (See **Machado, ABM**)

Mariana (Shafer)

Threadtail sp. *Neoneura mariana* **Williamson**, 1917

(See **Mary (Shafer)**)

Marianne

Hooktail sp. *Cornigomphus mariannae* **Legrand**, 1992

[Orig. *Paragomphus mariannae*]

Scarlet Skimmer ssp. *Crocothemis servilia mariannae* **Kiauta**, 1983

Maria (Marianne) Antonetta Johanna Elisabeth Kiauta, née Brink (b.1948) is the wife of the sub-species author Dr Milan Boštjan (Bastiaan) Kiauta (b.1937). (See **Kiauta, MAJE** & also see **Kiauta, MB** & **Bastiaan**)

Marie

Saint Marie Islander *Nesolestes mariae* **Aguesse**, 1968

This is a toponym; it refers to the locality Île Sainte-Marie off the east coast of Madagascar, where the species was discovered, which is also known as Nosy Boraha Island.

Marijan Matok

Pincertail sp. *Onychogomphus marijanmatoki* **Dow**, 2014

Pond Damselfly sp. *Papuagrion marijanmatoki* **Orr** & Richards, 2016

Marijan Matok (b.1972) of Ulm-Söflingen, Germany was honoured in the name in appreciation of his support of odonatological research in Malaysia through the International Dragonfly Fund.

Marikovsky

Small Spreadwing *Lestes virens marikowskii* **Belyshev**, 1961

[JS *Lestes virens vestalis* Rambur, 1842]

Professor Dr Pavel Iustinovich Marikovsky (1912–2008) was a Russian entomologist. After secondary school (1928), he took part in an expedition into the taiga beyond the Khekhstir Range as far as Ussuri, a trip that left a lasting impression for his whole life. At just sixteen he spent a year teaching in a village school then was a laboratory assistant for three years in the Far East Institute of Plant Protection, travelling all over the region, as a result of which he wrote his first book: *Beneficial Birds of the Far East Region and their Role in Agriculture* (1932). He spent the next four years studying and working as a laboratory assistant at the Far East Medical Institute in Khabarovsk, as well as a research assistant at the Anti-Plague Institute. Throughout this time, he continued to travel and collect and described his first species of flea (1935). After graduating he was sent to Blagoveshchensk as head of the Anti-plague Station, but soon returned to the Medical Institute when offered an assistant lecturing post. A further move led to a post at the Uzbec Institute of Microbiology and it was here he was conferred 'Candidate' (1941). He served in WW2, becoming a major by his discharge. He took the post of Head of the Entomology Laboratory at the Zoological Institute of Kazakhstan Academy of Sciences (1946–1950), which awarded his PhD (1950), and then became Professor of Zoology (1951). He became Head of the Entomology Laboratory at the Zoological Institute of Kirghiz Academy of Sciences (1954) and then Head of the Invertebrate Zoology Department at Tomsk State University (1956–1961). After this he moved to Kazakhstan as Head of the Laboratory of Entomology at the Zoological Institute at Ama-Ata (1961–1982) when he retired, devoting himself to writing. He wrote 160 scientific papers, 250 articles and 70 popular science books!

Marina Silva

Flatwing sp. *Philogenia marinasilva* **Machado**, 2010

Maria Osmarina Marina Silva Vaz de Lima (b.1958) is a Brazilian environmentalist and politician who has risen from illiterate rubber planter to international environmentalist. She was a colleague of Chico Mendes who was assassinated for defending the Amazon environment. Growing up, she survived five bouts of malaria, as well as hepatitis and metal

271

poisoning. Orphaned aged 16, she moved to the state capital, Rio Branco, to study and receive treatment for hepatitis. Nuns took her in at a convent and educated her. She became the first person in her family to learn to read and write. After leaving the convent, she went to work as a housemaid in exchange for lodging but continued to study. She graduated in history from the Federal University of Acre at 26 and became increasingly politically active. She helped create Acre's first workers' union (1984). She was elected a senator then became environmental minister (2003). The UN Environmental Program named her one of the Champions of the Earth (2007). She ran in the presidential elections (2010), earning nearly 20% of the vote for the Green Party and later ran as Socialist Party candidate, after the original candidate's assassination, in the presidential election (2014), but failed to make the second round.

Marirobi

Pond Damselfly sp. *Papuagrion marirobi* **Orr** & **Kalkman**, 2016

This is a joint dedication to Marius Seiler (b.1995) and Robert Seiler (b.1993), who are the sons of Mechtild and Klaus-Peter Seiler. The species was collected in Indonesian New Guinea and Mechtild & Klaus-Peter Seiler were donors among those who made the expedition possible.

Mariza

Emerald sp. *Aeschnosoma marizae* **Santos**, 1981

Professor Mariza Castello Branco Simões is a Brazilian entomologist at the Reserva Ecológica do Instituto Brasileiro de Geografia e Estatística (IBGE). When doing an aquatic insect survey in Brazil she found the type series of this species (November 1979). Among her written work is the paper: *Composição e flutuações sazonais das populações de Collembola (Apterygota) em serapilheira de cerrado* (1989). She is now in retirement.

Marquardt

Common Blue-tail *Ischnura elegans marquardti* **Schmidt**, 1939

[JS *Ischnura elegans* (Vander Linden, 1820)]

Karl Albert Marquardt (1864–1936) was an employee of the Dr Otto Staudinger (q.v.) & A Bang-Haas natural history dealership in Dresden (1887–1936). He was responsible for all the insect collection except Lepidoptera.

Marshall, G A K

Highland Dropwing *Misthotus marshalli* **Kirby**, 1905
[JS *Trithemis dorsalis* Rambur, 1842]
Marshall's Pinfly *Allocnemis marshalli* **Ris**, 1921
[Orig. *Chlorocnemis marshalli* A Blue Yellowwing]

Sir Guy Anstruther Knox Marshall (1871–1959) was an Indian-born British entomologist educated in England. Both his father and an uncle were naturalists who had written books on the birds and butterflies of the Indian sub-continent. He was sent by his father to South Africa (1891) where he first farmed sheep, then cattle, was in mining in Southern Rhodesia

(Zimbabwe) and worked in a law office in Salisbury (Harare). He was co-manager (1901) of Salisbury District Estates Company and built a store. He is recorded collecting there (1901–1902), contributing to scientific papers and he collected the types of the eponymous species (1903 & 1905). He was in London (1906) en route to Sarawak where he had been offered a post as museum curator, but illness prevented his departure. He was appointed (1909) as Scientific Secretary to the new Entomological Research Committee (Tropical Africa), which he quickly organised and expanded. He was responsible for founding the 'Bulletin of Entomological Research' (1909) and later (1913) the 'Review of Applied Entomology'. He was the first Director of the Commonwealth Institute of Entomology (1916), as a result of which he was awarded an honorary doctorate of science by Oxford University. He founded (1927) the biological control service 'Farnham House Laboratory' which eventually became the Commonwealth Institute of Biological Control, which he directed until retirement (1942). Among many honours, he was knighted (1930) and advanced to KCMG on retirement. In retirement, he worked on systematics at BMNH until a few weeks before his death. A reptile and a bird are also named after him.

Marshall, S A

Emerald sp. *Cordulisantosia marshalli* **Costa** & **Santos**, 1992
[Orig. *Santosia marshalli*]

Professor Dr Stephen Archer Marshall (b.1954) is a collections director, researcher and teacher at the University of Guelph, Canada. The University of Guelph awarded his BSc (1977) and PhD (1982), whereas Carleton University awarded his MSc (1979). He has worked there as assistant professor (1982-1987), associate professor (1987–1997) and full professor (1994–present). He has published over 250 papers, chapters, articles and books (1981–2014) such as: *Insects. Their Natural History and Diversity: A Photographic Guide to Insects of Eastern North America* (2007) and *Flies: The natural history and diversity of Diptera* (2012). Together with Costa (q.v.), he collected the holotype in Rio de Janeiro (March 1990). Dr Marshall said: "*I remember well collecting that specimen and giving it to Janira, (Janira Martins Costa) who described it first as Santosia new genus. Santosia was a preoccupied name, thus the currently valid replacement name.*"

Martens

Skimmer sp. *Orthetrum martensi* **Asahina**, 1978

Professor Dr Jochen Martens (b.1941) is a German zoologist who was teaching at the Johannes Gutenberg University in Mainz (1976–2006) and is now Professor Emeritus. He has undertaken (since 1969) virtually annual expeditions to the eastern Palearctic, including the former Soviet Union, Iran, India, Nepal Himalayas (1969–1970, 1973, 1974, 1980, 1988, 1995, 2001 & 2004) and, since 1996, on near-annual expeditions to China (in co-operation with the Chinese Academy of Sciences). During these expeditions, he collected over 2000 vertebrates (all groups except fish) and arthropods, especially soil-dwelling arachnids and insects. Beside evolution of Sino-Himalayan fauna in general, his particular interests include bird song as an evolutionary factor, molecular phylogenetics of birds and systematics and phylogeny of harvest spiders. More than 270 publications arose from his Himalayan collections with the footnote title *"Results of the Himalaya Expeditions of J. Martens"* (1971–2012). Two birds are also named after him.

Martha

Martha's Pennant *Celithemis martha* **Williamson**, 1922

Martha (Mattie) Wadsworth (1862–1943) was an amateur entomologist from Maine, USA. She published several papers in 'Entomological News', unusual for a woman of her era and was a long-term correspondent with the President of the American Entomological Society (1900–1915) Philip P Calvert (q.v.). Ill health precluded wider travel, but she thoroughly surveyed the Odonata of her own state. She collected the holotype.

Martin, L

Jewel sp. *Micromerus martinae* **Karsch**, 1891
[JS *Libellago semiopaca* (Selys, 1873)]
Limniad sp. *Amphicnemis martini* **Ris**, 1911

Ludwig Martin (1858–1924) was a German physician and entomologist. He lived and worked as a physician in Palu, Sulawesi, Indonesia. The type was taken in Sumatra (1882). He collected, mainly Lepidoptera, in Sumatra (1882 & 1895), South Celebes (1906), West Borneo (1909–1910) and Central Celebes. Among his published papers is: *Die Tagfalter der Insel Celebes* (1929).

Martin, P F

Shadowdamsel sp. *Palaemnema martini* Cowley, 1934

Paul Ferdinand Martin (1884–1935) was a German veterinarian. Just after he qualified (1919), he emigrated to Brazil (1920) and collected (especially Lepidoptera) in Peru, where he settled.

Martin, R

Furbelly *Archaeophlebia martini* **Selys**, 1896
[Orig. *Neophlebia martini*]
Hawker sp. *Anaciaeschna martini* Selys, 1897
[Orig. *Aeschna martini*]
Midget sp. *Mortonagrion martini* **Ris**, 1900
[Orig. *Argiocnemis martini*]
Parasol sp. *Neurothemis martini* **Krüger**, 1903
[JS *Neurothemis ramburii* Brauer, 1866]
Twinspot Tigertail *Synthemis martini* **Tillyard**, 1908
[JS *Synthemis leachii* Selys, 1871]
Duskhawker sp. *Gynacantha martini* Navás, 1911
[JS *Gynacantha adela* Martin, 1909]
Paddletail sp. *Sarasaeschna martini* **Laidlaw**, 1921
[Orig. *Jagoria martini*]
Spineleg sp. *Merogomphus martini* **Fraser**, 1922
[Orig. *Platygomphus martini*]

Clubtail sp. *Neurogomphus martininus* **Lacroix**, 1921
[Orig. *Oxygomphus martininus*]
Syrandiri Clubtail *Davidioides martini* Fraser, 1924
Martin's Islander *Nesolestes martini* **Schmidt**, 1951

René Martin (1846–1925) was a French naturalist and odonatologist. He attended the Lycee of Versailles (1859–1866) and then studied law in Paris (1866–1870). After marrying, he moved (1872) to Le Blanc where he practised as a solicitor (1872–1907). Shortly after arriving there he began studying natural history, particularly vertebrates and then Lepidoptera and Neuroptera finally specialising in Odonata. He devoted much of his life to the study of the Odonata of France and many other parts of the world, especially the Seychelles. He made a four day visit to Belgium just to see Selys' collections. He collected for himself in Italy (1894) and moved back to Paris in 1908. On Selys' death he took on the legacy of Selys, writing of monographs of Cordulines (1907) and Aeschnines (1908–1909, in three volumes) in the series *Collections zoologiques du baron Edm. de Selys Longchamps Catalogue systématique et descriptif.* He also submitted a manuscript of Calopterygines for the same series, but it remained unpublished due to the paper shortage during WW1. He had a wide-ranging interest and knowledge of wildlife and wrote a number of books such as: *Vertébrés sauvages du Département de l'Indre* (1894). His odonatological publications include descriptions of c. 170 new odonate species (of which nearly one third are synonyms) from many parts of the world, especially from the French colonies. His large collection was deposited at MNHN in Paris. He also named c. 20 odonate genera. Selys' dedication says a lot about their relationship: "…*Je dédie cette nouvelle espèce à mon excellent ami M. René Martin (au Blanc, Départment de l'Indre) trés bon Odonatologiste dont la collection qui s'accroît rapidement est sans doute la plus importante de France, et dont les travaux, notamment sur les odonates des Iles Séchelles, son aussi exacts que consienncieux.*" When his daughter moved to Chile (1920), Martin accompanied her and spent the rest of his life there, continuing to collect insects. His last paper was *Sur les Odonates de Chili* (1921). (Also see **Rene** & **Remartin**)

Martin Schorr

Fineliner sp. *Teinobasis martinschorri* Villanueva, 2010

Martin Schorr (b.1958) is a German entomologist. (See **Schorr**)

Martínez

Helicopter sp. *Mecistogaster martinezi* **Machado**, 1985

Antonio Martínez (1922–1993) was an Argentinian parasitologist and entomologist. He was Director of the Catedra de Microbiologia y Parasitologia of the Universidad de Buenos Aires and was a Principal Investigator with the Consejo Nacional de Investigaciones Cientificas. After retirement he worked as a volunteer with the Instituto de Investigaciones Entomologicas Salta (Argentina). He and A B M Machado (q.v.) often collected together and co-wrote a number of papers. He collected widely in Brazil and in other Latin American countries, initially with F S Pereira with whom he published a number of papers on Coleoptera such as: *Dois gêneros novos de Canthonini Americanos (Col. Scarabaeoidea, Scarabaeidae)* (1956), as well as with M A Vulcano and others. He took the type in Bolivia (1951). He wrote about 270 papers including a number on medical entomology, but the majority were on Coleoptera. He and his wife were killed in a road traffic accident while on a scientific trip to Bolivia. An amphibian is also named after him.

Martorell

Sombre Cruiser *Macromia martorelli* **Compte Sart**, 1967
[JS *Phyllomacromia melania* Selys, 1871]

Superb Sprite *Pseudagrion serrulatum martorelli* Compte Sart, 1967
[JS *Pseudagrion serrulatum* Karsch, 1894]

Brother Fernando Martorell collected (1961 & 1962) the type specimens of these species in Spanish Guinea (now Equatorial Guinea). He was a teacher in the catholic School of Arts and Crafts La Salle, Spanish Guinea. He was an enthusiastic naturalist and sent hundreds of dragonfly specimens collected by him and his helpers, Isidro Mora (q.v.) and Jose Mora (q.v.), to the author.

Mary (Laidlaw)

Ruby-tailed Hawklet *Epithemis mariae* **Laidlaw**, 1915
[Orig. *Amphithemis mariae*]

Mary Louisa Fortescue Folkard née Laidlaw (1912–fl.1994) was the author's daughter. Although no etymology was given, I am sure that this species was named after her. She married Laurence Folkard, the rector of Foxearth, a village in Essex. F F Laidlaw spent his last years in his daughter's home. (Also see **Laidlaw**, **Louisa**, **Ruby** and **Eric**).

Mary (Shafer)

Threadtail sp. *Neoneura mariana* **Williamson**, 1917

Miss Mary M Shafer (1889–1967) was the sister of Myrth Shafer (q.v.) and was honoured for her long service as bibliographer and assistant in the 'bug room'. (Also see **Myrth**)

Mary (Vick)

Mary's Longleg *Notogomphus maryae* **Vick**, 2003

Mary Christine Vick née Corben (1946–2005) was the wife of the author Graham Spencer Vick (b.1947). She was a botanist and biology teacher who helped set up the Cameroon Dragonfly Project with Graham Vick, David Chelmick and Otto Mesumbe (q.v.) (1996–2005); she did a lot of tropical odonatological field-work, especially in Cameroon, Malaysia, Australia and Brazil. (Also see **Vick**)

Masaki

Sylvan ssp. *Coeliccia flavicauda masakii* **Asahina**, 1951

Tsutomu Masaki (1907–1943) was a Japanese biologist and ethnologist. He worked at the Ishigakijima meteorological station (1929–1943) on the island where he was born. Here he also collected and studied the culture of the island as well as its biology. He collected the type series there, which he forwarded to the author (1939). He died in a 'shipwreck' following enemy action during WW2.

Masako

Cruiser ssp. *Macromia amphigena masaco* **Eda**, 1976

Masako Eda (née Katoh) (b.1940) is the wife of the author Dr Shigeo Eda (b.1932). (Also see **Eda**)

Mascagni

Salone Sprite *Pseudagrion mascagnii* Terzani & Marconi, 2004

Alessandro Francesco Mascagni (b.1945) is an Italian biologist particularly interested in limnology and entomology, with a focus on Coeloptera. He has been associated with Museo di Storia Naturale di Firenze, Florence for many years. He has written over 160 papers, often with Fabio Terzani such as: *Raccolte di Odonati in Trentino Alto-Adige (Insecta: Odonata)* (1983) and *Odonati del Trentino – Alto Adige e del Cadore (Italia settentrionale) (Odonata)* (2005).

Mason

Satinwing sp. *Euphaea masoni* **Selys**, 1879
Duskhawker sp. *Cephalaeschna masoni* **Martin**, 1909
[Orig. *Caliaeschna masoni*]

Professor James Wood-Mason (1846–1893) was a Scottish entomologist. After graduating from Queen's College Oxford he left to work at the Indian Museum, Calcutta (1869). He went to the Andaman Islands (1872), mostly studying marine fauna, but also collecting stick insects (describing two new species) and mantises. He was made Superintendent of the Indian Museum (1887). He collected the *Euphaea* in Burma and *Cephalaeschna* in Assam. He was aboard the Indian marine Survey ship HMS *Investigator*. For several years, he suffered from Bright's disease and left for England (April 1893), but never made it home, dying at sea. About ten marine species are also named after him.

Mathew

Wandering Glider *Orthetrum mathewi* Sing & Baijal, 1954
[JS *Pantala flavescens* (Fabricius, 1798), A. Globe Skimmer]

Dr Koshy Mathew's PhD was awarded by St John's College, Agra (1961). He collected the type in the Chakatha Range, Uttar Pradesh, India (1953) for the Zoological Survey of India. Among his publications is the co-authored *On a New Species of Ichthyurus (Chauliognathidae: Coleoptera) from Silent Valley* (1986).

Matile

Western Fairytail *Lestinogomphus matilei* **Legrand** & Lachaise, 2001

Loïc Henri Marcel Matile (1938–2000) was a French entomologist particularly interested in Diptera. Paris University awarded his PhD (1986). He worked (1990–2000) at the MNHN, rising to Professor of Entomology and Director of the Laboratoire d'Entomologie. He married Daniele Ferrero, a fellow entomologist. He wrote the two-volume: *Diptères d'Europe Occidentale* (2000). Lachaise was also second author of a paper that dedicated an Afrotropical diptera species to him, which Matile collected.

Matsuki

Forktail sp. *Macrogomphus matsukii* **Asahina**, 1986

Kazuo Matsuki (b.1948) is a Japanese entomologist who has collected in Taiwan and Thailand. He has published many papers on dragonflies and their larvae. He was co-founder (1983) of The Dragonfly Council of Kanto District to promote dragonfly research and information exchange among researchers and enthusiasts. He collected the holotype in Thailand (May 1985) and has provided most of the recent larval descriptions of the Thai and Taiwanese species. He has described two new Odonata species from Hong Kong.

Matsumoto

The genus *Matsumotopetalia* **Karube**, 2013

Dr Kouichi Matsumoto (b.1964) is a Japanese entomologist. He graduated from Tokyo University of Agriculture, Japan. He is interested in Coleoptera and Odonata as well as being a specialist of Psyllidae (Hemiptera). Among his papers is: *A New Species of the Genus Trioza (Homoptera, Psylloidea) Feeding on Ternstroemia gymuanthera from Japan* (1993) and the monograph: *Revision of the Scirtes flavoguttatus species-group* (2010). He is also a good friend of the author. (Also see **Ojisan**)

Matti

Featherleg genus *Matticnemis* Dijkstra, 2013
Demoiselle sp. *Noguchiphaea mattii* **Do**, 2008
Duskhawker sp. *Cephalaeschna mattii* **Zhang**, Cai & Liao 2013
Sylvan sp. *Coeliccia mattii* Phan & **Kompier**, 2016

Dr Matti Kalevi Hämäläinen (b.1947) is a Finnish entomologist. He discovered (2010), collected and described the peculiar new species *Platycnemis doi* from Vietnam, which Dijkstra placed in a new genus named after him. He also collected the first specimens of *Coeliccia mattii* in Vietnam (2006). (See **Hämäläinen**)

Mauffray

Pond Damselfly sp. *Metaleptobasis mauffrayi* Daigle, 2000
Swampdamsel sp. *Leptobasis mauffrayi* **Garrison** & von Ellenrieder, 2010
Pond Damselfly sp. *Calvertagrion mauffrayi* **Tennessen**, 2015

William Francis 'Bill' Mauffray (b.1945) is managing director (1996–present) of the International Odonata Research Institute that is based in Gainseville, Florida, US, which he joined as manager under M J Westfall (q.v.) (1992). He is also Curator of Odonata, Florida State Collection of Arthropods (1996–present). The Louisiana State University awarded his BS entomology (1969). His speciality is in Neotropical Odonata, having collected in Bolivia, Colombia, Ecuador, Belize, Nicaragua and Mexico. He is currently working on the Odonata of Ecuador; a comprehensive study of historical and published records, as well as new ones from a number of trips during the last 20 years. His publications include: *Oxyagrion tennesseni spec. nov. from Ecuador (Zygoptera: Coenagrionidae)* (1999).

Maurenbrecher

Shadowdamsel sp. *Protosticta maurenbrecheri* **van Tol**, 2000

Lucia Louis Angèle Maurenbrecher (1904–1975) was a Dutch amateur entomologist and collector, born in Java, the Netherland's East Indies. He went to school in The Hague and then studied Indology in Leiden, in preparation for a career in colonial administration. After completing his studies (1930), he was posted to the East Indies as Junior Controller in Yogyakarta, then (1933–1938) promoted to Controller on Banka Island. He was then trained and served (1938–1942) as controller for 'Javanese colonisation'; the relocation of people from overpopulated Java to remoter, less crowded parts of the archipelago. After the Japanese invasion (1942), during which he was interned on Celebes, he recorded his experiences in a diary. He went to the Netherlands after the war, but was appointed Assistant Resident in the Mandar district of Celebes, where he resumed the work of 'colonising' Javanese (1946–1948). He was then Assistant Resident in greater Batavia and later transferred to Purwakarta in the Krawang residency of Java to restore trade, the administration of justice and public health after a 'police action' (December 1948), remaining until after the transfer of sovereignty to the Indonesian Republic (1949). He was Assistant Resident (1950–1954) in Sorong, West New-Guinea, over which the Dutch had retained sovereignty. He returned to the Netherlands and worked for a metal company until retirement. He died during an excursion in the Idjen Highlands, east Java. He collected part of the type series in Celebes (now Sulawesi) when collecting for Lieftinck (1940–1941).

Maurits

Redbolt sp. *Rhodothemis mauritsi* Lohmann, 1984

Maurits Anne Lieftinck (1904–1985). (See **Lieftinck**)

Maxwell

Bayou Clubtail *Arigomphus maxwelli* Ferguson, 1950
[Orig. *Gomphus maxwelli*]

Richard E Maxwell collected the holotype in Texas (1940). I have failed to identify just who he was, although a Richard E Maxwell graduated at Bridgewater State College, Massachusetts (1950).

May

Dancer sp. *Argia mayi* **González-Soriano**, 2012

Dr Michael Love May (b.1946) is a leading American odonatologist who is a founding member of the World Dragonfly Association. Growing up in Gainesville, Florida he was an enthusiastic collector of insects and had accumulated an extensive butterfly collection by the time he left high school. His neighbour, and father of his school friend, was Dr Minter Westfall (q.v.) and May was taken on by him as an assistant in the early 1970s, working on Odonata. He was awarded his PhD (1974) and was a postdoctoral fellow at the University of Florida at the Smithsonian Tropical Research Institute (STRI) on Barro Colorado Island in Panama. He started (1978) at Rutgers University's School of Environmental & Biological Sciences, New Brunswick, rising to professor and continues to be a research associate. The focus of his current research is the taxonomy and phylogeny of Odonata and their migration. He has published widely on Odonata (1976-present) including major handbooks on North American Odonata, such as the co-authored: *Damselflies of North America* (1996), *A Manual of the Dragonflies of North America* (2000) and *Damselflies of North America, Color Supplement* (2007). He has described five new species and named three new genera of Odonata. He is a good friend of the author who "...*has made many contributions in the field of odonatology*".

Maya

Setwing sp. *Dythemis maya* **Calvert**, 1906
Darner sp. *Neuraeschna maya* **Belle**, 1989

The species are named after a people; the Maya Indians.

Mayné

Humped Hooktail *Paragomphus maynei* **Schouteden**, 1934
[Orig. *Mesogomphus maynei*]

Raymond Mayné (1887–1971) was a Belgian entomologist. He was the Government Entomologist in the Belgian Congo and a professor of zoology in Belgium. He studied the collection of insects from Central Africa at Tervuren, as well as qualifying in agriculture (1909) in Brussels: he also obtained a certificate, with distinction, in colonial studies (1910). He was appointed (1911) Entomologiste de la Colonie to the agricultural station at Congo da Lemba, responsible for an area covering fifty plantations, the majority of which grew cocoa. He combed the forest and plantations for insect specimens, which he sent back to Tervuren. He explored (1912) the potential of Yombe forests with Count J de Briey and spent a year in Belgium (1913) then returned (1914) to the Botanical Gardens in Eala (Equateur), becoming research director (1915). He spent time (1916) in London with his brother (a famous Arsenal football player), where he completed his most important published work: *Insectes et autres animaux attaquant le Cacaoyer au Congo Belge* (1917). He was in German East Africa (1917) using his entomological knowledge to help the Belgian troops combat the ravages of insect-borne diseases. He returned to Europe (1919) having sent Tervuren insect specimens and buffalo and antelope trophies. He became Professor of Zoology (1919–1959) at the State Institute of Agriculture, Gembloux, Belgium. He set up Belgium's first nature reserve (1942) at Torgny that is now named after him. He also had a collection of statues, spears and other souvenirs to remind him of the people he so admired, both for their art and their culture.

Mayoruna

Darner sp. *Neuraeschna mayoruna* **Belle**, 1989

Although the author did not give an etymology, this Peruvian species is surely named after the indigenous people of the same name that live in the area where it is found.

Mazu

Demoiselle sp. *Matrona mazu* Yu, Xue & **Hämäläinen**, 2015

Lin Moniang (960–987) is a legendary Chinese woman from Fujian, known as Mazu and worshipped as goddess of the sea. The species is endemic to Hainan Island. The etymology reads thus: "*Named after Mazu, a legendary Chinese woman Lin Moniang, who lived in Fujian in 960–987 during the Song Dynasty. In south-eastern coastal regions of China, including Hainan, Mazu is widely worshipped as a goddess of the sea, who protects fishermen and sailors*".

Mbarga

Green-eyed Citril *Argiocnemis umbargae* **Pinhey**, 1970
[JS *Ceriagrion annulatum* Fraser, 1955]

Father Augustin-René Mbarga was a missionary in Cameroon where he collected the type (1969).

McCleery

Ntchisi Yellowwing *Allocnemis maccleeryi* **Pinhey**, 1969
[Orig. *Chlorocnemis montana maccleeryi*]

Dr C H McCleery was District Medical Officer at Njombe, Tanzania (fl.1960s–fl.1980s), and is an amateur entomologist. He has a particular interest in butterflies and both bred and collected them in Africa, particularly Tanzania, Malawi, Sudan and Zambia. A number of his papers were published in the 'Bulletin of the Amateur Entomologists' Society', such as: *Sudanese butterflies on the edge of the Sahara* (1982) and at least one article was published in 'The Lancet', *Malaria Prophylaxis with Chloroquine* (1981). At least one butterfly is named after him (*Charaxes mccleeryi* Van Someren, 1972), which he discovered in the Uluguru Mountains, Tanzania (1966).

McClendon

Sanddragon sp. *Progomphus clendoni* **Calvert**, 1905

Jesse Francis McClendon (1880–1976) was an American chemist, zoologist and physiologist. The University of Texas awarded both his BSc (1903) and MSc (1904) and the University of Pennsylvania awarded his PhD in zoology (1906). He worked at the University of Texas (1903) and taught biology at Randolph-Macon College and the University of Missouri (1907–1910). He was an assistant instructor at Cornell University Medical College (1910–1914) and then worked at the Physiological Laboratory at the University of Minnesota Medical School (1914–1939) during which time he became Professor of Physiological Chemistry (1920). He left the University of Minnesota to become Research Professor of Physiology at the Hahnemann Medical School in Philadelphia and later was a research chemist at the Albert Einstein Medical Center, Philadelphia. He made substantial contributions in a number of fields of science including invertebrate zoology, nutrition, life processes of cell membranes and the role of Iodine in human health. He collected the species in Tuxpan, Jalisco State, Mexico (September 1903). (Also see **Clendon**)

McCulloch

Offshore Emerald *Anacordulia maccullochi* **Tillyard** 1926
[JS *Metaphya tillyardi* Ris, 1913]
Tiny Longlegs *Austrocnemis maccullochi* Tillyard, 1926
[Orig. *Agriocnemis maccullochi*]

Allan Riverstone McCulloch (1885–1925) was a noted systematic ichthyologist. He started his career at 13 as an unpaid assistant to Waite at the Australian Museum, Sydney. Encouraged by Waite to study, he became 'Mechanical Assistant' (1901) and later (1906–1925) Curator of Fishes there. He was a prolific collector, taking over 40,000 specimens in Queensland, Lord Howe Island, New Guinea, the Great Barrier Reef and a number of Pacific Islands, and wrote more than 100 papers, many of which he illustrated himself. He wrote: *Check List of Fishes and Fish-like Animals of New South Wales* (1922). He collected the holotypes of both species "*... during the period from November 1922 to January 1923, while exploring unknown regions of the central western part of Papua by boat and aeroplane, in company with Captain Frank Hurley.*" His health was poor and despite taking a year off he died while in Hawaii.

McGregor

Dryad sp. *Pericnemis mcgregori* **Needham** & Gyger, 1939

Richard Crittenden McGregor (1871–1936) was born in Australia, but moved to the USA with his American mother following the death of his father. He studied zoology at Stanford University, but his academic career was interrupted by a fish collecting expedition to Panama, delaying his bachelor's degree (1898) in philosophy. He was Ornithologist to the Manila Bureau of Science in the Philippines. Among his many publications are several articles on the birds of Santa Cruz County, California. He also wrote: *A Manual of Philippine Birds* (1909–1910), the first full treatment of the country's avifauna. He collected the type in Luzon (1930). Three birds, an amphibian and a reptile that he collected (1907) are also named after him.

McLachlan

Seychelles Islander *Allolestes maclachlanii* **Selys**, 1869
Tiger sp. *Gomphidia maclachlani* Selys, 1873
Forest Wisp *Agriocnemis maclachlani* Selys, 1877
Spotted Darner sp. *Boyeria maclachlani* Selys, 1883
[Orig. *Fonscolombia maclachlani*]
Pincertail sp. *Onychogomphus maclachlani* Selys, 1894
Emperor sp. *Anax maclachlani* **Förster**, 1898
Duskhawker sp. *Gynacantha maclachlani* Förster, 1899
Demoiselle sp. *Mnais maclachlani* **Fraser**, 1924
[JS *Mnais gregoryi* Fraser, 1924]

Robert McLachlan FRS (1837–1904) was a British biologist; he was primarily a botanist but then switched to entomology. When just sixteen his father died and he inherited his fortune when he came of age. He co-owned a ship and acted as broker but devoted most of his time to natural history. He made a voyage of thirteen months (1855–1856) to New South Wales and Shanghai, during which he collected a considerable herbarium. On his return, he began his written work with papers on Neuroptera and Lepidoptera. Hagen's publications further spurred his interest in Neuroptera, which became the chief interest of his life. (At that time Odonata were still considered Neuroptera.) He took almost annual visits to the Alps and Pyrenees and he amassed the largest collection of native and foreign Neuroptera in Britain.

He joined the Entomological Society of London (1858) and was secretary (1873–1875 & 1891–1904) as well as treasurer and president (1885–1886). He was among the first editors of the 'Entomologist's Monthly Magazine' (1864) and wrote numerous papers and several books such as: *A Catalogue of British Neuroptera* (1870) and his major work *Monographic Revision and Synopsis of the Trichoptera of the European Fauna* (1874). In his dragonfly papers, he described nearly 90 new species and eleven genera from all over the world, one third of them being calopterygoid species, in which he seemed to have a special interest. He was a close collaborator of Selys (from 1855). Later they met several times and also made joint short excursions. In his will Selys left McLachlan a considerable sum for reviewing and cataloguing Selys' Odonata collection, but due to his own ill-health he was unable to accept the bequest. Referring to Kipling's poem, P P Calvert (1904) ended his obituary: "*Twenty-four years younger than the great Belgian master* [Selys], *he has followed him within four years. The captains and the kings depart.*"

McLean

Pond Damselfly sp. *Melanesobasis mcleani* **Donnelly**, 1984

Dr John Alexander McLean (b.1943) is a New Zealand biologist and entomologist. The University of Auckland awarded his BSc (1965) and MSc (1968) and Simon Fraser University his PhD (1976), his thesis being: *Primary and Secondary Attraction in Gnathotrichus sulcatus (LeConte) (Coleoptera:Scolytidae) and their Application in Pest Management*. He was a high school teacher in New Zealand (1968–1969) and then taught biology at the University of the South Pacific, Suva, Fiji (1970–1973) where he collected the eponymous holotype (1973). He was appointed (1977) as the forest entomology lecturer in the Faculty of Forestry at the University of British Columbia. In his 30 years at UBC he also served as Associate Dean of Graduate Studies in the Faculty of Forestry including two years as acting dean. John retired (2007) but, as professor emeritus, has continued his interest in insect pheromones working with honey bees in New Zealand, especially those that are linked to lipids that are important nutrients and pheromone precursors for the bees. While John and his wife enjoyed ballroom dancing as a relaxing pastime in Vancouver, the maintenance of healthy bee populations is his major interest since returning to New Zealand (2009).

Medina

Rubyspot sp. *Hetaerina medinai* **Rácenis**, 1968

Dr Gonzalo Medina Padilla (1930–2009) was a biologist and lawyer who was Chief of Venezuela's Rancho Grande Biological Station, 'Henri Pitier', and curator of the museum there (1957–1981). He donated his own collection of birds to the museum when he took over. An amphibian is also named after him.

Mega Bin Luyog

Midget sp. *Mortonagrion megabinluyog* **Dow** & Choong, 2015

Mohammed Hassanal Mega bin Abdullah Luyog (b.1948) is 'an extremely able boatman' who acted as field assistant to the first author and took him to the locality where the type was taken (Brunei, 2013).

Megumi

Sylvan sp. *Coeliccia megumii* **Asahina**, 1984
[JS *Coeliccia kazukoae* Asahina, 1984]

Dr Megumi Hasegawa was the Chief of the Thai-Japan Cooperative Project 'Promotion of Provincial Health Service', Japan International Co-operation Agency in Chantaburi province (Thailand) at the time (1980) when Asahina visited Chantaburi twice and collected this new species there. In the same paper he also named another species, collected by himself at the same site, after Hasegawa's wife Kazuko (q.v.). Oleg Kosterin, who synonymized these two species (2011), used the 'ladies first' principle and demoted the name given after the husband in synonymy with the name after the wife. This may be the first case in the history of zoological nomenclature, when synonymy has been decided on this ground. (Also see **Kazuko**)

Melanson

Dryad sp. *Pericnemis melansoni* Villanueva, Medina & Jumawan, 2013

Fr. Louis Joseph Arthur Melanson (1879–1941) was a Canadian ordained (1905) priest who became Bishop of Gravelbourg, Saskatchewan (1933) and first Archbishop of Moncton, New Brunswick (1936). He founded Les Filles de Marie-de-L'Assomption, or the FMA congregation, which runs the Assumption College of Nabunturan. Milton N D Medina and Kim M Jumawan are based at the Research and Development Centre there, which is in Compostela Valley Province, Philippines.

Mell

Demoiselle sp. *Atrocalopteryx melli* **Ris**, 1912*
[Orig. *Calopteryx melli*]
Skimmer sp. *Libellula melli* **Schmidt**, 1948

Dr Rudolf Emil Mell (1878–1970) was a self-taught German naturalist who taught biology (until 1908). He then emigrated from Germany to China and founded the German-Chinese Secondary School in Canton (now Guangzhou). His original interests were entomology and herpetology, later developing interests in birds and mammals. He hired local collectors and undertook expeditions himself, such as to Ding Wu (north of Guangzhou) (1922). He discovered 76 new subspecies among the 431 bird species to be found in Kwantung province. He wrote on birds and Lepidoptera, such as: *Beiträge zur Fauna Sinica (II). Biologie und Systematik der Süd-chinesischen Sphingiden. Zugleich ein Versuch einer Biologie tropischer Lepidopteren überhaupt* (1922) and *Inventur und ökologisches Material zu einer Biologie der südchinesischen Pieriden* (1943) as well as: *Der Storch* (1930). Six birds and a number of Lepidoptera are also named after him.

*In the introduction to his paper Ris says that Herr C Mell collected the material he studied in Canton (1911). However, it does seem very likely that this initial was a mistake.

Mellis

Mellis's Sprite *Pseudagrion mellisi* **Schmidt**, 1951
Mellis's Rockstar *Tatocnemis mellisi* Schmidt, 1951

J V Mellis was a French anthropologist and insect collector who collected this species in Madagascar. He deposited a number of beetles with the BMNH that he collected (1934) and also collected anthropological artefacts. He wrote: *Nord et Nord-Ouest de Madagascar, Volamena et Vola (1938)* about the peoples of Madagascar.

Melville

Forest Watcher *Huonia melvillensis* **Theischinger** & **Watson**, 1998

This is a toponym; Melville Island is part of Australia.

Mendez

Rubyspot sp. *Hetaerina mendezi* **Jurzitza**, 1982

Bernabé Mendéz (1934–1968) was a National Park Ranger who was shot dead by poachers while on duty in the Iguazu National Park (14 April 1968). A waterfall in the park is also named after him.

Ménétriés

Snaketail sp. *Ophiogomphus? menetriesii* (**Selys**, 1854)
[*nomen dubium*]

Édouard Pétrovitch Ménétriés (1802–1861) was a French zoologist who collected in Brazil and Russia, where he settled. He studied in Paris under Cuvier (q.v.), among others, on whose recommendation he participated in an expedition in Brazil under Grigory Langsdorff (1821–1825). He was invited to become Conservator of Collections of the Russian Academy of Sciences in St Petersburg (1826). He explored and collected in the Caucasus (1829–1830) and wrote: *Catalogue Raisonée des Objects de Zoologie Recueillis dans un Voyage au Caucase et Jusqu'aux Frontiers Actuelles de la Perse* (1832). When the Zoological Museum of the Academy of Sciences was officially opened (1832), Ménétriés was designated curator of its entomological collections, a position he held for life. He studied the fauna of Siberia and also wrote one of the first works on the fauna of Kazakhstan, as well as many scientific papers. A reptile and four birds are also named after him, as well as the legendary tiger moth *Borearctia menetriesii*.

Menger

Dasher sp. *Micrathyria mengeri* **Ris**, 1919

E C Menger was an artist who was active (1866–1880) at the Institut Royal des Sciences Naturelles de Belgique. He drew illustrations of many of the species in Selys' collection. E B Williamson (q.v.) said "*Through the good offices of Dr. Ris, I obtained drawings from M. Menger of species in the de Selys collection.*" He also illustrated for B Y Sjöstedt (1866–1948) (q.v.).

Merina

Sprite sp. *Pseudagrion merina* **Schmidt**, 1951
Merina Wisp *Argiocnemis merina* **Lieftinck**, 1965

These species were named after a people; the Merina are the dominant 'highlander' Malagasy ethnic group in Madagascar.

Merton

Sylvan sp. *Idiocnemis mertoni* **Ris**, 1913

Dr Hugo Philip Ralph Merton (1879–1940) was a German zoologist and explorer. He studied at Bonn, Berlin and Heidelberg (1898–1905), culminating in the latter awarding his PhD (1905). He undertook research in Naples (1905–1906), returning to Heidelberg then joining an expedition (1907–1908) for the Senckenberg Nature Research Society to the Moluccas. He became Deputy Director of the Seckenberg Museum, Frankfurt (1909–1913). He was in the military during WW1 (1914–1918), later (1920–1935) becoming a professor at the University of Heidelberg, but was not allowed to continue when the Nazis took power due to his Jewish ancestry. He was visiting professor at the Institute of Animal Genetics, Edinburg (1937). He returned to Germany (1938), was interned and sent to Dachau concentration camp where he was seriously ill. He was allowed to emigrate, taking up the post of assistant at the Institute of Animal Genetics in Edinburgh (1939–1940) where he died, probably as a result of his time at Dachau. He collected the first specimen of the damselfly in the Aru Islands (1908). An amphibian, a reptile and two birds are also named after him.

Mesumbe

Skimmer genus *Mesumbethemis* **Vick**, 2000
[JS *Neodythemis* Karsch, 1889]
Cameroon Sparklewing *Umma mesumbei* Vick, 1996

Otto Mesumbe is a Cameroon entomological collector and guide. He guided the author and assisted him during his fieldwork in Cameroon. He discovered and collected many interesting species. He now lives in the USA where he works in hospital nursing.

Meurgey

Sylph sp. *Macrothemis meurgeyi* Daigle, 2007
[JS *Macrothemis celeno* Selys, 1857]

François Meurgey is a French biosystematist and entomologist. He is Curator of Insects at Museum d'Histoire Naturelle de Nantes (since 2000) where he works on the taxonomy, systematics and biogeography of Lesser Antillean insects, with a special emphasis on dragonflies and bees. He collected with J J Daigle in Guadeloupe (2007) where the type was taken and St Vincent (2010). Among his published papers are: *Contribution à l'étude de la faune des Odonates de Guadeloupe. Observation de Tholymis citrina (Hagen, 1876) et Tramea insularis Hagen, 1861* (2002), *Protoneura romanae spec. nov. from Guadeloupe (French West Indies)* (2006) and *Étude sur la répartition et l'écologie de Prontoneura romanae (Odonata, Zygoptera, Protoneuridae) Libellule endémique de Guadeloupe* (2007). More recently he co-wrote a book: *Les libellules des Antilles françaises : Ecologie, biologie, biogéographie et identification* (2011). (Also see **Romana**)

Meyer

Bombardier sp. *Lyriothemis meyeri* **Selys**, 1878
[Orig. *Calothemis meyeri*]

Dr Adolf Bernhard (Aron Baruch) Meyer (1840–1911) was a German anthropologist, entomologist and ornithologist. He studied at the Universities of Göttingen, Vienna, Zürich and Berlin. He was Professor of Anthropology and Ethnography and Director of Zoology at the Museum in Dresden (1874–1905). He was very interested in the evolution debate and corresponded with Wallace. He collected in the East Indies (Celebes) and New Guinea (1880–1883). He wrote: *The Birds of the Celebes* (1885), and made the first descriptions of a number of bird species from the East Indies. It was he who first recognised that the red male and green female King Parrot *Alisterus scapularis* constituted one species. Seventeen birds, one amphibian two mammals and two reptiles are also named after him, as are a number of fish.

Mézière

Blue-spotted Pricklyleg *Porpax mezierei* **Dijkstra** & Kipping, 2015

Nicolas 'Nico' Maximillien Armand Mézière (b.1980) is a French teacher and amateur odonatologist. He worked in Gabon for seven years (2008–2015) as a teacher, pursuing dragonflies in his spare time. He has also collected in Republic of Congo. He currently (2017) resides in French Guiana. The etymology says: "*Named in honour of our co-author Nicolas Mézière who, living in southeastern Gabon, has made many discoveries in this odonatologically unexplored part of Africa within a short time.*" He has co-described 21 new species of African Odonata (2015).

Mia

Adam's Gem *Libellago miae* **Lieftinck**, 1940
[JS *Libellago adami* Fraser, 1939]

No etymology is given.

Miao

Miao Flashwing *Vestalaria miao* Wilson & **Reels**, 2001
[Orig. *Vestalis miao*]

The Miao are an ethnic minority people of Vietnam in the area where the species is found.

Michael

Broad-Tailed Shadowdragon *Neurocordulia michaeli* Brunelle, 2000

Michael Erin Brunelle (b.1991) is the son of the author Paul-Michael Brunelle (b.1952).

Michalski

Fineliner sp. *Teinobasis michalskii* **Theischinger** & **Kalkman**, 2014

John Charles Michalski (b.1963) is an American entomologist and odonatologist who collected Odonata and other insects as a youngster. He is also a traveller and writer of travel nonfiction, under the pseudonym Charles McAllister. He studied biology and entomology at Rutgers University for his BSc (1991), learning the basics of odonatology under Dr Michael May (q.v.). He spent a great deal of time (1983–1996) on the Caribbean island of Trinidad. Marygrove College awarded his MSc in teaching (1994). He was a Curatorial Assistant at Newark Museum (1993–1994), where he curated their insect collection. He has been teaching biology in New Jersey public schools since 1995. His research is strongly influenced by the instruction and philosophy of Dr T W Donnelly (q.v.) of Binghamton, NY. He has written many papers (including with Gunther Theischinger), such as: *A Catalogue and Guide to the Dragonflies of Trinidad (Order Odonata)* (1988). He has written often with

Steffen Oppel, an example being: *Two new species of Argiolestes from Papua New Guinea (Odonata: Megapodagrionidae)* (2010). He wrote the books: *A Manual for the identification of the Dragonflies and Damselflies of New Guinea, Maluku & the Solomon Islands* (2012) and *The Dragonflies & Damselflies of Trinidad and Tobago* (2015). He has described eight new Odonata species from New Guinea.

Mielke

Emerald sp. *Navicordulia mielkei* **Machado** & **Costa**, 1995

Professor Dr Olaf Hermann Hendrik Mielke (b.1941) is a Brazilian zoologist. The State University of Guanabara awarded his BSc (1964) and MSc (1971) and the Federal University of Paraná his PhD (1982). He was president (1994–2002) and editor (1992–1996) of the Brazilian Society of Zoology, vice-president of the Brazilian Society of Entomology (2005–2007) and vice-president of the Lepidopterists' Society–USA (1975 & 1986). He has been a visiting researcher in the largest scientific collections in Brazil, South and North America and Europe. He has been a member of the Board of the Association of Tropical Lepidoptera, USA (1982–2008), the Brazilian Society of Zoology, Brazilian Society of Entomology, Entomological Society of Brazil, Lepidopterists' Society, Union Des Belges Entomologistes, Association for Tropical Lepidoptera and Sociedad Hispano Luso American lepidopterology. He was full professor at the Federal University of Paraná and researcher at the Brazilian National Council for Scientific and Technological Development. His particular interests are the systematics, bionomics, morphology and biodiversity of Neotropical Lepidoptera. He collected the holotype in Brazil (October 1987). The authors wrote of him: "*...whose wanderings throughout South America in search of butterflies, have also yielded many valuable dragonfly species*".

Miers

Emerald sp. *Navicordulia miersi* **Machado** & **Costa**, 1995

Herbert Willy Miers (1928–2009) was an amateur Brazilian entomologist. The authors wrote: "*We dedicate this species to Mr Herbert Miers, lepidopterist from Joinville, who collected the holotype and has made an important contribution to the odonatological studies of one of the authors* (Machado)." He often collected with visiting entomologists. In another etymology, he was thanked for his: "*...innate enthusiasm in helping scientists for over 30 years*". A number of butterflies are also named after him.

Mildred

Bluetail sp. *Ischnura mildredae* **Fraser,** 1927

Cascader ssp. *Zygonyx iris mildredae* Fraser, 1926

Mildred Wall was a keen collector and the wife of the physician and herpetologist Colonel Frank Wall (1868–1950). He collected these species in Upper Burma. (Also see **Wall**)

Miles

Oread sp. *Calicnemia miles* **Laidlaw**, 1917
[Orig. *Calicnemis miles*]

This was a replacement name for a species from Burma, which had earlier been misidentified with several different names. In Latin 'miles' means a common soldier (private).

Millard

Duskhawker sp. *Gynacantha millardi* **Fraser**, 1920

Walter Samuel Millard (1864–1952) was a British entrepreneur and naturalist. He went to India (1884) as an assistant in a wine business run by the then secretary of the Bombay Natural History Society, who also made Millard assistant editor of their journal. He was the long-time Honorary Secretary of the Bombay Natural History Society and took over as their journal editor (1906–1920). He was a very keen gardener and wrote: *Some Beautiful India Trees* (1937) as well as enjoying ornithology and being a pioneer conservationist. He was also responsible for the mammal survey of the sub-continent carried out by the society (1911–1923). He left India on retirement (1920), but managed the Bombay Natural History Society's business in London for many years. No etymology was given in the duskhawker account, but undoubtedly the species was named after him.

Miller

Firetail sp. *Telebasis milleri* **Garrison**, 1997

Dr Peter Lamont Miller (1931–1996) was a British biologist and zoologist and at the time of his death one of the world's leading odonatologists. After national service, he began studying to be a veterinarian (1951), but physiology study led him to focus on insects and other invertebrates and he changed course to read natural sciences at Downing College, Cambridge. He graduated with first class honours and won the Frank Smart Prize (1955) as the best zoologist of the year; he went on to complete his PhD there (1958). He was a Research Fellow at Downing College (1956–1959) after which he married a fellow dragonfly enthusiast and they went to Uganda where he lectured in zoology at Makerer College, Kampala (1959–1962). He moved back to England as a lecturer, tutor and research fellow at Queen's College, Oxford (1962–1994) until taking early retirement when he returned to lecture in Uganda and study the Odonata. He studied physiology of Orthoptera, Hemiptera, Coleoptera, Diptera and Opiliones, but narrowed his focus (from 1981) on Odonata. His early work followed from his PhD studies on respiration, including work on neural control of spiracle movement in Odonata; but he widened his field to cover their locomotion and behaviour, copulatory mechanisms, oviposition, territorial strategies and courtship. He published over 80 papers on Odonata and 47 others (1961–1996). Among his most influential works were the books: *Dragonflies* (1987, 2nd ed 1995) and *East African dragonflies* (2003) co-authored by his wife. During his time at Oxford he travelled to Kenya, Uganda, and India. His plans to write and study, taking trips there every year, were cut short by a sudden fatal illness. (See also **Peter Miller**)

Millot

Genus *Millotagrion* **Fraser**, 1953

Proto sp. *Protolestes milloti* Fraser, 1949
Lemur Skimmer *Orthetrum milloti* Fraser, 1949
[JS *Orthetrum lemur* Ris, 1909]

Professor Jacques Millot (1897–1980) was a French physician and arachnologist at the Muséum National d'Histoire Naturelle, Paris. As well as publishing on spiders, he wrote *Biology of the Human Races* (1952). He spent many years in Madagascar, where the type was taken (1946), in charge of the Scientific Institute of Madagascar and (1953) started the periodical 'Le Naturaliste malgache'. An amphibian and a reptile are also named after him.

Milne

Phantomhawker sp. *Planaeschna milnei* **Selys**, 1883
[Orig. *Aeschna milnei*]

John Milne (1850–1913) was a British geologist and mining engineer who trained at the Royal School of Mines. He was hired in the search for coal in Newfoundland and Labrador (1873 & 1874) but his natural history leanings were already in evidence as at that time he wrote a paper on the newly extinct Great Auk. He also went to Arabia (1873) on a mapping expedition and collected fossils while there, which he sent to the BMNH. He was hired by the Japanese government as an advisor on mining and geology at the Imperial College of Engineering, Tokyo (1875–1895). He spent much time training Japanese students and studying earthquakes there, so much so that he was given an honour by the Emperor and a pension when he returned to England. He also took an interest in anthropology while in Japan. On his return to England with his Japanese wife he was made Professor Emeritus of Tokyo Imperial University. He created a worldwide network of seismic observatories and his was the world headquarters for earthquake seismology. He wrote: *Earthquakes and Other Earth Movements* (1898). When he died of Bright's disease, his wife returned to Japan. Selys dedicated this Japanese species to him for his contributions in developing science in Japan.

Mina

Mina Net-winged Darner *Neuraeschna mina* **Williamson** & Williamson, 1930

Mina Louise Winslow (1890–1982) was an American malacologist who was the first Curator of Molluscs at the Museum of Zoology, the University of Michigan (1914–1929), until her resignation. She collected in southern Africa (1924–1925) on a fully paid sabbatical year. She published a number of papers and longer works such as: *Two new freshwater snails from Michigan* (1923) and *African Adventure 1924* (1980); the latter was written for a writing class. She was also an accomplished artist and illustrator.

Minck

Skimmer sp. *Dasythemis mincki* **Karsch**, 1890
[Orig. *Malamarptis mincki*]

Max Minck was a German merchant and entomologist, who served for many years as the Secretary of the Entomological Society of Berlin. He was also a keen lepidopterist with at least one species named after him.

Minjerriba

Dune Ringtail *Austrolestes minjerriba* **Watson**, 1979

This is a toponym; Minjerriba is the Aboriginal name for North Stradbroke Island, Australia.

Minter

Flatwing sp. *Philogenia minteri* **Dunkle**, 1986
Pond Damselfly sp. *Metaleptobasis minteri* Daigle, 2003

Minter Jackson Westfall Jr (1916–2003) (See **Westfall**)

Mirna

Pond Damselfly *Oxyagrion mirnae* **Machado**, 2010

Professor Dr Mirna Martins Casagrande (b.1953) is a Brazilian entomologist at the Laboratório de Estudos de Lepidoptera Neotropical, Departamento de Zoologia, Universidade Federal do Paraná, Curitiba, Paraná, Brazil. He is editor of the journal of the Sociedade Brasileira de Zoologia, of which Olaf Mielke (q.v.) is president. They have written often together, especially on Lepidoptera such as: *Notas taxonômicas em Hesperiidae neotropicais, com descrições de novos taxa (Lepidoptera)*. (2002). Together with Bonatto they compiled the D*iretório de zoólogos do Brasil* (1997). Casagrande collected the type in Brazil (2010).

Misra

Crimson Dropwing *Crocothemis misrai* Baijal & Agarwal, 1956
[JS *Trithemis aurora* (Burmeister, 1839)]

Dr S D Misra was Professor of Zoology, University of Jodhpur. He collected the holotype and co-wrote: *A Systematic Catalogue of the Main Identified Entomological Collection at the Forest Research Institute, Dehra Dun* (1951).

Mitwaba

Katanga Yellowwing *Allocnemis mitwabae* **Pinhey**, 1961

This is a toponym; Mitwaba is a territory of Democratic Republic of the Congo.

Miyashita

Skydragonfly sp. *Chlorogomphus miyashitai* **Karube**, 1995

Tetsuo Miyashita (b.1947) is a Japanese commercial insect supplier. He collected the type specimens in Laos (1993).

Mjöberg

Pimple-Headed Hunter *Austrogomphus mjoebergi* **Sjöstedt**, 1917
[Orig. *Austrogomphus mjöbergi*]
Tropical Evening Darner *Telephlebia mjöbergi* Sjöstedt, 1917
[JS *Telephlebia tillyardi* Campion, 1916]

Dr Eric Georg Mjöberg (1882–1938) was a naturalist, entomologist, ethnographer and explorer. He took his initial degree at Stockholm University (1908) and his master's at Lund University (1912). He held various jobs in Sweden, including working at the Naturhistoriska Riksmuseet, Stockholm, and teaching in high schools (1903–1909). He led Swedish scientific expeditions in northwestern Australia (1910–1911) and Queensland

(1912–1913). He worked in Sumatra at an experimental station (1919–1922), combining the duties with those of being Swedish Consul. He was Curator of the Sarawak Museum (1922–1924) and led a scientific expedition to Borneo (1925–1926). Sjöstedt wrote: *Results of Dr. E. Mjöberg's Swedish scientific expeditions to Australia 1910–1913, Odonata.* (1917). Two birds, two reptiles and three amphibians are also named after him.

Mocsáry

Paddle-Tipped Duskhawker *Gynacantha mocsaryi* **Förster**, 1898

Alexander 'Sándor' Mocsáry (1841–1915) was a Hungarian entomologist specializing in Hymenoptera. He studied at a local gymnasium, but tuberculosis forced him to withdraw and he continued to study at home (1859). He later finished his education at the University of Vienna where he graduated from the Faculty of Natural Sciences. He was an assistant (1870) rising to curator and director (1910), until retirement (1914), of the Hungarian National Museum, where he created a large collection of Hymenoptera. He published more than 200 papers and longer works. He confined his collecting activities to Hungary, but a great deal of material came to him from other parts of the world, especially the East Indies and Australia. More than fifty insect species are named after him.

Modigliani

Leaftail sp. *Oligoaeschna modiglianii* **Selys**, 1889
Satinwing sp. *Euphaea modigliani* Selys, 1898

Elio Modigliani (1860–1932) was an Italian zoologist and anthropologist who collected in Sumatra and nearby islands (1886–1894). His first trip was to Nias (1886), following which he wrote the book *Un viaggio a Nias*. He travelled to Enggana Island (1891) and wrote about it and its apparently female dominated society in *L'Isola delle Donne* (The Island of Women, 1894). He sent many artefacts back to the Ethnographic Museum of Florence. Three birds and five reptiles are also named after him, as is an amphibian that was only described after the museum specimens were close to 100 years old!

Mohan

Hooktail sp. *Acrogomphus mohani* Sahni, 1965
[JS *Anisogomphus bivittatus* Selys, 1854]

No etymology was given. However, it almost certainly refers to the Indian poet, Maulana Hasrat Mohani née Syed Fazl-ul-Hasan (1875–1951), who was also a journalist, politician and part of the independence movement. However, his adopted name Mohani also refers to his place of birth – Mohan; a town in Uttar Pradesh so this could, in a sense, be regarded as a toponym too.

Mohéli

Sprite sp. *Pseudagrion mohelii* **Aguesse**, 1968

This is a toponym; it refers to the locality Mohéli Island in the Comores where the species was found.

Mojca

Mojca's Shadowdamsel *Ceylonosticta mojca* Bedjanic, 2010
[Orig. *Drepanosticta mojca*]

Mojca Bedjanič (b.1973) is the wife of the author Matjaž Bedjanič (b.1972).

Moluam

Guangdong Hooktail *Mellligomphus moluami* **Wilson**, 1995
[JS *Melligomphus guangdongensis* Chao, 1994]

Moluam Cook (b.1988) is the son of the author's friend, David Cook, who collected the type in Hong Kong (1993). The etymology reads: *"The author and Mr David Cook with his son Moluam spent a considerable amount of time in the New Territories trying to obtain larvae and adult specimens of this species. In the end, the male type specimen was finally obtained in a stream a few score metres from Mr Cook's flat on Hong Kong Island! The author is pleased to name the new species after Mr Cook's son Moluam."*

Monard

Genus *Monardithemis* **Longfield**, 1947
[JS *Micromacromia* Karsch, 1890]

Southern Fluttering Dropwing *Trithemis monardi* **Ris**, 1931
Angola Sprite *Pseudagrion monardi* Longfield, 1947
[JS *Pseudagrion angolense* Selys, 1876]
Woodland Skimmer *Orthetrum monardi* **Schmidt**, 1951

Firebelly ssp. *Eleuthemis buettikoferi monardi* Schmidt, 1951
[Orig. *Eleuthemis büttikoferi monardi*]

Professor Dr Albert Monard (1886–1952) was a Swiss zoologist, naturalist and explorer who made six expeditions to Africa (1928–1947). He taught at a high school in La Chaux-de-Fonds and was Curator of its Natural History Museum (1920–1952). His best-known work was: *The Little Swiss Botanist* (1919), still used as a school textbook in the French-speaking cantons of Switzerland. Two mammals and a reptile are also named after him.

Monastyrski

Clubtail sp. *Davidius monastyrskii* **Do**, 2005

Dr Alexander L Monastyrskii (b.1954) is a Russian entomologist specializing in Lepidoptera, who was living in Vietnam when he collected the holotype (April 1997). He has worked at the Vietnam Forest Museum of the Forest Inventory and Planning Institute, Hanoi, Vietnam and the Ecology Department of the Vietnam-Russia Tropical Centre, Hanoi, Vietnam. He has written five books on Vietnamese butterflies, including three volumes of the series: *Butterflies of Vietnam* (2005, 2007, 2011), as well as continuing to describe Vietnamese butterflies.

Montague

Tigertail sp. *Synthemis montaguei* **Campion**, 1921

Lieutenant Paul Denys Montague (1890–1917) was a Greek-born British zoologist, ethnologist, explorer and pilot in the Royal Flying Corps, known as 'the Birdman of Salonika'. He explored in northwest Australia (1912). He took part in the 'Salonika Campaign' during WW1 but died when his plane was shot down. He collected the type in New Caledonia (1914) among a large number of other insects. Campion wrote: "*I have honour of dedicating this very fine species to the memory of its discoverer, who afterwards gave his life in the course of freedom on the battlefields of Macedonia.*" A bird is also named after him.

Montandon

Blue-eyed Goldenring *Cordulegaster montandoni* St. Quentin, 1971
[JS *Cordulegaster insignis* Schneider, 1845]

Dr Arnold Lucien Montandon (1852–1922) was a French naturalist, primarily an entomologist, who settled in Romania (1873) where he collected the type. He was at the Natural History Museum, Bucharest as assistant to Dr Grigore Antipa, after whom the museum is now named, helping to set it up (1896–1907). He wrote: *Espèces d'Hémiptères-Hétéroptères d'Algérie et de Tunisie* (1897). An amphibian is also named after him.

Monteil

Reedling sp. *Lestes monteili* **Navás**, 1935
[JS *Indolestes extraneus* Needham, 1930]

Paul Monteil was the Director of the Catholic missionary hospital, St Louis Hospital, in Nanchang, Kiangsi, China. He collected one of the type series in Kuling, Kiangsi (September 1934).

Monteiro

Giant Skimmer *Thermorthemis monteiroi* **Kirby**, 1900
[JS *Orthetrum austeni* (Kirby, 1900)]

Joaquim (Joachim) John (João José) Monteiro (1833–1878) was a Portuguese mining engineer who had English ancestors (hence 'John' and alternatives to it). He collected natural history specimens in Angola (1860–1875) with his wife, Rose Monteiro née Bassett (1840–1897), who collected the type. He wrote: *Angola and the River Congo* (1875). Five birds and at least one fish are also named after him.

Montezuma

Band-winged Dragonlet *Trithemis montezuma* **Calvert**, 1899
[JS *Erythrodiplax umbrata* (Linnaeus, 1758)]

There were two Aztec emperors called Montezuma, Montezuma I (c.1398–1469) and Montezuma II (c.1466–1520).

Montgomery

Bannerwing sp. *Chalcothore montgomeryi* **Rácenis**, 1968
[Orig. *Euthore montgomeryi*]

Professor Basil Elwood 'Monty' Montgomery (1899–1983) was an entomologist and authority on dragonflies. He graduated BSc from Oakland City College (1922) and was awarded his

MA by Purdue University (1925) and his PhD by Iowa State University (1936). He was a schoolteacher (1918–1928) and then spent most of his professional career as a faculty member at Purdue University (1928–1955), thereafter Professor Emeritus and a Professor at Marian College, Indianapolis and Frostberg State College, Maryland. He is most remembered by fellow odonatologists for the first worldwide colloquium on Odonata that he organised at Purdue University (1963) and for initiating and editing the newsletter 'Selysia'. These initiatives were the first steps in building a worldwide odonatological community. He donated a large collection (20,000 of 1000 species) of dragonflies to the National Museum of Natural History as well as his Odonata library. He wrote more than 100 articles and papers on Odonata, bees and the history of entomology. His major publications include: *Studies in the Polythoridae. 1. A synopsis of the family, with keys to genera and species, information on types, and the description of a new species* (1967) and *The family and genus-group names of the Odonata. I. Calopterygoidea* (1967). Other works included: *Odonatological bibliography of Frederick Charles Fraser* (1988). He married Esther Barrett (q.v.) (1930) and they had a daughter Emily Joan Alward (q.v.). He made a number of study trips to Alaska and one to New Zealand, as well as visiting all the important museum collections in Europe. He was a keen collector of stamps with insect images on them. He described seven new species of odonata. (Also see **Emily** & **Esther (Montgomery)**)

Moor

Zambezi Hawker *Pinheyschna moori* **Pinhey**, 1981
[Orig. *Aeshna moori*]

Dr Ferdinand 'Ferdy' Cornelis de Moor (b.1947) is a Dutch aquatic entomologist and ecologist. Witwatersrand University, Johannesburg awarded his BSc biological science (1971), his BSc zoology (1972) and his PhD (1983). His career started as Assistant Keeper of Entomology at the National Museums of Rhodesia (1972–1975) (during which time he collected the type), becoming Curator of Invertebrates and Senior Keeper of Entomology (1975–1977). He was then (1977–1981) Research Fellow with National Institute for water research (NIWR) stationed in Warrenton. He became Senior Research Officer (1981–1983) and Chief Research Officer (1983–1984) at CSIR (NIWR) Pretoria, before becoming (1984–1994) Principal Natural Scientist & Curator Freshwater Invertebrates and then (1994–1996) Specialist Scientist, Aquatic Invertebrates, Albany Museum Grahamstown. Since then he has been Senior Specialist Scientist and Curator, Department of Freshwater Invertebrates, Makana Biodiversity Centre, Albany Museum, and Department of Zoology and Entomology, Rhodes University, South Africa. He specialises in biology and identification of aquatic macro-invertebrates; river ecology; assessment of water quality, conservation status of aquatic ecosystems; systematics and biogeography of Trichoptera and Simuliidae (Diptera). He has published over 300 papers, reports, reviews and articles. He collected the holotype on a Bulawayo Museums collecting expedition to NW Zambia (April 1972) and said: "*I remember catching the aeshnid in Zambia next to the Kamenkundju River in an open patch in the forest. I also photographed the specimen to capture the live colouration.*"

Moore, F

Cruiser sp. *Macromia moorei* **Selys**, 1874

Frederic Moore (1830–1907) was a British entomologist who became the Director of the East India Company Museum, London. He was appointed assistant at the museum (1848) and became a temporary writer and then assistant curator until retiring (1879). He began compiling the 10-volume: *Lepidoptera indica* (1890–1913), a major work on the butterflies of South Asia, completed after his death by Charles Swinhoe. His artist son, F C Moore, drew many of the plates and the books described many new species. His other works included: *A catalogue of the birds in the museum of the East-India Company* (1854–1858) and *The Lepidoptera of Ceylon* (1880–1887). Although Selys' etymology named him as Dr Moore, there seems to be no evidence that he completed a doctorate although he was a Fellow of the Zoological Society.

Moore, N W

Citril sp. *Ceriagrion moorei* **Longfield**, 1952
Shadowdamsel sp. *Drepanosticta moorei* **Van Tol** & Müller, 2003
Threadtail sp. *Neoneura moorei* **Machado**, 2003
Large Longleg *Notogomphus moorei* **Vick**, 2003

Dr Norman Winfrid Moore (1923–2015) was a British conservationist described as one of the most influential figures in conservation over half a century. He worked extensively on studies of dragonflies and their habitats and was one of the first people to observe and warn of the adverse effects of DDT and other organochlorine pesticides on wildlife. He graduated from Cambridge during WW2 and served as a Lieutenant in the Royal Artillery (1944–1946) in Holland and Germany, where he was wounded and taken prisoner. He studied at, and was awarded his PhD by, Bristol University (1954) with a thesis entitled: *On the behaviour and ecology of dragonflies (Odonata, Anisoptera)*. He worked as zoology lecturer at Bristol University (1948–1953) and became a Scientific Officer at the Nature Conservancy Council (1953–1983), rising to Chief Advisory Officer. He was Head of the Toxic Chemicals and Wildlife Division at Monks Wood Experimental Station where he studied the effects of toxic chemicals on wildlife (1960–1974). He was also Visiting Professor of Environmental Studies at University of London (1979–1983). He was a founding member and one time chairman of the Farming and Wildlife Advisory Group (FWAG) and later a vice-president of the British Association of Nature Conservationists. A three-month expedition to the Gambia (1947) reignited his interest in dragonflies, which had started in his teen years. His work on dragonflies and conservation led to him coining the term 'the birdwatcher's insect'. He founded the Odonata Specialist Group of the IUCN Species Survival Commission (1980) and chaired it (until 1999). In this capacity, he wrote *Dragonflies: Status Survey and Conservation Action Plan* (1997). He has contributed to, written or co-written a number of books including his award-winning professional autobiography: *The Bird of Time* (1987) and *Oaks, Dragonflies and People* (2002). He was also co-author of *Dragonflies* (1960), a book in The New Naturalist series by Collins. He was honoured for his contributions to the conservation of Odonata. He was a polite, gentle and humble man and never used his title of the 3rd Baronet of Hancox (1959). On retirement, he was offered an OBE, but he considered that this was inappropriate for a civil servant 'just doing his job'.

Moremi

Angola Bluet *Enallagma moremi* **Balinsky**, 1967
[JS *Pinheyagrion angolicum* Pinhey, 1966]

This is a toponym; it is named after Moremi Game Reserve in Bechuanaland where this species was found.

Morin

Mealy Threadtail *Elattoneura morini* **Legrand**, 1985

Claude Morin is a naturalist who was in Brazzaville with Legrand where they collected the type specimens together (April 1979). He co-wrote: *Contribution à la faune de la république populaire du Congo: Les collections de l'Université de Brazzaville* (1974).

Morrison, H K

Great Basin Snaketail *Ophiogomphus morrisoni* **Selys**, 1879

Herbert Knowles Morrison (1854–1885) was an American entomologist. He was one of the founders (1874) of the Cambridge Entomological Club (Harvard) when only 19 years old, having begun collecting and studying at 12. His experiences with the club convinced him to become a professional insect collector and he collected across the US (1875–1885). He died in Key West Florida after an attack of dysentery. The doctor who attended him during this fatal illness said he had the finest physique of any man he ever saw, perhaps as he spent most days walking up to forty miles a day in pursuit of insects! He is commemorated in the names of other taxa such as the moth Morrison's Sallow *Eupsilia morrisoni*.

Morrison, J D

Rapids Dropwing *Trithemis morrisoni* **Damm & Hadrys**, 2009

James Douglas 'Jim' Morrison (1943–1971) was an American singer-songwriter and poet, most famous for being the lead singer of 'The Doors' rock band. He was the self-proclaimed 'King of Orgasmic Rock'. He became alcohol dependent, which probably led to his death. Although some allege he overdosed on heroin, there was no autopsy. Morrison was well known for often improvising passages of spoken words while the band played live. The etymology says: *"Named after the poet James Douglas Morrison and his passion for deserts and the hidden mysteries of nature"*.

Mortimer

Harpoon Clubtail *Gomphus mortimer* **Needham**, 1944
[JS *Gomphus descriptus* Banks, 1896]

Dr Mortimer Demarest Leonard (1890–1975) was an American entomologist. He was also the author's friend and fieldwork associate. He attended Cornell University (1909), leading to his BSc (1913). He stayed to study for his PhD whilst working in nurseries and laboratories until he was awarded his PhD (1921). He was offered a post directing the field work of the Bowker Chemical Company in the eastern US (1921–1923) and was then acting New York State Entomologist (1923–1924), spending the next year back at Cornell (1924–1925). He worked for the Florida Agricultural Supply Company (1925–1927) after which he again returned to Cornell. He became Chief of the Division of Entomology at Puerto Rico's Insular Experimental Station (1930–1932). There followed a company post in pest control in Florida (1933–1939), various other companies (1939–1940, 1940–1941) and then he worked as a business analyst (1941–1945). He joined the US Bureau of Entomology (1945–1947) and then they followed another period of various company employments until retirement (1948–1961). He wrote, among other works: *A list of the insects of New York* (1928), and co-wrote: *Manual of Vegetable Garden Insects* (1918).

Morton, E S

Narrow-winged Skimmer sp. *Cannaphila mortoni* **Donnelly**, 1992

Dr Eugene Siller Morton (b.1940) is an American ornithologist. He attended Denison University (BS), the University of the Pacific, Cornell University (MS) and completed his PhD at Yale University, with predoctoral and postdoctoral fellowships at the Smithsonian Tropical Research Institute in Panama. He was professor at the University of Maryland and senior scientist at the Smithsonian Institution until his retirement (2005) where he is now Senior Scientist Emeritus at the Migratory Bird Center and Adjunct Professor at York University, in Toronto and Director of the Hemlock Hill Field Station near Cambridge Springs, PA. He has written, co-written or edited several books including: *Migrant Birds in the Neotropics* (1982), *The Smithsonian Book of Birds* (1990), *Animal Talk* (1991) and *The Behavioral Ecology of Tropical Birds* (2001). He collected part of the type series of this species in Panama. The etymology says that he: *"...took time from his ornithological studies to assemble an extensive collection of odonates and to pursue studies on many of these insects."* (Also see **Letitia**)

Morton, K J

The Genus *Mortonagrion* **Fraser**, 1920
Sylph sp. *Macrothemis mortoni* **Ris**, 1913
Oread sp. *Calicnemia mortoni* **Laidlaw**, 1917
[Orig. *Calicnemis mortoni*]
Bombardier sp. *Lyriothemis mortoni* Ris, 1919
Shadowdamsel sp. *Protosticta mortoni* Fraser, 1924
[JS *Protosticta gravelyi* Laidlaw, 1915]
Skydragon sp. *Chlorogomphus mortoni* Fraser, 1936
Rifle Sprite *Pseudagrion mortoni* **Schmidt**, 1936
[JS *Pseudagrion sublacteum* Karsch, 1893]

Common Bluetail. *Ischnura elegans mortoni* Schmidt, 1938
[JS *Ischnura elegans* (Vander Linden, 1820)]

Kenneth John Morton (1858–1940) was an amateur Scottish entomologist whose particular interests included Odonata. He worked for the British Linen Bank in Glasgow (1874–1897), then Edinburgh (1897) until retirement (1922) rather than accepting the promotion he was offered because he wanted to devote himself to entomology. He collected insects as a boy building a collection of Lepidoptera, but around the age of 18 turned to Odonata. However, he was also interested in Trichoptera, Plectoptera, Ephemeroptera and Neuroptera. He mainly collected in Scotland along the River Clyde, often with his friend Robert McLachlan (q.v.), but also often visited Europe including France, Switzerland, Norway, Italy, Corsica, Austria and Spain. He was closely associated with Ris (q.v.) and McLachlan, exchanging more than 750 letters with the latter and some 250 with the former. His collections are now in the National Museum of Scotland. Among other works (including around forty-five papers on Trichoptera 1883–1940 and fifty on Odonata) he wrote: *On the Oral Apparatus of the Larva of Wormaldia, a Genus of Trichoptera* (1887) and *Notes on the Odonata of Yunnan, with Descriptions of New Species* (1928). He described 18 new Odonata species and named two new genera. He also described many other new insect species, especially Trichoptera and is honoured in the names of several other insects.

Moulds

Kimberley Hunter *Austrogomphus mouldsorum* **Theischinger**, 1999
Striped Threadtail *Nososticta mouldsi* Theischinger, 2000

Dr Maxwell Sydney Moulds (b.1941) is an Australian entomologist and museum curator; he is married to Barbara J Moulds (b.1944). After his schooling (1958) he became a primary school teacher which allowed him long holidays to collect insects. At that time, it was difficult to buy collecting equipment in Australia so he set up (1962) a business 'Australian Entomological Supplies'. He gave up teaching (1968) and devoted his days to insect research at the museum, where he was made honorary research associate, running the mostly mail order business in the evenings. He sold his business (1972) and that year set up the Australian Entomological Press and he was the founding Editor (1972–1988) of the 'Australian Entomological Magazine'. He published a bibliography of Australian butterfly literature (1977). He took over looking after the invertebrate collection at the Australian Museum (1989) when the technical officer had a stroke and when the entomology collection manager position was made permanent (1990) he was offered the post, which he held until retirement (2003). He remains an active senior research fellow. He was awarded his MSc by Macquarie University (1995) and his PhD by Sydney University (1999). He collected in Papua New Guinea (1996). His principal interest is in Cicadas and among his written works is: *Australian Cicadas* (1990). The dedication for *mouldsorum* refers to the collectors M S and B J Moulds. (See **Barbara**)

Moulton

Threadtail sp. *Disparoneura moultoni* **Laidlaw**, 1912
[JS *Prodasineura hyperythra* Selys, 1886]
Jewel sp. *Rhinocypha moultoni* Laidlaw, 1915

Major John Coney Moulton (1886–1926) was a British army officer, ornithologist and entomologist. He sent material to Laidlaw, much of which he had collected himself. He was Curator of Sarawak Museum (1908–1915), leaving to serve with the Wiltshire Regiment in India and Singapore (1915–1919). He was Director of Raffles Museum, Singapore (1919–1923) but returned to Sarawak (1923) as Chief Secretary to Charles Vyner Brooke (q.v.), the second 'White Rajah'. He collected the types in Borneo (1911 & 1913). Two birds are also named after him.

Mound

Leaftail sp. *Phyllogomphus moundi* **Fraser**, 1960

Dr Laurence Alfred Mound (b.1934) is a British-born Australian entomologist. The University of London awarded his BSc in zoology (1957) and Imperial College his economic entomology diploma (1958). ICTA, Trinidad awarded his diploma in tropical agriculture and University of London his DSc (1975). He was an entomologist to the Nigerian Federal Government (1959-1961), undertaking research into whitefly vectors and crop diseases at Ibadan. He was then (1961-1964) entomologist to a cotton corporation in Sudan before joining the BMNH (1964), until retirement (1992), where he became Keeper of Entomoloy (1981-1992) whilst also being Consultant Director CAB International Institute of Entomology (1976-1992). He wrote: *Thrips of Central & South America* (1992-1994) and then moved to Australia where he was a research fellow at CSIRO (1995-1996) and thereafter honorary research fellow at CSIRO Ecosystem Sciences at the Black Mountain Campus. He has written more than 330 publications (1963-2013). He collected the type at Ibadan in Nigeria (1960).

Moura

East Coast Citril *Ceriagrion mourae* **Pinhey**, 1969

Dr Armando Reis Moura is a Portuguese palaeontologist and geologist working in Mozambique who is also interested in Odonata. He collected the holotype there (August 1964). He has a great many publications (1960s-1980s) mostly on marine fossils but also including *Barcos do litoral de Moçambique* (1986).

Moure

Pond Damselfly sp. *Aceratobasis mourei* **Santos**, 1970
[Orig. *Telagrion mourei*]

Professor Father Jesus Santiago Moure (1912-2010) was a Brazilian priest. After attending seminary (1929-1932) he took a degree in theology at Seminario Maior Claretiano, Curitiba (1933-1936). While there he began studying insects as evidenced by his correspondence with the Paulista Museum. He was ordained (1937) but studied bees at the museum. He started collecting, but also employed a full-time collector (Claudionor Elias q.v.). He returned to Curitiba (1938) as Professor of Natural Sciences at the seminary. He was involved with the foundation of the Universidade do Parana, assuming the chair of zoology and was later Director of the Museu Paranaense (1952-1954). He travelled to Argentina (1953) and Chile (1954), then to the University of Kansas and across the USA (1956-1957) followed by study for a year in Europe (1957). Over the next four decades he taught, travelled and studied in the USA and Europe. He published over 220 articles, mostly on bees, and three books culminating in *Catalogue of Bees in the Neotropical Region* (2007). He amassed a collection of over 5 million insects!

Mudginberri

Top End Archtail *Nannophlebia mudginberri* **Watson** & **Theischinger**, 1991

This is a toponym; Mudginberri Station is a locality in Northern Territory, Australia.

Mukherjee

Oread sp. *Calicnemia mukherjeei* Lahiri, 1976

Professor Durgadas Mukherjee (fl.1924–d.>1976) was an entomologist at Calcutta University. The author collected the paratypes (1973) and holotype (1974) from the Khasi Hills in Maghalaya, India. Lahiri's etymology says: *"The new species is named after the eminent entomologist, the late professor Durgadas Mukherjee of Calcutta University."*

Müller, L

Carnarvon Darner *Austroaeschna muelleri* **Theischinger**, 1982

Leonard Müller (b.1942) of Berowra, Sydney, NSW has participated in collecting trips across Australia with the author, including to northeastern New South Wales (1998) and Western Australia (1981). The etymology reads: *"I also wish to express my special gratitude to my friends Mr L Müller (Berowa) and Dr J.A.L. Watson (Canberra) who supported my work in many ways."* Theischinger has also named species after Leonard's wife Elke. (Also see **Leonard**, & **Elke**)

Müller, RA

Roland Albert Müller (1936-2016) (See **Roland Müller**, also see **Pistor**)

Muller

Flatwing sp. *Argiolestes muller* **Kalkman**, Richards & Polhemus, 2010

This is a toponym; it is named after the Muller Mountain Range, which was named after Major George Muller, so is an eponym once removed. Employed by the Dutch Government, he tried to explore Kalimantan east to west (1824) but died (1825) in the attempt, killed by Dayak people.

Mumford

Emerald sp. *Hemicordulia mumfordi* **Needham**, 1933

Dr Edward Philpott Mumford (1902–1977) was a British entomologist. Cambridge University awarded his BSc (1926) and Victoria College, Toronto, Canada awarded his MSc (1929). At the conclusion of WW2, he returned to England and continued his studies on biogeography at Jesus College, Oxford, where he was awarded his PhD (1948). He was put in charge of the Pacific Entomological Survey when it was formed (1927). It was funded for a 5-year period, through a cooperative agreement between the Hawaiian Sugar Planters' Association, the Association of Hawaiian Pineapple Canners, and the Bernice Pauahi Bishop Museum. The original director of the Survey was C F Baker (q.v.) from the Philippines. However, Baker's untimely death meant Mumford was put in charge. The survey focused on the Marquesas Islands where the holotype was collected. He worked most of the time in California, but he had an extended sojourn in Hawaii (1932–1933), when he and Dr Martin Adamson, who had been studying the fauna of the Marquesas Islands with him, worked on their accumulated collections for several months at the Bishop Museum. His health was undermined by the harsh climate and conditions of his fieldwork and he was forced to seek recuperation in the cool, dry climate of California, where he remained. After working with the Pacific Entomological Survey, he took up residence in Palo Alto and extended his studies, under the aegis of Stanford University, to include the world distribution of pathogenic parasites of man. These studies led him to become the principal promoter of

Stanford's Pacific Islands Research War Project and to contribute a number of publications to the project including *Preliminary Report on Parasitic and Other Infectious Diseases of the Japanese Mandated Islands and Guam* (1943). This was followed by his most significant work: *Manual on the Distribution of Communicable Diseases and their Vectors in the Tropical Pacific Islands* (1944). He joined the faculty of Dominican College, San Rafael, California, as Professor of Biology and Department Chairman until his retirement (1971), then as Research Scholar in Residence until his death. Needham did not include an etymology in his description but clearly this is the man he had in mind.

Muzon

Threadtail sp. *Peristicta muzoni* Pessacq & **Costa**, 2007
Threadtail sp. *Drepanoneura muzoni* von **Ellenrieder** & **Garrison**, 2008

Dr Javier Muzón (b.1961) is an Argentine aquatic entomologist and odonatologist also interested in systematics, biogeography and conservation biology. His BSc was awarded by Universidad Nacional de La Plata (1986) and his PhD by Universidad de Buenos Aires (1993) (his home town). He undertook research and teaching (1987–2004) at various institutions, as well as managing a number of projects. He became Head of the Entomology section of the Instituto de Limnología, Dr. Raúl A Ringuelet, National University of La Plata, Argentina (2004–2008). He is currently (since 2015) Director, Laboratorio de Biodiversidad y Genética Ambiental (BioGeA), Universidad Nacional de Avellaneda, Researcher of the Consejo Nacional de Investigaciones Científicas y Tecnológicas (CONICET), Argentina (Category: Independiente), Professor of Invertebrate Zoology II Arthropods (Universidad Nacional de La Plata) and Professor of the Masters in Entomology (Universidad Nacional de Tucumán. He was recently (2016) elected President of the Sociedad de Ondonatologia Latinoamericana (SOL). Among his recent written or co-written works are: *The Odonata (Insecta) of Patagonia: A synopsis of their current status with illustrated keys for their identification* (2014) and *Odonata Diversity and Synantrophy in Urban Areas: A Case Study in Avellaneda City, Buenos Aires, Argentina* (2016). He is author or co-author of nine new Odonata species.

Myers

Tepui Shinywing *Iridictyon myersi* **Needham** & Fisher, 1940

Dr John Golding Myers (1897–1942) was a British entomologist. His parents emigrated (1911) to New Zealand and he attended Wellington University, attaining his BSc and MSc after WW1 service in Europe. He was employed (1919–1924) as entomologist in the Biological Division of the New Zealand Department of Agriculture, where he worked on the cattle tick and other pests. He won a scholarship (1924) to Harvard which awarded his DSc (1925). After this he went to London and joined the staff of the Imperial Institute of Entomology. His first task was undertaking research in France, studying the enemies of the Pear Leaf-curling Midge then studying other pest species and their predators. The task included organising the breeding of parasites of pest species, which were exported as biological controls to Australia (which he visited), New Zealand and elsewhere. He went to Trinidad (1928) and then on to British Guiana and Suriname. He visited Venezuela and then Brazil (1932 & 1933), first discovering a parasitic fly (Amazon Fly), then collecting some for introduction to British Guiana sugar plantations to parasitize 'cane-borers'. In 1934 he joined the staff of the Imperial College of Tropical Agriculture in Trinidad. He was appointed economic botanist to the government of Anglo-Egyptian Sudan to survey economic opportunities in its southern province, Equatoria. He never finished his report because he was killed there in a road traffic accident. He wrote a book on Cicadas: *Insect Singers: A Natural History of the Cicadas* (1929). He collected the shinywing from Mount Roraima in British Guiana (1932).

Myrian

Firetail sp. *Telebasis myrianae* **Machado**, 2010

Myrian Morato Duarte (b.1969) is a biologist and talented scientific illustrator. Machado wrote: "*Named myrianae in honor of my good friend the biologist Myrian Morato Duarte who for 17 years has been illustrating my papers on dragonflies.*"

Myrth

Threadtail sp. *Neoneura myrthea* **Williamson**, 1917

Miss Myrth V Shafer (1890–1967) was the sister of Mary M Shafer (q.v.), and was honoured for her long service as bibliographer and assistant in the 'bug room'. (Also see **Mariana (Shafer)**)

Mzymta

Caucasian Goldenring *Cordulegaster mzymtae* **Bartenev**, 1929

This is a toponym; it is named after the type locality Mzymta River in the western Caucasus.

N

Nagamine

Phantomhawker ssp. *Planaeschna ishigakiana nagaminei* **Asahina**, 1988

Kunio Nagamine is a Japanese entomologist with a focus on Lepidoptera, Coleoptera and Odonata. He worked as a science teacher in junior high school. He collected the type in Japan (1986–1987) and collected other Odonata types, such as on Tokashiki Island (1993). Among his papers is: *Host-plants of butterflies in Okinawa Island. Satsuma* (1963) and *Cloning, phylogeny, and expression analysis of the Broad-Complex gene in the longicorn beetle Psacothea hilaris* (2014).

Nakamura

Skydragon sp. *Chlorogomphus nakamurai* **Karube**, 1995
Clinchtail sp. *Amphigomphus nakamurai* Karube, 2001

Shin-ichi Nakamura (b.1951) is a Lepidoptera researcher. He often helps the author in his studies. He collected the holotype male of the Clinchtail on Mount Tamdao, Vinh Phuh Province, Vietnam (1993).

Nancy

Nancy's Shadowdamsel *Ceylonosticta nancyae* Priyadarshana & Wijewardana, 2016

Nancy Elizabeth van der Poorten (née Chaprin) (b.1953) is a Canadian entomologist. She was awarded her BSc in botany by the University of Guelph (1977). She worked for a few years as a laboratory assistant and research assistant in the Department of Botany and the Department of Crop Science. Over the next 25 years, she worked at various (non-entomological) jobs, but maintained her interest in insects, particularly butterflies and odonates. She was an active member of the Toronto Entomologists' Association in Canada and its president (1998–2004). Since retirement, she has devoted her time to the study of the Lepidoptera (mostly butterflies but also moths) and Odonata of Sri Lanka. She has described two new species and one new subspecies of Odonata from Sri Lanka, and has authored or co-authored 13 papers on the Odonata of the island. She has mentored many upcoming odonatists in Sri Lanka, helping them with identification and writing manuscripts. She was co-author of the 2012 Red Listing project in Sri Lanka: *The Taxonomy and Conservation Status of the Dragonfly Fauna of Sri Lanka. In: The National Red List 2012 of Sri Lanka; Conservation Status of the Fauna and Flora*) and contributed a chapter on Odonata to the book *Horton Plains: Sri Lanka's Cloud-Forest National Park* (2012). She co-authored the book, *Dragonfly fauna of Sri Lanka: distribution and biology, with threat status of its endemics* (2014). She reviews papers for several journals and is president-elect (2015–2017) of the Worldwide Dragonfly Association. In addition to her dragonfly work, she has co-authored 9 papers on the butterflies of Sri Lanka with her husband, George, with whom she also co-authored the book *The Butterfly Fauna of Sri Lanka* (2016).

Natalia

Grisette sp. *Amphipteryx nataliae* **González-Soriano**, 2010
Pond Damselfly sp. *Mesamphiagrion nataliae* Bota-Sierra, 2013

Dr Natalia von Ellenrieder (b.1972) is a biologist, entomologist and odonatologist. The Universidad Nacional de La Plata, Argentina, where she was a teaching assistant (1994–2000), awarded her Licenciatura (equivalent to bachelor's and master's degrees) (1996) and PhD (1999). She was a postdoctoral fellow at the Staatliches Museum für Naturkunde Stuttgart, Germany (2000), a research associate at Los Angeles County Museum of Natural Science of (2001–2005) and an associate research biologist for Chico Research Foundation and California Department of Fish and Game (2004–2005). She was assistant researcher (2005–2007) then adjunct researcher (2007–2009) for CONICET. She is Senior Insect Biosystematist (since 2012) at the Plant Pest Diagnostics Branch, California Department of Food & Agriculture, having been associate (2010–2012). She has worked in the field in Argentina (1997, 1999, 2005–2009, 2012) Bolivia (2000), Chile (1999), Costa Rica (2013), Ecuador (2009), Guyana (2014), Mexico (2009), Namibia (2007), Suriname (2010), Vietnam (2014) and USA (Texas 2001, Arizona 2001, 2002, West Virginia 2002, California 2003 & Florida 2012) as has examined museum collections in South America, North America and Europe. She has written more than 110 papers or book chapters and co-wrote: *The dragonfly genera (Odonata: Anisoptera) of the New World: An illustrated and annotated key* (2006); *Dragonflies of the Yungas. A Field Guide to the Species from Argentina* (2007); and *Damselfly Genera of the New World. An Illustrated and Annotated Key to the Zygoptera* (2010). Her publications include descriptions of 74 new species and 5 new genera of Odonata.

Nathalia

Pond Damselfly sp. *Aceratobasis nathaliae* **Lencioni**, 2004
[Orig. *Telagrion nathaliae*]
Pond Damselfly sp. *Angelagrion nathaliae* Lencioni, 2008

Nathalia Rodrigues Lencioni (b.2000) is the 'beloved' daughter of the author Frederico Augusto de Atayde Lencioni (b.1970). (Also see **Lencioni**)

Navás

Navás' Darter *Trithetrum navasi* **Lacroix**, 1921
[A Fiery Darter]

Reverendo Padre Longínos-Blas Navàs Ferrer Homdedeu Marco (1858–1938), generally known as Longínos Navàs, was a Spanish entomologist. He entered a Jesuit seminary in France because they were banned in Spain at that time (1875). There he studied classical language and philosophy. He was at college in Manressa (ordained, 1890) and later (1892) taught natural history at Colegio del Salvador de Zaragoza. He was also responsible for the museum there where over the years he built up the insect collection and traded with other museums, particularly in Latin America. He took a degree in natural science (1904).

He was one of the founders of the Spanish Entomological Society (1918). He produced 620 papers in entomology and described over 3000 species and genera, mostly Neuroptera (c.120 dragonflies), although the majority of them are synonyms or *nomina dubia*. His collection suffered considerable damage and only around 400 specimens could be saved, much of what he sent to others is scattered around the world and is in need of revision. Later entomologists have found his descriptions imprecise, giving rise to much confusion. A biography has been published about him: *Longínos Navàs, científico jesuita* (1989).

Nawa

Demoiselle sp. *Mnais strigata nawai* Yamamoto, 1956
[JS *Mnais costalis* Selys, 1869]

Yasushi Nawa (1857–1926) was a Japanese entomologist. He was interested in insects even as a child and this continued through his later education and training. He attended the Gifu Agricultural Learning Centre (now the Gifu Prefectural Agricultural High School), staying on at the school as an assistant after he graduated (1882). The following year (1883) he discovered a new butterfly in the city of Gero. He went to Tokyo Imperial University (1886) and earned his junior high school teaching certificate in just six months, after which he began teaching at junior high and elementary schools in Gifu Prefecture. After a decade of teaching, he founded the Nawa Insect Research Centre (1896) where he studied protection of beneficial insects and combatting harmful ones, particularly termite control. Later (1919), the centre become the Nawa Insect Museum in Gifu, the first such museum in Japan.

Neblina

Clubskimmer sp. *Brechmorhoga neblinae* **De Marmels**, 1989
Pond Damselfly sp. *Tepuibasis neblinae* De Marmels, 1989
[Orig. *Aeolagrion neblinae*]

This is a toponym; it is a location in Pico da Neblina National Park in northern Brazil.

Needham

Phantom Darner *Triacanthagyna needhami* **Martin**, 1909
[JS *Triacanthagyna trifida* (Rambur, 1842)
Bluet sp. *Coenagrion needhami* **Navás**, 1933
[JS *Ischnura forcipata* **Morton**, 1907]
Needham's Skimmer *Libellula needhami* **Westfall**, 1943
Duskhawker sp. *Cephalaeschna needhami* **Asahina**, 1981

Dr James George Needham (1868–1956) was an American entomologist who popularised the study of Odonata and made major contributions to the understanding of their larval stages. His BSc and MSc degrees were awarded by Knox College, Illinois, where he also taught (1894–1896) and wrote (1895) *Elementary lessons in Zoology*. This impressed Professor J H Comstock and resulted in a scholarship to Cornell University (1896–1898), which awarded his PhD. He was then Professor of Biology at Lake Forest University (1898–1907), returning to Cornell (1907) as Assistant Professor of Limnology. When Professor Comstock retired (1914), he succeeded him as head of the Department of Entomology until his own retirement (1935). He travelled widely, including almost a year (1927–1928) studying insects in China, resulting in: *A manual of the dragonflies of China* (1930). He published c.350 scientific articles, educational papers, and textbooks, mostly on water-inhabiting insects, including *A Handbook of Dragonflies of North America* (1929) with Hortense Butler Heywood and *A Manual of the Dragonflies of North America (Anisoptera)* (1955) with Minter Westfall (q.v.). He described over 150 new species of Odonata (of which one quarter are synonyms) and named over 25 new genera. He also wrote poetry and philosophical writings. It has been said that teaching was his life's main calling. He used to say that nature should be studied outdoors, not from books. He enjoyed teaching young children on the wonders of nature, and at a time when women students were often ignored, he taught and mentored several and published with them. He is also remembered for the Comstock-Needham system of naming insect wing veins, based on Needham's now overturned 'pretracheation' theory.

Nehalenni

The damselfly genus *Nehalennia* **Selys**, 1850

Nehalenni was a Germanic or Celtic goddess associated with the mouth of the Rhine.

Nemésio

Flatwing sp. *Philogenia nemesioi* **Machado** 2013

Professor André Nemésio de Barros Pereira (b.1971) is a biologist, ornithologist and entomologist at the Universidade Federal de Uberlândia, Brazil. The Federal University of Minas Gerais awarded his bachelor of vertebrate zoology (1997) and his degree in biological science (1999). He later returned and took his masters in ecology, conservation and wildlife management (2004), which he followed with a PhD (2008). He was substitute Professor of Vertebrate Zoology there (2004–2006) and is now undertaking postdoctoral research in molecular genetics at the Federal University of São Carlos. He is a good friend of the author who also critically examined the manuscript of the paper in which the species was named. His interests include Euglossini (Orchid bees) and conservation. He collected the type in Peru (2012). He has written three books, articles, chapters and 65 papers (1996–2014) such as: *Orchid bees (Hymenoptera: Apidae) of the Brazilian Atlantic Forest* (2009). He has also edited or co-edited three journals. In his descriptions of new bee taxa, his sense of humour and love of football is revealed as he is fond of using unusual names such as: 'Atletico', 'Ronaldinho' and *Euglossa bazinga* in reference to the character Sheldon Cooper, of the television series 'The Big Bang Theory'.

Netta

Pretty Tigertail *Eusynthemis netta* **Theischinger**, 1999

Ms Annette 'Netta' Smith (b.1955) is an American biologist currently (2014) studying colour vision in a molecular biology laboratory. She is married to Dennis Paulson (q.v.) She has collected in seventeen US states so far (1993–2017), British Columbia (1997–2006), Mexico (1999-2006), Belize (1987), Costa Rica (2005), Peru (2002), Venezuela (2000), Australia (1988–2003), and southern Africa (1997) among other places. With her husband, she was the co-collector of this species at Mount Lewis, Queensland (1998). She now mostly photographs dragonflies while on collecting trips.

Newton, AH

Newton's Sprite *Pseudagrion newtoni* **Pinhey**, 1962

[A. Harlequin Sprite]

Dr Arthur Henry 'Harry' Newton (1885–1966) was a British physician and amateur entomologist. He sent 'a long series of specimens from Zululand' to Pinhey. The holotype series was taken at Nqutu, KwaZulu-Natal, South Africa (1961).

Newton (Dias dos Santos)

Sylph sp. *Macrothemis newtoni* **Costa**, 1990

Emerald sp. *Cordulisantosia newtoni* Costa & **Santos**, 2000

[Orig. *Santosia newtoni*]

(See **Santos**)

Newton Santos

Skimmer sp. *Elga newtonsantosi* **Machado**, 1992

(See **Santos**)

Nicéville

Common Blue Skimmer *Orthetrum nicevillei* **Kirby**, 1894

[JS *Orthetrum glaucum* (Brauer, 1865)]

Charles Lionel Augustus de Nicéville (1852–1901) was an entomologist particularly interested in Lepidoptera who was born in England and died in India. After schooling in Brighton, England, he left for India (1870) becoming a clerk in a Calcutta court, but (from 1871) he devoted all his spare time to entomology, during which time he made several trips to Sikkim and (1887) to the Baltistan glaciers. Among his published papers was one cataloguing 631 butterflies in Sikkim; others included butterflies found in Darjeeling, Buxa and Bhutan, areas contiguous with Sikkim state. He collected the type in Upper Burma. He was appointed as Imperial Entomologist (1901).

Nicola

Rock Skimmer sp. *Paltothemis nicolae* Hellebuyck, 2002

Nicole Hellebuyck née Chiasson was the wife of the author, Por Victor J Hellebuyck (1949–2006), who thanked his wife for her patience and support, especially during his numerous trips away from home. She is associated with the Université de Sherbrooke, Quebec, Canada.

Nielsen

Small Red Damsel *Ceriagrion tenellum nielseni* **Schmidt**, 1953

[JS *Ceriagrion tenellum* (De Villers, 1789)]

Dr Cesare Nielsen (1898–1984) was an Italian dentist, physician and amateur entomologist in Bologna. He studied medicine in Switzerland at the Universities of Bern and Geneva, qualifying as a dentist (1922). To practice in Italy, he needed a further medical degree, which was awarded by the University of Bologna (1925). Throughout his career, he continued to pursue his interest in entomology and dragonflies in particular, and became an honorary curator at The Bologna Entomological Institute, which enabled him to take part in scientific expeditions. A severe digestive disorder and surgical intervention led him to retire from dentistry (1966). He published about 20 papers on Italian and African dragonflies and wrote: *Odonati della Venezia Tridentina* (1932). Jointly Nielsen and Cesare Conci (q.v.) published his most significant book: *Odonata* in the *Fauna d'Italia* series (1956). He described four new odonate species, including the European *Somatochlora meridionalis* (Balkan Emerald) (1935). (Also see **Caesar**, **Cesare** & **Conci**)

Nietner

Nietner's Grappletail *Heliogomphus nietneri* **Hagen**, 1878
[Orig. *Gomphus? nietneri*]
Nietner's Shadowdamsel *Ceylonosticta nietneri* **Fraser**, 1931

Johannes Werner Theodor Nietner (1828–1874) was a Prussian born British entomologist who worked on the insects of Ceylon (now Sri Lanka). He worked at the botanical gardens in Paradeniya and collected plants and insects across Ceylon. He also purchased a small coffee plantation and ran it for more than a decade, which afforded him leisure time to collect and to develop part of his estate as a herbarium. He died of dysentery when embarking for a trip back to Europe. A number of beetles and many deciduous shrubs and liverworts are also named after him. He sent most of the insects he collected for study to institutions in Britain, France and Germany.

Nihar

Threadtail sp. *Elattoneura nihari* Mitra, 1995

Nihar Ranjan Mitra was the father of the author Tridib Ranjan Mitra (1942–2012).

Niisato

Paddletail sp. *Sarasaeschna niisatoi* **Karube**, 1998
[Orig. *Oligoaeschna niisatoi*]

Tatsuya Niisato (b.1957) is a Japanese entomologist whose focus is Coleoptera. He currently works at the Bio-indicator Company, Shinjuku, Tokyo. He wrote: *Two new taxa of the Japanese Clytini (Coleoptera, Cerambycidae)* (2005) and co-wrote: *A revision of the genus Amamiclytus Ohbayashi from Taiwan and the Ryukyu Islands (Coleoptera, Cerambycidae)* (2011) among other papers.

Noguchi

Genus *Noguchiphaea* **Asahina**, 1976

Yoshiko Noguchi (1926–1976) was a Japanese medical entomologist at the Department of Medical Entomology, National Institute of Health, Tokyo who was a faithful collaborator of Asahina. In the 1960s, she wrote or co-wrote a number of papers, mostly on the control of flies and mosquitos, including: *Ommatidial number as a diagnostic character for Japanese Autogenus Culex molestus* (1966). (Also see **Yoshiko**)

Nomura

Darter sp. *Sympetrum nomurai* **Asahina**, 1997

Dr. Shuhei Nomura (b.1962) is a Japanese zoologist who is Senior Curator at the Department of Zoology, Division of Terrestrial Invertebrates, National Museum of Nature and Science, Tokyo (NMNS) (called NSMT before 2007). He is also a vice-president of the Coleopterological Society of Japan (2013–2014). Kyushu University awarded his BS (1985), MS (1987) and PhD (1990) and he was assistant professor there (1990–1995). He moved to the National Science Museum, Tokyo (NSMT) as a curator in the Zoology Department (1995), becoming Senior Curator (1988). His research specialties are taxonomy, phylogeny and biodiversity of beetles (Insecta, Coleoptera) and he is also working on morphology of beetles' microstructures. He has given scientific names to more than 280 new taxa (genera, subgenera, species and subspecies) of beetles, most of which belong to the family Staphylinidae. He is one of the editors of: *The Insects of Japan*, an on-going series of books from the Entomological Society of Japan and is the author of more than one hundred original papers in scientific journals (1986–2014). He provided the material to Syoziro Asahina.

Nourlangie

Cave Duskhawker *Gynacantha nourlangie* **Theischinger** & **Watson**, 1991

This is a toponym; Nourlangie Creek is a locality in West Arnhem Land, Northern Territory, Australia.

Noval

Clubtail sp. *Tibiagomphus noval* Rodrigues Capitulo, 1985
[Orig. *Cyanogomphus noval*]

Nora Gomez (b.1957) and Valeria Rodrigues Capitulo (b.1983) are jointly honoured, being wife and daughter of the author Alberto Rodrigues Capitulo (b.1953).

Nurse

Pond Damselfly sp. *Rhodischnura nursei* **K J Morton**, 1907
[Orig. *Ischnura nursei*]

Lieutenant Colonel Charles George Nurse (1862–1933) of the Indian Service Corps was a naturalist, especially ornithologist and entomologist who divided his time between northwest India and Aden (which garrison was part of the Western Army Corps of India). A linguist with a sound knowledge of Russian and several Oriental languages, he rapidly advanced his career after being commissioned (1881). He went to India with his regiment (1881–1884) and was then in action in the Sudan Expedition (1884). Back in India he was promoted to First Lieutenant (1885) and, after service in Somalia (1890), he was promoted to Captain (1892). He became Major (1901) and Lieutenant Colonel (1907) and at retirement (1909) he was commander of the 33rd Punjabi Regiment. He was re-appointed from retirement and attached to the 3rd Battalion Bedfordshire Regiment in Flanders (1915). He had a lifelong interest in birds, collected butterflies and moths and then became interested in hymenoptera. At this time, Indian Hymenoptera were neglected. Nurse described many new species and added greatly to knowledge of the group. He sold part of his private collection of Lepidoptera to John James Joicey (1919) but the bulk was bequeathed to the BMNH. He wrote papers on birds, insects and about his travels such as: *A Journey Through Part of Somali-Land, Between Zeila and Bulhar* (1891) and *New and little known Indian Bombyliidae* (1922). A reptile is also named after him.

O

Oberthür

Jewelwing sp. *Atrocalopteryx oberthuri* **McLachlan**, 1894
[Orig. *Calopteryx oberthüri*]

René Oberthür (1852–1944) was a French entomologist with an interest in Coleoptera. He published a number of papers including: *Coleopterorum Novitates – Recueil spécialement consacré à l'étude des Coléoptères* (1883) and the longer *Etudes d'Entomologie* (1880). He worked in a print business founded and owned by his father François-Charles Oberthür (1818–1893), who was an amateur entomologist with a strong interest in Lepidoptera. René's brother was Charles Oberthür (1845–1924), who was also an entomologist with an interest in Lepidoptera. He too wrote several papers, including: *Faune entomologique armoricaine. Lépidoptères* (1912). McLachlan was sent the holotype in a collection made in western China by René Oberthür, who was his friend. His collections are conserved by MNHN (5 million specimens in 20,000 cases), Museum Koenig and Museo Civico di Storia Natureale de Genova.

Obiri

Cave Reedling *Indolestes obiri* **Watson**, 1979

This is a toponym; Obiri (Obirie) Rock is a cave habitat in Australia.

O'Brien

Pond Damselfly sp. *Homeoura obrieni* von **Ellenrieder**, 2008

Mark Francis O'Brien (b.1956) is an American entomologist. SUNY College of Environmental Science & Forestry awarded his MS in entomology and his MS in science was awarded by Syracuse University (1981). He has been Collections Manager, Insect Division, at the UMMZ since 1981. He is responsible for collection maintenance, curating, specimen preparation, loans, exchanges and Insect Division data resources. His expertise is insect natural history, biology and distribution of Hymenoptera and Odonata in Michigan. He served as editor of the 'Great Lakes Entomologist' for 10 years, and is the coordinator of the Michigan Odonata Survey. Among his published works are the papers: *An annotated list of the name-bearing types of species-group names in Odonata preserved in the University of Michigan Museum of Zoology* (2003) and *Odonatological History in Michigan – 1875–1996* (2009). The etymology reads: "*Named after my colleague Mark O' Brien, in gratitude for his manifold assistance to students interested in the rich dragonfly collection in UMMZ.*"

Odobenus

Damselfly sp. *Igneocnemis odobeni* **Hämäläinen**, 1991
[Orig. *Risiocnemis odobeni*]

Not an eponym but a reference to another animal. The posterior view of the ventral processes of the superior appendages gives an image just like that of the upper canine teeth of the walrus (*Odobenus*).

O'Donel

Clubtail sp. *Asiagomphus odoneli* **Fraser**, 1922
[Orig. *Gomphus odoneli*]
Duskhawker sp. *Gynacantha odoneli* Fraser, 1922
Threadtail sp. *Prodasineura odoneli* Fraser, 1924
[Orig. *Caconeura odoneli*]

H V O'Donel (b.c.1887–1936) was a naturalist and collector operating in Vietnam, then India (Bihar, Bengal & Sikkim; c.1913–c.1936). He wrote a number of articles published in the 'Journal of the Bombay Natural History Society', on birds, mammals and insects, such as: *Notes on the Chestnut-headed Shortwing* (1913), *Notes on the Burmese Ferret-Badger (Helictis personata)* (1916) and *The Indian Cuckoo* (1936). Fraser said of him (1933): "... *Mr H V O'Donel collected for me in the Duars, finding several new species*" (1933) and "*I hear with great regret of the early death of H V O'Donel in Calcutta. He contributed much of the material from Bengal and Bihar described in this work; not only Entomology but also Ornithology has lost a valuable field-worker who will be hard to replace.*" (1936) (Also see **Dorothea (O'Donel)**)

Odzala

Congo Threadtail *Elattoneura odzalae* **Aguesse**, 1966
[Orig. *Disparoneura odzalae*]

This is a toponym; d' Odzala, in Congo-Brazzaville is the type locality.

O'Farrell

Genus *Tonyosynthemis** **Theischinger**, 1995

Slender Tigertail *Tonyosynthemis ofarrelli* Theischinger & **Watson**, 1986
[Orig. *Synthemis ofarrelli*]

Professor Anthony 'Tony' Frederick Louis O'Farrell (1917–1997) was an Australian entomologist and zoologist born in India to Irish parents who moved to England. Imperial College of Science awarded his first-class BSc (1939) and he was awarded the Forbes Memorial Medal in Entomology. Declared unfit for active service in WW2, he joined the Home Guard at the same time as being Entomologist with the UK Ministry of Agriculture & Food (1940–1946). He was posted to Belfast where he met his wife being married in 1944. They emigrated to Australia (1947), where he was appointed Lecturer (1947–1955) then Professor of Zoology (1955–1982) at New England College (later University), Armidale, New South Wales. He was very interested in limb regeneration in insects, particularly cockroaches and was always interested in Odonata, collecting in many parts of eastern Australia, although he only published 15 papers on them. Perhaps his best work was the Odonata chapter in *The Insects of Australia* (1970). His is co-author of four new species and one subspecies. He founded a laboratory at Arrawarra Headland that was later named after him. The dedication for the tigertail reads "…*in honour of his 70th birthday (9 January, 1987) and in recognition of the great contribution he has made to the knowledge of the Australian Odonata.*" The etymology for the genus reads for: "*AFL (Tony) O'Farrell (1917–1997) and Dr JAL (Tony) Watson (1935–1993), two unforgettable friends and outstanding odonatologists.*" (Also see **Tony (O'Farrell)**)

*Jointly with Dr JAL Watson (q.v.)

Ogata

Shadowdamsel sp. *Sinosticta ogatai* **Matsuki** & **Saito**, 1996

[Orig. *Drepanosticta ogatai*]

Seiji Ogata (1948–2012) was a Japanese amateur lepidopterist living and working in Hong Kong. He collected the type series there (1994–1995).

Oguma

Clubtail sp. *Trigomphus ogumai* **Asahina**, 1949

[Orig. *Gomphus ogumai*]

Kan (Mamoru) Oguma (1886–1971) was a Japanese entomologist who was the founding president of the National Research Centre of Genetics (1949). He published several papers on Japanese dragonflies (1913–1932), including a chapter on Odonata in the book: *Iconographia Insectorum Japonicorum* (1932). He described 24 new Odonata species, two thirds of which are valid, and named the genus *Nihonogomphus*.

Ojisan

Skydragon sp. *Watanabeopetalia ojisan* **Karube**, 2013

Ojisan is the familiar name (nickname) of Dr Kouichi Matsumoto. (See **Matsumoto**)

Oláh

Skimmer sp. *Diplacina olah* Kovács & **Theischinger**, 2015

Professor Dr János Oláh (b.1942) is a noted Hungarian expert of world Trichoptera (caddisflies) and a supporter of Tibor Kovács' research in Batanta in Indonesian Papua New Guinea. His interests cover conservation, zoology, landscape ecology, minerology and ornithology and he was a co-founder of the birdwatching tour company 'Sakertours'. He has written fifteen books and over 300 scientific papers, including ones co-authored with Kovács, such as *New species and records of Balkan Trichoptera II* (2013) and has described many species.

Olive

Delicate Tigertail *Choristhemis olivei* **Tillyard** 1909

[Orig. *Synthemis olivei*]

Edmund Abraham Cumberbatch Olive (1844–1921) was an English born Australian businessman, auctioneer and horse and cattle salesman during the early days of the Palmer River gold rush (1875). He lived out his life in Cooktown. In his later natural history collecting he relied heavily on the knowledge and assistance of Aboriginal people, particularly one man who collected for him, known as Billy Olive. In his etymology, Tillyard says he named the species thus: "*Dedicated to my friend, Mr. E.A.C. Olive, of Cooktown.*" Two birds and four reptiles are also named after him.

Olsufyev

Olsufieff's Rockstar *Tatocnemis olsufieffi* **Schmidt**, 1951

Olsufieff's Sprite *Pseudagrion olsufieffi* Schmidt, 1951

Professor Dr Grigoriy Vasilyevich Olsufyev (Gregor von Olsufieff) (1875–1957) was a Russian parasitologist. He collected in Madagascar (1930s), including the type (February 1934) and other species that Schmidt described (1950s), such as *Platycnemis sanguinipes* (1931). He wrote mostly about pests and parasites, particularly ticks as disease vectors and treatments, such as: *A study on flies parasitic on the Asiatic Locust (Locusta migratoria, L.) and their superparasites* (1929) and *Sensitivity to macrolide antibiotics and lincomycin in Francisella tularensis holarctica* (1978). He is named in other taxa, such as *Tabanus olsufievi* Bogatshev and Samedon (1949).

Olthof

Fineliner sp. *Teinobasis olthofi* **Lieftinck**, 1949

Jan Olthof (d.1945) was a Dutch entomologist in Java, Indonesia. He took part in the third Archbold expedition (1938–1941) to the type locality, New Guinea, as assistant to Dr L J Toxopeus. He became Assistant Entomologist, then Curator of Entomology at the Buitenzorg Museum and was described as a young but zealous collector and breeder of Macrolepidoptera. He died from the consequences of imprisonment (1945) during the Japanese occupation. His collections are in the Bultenzorg Museum.

Ono

Sylvan sp. *Coeliccia onoi* **Asahina**, 1997

[JS *Coeliccia cyanomelas* Ris, 1912]

Dr Hirotsugu Ono (b.1954) is an arachnology researcher who is Senior Curator (since 1983) in the Department of Zoology, Division of Terrestrial Invertebrates at National Museum of Nature & Science, Tokyo. His first degree was in law (1976) at Gakushuin University, and his doctorate was from Kyoto University. Before joining the staff in Tokyo, he was a research assistant (1977–1983) at Gutenberg University, Mainz, Germany. He collected the holotype on the second National Science Museum expedition to Vietnam (1997). Among his many papers is the monograph: *A revisional study of the spider family Thomisidae (Arachnida, Araneae) of Japan* (1988) and the longer work *The Spiders of Japan* (2009).

Ophelia

Pixie sp. *Brachygonia ophelia* **Ris**, 1910

Ophelia is a fictional character in the play 'Hamlet' by William Shakespeare.

Ordos

Bluetail sp. *Ischnura ordosi* **Bartenev**, 1911

This is a toponym; it is named after the type locality Ordos in Inner Mongolia.

Oreades

Demoiselle sp. *Matrona oreades* Hämäläinen, Yu & Zhang, 2011

In Greek mythology, the Oreades or Orestiad were mountain nymphs that are known according to their dwelling, i.e. the Idae were from Mount Ida, Peliades from Mount Pelia etc.

Orr

Jewel sp. *Libellago orri* **Dow** & **Hämäläinen**, 2008
Flatwing sp. *Celebargiolestes orri* **Kalkman**, 2016

Dr Albert George Orr (1953) is an Australian entomologist, currently (2017) Associate Editor of 'Odonatologica' and Subject Editor of 'International Journal of Odonatology'. His BSc (entomology, mathematics) was awarded by the University of Queensland (1974), his Diploma in Education by Charles Stuart University (1982), his BSc (Hons) by the School of Australian Environmental Studies, Griffith University (1980) and his PhD by Griffith University (1988). His doctoral thesis was entitled: *Mate conflict and the evolution of the sphragis in butterflies*. He is Adjunct Research Fellow, Griffith University and formerly lecturer in biology, University of Brunei Darussalam (1990–1996). He has written papers on diverse entomological topics, focussed mainly on Macrolepidoptera and Odonata. He has described or co-described 37 new odonate species as well as four new butterflies. He has also authored or co-authored several award-winning books including: *A guide to the dragonflies of New Guinea, their identification and biology* (2003), *The butterflies of Australia* (2010), *Field Guide to damselflies of New Guinea* (2013) and *Field Guide to dragonflies of New Guinea* (2016). He also wrote: *Dragonflies of Peninsular Malaysia and Singapore* (2005) and *Metalwing Demoiselles of the Eastern Tropics* (2007), He is a Fellow of the Royal Entomological Society. His most important contributions have been in the areas of reproductive biology of butterflies, moth biodiversity, behaviour of chlorocyphid damselflies and the promotion of popular interest in entomology. His semi-popular guide books have stimulated local interest in Odonata in Southeast Asia and New Guinea and he is presently collaborating with Vincent Kalkman (q.v.) on a guide to Odonata of Bhutan. He is a noted entomological artist, producing both technical illustrations and depictions of living insects in their natural habitats.

Ortiz

Clubtail sp. *Diastatomma ortizi* Compte Sart, 1964
[JS *Diastatomma selysi* Schouteden, 1934]

Eugenio Ortiz de Vega (1919–1990) was a Spanish naturalist. Madrid University awarded his masters (1948) and his doctorate (1956). He was a research fellow at the Max Planck Institute in Germany while completing his PhD and an associate professor on Arthropods at Madrid (1952–1959) and then Professor of Genetics there (1959–1968). He was also a research worker in the Consejo Superior de Investigaciones Cientificas in Madrid (1959–1985) and became Director of the Museo Nacional de Ciencias Naturales Madrid (1975–1984). He collected the holotype in Spanish Guinea (Equatorial Guinea) when part of an expedition organised by the museum (July 1948) along with fellow naturalist Emiliano Aguirre.

Osvalda

Sombre Dropwing *Trithemis osvaldae* D'Andrea & Carfi, 1997

Osvalda Andrei (b.1949) is an Italian associate professor of the University of Siena; in private life, she is the wife of the first author Marcello D'Andrea (b.1956). After graduating in Greek language and literature at the University of Florence, she became a research fellow at Siena (1975–1981) where she became assistant professor (1981–2004) and Associate Professor (from 2005) of the History of Christianity. Her main publications concern this area of study, principally Christian chronography and historiography.

Othello

Black Knight *Camacinia othello* **Tillyard**, 1908

Othello is, of course, the title character of one of Shakespeare's tragedies who, mistakenly believing he has caused his wife's death, kills himself. Significantly, given the etymology, the subtitle of the play is 'The Moor of Venice'. Tillyard (1908) stated that the Odonata name was chosen to reflect was the primary character of Shakespeare's hero-its dark pigmentation on the wings.

Otto

Threadtail sp. *Epipleoneura ottoi* Pessacq, 2014

Raúl Adolfo 'Otto' Pessacq (b.1942) is the father of the author Pablo Pessacq (b.1973).

Oubaji

Clubtail sp. *Gomphus ubadschii* **Schmidt**, 1953

Hamdi Oubaji from Syria collected the type.

Ouvirandra

Striped Junglewatcher *Oreoxenia ouvirandrae* Förster, 1899
[JS *Neodythemis hildebrandti* Karsch, 1889]
Pied-spot sp. *Hemistigma ouvirandrae* Förster, 1914

These species from Madagascar were named after the Madagascarian aquatic plant *Ouvirandra fenestralis* (Madagascar Laceleaf, Lattice Leaf or Lace Plant), the present scientific name of which is *Aponogeton madagascariensis*. This plant is characterized by its uniquely porous leaves. Its native Malagasy name is 'ouvirandrano', literally meaning 'yam of the water'. The plant is popularly sold and it is endangered in the wild, where it inhabits fast flowing, shady streams. The thorax of the *Oreoxenia* has a reticulate colour pattern, the reason for the name of this species.

Overlaet

Clubbed Cruiser *Phyllomacromia overlaeti* **Schouteden**, 1934
[Orig. *Macromia overlaeti*]

François 'Frans' Guillaume Overlaet (1887–1956) was a Belgian entomologist. He collected (1930s–1950s) in the Belgian Congo (now Congo Democratic Republic) for the Belgium

Herbarium and Musée Royal de l'Afrique and Musée du Congo. He wrote among other works: *Formes nouvelles ou peu connues de Nymphalides africains* (1952).

Owada

Skydragon sp. *Chloropetalia owadai* **Asahina**, 1995
[Orig. *Chlorogomphus owadai*]
Flatwing sp. *Rhipidolestes owadai* Asahina, 1997
Phantomhawker sp. *Planaeschna owadai* **Karube**, 2002

Dr Mamoru Owada is a zoologist and Curator at the Department of Zoology, National Natural History Museum, Tokyo. He made an expedition to Vietnam (1995), where he collected the type. The author wrote "...*I thank Dr. M. Owada and M. Tomokuni of the National Science Museum (Nat. Hist.), Tokyo, who gave me an opportunity to study their valuable specimens.*" He has published a great many papers, particularly on moths such as co-writing: *Notes on some chalcosiine moths from the Indo-Chinese peninsula, with descriptions of two new species and two new subspecies* (1999).

P

Pagenstecher

Jewel sp. *Rhinocypha pagenstecheri* **Förster**, 1897

Dr Arnold Andreas Friedrich Pagenstecher (1837–1913) was a German entomologist, particularly interested in Lepidoptera. He studied medicine at Wurzburg, Berlin and Utrecht, then worked as an assistant to his cousin Alexander Pagenstecher (1828–1879) at his ophthalmology clinic in Wiesbaden where established a general practice there (1863). He became (1876) a *Sanitätsrat* (medical officer), followed in 1896 by an appointment as *Geheimen Sanitätsrat* (privy medical counsellor). He was also Director of the Naturhistorischen Museum in Hamburg. He is best known for his extensive studies of Lepidoptera species of the Malay Peninsula and he wrote: *Die geographische Verbreitung der Schmetterlinge* (1909) looking at the causes of geographical distribution of Lepidoptera. He provided the Jewel species from Sumba Island (collected by H Fruhstorfer) for the author to study.

Pahyap

Satinwing sp. *Euphaea pahyapi* **Hämäläinen**, 1985

Pah-yap Kamnerdratana (since 1991, his first name has been Songphol) (b.1932) was, at the time the eponym was published, Senior Lecturer of Forest Entomology at Kasetsart University (Bangkok). He accompanied the author during his first four visits to Thailand collecting dragonflies. All the chiefs of the National Parks and Wildlife Sanctuaries they visited were either his friends or students. This greatly facilitated the field work and made some of the visits unforgettable to the 'farang' (foreigner). The satinwing was found in Khao Phanom Bencha in Krabi (1982), a place very few foreigners had visited before.

Pamela

Bluetail sp. *Ischnura pamelae* **Vick** & **Davies**, 1988
Tigertail sp. *Synthemis pamelae* Davies, 2002

Pamela Tobin (d.1988) was author Allen Davies' secretary and companion and with him co-wrote the two-volume: *The Dragonflies of the World: a systematic list of the extant species of Odonata* (1984–1985). She had accompanied Davies when these new species were found in New Caledonia (1981 & 1983) and was the first person to find a specimen of the Tigertail. (Also see **Allen (Davies)** & **Davies**)

Paprzycki

Shortwing sp. *Perissolestes paprzyckii* **Kennedy**, 1941

Pedro Paprzycki (1893–1959) was a commercial collector from Poland. He collected in Peru for several years (1936–1942), living in Satipo.

Paraense

Firetail sp. *Telebasis paraensei* **Machado**, 1956

Professor Dr Wladimir Lobato Paraense (1914–2012) was a Brazilian physician, teacher and zoologist. He graduated in medicine at the School of Medicine and Surgery of Pará (1931–1934) and School of Medicine of Recife (1934–1937). He began his career specialising in pathological anatomy at the University of São Paulo School of Medicine, then joined the Oswaldo Cruz Institute (IOC) (1942). He then went to the Research Center in Belo Horizonte (1952–1954), studying skin diseases and was there again some years later (1960). Around this time, he began to study freshwater snails, both in relation to the diseases they carry and ordering their systematics. He discovered several new species and became a leading malacologist and continued that research for the rest of his life. He was Professor of Parasitology in Brasilia (1968–1976). He published over 200 scientific papers and contributions to longer works. He is also honoured in the names of a flatworm, a fly and a protozoon.

Pareci

Firetail sp. *Telebasis pareci* **Machado**, 2010
[JS *Telebasis lenkoi* Machado, 2010]

The name refers to the Pareci indigenous people that inhabit the region of Utiariti in the State of Mato Grosso where the holotype was collected.

Pariwono

Shadowdamsel sp. *Protosticta pariwonoi* **Van Tol**, 2000

Scipio S Pariwono (formerly Liem Swie Liong) (b.1919) was an Indonesian guide and collector. Born in Gorang Gareng, Eastern Java, he mainly lived in Bogor, Indonesia. He was an assistant to A M R Wegner at Nongkodjadjar and A Diakonoff in Bogor, Java. Through numerous field trips he contributed significantly to knowledge of the Odonata of Indonesia, especially of Java, Borneo and Sulawesi. He made collecting and reconnaissance trips to Sulawesi with M A Lieftinck (q.v.) (1982) and with J van Tol (q.v.) (1989), where he collected some of the paratypes of this new species (1983).

Parzefall

Clear-tipped Amberwing *Perithemis parzefalli* **J Hoffmann**, 1991

Professor Dr Jakob Parzefall (b.1938) is a German biologist at the Zoological Institute of the University of Hamburg. His interests range from plant and animal science to anthropology, genetics and speleology. He has written or co-written a number of scientific papers, and contributions to longer works such as *Behavioural Aspects in Animals Living in Caves* in *The Natural History of Biospeleology* (1992). He was the author's teacher and inspired him to make his first trip to Peru.

Pasquini

Ethiopian Threadtail *Elattoneura pasquinii* Consiglio, 1978

Professor Pasquale Pasquini (1901–1977) was Professor of Zoology and Comparative Anatomy and Fellow of the Accademia Nazionale dei Lincei in Rome. He wrote numerous papers on zoology and morphology, especially experimental morphology, as well as several more general works such as: *General Biology* (1940) and *Animal Life* (1971).

Pataxo

Firetail sp. *Telebasis pataxo* **Machado**, 2010

The species is named after a people; the Pataxo Indians inhabit the region of Bahia State where the holotype was collected.

Patricia (Beldade)

Damselfly sp. *Idiocnemis patriciae* **Gassman** & Richards, 2008

Dr Patricia Margarida do Ó de Oliveira Beldade (b.1972) is Assistant Professor of Biology at Leiden University. Her degree in biology is from the University of Lisbon and the University of Leiden awarded her PhD in evolutionary and developmental genetics. She has already published around 30 papers. Her research focuses on genetic and developmental basis of variation in adaptive phenotypes. The etymology reads thus: "*It is the first author's pleasure to name the new species after his dear colleague Dr Patricia Beldade of Leiden University.*"

Patricia (Tillyard)

Waterfall Redspot *Austropetalia patricia* **Tillyard**, 1909

[Orig. *Phyllopetalia patricia*]

Patricia 'Pattie' Tillyard MBE née Cruske (1880–1971) was the author's wife (1909). Born in England, she attended Newnham College, Cambridge, where she completed her natural sciences (botany) tripos with second-class honours because no Universities awarded degrees to women at that time – she was awarded a belated MA (Cantab) when Cambridge changed its policy (1921). She became Headmistress of Hitchen Girls Grammar Schools. She studied for a diploma in health then spent three months in the Nile Delta with her brother before moving to Australia to marry Tillyard. Throughout the rest of her life she was a community leader and campaigner for women's rights. Fiercely independent, she only relinquished her driving licence at the age of 89. (Also see **Tillyard**)

Patricia

Spreadwing sp. *Lestes patricia* **Fraser**, 1924

Patricia is a feminine forename, but not a specified one. The author gave no etymology, so it is not possible to know for sure if this was named after a particular person. However, the lack of a Latin ending suggests it is not an eponym, as does Fraser's frequent use of apparent forenames. On the other hand, Fraser often failed to identify persons honoured in his etymologies. Fraser collected the damselfly in Coorg (aka Kodagu) Karnataka, India (June 1923).

Paul, E

Orange-tipped Yellowtail *Allocnemis pauli* **Longfield**, 1936
[Orig. *Chlorocnemis pauli*]

Eric Paul was a coffee plantation owner from Buzirasagam, Uganda. He and Jean Paul had a coffee farm on the edge of Kibale Forest, but they had turned their large comfortable house into a guesthouse. Longfield stayed there for 10 days (February–March 1934) and collected this damselfly in Kibale Forest.

Paul (Preuss)

Greater Double-spined Cruiser *Phyllomacromia paula* **Karsch**, 1892
[Orig. *Macromia paula*]

Paul Preuss (1861–1926). (See **Preuss**)

Paulian

Genus *Paulianagrion* **Fraser**, 1941
[JS *Pseudagrion Selys*, 1876]

Comoro Islander *Nesolestes pauliani* Fraser, 1951
Rusty Junglewatcher *Neodythemis pauliani* Fraser, 1952

Dr Renaud Maurice Adrien Paulian (1913–2003) was a French zoologist considered to be one of the greatest European entomologists of the 20th century. He was particularly interested in Coleoptera. He was Deputy Director of the Institut de Recherche Scientifique de Madagascar (1947–1961), then became Director of the Institut Scientifique de Congo-Brazzaville (1961–1966) and head of the local university. He was Head of the Université d'Abidjan in the Ivory Coast (1966–1969) before returning to France where he became Rector of the Academy of Amiens and then of the Academy of Bordeaux. He was elected correspondent of the Academy of Sciences (1975). He initiated (1956) the important series, *Faune de Madagascar*, over 90 volumes of which have been published. He wrote over 350 papers and a number of books, including: *Madagascar, un sanctuaire de la Nature* (1981). He sent Fraser material to study and collected the holotype of *Neodythemis*. Among other taxa, two reptiles, two amphibians, a mammal and a bird are also named after him. (Also see **Renaud**)

Paulina

Sooty Saddlebags *Tramea paulina* **Förster**, 1910
[JS *Tramea binotata* Rambur, 1842]

Etymology was not given. However, since the species had been collected in Sao Paulo, the name may refer to this locality. Supporting evidence to this speculation that it is a toponym is that, in the same paper, Förster named a form of *Uracis fastigiata* as *machadina* that had been collected in Rio Machados.

Paulinus

Cape York Tiger *Ictinogomphus paulini* **Watson**, 1991

Paulinus (d.644) was the first Archbishop of York (627–633). He was sent by the Pope to the Kingdom of Kent (601) as a missionary to convert the Anglo-Saxons from paganism. The Australian species may have been dedicated because its "'...*distribution appears to be confined to the northern part of Cape York Peninsula*".

Paulo

Darner sp. *Rhionaeschna pauloi* **Machado**, 1994
[Orig. *Aeshna pauloi*]

Paulo Augusto Ribeiro Machado (b.1968) is the son of the author. The etymology says: *"This beautiful species is dedicated to my companion of odonatological excursions, my son, Paulo**. (Also see **Machado**)

Paulson

Paulson's Knobtail *Epigomphus paulsoni* **Belle**, 1981
Scarlet Pigmyfly *Nannophya paulsoni* **Theischinger**, 2003
Pond Damselfly sp. *Phoenicagrion paulsoni* von **Ellenrieder**, 2008
Tropical King Skimmer sp. *Orthemis paulsoni* von Ellenrieder, 2012

Dr Dennis Roy Paulson (b.1937) is an American zoologist and naturalist and a world authority on Odonata. His BSc (1958) and PhD (1966) in zoology were both awarded by the University of Miami, and he was assistant curator of their vertebrate research collection (1954–1964). He became an instructor in zoology at the University North Carolina (1964–1965), then US Public Health Service fellow there (1966). After that he was research associate (1966–1969, 1974–1976) and assistant professor (1969–1974) at the Department of Zoology, University Washington, then Affiliate Curator of Vertebrates at its Burke Museum (1976–1990). He was Director, Slater Museum of Natural History, University of Puget Sound (1990–2004), and Emeritus Director since then. He has also been an instructor for the Organization for Tropical Studies, Universidad de Costa Rica (1967, 1969, 1970, 1975) and Visiting Instructor at The Evergreen State College (1976–1977). He has collected Odonata widely in the New World tropics from Mexico to Argentina, with special attention to Costa Rica, as well as in Africa, Asia and Australia. He wrote: *Dragonflies and Damselflies of the West* (2009) and *Dragonflies and Damselflies of the East* (2011) and books on other taxa such as *Shorebirds of the Pacific Northwest* (1993) and *Shorebirds of North America: The Photographic Guide* (2005). He has published 52 papers on Odonata, among which six new species are described, 37 papers on birds, four papers on reptiles, and one on beetles. A lizard, a frog, two mammals and two snails are also named after him.

Pavie

Common Spineleg *Merogomphus pavici* **Martin**, 1904

Auguste Jean Marie Pavie (1847–1925) was a French colonial civil servant, explorer and diplomat who was instrumental in establishing French colonial control over Laos. He was a civil servant in Cambodia and Vietnam before becoming the first Governor-General for the colony of Laos (1885). He joined the army at 17 (1864) and was posted to Indo-China in the infantry (1869) but recalled to fight in the Franco-Prussian War, during which he reached the position of sergeant major, then (1871) returned to Cambodia in charge of a small telegraph office. This posting allowed him to develop a deep understanding of the local culture and language. He 'went native' and wandered dressed in local garb, recording everything he saw. Brought to the attention of the governor (1879), he was entrusted with a five-year mission to explore the Gulf of Siam. These 'Missions Pavie', which were extended

(1879–1895), were so successful that he was appointed as vice-consul in Luang Prabang, then consul (1889) and eventually consul general (1891). The missions gathered vast amounts of scientific data and specimens, from archaeology to entomology. On his return to France, he wrote a multi-volume work: *La mission Pavie, A la conquête des coeurs* and *Contes du Cambodge, du Laos et du Siam* (1898–1921). Other taxa, including insects, gastropods and at least one fish are also named after him.

Paya

Threadtail sp. *Neoneura paya* **Calvert**, 1907

The species is named after a people; the Paya (now known as Pech) are an Indian tribe in northeast Honduras.

Peacock

Flatwing sp. *Philogenia peacocki* Brooks, 1989

Harold J Peacock (d.1990) was a Costa Rican landowner. He *"...very kindly allowed his ranch to be purchased by Guanacaste National Park"* where the species was first found. (See also **Baltodano, Burgos, Echeverri**).

Pearson, DL

Clubtail sp. *Peruviogomphus pearsoni* **Belle**, 1979

David Leander Pearson (b.1943) is an American ornithologist and ecologist. The Pacific Lutheran University, Tacoma, Washington awarded his bachelor's degree (1967), Louisiana State University his masters (1969) and University of Washington his PhD (1973). He was Assistant Professor of Zoology at Washington University (1973–1974) then Assistant Professor of Biology at Pennsylvania State University (1974–1980) and Associate Professor of Biology there (1980–1988) then becoming Research Professor of Biology at Arizona State University (1988–2008). He has published around 100 papers, many of which he has presented at international meetings. He collected the holotype on the lake edge at Limoncocha, Napo Province, Ecuador (October 1971).

Pearson, J

Rivulet Tiger *Gomphidia pearsoni* **Fraser**, 1933

Dr Joseph Pearson (1881–1971) was an English born zoologist and marine biologist. He was educated in Liverpool and attended Victoria University that awarded his BSc (1902). He became Naturalist at the Ulster Fisheries and Biology Association (1903) and later (1903–1910) lectured at Liverpool and Cardiff Universities. Manchester University awarded his DSc (1922). He then became Director of the National Museum of Colombo, Ceylon (now Sri Lanka) (1910–1933) and taught biology at Ceylon Medical College (1910–1921) (apart from military service for two years in France during WW1) and zoology at University College, Colombo (1921–1924) and he was also Director of the Department of Fisheries for Ceylon (1920–1933). He then became Director of the Tasmanian Museum (1934) and a member of the Tasmanian Fisheries Board. A mammal is also named after him.

Pecchioli

Blue Featherleg *Platycnemis pennipes pecchioli* **Selys**, 1863
[JS *Platycnemis pennipes* (Pallas, 1771)]

Vittoria (Victor) Pecchioli (1790–1870) was an Italian geologist and malacologist. He made an important collection of molluscs, now in the Museo di Storia Naturale dell'Università di Firenze, Italy. He collected the type in Italy too. He was also a noted translator and translated the works of Sir Walter Scott into Italian.

Pechuman

Clubtail sp. *Epigomphus pechumani* **Belle**, 1970

Dr Laverne 'Verne' Leroy Pechuman (1913–1992) was an American entomologist. He took a year out after secondary education (1930), then attended Cornell University, which awarded his bachelor's degree in entomology (1935). Almost immediately he was appointed an assistant in entomology for the Dutch elm disease investigation being conducted jointly by the Departments of Entomology and Plant Pathology. His continued study at Cornell led to his masters (1937) and doctorate (1939). He started work for the Chevron Chemical Company as a field representative (1939–1945), branch manager (1945–1947), district manager (1947–1961) and finally Senior Research Scientist (1961–1962). He returned to Cornell's Entomology Faculty as an associate professor (1962–1972) and Curator of the insect collection, becoming Professor of Entomology (1972–1982) and afterwards Professor Emeritus. His special area of interest was in blood-sucking flies. He collected mainly in New York State. He wrote or co-wrote around 100 papers and monographs and described 22 species; 25 other insects are named after him. He died of cancer.

Pedro

Emerald sp. *Neocordulia pedroi* **Costa**, Carriço & Santos, 2010

Pedro Martins Costa Mattos (b.2007) is the grandson of the first author Dr Janira Pedreira Martins Costa (b.1941). (See **Costa**)

Peitho

Jewel sp. *Rhinocypha peitho* **Hämäläinen**, 2017

In Greek mythology, Peitho was a goddess of persuasion and seduction.

Pelops

Jewel sp. *Rhinocypha pelops* **Laidlaw**, 1936

In Greek mythology, Pelops was a king of Pisa in Peloponnesus.

Pemón

Flatwing sp. *Heteragrion pemon* **De Marmels**, 1987

Named after the Pemón Indigenous people, whose native land is in the famous Tepuis (table mountains) in Venezuela. The damselfly was found on Gran Sabana highland plateau on which these mountains stand.

Pendlebury

Slendertail sp. *Leptogomphus pendleburyi* **Laidlaw**, 1934
Coraltail ssp. *Ceriagrion fallax pendleburyi* Laidlaw, 1931
[Orig. *Ceriagrion pendleburyi*]

Henry Maurice Pendlebury (1893–1945) was a British entomologist and lepidopterist. He was an army Captain during WW1 (1914–1919) and during WW2 the Japanese occupational army imprisoned him in Changi Prison. He was Curator of the Federated States of Malaya Museum, Kuala Lumpur and collected across Malaya, including the clubtail holotype. He co-wrote, among other works, *The Butterflies of the Malay Peninsular* (1934). He is also honoured in the names of other insects such as a fly and also in the name of a bat, both of which he collected. Sadly, he died in Bangalore on route for England from the POW camp in Singapore.

Pereira

Threadtail sp. *Epipleoneura pereirai* **Machado**, 1964

Francisco Silvério Pereira (1913–1991) was a Brazilian priest and entomologist. His specialist interest was Scarabid beetles and he collected them in Brazil and other parts of Latin America. He worked at the Museu de Zoologia, São Paulo and at the Instituto Biológico there (1960–1983). He published many papers describing new genera and species of insects, especially *Scarabaeidae*. He was also a friend of the author and accompanied him on many collecting trips.

Péringuey

Rock Malachite *Ecchlorolestes peringueyi* **Ris**, 1921
[Orig. *Chlorolestes peringueyi*]

Dr Louis Albert Péringuey (1855–1924) was a French entomologist and naturalist. He left France (1879) for South Africa, where he became a scientific assistant at the South African Museum (1884), was Curator in charge of the Invertebrates Collection (1885), and became the museum's Director (1906–1924). He dropped dead while walking home from the museum. Two reptiles, two birds and a fish are also named after him. Ris wrote: "*The origin of the present paper is a request by Dr. L. Peringuey to the author to write a paper which would help the resident entomologist to get a reliable knowledge of the South African dragonflies. In 1908, the writer had published… …an annotated catalogue of the fauna herein discussed, and this catalogue, in the opinion of Dr. Peringuey, would have to be modified for the purpose above mentioned.*"

Peris

Guinea Threadtail *Guineagomphus perisi* **Compte Sart**, 1963
[JS *Phyllogomphus coloratus* Kimmins, 1931]
Guinea Threadtail *Elattoneura perisi* Compte Sart, 1964
[Orig. *Prodasineura perisi*]

Dr Salvator Vicente Peris Torres (1922–2007) was a Spanish zoologist. He studied for his first degree at the University of Madrid (1941) which also awarded his masters (1985) and became assistant entomologist in the Section of Epidemiology, Institute of Colonial Medicine

(1943–1946). He spent two years in London (1947–1949) to research at BMNH and work with other entomologists. Following his PhD from Madrid (1950), he was appointed partner of the Department of Entomology at the Aula Dei Experimental Station, Zaragoza (1950–1953), then senior scientist there (1953–1960). He was Secretary of the Institute of Soil Science (1960–1966) then became Professor of Zoology at the University of Seville (1966–1969) before moving to the same chair at the University of Madrid (1969–1987) until retirement when he became Professor Emeritus (1987–1992). He was also Director of the Spanish Entomology Institute (1978–1985). He took part in a number of scientific expeditions to the Iberian Peninsula and elsewhere, such as Annobón Island (1959), the Gulf of Guinea (1961) and Northwest Territory Sahara (1973). He published around 100 papers on insects and arthropods and described around 50 Diptera; 21 other insect species are also dedicated to him.

Perrens

Darner sp. *Coryphaeschna perrensi* **McLachlan**, 1887
[Orig. *Aeschna perrensi*]

Richard (d.c.1901), William and Thomas Perrens were Englishmen who settled in Argentina on the banks of the Corrientes River, in the province of the same name. They collected insects and other taxa, which they sent to various friends and institutions in England. McLachlan's etymology reads: "*I dedicate the species to Mr. Perrens, who on several occasions has sent me extensive consignments of Odonata, &c., from Corrientes.*" And his notes mention… "*1 (slightly immature and somewhat crushed) specimen, collected by Mr. Perrens.*" There are butterflies named after one of them, probably not Richard, who probably collected the darner as they share the name *perrensii*. Richard certainly made collections of mammals (1876–1894), which he sent to England, one containing the Goya Tuco-tuco *Ctenomys perrensi* named after him.

Persephone

Skimmer sp. *Diplacina persephone* **Lieftinck**, 1933

Persephone was a character from Greek mythology, the daughter of Zeus and Demeter and queen of the underworld.

Pesechem

Pond Damselfly *Papuagrion pesechem* **Lieftinck**, 1949

The species is named after a people; the Pesechem are a tribe of Papuans in the vicinity of the type locality in New Guinea.

Peter Miller

Shortleg sp. *Stenagrion petermilleri* **Hämäläinen**, 1997

Dr Peter Lamont Miller (1931–1996). (See **Miller**)

Peters

Subgenus *Petersaeschna* **Theischinger**, 2012

Günther Peters (b.1932). (See **Günther Peters**)

Petersen

Pond Damsel sp. *Andinagrion peterseni* **Ris**, 1908

[Orig. *Oxyagrion peterseni*]

Peter Esben-Petersen (1869–1942) was a Danish amateur entomologist, particularly interested in Neuroptera as well as the Orthoptera and Ephemeroptera of Denmark. He was a teacher in Silkeborg but became expert in entomology and was awarded an honorary doctorate by the University of Copenhagen, which museum now houses his collection. He was also a local politician and became a town mayor. He wrote several books in the series *Denmark Fauna*.

Pethiyagoda

Forest Shadow-Emerald *Macromidia donaldi pethiyagodai* **Nancy** van der Poorten, 2012

Tilak Rohan David Pethiyagoda (b.1955) (abbreviated to Rohan Pett by deed poll in 2010) is a Sri Lankan biologist. Kings College, London awarded his BSc in electronics (1977) and Sussex his M Phil in biomedical engineering (1980). He worked as an engineer in the Division of Biomedical Engineering of the Ministry of Health in Sri Lanka (1981–1987), becoming director (1982). He then resigned government service to research freshwater fish, leading to his first book: *Freshwater fishes of Sri Lanka* (1990). With colleagues from the Wildlife Heritage Trust that he founded, he has discovered over 100 vertebrates in Sri Lanka including fish, amphibians and reptiles. Concerned by the rapid loss of montane forest in Sri Lanka, he began (1998) an ongoing project to restore abandoned tea plantations to natural forest. He has received many honours and sits on many national and international committees. A reptile and several fish are also named after him.

Pfeiffer

Western Demoiselle *Agrion splendens pfeifferi* **Götz**, 1923

[JS *Calopteryx xanthostoma* (Charpentier, 1825)]

Ernst Pfeiffer (1893–1955) was a German amateur entomologist with a focus on Lepidoptera. By profession he was a publisher and bookseller. He collected frequently and widely in Hungary, Dalmatia, Bulgaria and Persia. He published many papers, principally on butterflies. He also co-edited: *Lepidopteren-Fauna von Marasch in türkisch Nordsyrien*. His collection is now in the zoological museum in Munch.

Philip

Pond Damsel sp. *Aeolagrion philipi* **Tennessen**, 2009

Tropical King Skimmer sp. *Orthemis philipi* von **Ellenrieder**, 2009

Professor Dr Philip Steven Corbet (1929–2008). (See **Corbet**)

Phillips

Abbott's Skimmer *Orthetrum phillipsi* **Kirby**, 1896

[JS *Orthetrum abbotti* Calvert, 1892]

Ethelbert Edward Lort-Phillips (1857–1943) was a British architect, traveller and hunter who shot big game around the world. He was also a collector of natural history specimens, particularly mammals and birds. He was in East Africa (1884–1895) and, with his wife Louisa and a party of friends, explored parts of Somaliland (Somalia) and collected the type (1895). This must have been a grand Victorian odyssey, as the party included both Mrs Phillips and a Miss Gillett, with her brother, Ethelbert's nephew, and possibly his heir, Frederick William Alfred Herbert Gillett (1872–1944). He later (1926) assumed the name Lort-Phillips. Phillips is also remembered in Norway as the man who developed an estate called Vangshaugen on Lake Storvatnet, where he planted a rhododendron garden, something virtually unknown in Norway at that time. He became Vice-president of the Zoological Society of London. Three reptiles, two birds and two mammals are also named after him. (Also see **Lort**)

Phoebe

Skimmer sp. *Diplacina phoebe* **Ris**, 1915

Phoebe is a character in Greek mythology like many other new names in the same paper. She was one of the original Titans, the sons and daughters of Uranus and Gaia.

Phyllis

Flutterer sp. *Rhyothemis phyllis* Sulzer, 1776
[Orig. *Libellula phyllis*]

Phyllis was a character from Greek mythology who married Demophon, the King of Athens.

Pictet

Common Spreadwing *Agrion picteti* **Fonscolombe**, 1838
[Syn. of *Lestes sponsa* Hansemann, 1823]
Dark Spreadwing *Lestes picteti* **Selys**, 1840
[Homonym, syn. of *Lestes macrostigma* (Eversmann, 1836)]

Professor François Jules Pictet-De la Rive (1809–1872) was a Swiss zoologist and palaeontologist. He studied at Muséum National d'Histoire Naturelle, Paris, under Cuvier (1829–1830), before becoming Professor of Zoology, Université de Genève (1835). He retired from teaching (1859) to devote his time to the Museum d'Histoire Naturelle, Geneva. He also has a reptile named after him.

Piel

Common Spineleg. *Anisogomphus pieli* **Navás**, 1932
[JS *Merogomphus pavici* Martin, 1904]
Pond Damselfly sp. *Agriocnemis pieli* Navás, 1933
[taxonomic status unverified]
Demoiselle sp. *Mnais pieli* Navás, 1936
[JS *Mnais tenuis* Oguma, 1913]

Father Octave Piel (1876–1945) was a Jesuit priest and entomologist, resident (<1918–c.1943) in Shanghai, China. He had been resident in China from at least 1918 as he is recorded collecting insects there at that time. The Jesuit priest Pierre Marie Heude founded the Museum at Xujiahui, Shanghai (1868) and Piel was director for the rest of his life (1902) when the name was changed to honour him. Piel visited the US (1930) and spent time at the US National Museum helping to identify oriental bees. He was (1931–c.1943) the Director of the Musée Heude, which was taken over (1949) after the revolution and closed (1952) when all the collections were transferred to the Chinese Academy of Science (at which time the Shanghai Natural History Museum was founded). He was particularly interested in bees and is commemorated in other taxa, such as a scorpionfly genus and species, as well as several bees.

Pierrat

Featherleg sp. *Platycnemis pierrati* **Navás**, 1935
[JS *Copera marginipes* (Rambur, 1842)]

Père Charles-J B Pierrat was a Jesuit Priest Resident in China as a missionary. He was in Canton (c.1914–c.1934).

Pijpers

Sanddragon sp. *Progomphus pijpersi* **Belle**, 1966

H P Pijpers was a corporal in the army of Suriname where he collected the sanddragon species (August 1963). He is also recorded as having collected other taxa there, such as fish, during the Sipaliwini & Wilhelmina Mountains Expeditions (1960–1962) with among others D C Geijskes (q.v.), Director of the Surinam Museum. He is mentioned in several papers as part of collecting expeditions, but his status is not spelt out. A number of fish, including the sub-species that he collected, *Curimatus (Hemicurimata) esperanzae pijpersi,* and the catfish species, *Imparfinis pijpersi,* are also named after him.

Pilbara

Pilbara Threadtail *Nososticta Pilbara* **Watson**, 1969

This is a toponym; Pilbara is an area of Western Australia.

Pindrina

Pilbara Billabongfly *Austroagrion pindrina* **Watson**, 1969

This is a toponym; Pindrina Spring is a locality in Western Australia.

Pinhey

The Bluet Genus *Pinheyagrion* **May**, 2002
The Hawker Genus *Pinheyschna* **Peters** & **Theischinger,** 2011
Tanganyika Jewel *Platycypha pinheyi* **Fraser**, 1950
Eastern Horntail *Nepogomphoides pinheyi* Fraser, 1952
[JS *Nepogomphoides stuhlmanni* **Karsch**, 1899]
Pinhey's Spreadwing *Lestes pinheyi* Fraser, 1955
Pinhey's Wisp *Agriocnemis pinheyi* **Balinsky**, 1963
Pinhey's Bluet *Pinheyagrion angolicum* Pinhey, 1966
Eastern Snorkeltail *Mastigogomphus pinheyi* **Cammaerts**, 1968
[Orig. *Neurogomphus pinheyi*]
Inland Darner *Austroaeschna pinheyi* Theischinger, 2001
Emerald-striped Slim *Aciagrion pinheyi* Samways, 2001

[A. Green-striped Slim]
Flatwing sp. *Priscagrion pinheyi* **Zhou** & **Wilson**, 2001
Pinhey's Horntail *Tragogomphus ellioti* **Legrand**, 2002
Unicorn Darner ssp. *Austroaeschna unicornis pinheyi* Theischinger, 2001

Dr Elliot Charles Gordon Pinhey (1910–1999) was an entomologist who worked in Africa and specialised in African Lepidoptera and Odonata. He is regarded as the doyen of African Odonatology. Born of British parents who were on holiday in Belgium, Pinhey made major contributions in entomology to the knowledge of butterflies, moths and dragonflies. The University of London awarded his BSc in mathematics, physics, chemistry and biology (1934), after which he taught as a science master and in 1962 he was granted a DSc. His chronically poor health led him to emigrate to the better climate in Southern Rhodesia (Zimbabwe) (1939). After working in the RAF Meteorology Department, he joined the Agriculture Department as economic entomologist. He joined the Transvaal Museum, South Africa, as Assistant Professional Officer in Entomology, becoming an Odonata specialist. He worked at the Coryndon Museum, Nairobi (1949–1955) under Louis Leakey and collected extensively in East and Central Africa. He was invited (1955) to become Keeper of Invertebrate Zoology at the National Museum in Bulawayo and later was President of the Entomological Society of Southern Africa (1974–1975) and an active member of the Societas Internationalis Odonatologica (SIO). He made an exciting overland expedition (1958) crossing Africa by Land Rover, travelling through the Belgian Congo to Cameroon and Nigeria. He published (1949–1985) over 130 papers on African Odonata, including many generic revisions and regional synopses, and described over 190 new species or subspecies (of which about half represent valid full species) and named 10 new genera. His Odonata publications include: *The dragonflies of Southern Africa* (1951), *Dragonflies (Odonata) of Central Africa* (1961), *A survey of the dragonflies (Order Odonata) of Eastern Africa* (1961), *A descriptive catalogue of the Odonata of the African continent (up to December 1959), Parts 1 and 2* (1962), and *Monographic study of the genus Trithemis Brauer (Odonata; Libellulidae)* (1970). He also published papers and books on African butterflies and moths and a guide book: *Introduction to Insect Study in Africa* (1968). (Also see **Elliot**)

Pinratana

Cruiser sp. *Macromia pinratani* **Asahina**, 1983
Longleg sp. *Anisogomphus pinratani* **Hämäläinen**, 1991
Spectre sp. *Petaliaeschna pinratanai* Yeh, 1999
[JS *Petaliaeschna flavipes* Karube, 1999]

Amnuay Pinratana (b.1929) is a Thai Catholic lay brother, amateur entomologist and originator of many books on Thai insects. He was educated in India (1949–1953) and then worked as an English teacher in various Catholic schools in Thailand. He was Provincial Superior (head of Catholic brothers in Thailand) (1965–1973) and became Director (1974–1977) of Saint Gabriel's College in Bangkok, a renowned primary and secondary school for boys. Thereafter, he remained teaching English and moral principles until retirement. His early interest (1970s) in butterfly collecting intensified and he began to make collecting trips to various parts of the country, accompanied by his friends, helpers and numerous foreign experts. Besides butterflies, he collected moths, dragonflies (since 1982) and various showy beetle groups. This activity continued up to c.2003. By that time his insect collections (now maintained as a special private museum at Saint Gabriel's College) had grown to be the most diverse in the whole country. With the help of numerous foreign entomologists, he has published (usually as a co-author or editor) numerous books on Thai insects, such as: *Butterflies in Thailand* (six volumes) and *Moths of Thailand* (six volumes). He co-authored Hämäläinen's: *Atlas of the dragonflies of Thailand* (1999). He has guided numerous visiting foreign odonatologists in field trips to different parts of the country. Bro Amnuay (also known as Brother Philip) is known as a very humble, friendly and helpful man. At least 24 insects are named after him.

Piper

Pond Amberwing *Perithemis piperi* **Hoffmann**, 1987
[JS *Perithemis mooma* Kirby, 1889]

Werner Piper (b.1957) is a German entomologist who visited (1990) Peru with Hoffmann, collecting at different places along the coast up into the rainforest on the Madre de Dios. He has worked at irregular intervals (since 1993) in Brazil.

Pipila

Dancer sp. *Argia pipila* Calvert, 1907

The species is named after a people; the Pipila are a Guatemalan Indian tribe in the region where the species was found.

Pirassununga

Dasher sp. *Micrathyria pirassunungae* **Santos**, 1953

This is a toponym; Pirassununga, near Sao Paulo, is the type locality of this species.

Pistor

Damselfly sp. *Igneocnemis pistor* **Gassmann** & **Hämäläinen**, 2002
[Orig. *Risiocnemis pistor*]
Shadowdamsel sp. *Drepanosticta pistor* **Van Tol**, 2005

This is one of the 'clever' names that taxonomists are so keen on. Pistor is a translation into Latin of the German name of the collector, Müller, which in English translates as Miller. As the etymology puts it: *"We are pleased to honour Roland A Müller by naming the present species pistor ([Lat.] = Müller [Germ.])."* (See **Müller** & **Roland Müller**).

Plateros

Tiger sp. *Gomphidia platerosi* **Asahina**, 1980
[JS *Gomphidia kirschii* Selys, 1878]

Professor Dr Cristobal G. Plateros (d.2013) held the Chair of Biology (1977–1983) at the College of Arts & Science at the University of San Carlos, Philippines. He collected the type there at Leyte (1970) which he had visited a number of times on USC expeditions. He took part in several USC expeditions (1954, 1957, 1958, 1969, 1960, & 1976).

Plaumann

Brown-and-Red Skimmer *Orthemis plaumanni* **Buchholz**, 1950
[JS *Orthemis ambinigra* Calvert, 1909]

Dr Fritz Plaumann (1902–1994) was a German (in modern day Lithuania) born amateur entomologist. He moved to Brazil (1924) where he farmed and taught, but also studied orchids and insects for the next 70 years. He amassed c.80,000 specimens of c.17,000 insect species, including c.1,500 new to science. His library was built up by sending insects to others who sent him books in return and he also sent insects to many of the great museums of Europe. He is considered by many to be the greatest collector of insects in Latin America in the twentieth century. The Entomological Museum Fritz Plaumann in the district of Nova Teutonia, Seara, Santa Catarina was opened in his honour (1988) and also houses his collection. A biography was written of him based on his own diaries: *The Diary of Fritz Plaumann* (2001) and a TV documentary made about him. He collected the specimens of this skimmer. Around 150 other insect species, especially beetles, bear his name.

Plebeja

Damselfly sp. *Igneocnemis plebeja* **Hämäläinen**, 1991
[Orig. *Risiocnemis plebeja*]

Although no etymology was given, there is jocular story behind this name, which almost qualifies as an eponym. The author described three new *Risiocnemis* species from Sibuyan Island in the Philippines: *R. kiautai* (see **Kiauta**), *R. rolandmuelleri* (see **Roland Müller**) and *R. plebeja*. Since the last species is rather plainly coloured, it was named *plebeja*, a form of the Latin adjective 'plebeius', (from the noun 'plebs'), used as an adjectival noun. In ancient Rome either the noun, or the adjective used as a noun, meant the ordinary citizens, as distinct from the elite.

Pluot

Senegal Threadtail *Elattoneura pluotae* **Legrand**, 1982

Dr Dominique Pluot-Sigwalt (b.1941) is a French entomologist who is Attachée Honoraire at the Département Systématique & Evolution, MNHN, Paris. There she was a colleague of Jean Legrand working on Heteroptera. She has spent several months in Senegal undertaking fieldwork. She co-wrote, among other works: *Insecta Hemiptera Heteroptera Coreidae* (2012).

Pocahontas

Slough Amberwing *Perithemis pocahontas* **Kirby**, 1889
[JS *Perithemis domitia* (Drury, 1773)]

Pocahontas (born Matoaka, known as Amonute and later known as Rebecca Rolfe) (c.1595-1617) was a Virginia Indian famous for being held captive, converting to Christianity and marrying an Englishman (1614). She came with him to England (1616), being feted as a 'civilized savage', but died just before setting sail back to Virginia.

Pocomán

Dancer sp. *Argia pocomana* **Calvert**, 1907

The species is named after a people; the Pocomán are Kekchian Maya people in Guatamala, where the species was found.

Poey

Three-Striped Dasher *Mesothemis poeyi* **Scudder**, 1866
[JS *Micrathyria didyma* Selys, 1857]

Professor Felipe Poey y Aloy (1799–1891) was a Cuban zoologist, naturalist, and artist. He was brought up in France (1804–1807) and later in Spain where he qualified as a lawyer in Madrid, but his ideas were too liberal for the age and he was forced to return to Cuba (1823). He concentrated on natural history, describing 85 species of Cuban fish. His *Memorias sobre la historia natural de la isla de Cuba, acompañadas de sumarios latinos y extractos en frances,* (1858) depicts mainly fish and snails, but also some mammals, Hymenoptera and Lepidoptera; the drawings are all done by Poey. The Felipe Poey Natural History Museum in Havana (created 1842) merged (1849) with the University of Havana, becoming the first such museum in Cuba; it was named in his honour and he was its first director. He also became the first Professor of Zoology and Comparative Anatomy at the University of Havana. Poey, who also supplied Cuvier in Paris with Cuban fish specimens, prepared many of the museum's exhibits, particularly fish. He is commemorated in the scientific names of other taxa, especially fish, such as Poey's Scabbardfish *Evoxymetopon poeyi*, but also moths such as *Nannoparce poeyi* and a mammal. His 175th anniversary was celebrated with a set of Cuban stamps.

Polhemus

Sylvan sp. *Idiocnemis polhemi* **Gassmann**, 2000

Dr Dan Avery Polhemus (b.1958) is an American biologist at the Hawaii Biological Survey. The Colorado State University awarded his bachelor's degree and the University of Utah awarded his PhD. He was a postdoctoral fellow at the Smithsonian Institution (1988–1990), then an associate entomologist at the Department of Natural Sciences, Bishop Museum, Hawaii (1990–1996). He returned to the Smithsonian as Research Entomologist (1996–2006) and then took a post as administrator in the Division of Aquatic Resources at the Department of Land and Natural Resources, Hawaii (2005–2010) before his present job as Program Manager at the Aquatic Ecosystem Conservation Program, Pacific Islands Fish & Wildlife Office also in Hawaii. He has at different times undertaken field work in Southeast Asia, Madagascar, New Guinea, Australia, South America and other US states, during one of which he collected the type of the damselfly in Papua New Guinea. One of his principal interests is in the systematics, zoogeography, and ecology of true bugs (Insecta: Heteroptera) and damselflies (Insecta: Odonata: Zygoptera). He has written or co-written more than 170 papers (1981–2013) and longer works, including: *Hawaiian Damselflies – a field identification guide* (1996). He has described or co-described nine new Odonata species.

Poll

Phasm sp. *Linaeschna polli* **Martin**, 1909

Jacobus Rudolphus (Jacob Rudolph) Hendrik Neervoort van de Poll (1862–1924) was a Dutch entomologist and collector of insects who specialised in Coleoptera, many of which he purchased from dealers. No etymology was given, but we believe it to be named after this man. Our thinking is informed by the fact that Poll's collection was sold on his death and specimens are now found in many museum collections. Other collectors, who donated them to museums or allowed them to be studied, had bought some. The hawker is a Bornean dragonfly and came from the collection of Herman Willem van der Weele (1879–1910) (q.v.). Most of Poll's papers are descriptions of insects in his own collection, particularly beetles, and a number of species are named after him such as a butterfly *Troides vandepolli*.

Pollen

Black-splashed Elf *Tetrathemis polleni* **Selys**, 1869
[Orig. *Neophlebia polleni*]

François Paul Louis Pollen (1842–1886) was a Dutch naturalist and merchant, who made major contributions to the study of Malagasy fauna. He went to study medicine at Leiden (1862) but was encouraged by Hermann Schlegel to study zoology instead. He collected in Madagascar (1863–1866) for the Leiden Museum financing the trip and the fieldwork of others from his trading and inherited wealth. He sent libellulid specimens to Selys for study (1867). He also visited the Comoros Islands and Reunion. He wrote: *Récherches sur la Faune de Madagascar et de ses Dépendances – d'après les découvertes de F.P.L. Pollen et D.C. van Dam* (1868). Two birds and two reptiles are also named after him.

Ponce

Threadtail sp. *Prodasineura poncei* Villanueva & Cahilog, 2013

Dr Dennis V Ponce de Leon was a physician who became Municipal Health Officer of the Municipality of Balabac, Palawan, Philippines and was honoured for the help he gave the Balabac survey. Cahilog visited Balabac Island in the Philippines (March 2013), where he discovered this new species. The survey needed practical aid and political support and Dr Ponce facilitated meetings with the local mayor etc.

Popoluca

Dancer sp. *Argia popoluca* **Calvert,** 1902

The species is named after a people; the Popolucas were a tribe formerly inhabiting Southern Mexico.

Portia

Portia Widow *Palpopleura portia* Drury, 1773
[Orig. *Libellula portia*]

This is not an eponym in the true sense as there is no particular person honoured, but from antiquity, the classical form would be Porcia, a female member of the Porcii clan. Drury's dragonfly names almost always came from Roman antiquity.

Poungyi

Sylvan sp. *Coeliccia poungyi* **Fraser**, 1924

This is not an eponym; Burmese Buddhist monks are called poungyi. 'Poun Gyi' means an 'elder monk'.

Possompes

Common Bluet *Enallagma cyathigerum possompesi* Heymer, 1968
[JS *Enallagma cyathigerum* (Charpentier, 1840)]

Bernard Possompes (1912–1975) was a French entomologist at the Laboratoire de Zoologie, Faculty des Sciences at the University of Paris. Among his published papers is: *Niveau de Rupture de la Patte des Insectes* (1963).

Poyarkov

Black Darter *Sympetrum danae pojarkovi* **Grigoriev**, 1905
[JS *Sympetrum danae* (Sulzer, 1776)]

Erast Fedorovich Poyarkov (sometimes Pojarkov) collected the type in Kirghizia (1903).

Prakriti

Sylvan sp. *Coeliccia pracritiae* Lahiri, 1985
[Orig. *Coeliccia prakritii*]

Prakriti Lahiri was the mother of the author Ashok Ranjan Lahiri (d.2012). The etymology states: "*The species is named after and in memory of my mother.*" (See **Chittaranji**)

Pramote

Leaftail sp. *Sarasaeschna pramoti* Yeh, 2000
[Orig. *Oligoaeschna pramoti*]

Pramote Saiwichian (b.1935) is a Thai forester. He was a chief of several national parks in southern Thailand, currently living in Nakhon Ratchasima. A friend of Amnuay Pinratana (q.v.) he accompanied him on many field trips in different parts of Thailand and facilitated his work in many ways. It was on the field trip to Doi Inthanon National Park, that this new species was found (May 1999). Pinratana and Matti Hämäläinen (q.v.) were also there and saw Mr Somnuk Panpichit catch the two specimens of this new fish. The specimens were sent for study by Wen-Chi Yeh, who works at the Forest Protection Division of the Taiwan Forestry Research Institute. Both Pramote and Pinratana once visited Matti Hämäläinen in Finland. It was exceptionally warm in Finland at that time and the guest room where the visitors slept was hotter than they were used to back home in Thailand!

Prater

Glider sp. *Pseudotramea prateri* **Fraser**, 1920

Stanley Henry Prater (1890–1960) was a British naturalist. He took part in the Bombay Natural History Society's Mammal Survey (1911–1923), during which he was seriously wounded when accidentally shot in the thigh. He trained in taxidermy in England (1923) and became Curator of the Bombay Natural History Society and of the Prince of Wales Museum of Western India (1923–1947), which he first joined when 17 (1907). He also visited several US museums (1927) to learn more about museum exhibition techniques. He wrote: *The Book of Indian Animals* (1948) since updated and republished several times. He was also a parliamentarian, President of Anglo-Indian and Domiciled European

Association, and their representative in the Bombay legislative assembly (1930–1947). After independence he represented that community in the Indian Constituent Assembly. He emigrated with his family to England (c.1950) and died there a decade later after a very long and debilitating illness. A bird and a mammal are also named after him.

Preuss

Swamp Junglewatcher *Neodythemis preussi* **Karsch**, 1891
[*S. Allorrhizucha preussi*]

Professor Dr Paul Rudolf Preuss (1861–1926) was a Poland-born German naturalist, botanist and horticulturist. He collected in West Africa (1886–1898) and was a member of Zintgraff's (1888–1891) military expedition to explore the hinterland of Cameroon, which was then a German colony. Whilst storming a native village, the troop commander was killed and the second-in-command severely wounded. Preuss took over command and led the remaining troops back to the coast. He collected Odonata around Barombi-Station, which were examined by Karsch, revealing many new species. He constructed the botanical gardens of Victoria (Limbe), Cameroon (1901) while employed by the colonial government. He later collected in New Guinea (c.1903) and again in West Africa (1910). Six birds, three mammals, an amphibian and a reptile are also named after him. (Also see **Paul**)

Priydak

Dancer sp. *Onychargia priydak* **Kosterin**, 2015

Natalya Vladimirovna Priydak (b.1976) from Novosibirsk, Russia, is the wife of the author, Dr Oleg Engelsovich Kosterin (q.v.). She studied chemistry for her first degree at the Natural Science Faculty of Novosibirsk State University (1998), then was awarded her MS at the Philosophy Faculty at the same university (2001). She works (2000–present) as a botanical artist at the Central Siberian Botanical Garden, Novosibirsk. She has (since 1999) accompanied and helped the author on a number of his entomological expeditions. When pregnant (2010), she inspired him to undertake a joint expedition to Cambodia, the first of many trips investigating Cambodia's Odonata. She is (since 2015) an environmental activist in Novosibirsk.

Procter

Pincertail sp. *Onychogomphus procteri* **Chao**, 1953
[*nomen nudum*; JS *Ophiogomphus sinicus* (Chao, 1954)]

Dr William Procter (1872–1951) was an American businessman, philanthropist, biologist, zoologist and entomologist. Yale awarded his PhB (1894) in chemistry and business. He took an extended trip around the world (1895–1897) visiting Japan, China and India among many other places. In between times he also graduated from the Sorbonne. He spent the next two decades in business, mostly in securities and the railroad, but latterly as a director of Procter & Gamble, which his grandfather had founded. He gradually relinquished his business career (1917–1920) to do postgraduate work in zoology at Columbia University. From his teenage years, he had spent his summers on Mount Desert Island, Maine at Bar Harbor. He established a research facility there (1921), eventually establishing his own laboratory on his estate at Corfield, Frenchman's Bay, studying marine fauna. His work there led to him taking an examination at Montreal University, which awarded him his DSc (1936). He was also a research associate of the Philadelphia Academy of Science (1928–1936). He switched (1932) his studies on the island more towards its insects and there followed a series of volumes about the insects, of which he wrote volumes two to seven. Over this period, he amassed records on the island of no less than 6,378 species of insect! Among his many voluntary posts and memberships, he was a trustee of AMNH (1931–1951). His collection was bequeathed to the University of Massachusetts.

Pryer

Clubtail sp. *Asiagomphus pryeri* **Selys**, 1883
[Orig. *Gomphus pryeri*]
Petaltail sp. *Tanypteryx pryeri* Selys, 1889
[Orig. *Tachopteryx pryeri*]
Paddletall sp. *Sarasaeschna pryeri* **Martin**, 1909
[Orig. *Jagoria pryeri*]

Henry James Stovin Pryer (1850–1888) was a British lepidopterist who went to China (1871) and shortly afterwards settled in Japan, where he collected the type specimens of all three species. He devoted his spare time to collecting natural history specimens, later donating them to the British Museum. The Petaltail was found in Gifu (May 1886). A few days earlier he had found a pair of the unique Anisozygopteran *Epiophlebia superstes* in the same location. Selys described both species in the same paper, which contains a fine colour plate by Guillaume Séverin (q.v.) of '*Palaophlebia superstes*'. Pryer died of pneumonia in Yokohama. He also has a reptile and four birds named after him.

Psyche

Cup Ringtail *Austrolestes psyche* **Hagen**, 1862

Psyche is a character from Greek and Roman mythology. She was a mortal princess. She was so beautiful and graceful that people compared her beauty to Aphrodite, the goddess of love and beauty and all the gods and men who once loved Aphrodite turned towards Psyche. This made Aphrodite very jealous so she asked her son, Eros, to make Psyche fall in love with the ugliest man on earth by shooting his golden arrows that make people fall in love. He accidentally pricked himself with one of his own arrows and fell in love with Psyche himself.

Pujol

Western Blacktail *Nesciothemis pujoli* **Pinhey**, 1971

Raymond Pujol (b.1927) is an agronomist and ethno-zoologist. He is Professor Emeritus of Ethno-Zoology at Muséum National d'Histoire Naturelle, Paris, where he was originally employed in the entomological department before working in West Africa; he collected in the Bangui area of the Central African Republic (1965–1969) including the type. He also led an expedition to Sérédou, part of Guinea (1957–1958). He was put in charge of the ethno-zoological department, MNHN in 1966. He taught at the University of Paris (1981–1998).

He co-wrote: *Dictionnaire raisonné de biologie* (2003). An amphibian and a bird are also named after him.

Purépecha

Sinuous Snaketail *Ophiogomphus purepecha* González-Soriano & Villeda-Callejas, 2000

The Purépecha are an indigenous people mostly found in the northwestern region of the Mexican state of Michoacán.

Q

Quandt

Pond Damselfly sp. *Palaiargia quandti* **Orr**, **Kalkman** & Richards, 2014

Ludwig Quandt (b.1961) is a musician and presently First Principal Cello of the Berlin Philharmonic. He was honoured for his generous support of Odonata research in New Guinea through the International Dragonfly Fund (IDF).

Quarré

Quarré's Tiger *Gomphidia quarrei* **Schouteden**, 1934
[Orig. *Diastatomma quarrei*, A. Quarre's/Southern Fingertail]

Paul Quarré (1904–1980) was a Belgian botanist and plant collector (1924–1946) who had an interest in entomology. He collected in the Congo/Zaire (1930s) and in Vietnam, as well as Europe. He is also commemorated in the names of a variety of plants.

Quirk

Pincertail sp. *Onychogomphus quirkii* **Pinhey**, 1964
[JS *Onychogomphus seydeli* (Schoudeten, 1934)]

Paul Quirk is a businessman. He sponsored and participated in Pinhey's expedition (April–May 1963) to the northern parts of Rhodesia (now Zambia) where the species was found.

R

Rácenis

The Darner Genus *Racenaeschna* **Calvert**, 1958

Sanddragon sp. *Progomphus racenisi* **De Marmels**, 1983
Firetail sp. *Telebasis racenisi* **GH Bick** & **JC Bick**, 1995

Dr Janis Rácenis (1915–1980) was a Latvian-born Venezuelan entomologist and ornithologist. The University of Latvia awarded his first degree (1943) and he began to study ornithology. He moved to Germany and finished his education, being awarded his PhD by the University of Erlangen (1947), where he married Gaida (q.v.). He left (1948) to teach in the Department of Science at the Central University of Venezuela, Caracas. He stayed until retirement (1976). During his time there, he served in various management positions, including in the School of Biology and the Biology Museum of the Central University of Venezuela (MBUCV), becoming its founding director (1949). He was also founder editor and director of the journal 'Acta Biologica Venezuelica' and founding director (1965) of the Institute of Tropical Zoology. In his 22 odonatological papers, he described 35 new species and five new genera. His papers include: *Notas taxonômicas sobre la familia Megapodagrionidae (Odonata: Zygoptera) con sinopsis de las especies venezolanas* (1959) and *Lista de los Odonata del Peru* (1959). He was also honoured in the names of a spider, a crustacean, a scorpion, a fish, an amphibian, a reptile and several insects. Sadly, he died on his 65th birthday.

Racleay

Southern Giant Darner *Telephlebia racleayi* **Martin**, 1909
[JS *Austrophlebia costalis* Tillyard, 1907]

The species epithet *racleayi* is a misspelling; it should have been *macleayi*. (See **Macleay**)

Radama

Radama Duskhawker *Gynacantha radama* **Fraser,** 1956
Radama Islander *Nesolestes radama* **Lieftinck**, 1965

Radama I (1793–1828) and Radama II (1829–1863) were sovereigns of Madagascar.

Rainey

Swampdamsel sp. *Leptobasis raineyi* **Williamson**, 1915
Skimmer sp. *Oligoclada raineyi* **Ris**, 1919
[JS *Oligoclada abbreviata* (Rambur, 1842)]

B J Rainey collected the swampdamsel in a small swamp near Cumuto, Trinidad (1912). He sent many specimens to E B Williamson (q.v.) having collected with him and his father in

Guatemala and British Guiana. In naming the swampdamsel Williamson wrote: "*Described from a single male in my collection taken... ...by Mr B J Rainey, to whom I am indebted for this and many dragonflies, and for whom I take pleasure in naming the species.*" Ris described the second species from a specimen taken in British Guiana that was in Williamson's collection.

Rambur

Keeled Skimmer ssp. *Libellula ramburii* **Selys**, 1848
[JS *Orthetrum coerulescens anceps* Schneider, 1845]
Rambur's Forktail, *Ischnura ramburii* Selys, 1850
[Orig. *Agrion ramburii*]
Parasol sp. *Neurothemis ramburii* **Brauer**, 1866
[Orig. *Polyneura ramburii*]
Violet Dropwing *Trithemis ramburii* **Kirby**, 1890
[JS *Trithemis annulata* (Palisot de Beauvois, 1805)]
Black Percher *Diplacodes ramburii* Kirby, 1890
[JS *Diplacodes lefebvrii* (Rambur, 1842)]
Threadtail sp. *Caconeura ramburi* **Fraser**, 1922
[Orig. *Indoneura ramburi*]
Hybrid Bluetip *Coenagriocnemis ramburi* Fraser, 1950

Jules Pierre Rambur (1801–1870) was a French physician and early entomologist who studied the insects of Corsica and Andalusia. He studied in Tours for his first degree (1821) and then medicine at Montpelier, qualifying MD in 1827. He returned to live in Tours but began travelling to the Alps, collecting around Paris, then to Corsica with his school friend Adolphe de Graslin (1829) publishing *Catalogue des Lépidoptères de L'Ile de Corse* (1832). He spent several years in Paris and was a founding member of the French Entomological Society (1832). He undertook a tour of Europe (1834–1835), culminating in southern Spain where he was accused by the British of spying when observing the Barbary apes of Gibraltar. He grew tired of practising medicine and settled back in Tours, having married (1841). His best-known work was: *Histoire naturelle des insectes Névroptères* (1842), in which he described over 230 new Odonata species of which, however, c.100 were synonyms. He also named c.15 new genera. His other publications include: *Faune entomologique de l'Andalousie* (1837–1840) and *Catalogue systématique des Lépidoptères de l'Andalousie* (1858–1866). He moved to Geneva and later died there from dysentery he had contracted in Barcelona.

Ram Mohan

Duskhawker sp. *Gynacantha rammohani* **Mitra** & Lahiri, 1975

Ram Mohan Roy (1772–1833) was an Indian religious, social and educational reformer, who challenged traditional Hindu culture and advocated societal progress under British rule, especially the abolition of such practices as 'sati' where a widow would throw herself on her husband's funeral pyre. He has been called the 'father of modern India' and, in particular, Bengali society. The authors wrote: "*The species has been named in honour of Raja Ram Mohan Roy (1772–1833) for his sincere advocacy (1831) in the formation of Supra National Organisation for settling all disputes among nations and for furthering the cause of peace in the world.*"

Ranavalona

Cascader sp. *Zygonyx ranavalonae* **Fraser**, 1949
Gilded Islander *Nesolestes ranavalona* **Schmidt**, 1951

Ranavalona I (born Rabodoandrianampoinimerina and also called Ramavo aka Ranavalo-Manjaka I) (c.1778–1861) was sovereign (1828–1861) of the Kingdom of Madagascar, where the two species are found. Following the death of her husband Radama I, she pursued a policy of self-sufficiency and isolation, reducing the political and economic ties with Europe, including repelling a French attack. She also suppressed the Christian movement and re-established traditional practices. Recent academic research has recast Ranavalona's actions as those of a queen attempting to expand, her empire while protecting Malagasy sovereignty against the encroachment of European cultural and political influence.

Ransonnet

Ransonnet's Skimmer *Orthetrum ransonnetii* **Brauer**, 1865
[A. Desert Skimmer, Orig. *Libellula ransonnetii*]

Baron Eugen von Ransonnet-Villez (1838–1926) was an Austrian diplomat, painter, lithographer, biologist and explorer. He studied law in Vienna (1855–1858) and began his diplomatic career when he became Ministerialoffizial (1858) at the Imperial and Royal Ministry of Foreign Affairs. In his spare time, his love was for the sciences, photography, painting, and especially colour lithography. His diplomatic postings (1860 onwards) took him to Palestine, Egypt, India, Ceylon (Sri Lanka) and Japan. He built himself a diving bell, which he used to make drawings of underwater scenes. Among other works, he published (1867) *Sketches of the inhabitants, animal life and vegetation in the lowlands and high mountains of Ceylon, as well as the submarine scenery near the coast, taken in a diving bell.* He donated (1892) his underwater paintings and more than 5,000 zoological specimens to the Vienna Natural History Museum. Ransonnet gave all the dragonflies and Neuroptera collected during his travels to Brauer to study. Georg Ritter von Frauenfeld collected the skimmer at El Tur in Sinai.

Raphael (Mpala)

Bullseye Hawker ssp. *Pinheyschna rileyi raphaeli* **Pinhey**, 1964
[Orig. *Aeshna rileyi raphaeli*]

Raphael Mpala (presently Raphael Chahwanda) is a technical officer of the National Museum, Bulawayo. He collected the holotype in Northern Rhodesia (Zambia) (April 1963).

Raphael (Selys)

Flatwing sp. *Philogenia raphaella* **Selys**, 1886

This could honour Selys' eldest son, Raphäel of Selys Longchamps (1841–1911), who became an officer and was a noted early photographer. However, this could also be an allusion to the archangel Raphael.

Rappard

Pincertail sp. *Onychogomphus rappardi* **Lieftinck**, 1937

Frederik Willem Rappard (1907–1994) was a Dutch forester who, after agricultural college in Wageningen, became a forest officer (1934) in the Dutch East Indian Forest Service, first stationed at Lahat, Benkoelen, South Sumatra and then Tjepoe, Java (1937). During WW2, he was a POW and was evacuated to Holland afterwards. He returned to Java (1947), stationed at Banjoewangi. He was appointed (1954) Head of the Forestry Division at Hollandia, New Guinea, until retirement (1957), when he returned to The Hague, the city of his birth. He collected the holotype (December 1936). He wrote some botanical papers and was a good photographer. At least one plant, *Dendrobium rappardii,* is named after him.

Rasoherina

Madagascar Demoiselle *Phaon rasoherinae* **Fraser**, 1949

Queen Rasoherina (born Rabodozanakandriana) (1814–1868) was sovereign of Madagascar (1863–1868).

Raychaudhuri

Emerald Spreadwing sp. *Megalestes raychoudhurii* Lahiri, 1987

Professor Dr Dhirendra Nath Raychaudhuri (1924–1981) worked at the Entomology Laboratory, Department of Zoology, Calcutta University. After initial schooling, he understood his intermediate course at St. Xaviers' College, Kolkata, and graduated in zoology from University of Calcutta and also obtained the MSc there (1947). He joined St. Xaviers' College as a faculty member to teach biology. Simultaneously he started serving City College and Charuchandra College. He went (1954) to Leiden, The Netherlands, to join Professor Hille Ris Lambers and subsequently obtained his DSc. Then he returned to St. Xaviers' College where he initiated aphid studies. He established the Zoology Department of Charuchandra College and became Vice-Principal. He joined the Department of Zoology, University of Calcutta (1961) and initiated research in entomology. He continued his work on aphids as well as the physiology of insects. Together with his students, he had over three hundred publications, notably, *Aphids of Northeast India and Bhutan* (1980).

Raymond (Compte Porta)

Black-banded Duskhawker *Heliaeschna raymondi* **Compte Sart**, 1967
[JS *Heliaeschna fuliginosa* Selys, 1883]

Ramón Compte Porta (1909–1991) was the father of the author Arturo Compte Sart (b.1933). He was self-educated and an extremely competent amateur astronomer. He was founder (1950) of the Agrupación Astronómica Balear and a delegate in the Balearic Islands for the Sociedad Astronomica de España y America (1950–1970). He was also a founder of the astronomical observatory in Palma de Mallorca and an observer correspondent for NASA during the Apollo XI–XVII programme. He wrote hundreds of articles and several books on astronomy including: *L'astronomia a Mallorca* (1991). (See also **Margarita**)

Raymond (Straatman)

Skimmer sp. *Phyllothemis raymondi* **Lieftinck**, 1950

Raymond Straatman (1917–1987) was a Dutch amateur entomologist who was interested in butterflies, particularly the Birdwing species. He moved to Indonesia (late 1940s) and settled in Sumatra, where he worked for a Dutch company on their rubber and copra estate. When the political situation in Indonesia made it difficult for the Dutch, he migrated to Australia (1959). He then (1960) went to Papua New Guinea to work as a rubber planter near the Veimauri River. He later (1965–1967) worked for the Bishop Museum, Hawaii, collecting for them throughout Papua New Guinea, Bougainville and the Solomon Islands. He not only collected and studied butterflies, but also bred them and became an independent collector (1970). The Papua New Guinea Fishery and Wildlife Department charged him (1978) with the illegal catching of, and dealing in, protected birdwings. Found guilty, he was ordered to leave so settled in Cairns, Queensland and later built a house near Kuranda. He continued to visit Indonesia including West New Guinea's Arfak Mountains, and travel to Europe and the US. He visited Leiden (1981). He collected the type in Sumatra (1948). He published a number of papers on butterflies.

Rebecca

Peri sp. *Archibasis rebeccae* Kemp, 1989

Rebecca Louise Kemp (b.1985) is the daughter of the author Robert Graham Kemp (b.1953). Four male specimens were found in Pahang, Peninsular Malaysia, in February 1988, when Rebecca was two years old. She gained degrees in medicine, surgery and biomedical science (2006 & 2008) at Nottingham University and currently works as a general medical practitioner. She maintains a general interest in dragonflies and has accompanied her father on numerous foreign trips looking for them. The author wrote: "*I have pleasure in naming this new taxon after my daughter who accompanied me at the time when the specimens were taken.*"

Reels

Skimer sp. *Atratothemis reelsi* **KDP Wilson**, 2005
Phantom sp. *Bornargiolestes reelsi* **Dow**, 2014

Graham Thomas Reels (b.1964) is a British animal ecologist and conservationist. The University of Hertfordshire (then Hatfield Polytechnic) awarded his BSc in environmental studies (1988) and the Department of Zoology at Hong Kong University his MPhil (1994). He was then employed by HKU to conduct aquatic invertebrate surveys of freshwater wetlands across the whole territory of Hong Kong, and to work on a territory-wide biodiversity survey of Hong Kong, with responsibility for insects, until the project's completion (1998). Four staphylinid beetles and one geometrid moth were named after him as a direct consequence of his field collections. He collected the holotype of *A. reelsi* when employed to undertake rapid biodiversity surveys of Guangxi, Guangdong and Hainan in tropical southern China (1998–2000), while working for Kadoorie Farm & Botanic Garden, a Hong Kong-based conservation charity. He worked as an ecological consultant in Hong Kong, continued surveying Chinese Odonata for KFBG on a voluntary basis, and was lead author of *A Field Guide to the Dragonflies of Hainan* (2015). He has co-authored several odonatological papers with Keith Wilson (q.v.) and Rory Dow (q.v.). He has an interest in the dragonflies of Sarawak, having made numerous field visits (2004–2016) to help discover or rediscover many more species, and is currently collaborating with Rory Dow on a long-term project to write a field guide to that fauna. He now lives in England and is a co-author of 20 new species of Odonata.

Reeves

Queensland Pin *Eurysticta reevesi* **Theischinger**, 2001

Deniss Michael Reeves (b.1933) is a pharmacist and amateur Australian naturalist. He is a well-known member of the Queensland Naturalists' Club. He has been a member for over three decades (1968) and is very knowledgeable about Lepidoptera and Odonata. He has participated in several scientific expeditions organised by the Royal Geographical Society of Queensland and has conducted surveys of the Odonata and Lepidoptera, resulting in numerous scientific publications. He has done much to raise community awareness and understanding about the Odonata. He was awarded the Queensland Natural History Award (2010) and is also a widely-published wildlife photographer. He donated a collection of 408 Tasmanian butterflies to the CSIRO Australian National Insect Collections (1999). He was the first person to draw attention to the existence of a species of *Eurysticta* in Queensland.

Regina

Two-banded Cruiser *Macromia reginae* **Le Roi**, 1915
[JS *Phyllomacromia contumax* Selys, 1879]

Margarethe Koenig née Westphal (1865–1943) was the wife of Dr Alexander Ferdinand Koenig (1858–1940) who funded and participated in a number of expeditions to Egypt and the Sudan, including the one which collected the type. She accompanied him on most of these trips. The name 'Regina' is used to infer 'queen' as 'Koenig' means 'King'… so the wife of a king is a queen.

Regis Albert

Congo Tigertail *Ictinogomphus regisalberti* **Schouteden**, 1934
[Orig. *Ictinus Regis-Alberti*]
Regal Cascader *Zygonyx regisalberti* Schouteden, 1934
[Orig. *Pseudomacromia Regis-Alberti*]

Albert Léopold Clément Marie Meinrad (Albert I) (1875–1934) was King of Belgium (1909–1934). He died in a mountaineering accident while climbing alone in the Ardennes region of Belgium. There are two possible explanations for his death: the first was he leaned against a boulder at the top of the mountain which became dislodged; or second, the pinnacle to which his rope was belayed had broken, causing him to fall about sixty feet. His grandson, Albert Félix Humbert Théodore Christian Eugène Marie Prince Albert of Belgium (born June 1934) reigned later as Albert II, King of the Belgians (1993–2013).

Rehn

Pond Damselfly sp. *Bromeliagrion rehni* **Garrison**, 2005

Dr Andrew Charles Rehn (b.1970) is an aquatic entomologist who collected the holotype in Ecuador (2001). The University of California, Davis, awarded his PhD (2000). He works (since 2000) at the California Department of Fish and Wildlife Aquatic Bio-assessment

Laboratory. His career has focused mostly on the development of biological indices to assess stream health based on benthic invertebrates, aiming to provide a scientific foundation for management policies that protect aquatic life. His papers in this field include: *Benthic macroinvertebrates as indicators of biological condition below hydropower dams on west slope Sierra Nevada streams, California, USA* (2009) and *Bioassessment in complex environments: designing an index for consistent meaning in different settings* (2016). He is also an expert on Odonata, publishing such as: *Phylogenetic analysis of higher-level relationships of Odonata* (2003). The etymology states that *"...this species is named for my friend Dr. Andrew Rehn who collected the types and graciously allowed me to describe this species."*

Reinhard

Signaltail sp. *Rhinagrion reinhardi* **Kalkman** & Villanueva, 2011

Dr Reinhard Jödicke (b.1948) is a German biologist with a focus on zoology, conservation and molecular genetics. After a doctoral dissertation on Cladocera at the University of Cologne his professional focus was on Odonata and he was co-editor of the journals 'Odonatologica', 'Libellula' and 'Notulae odonatologicae'. He became Executive Editor of the 'International Journal of Odonatology', which he successfully managed for a decade (2001–2010). Many of his studies are connected with the Odonata of Spain, where he spent much of his lifetime, and he edited (1996) the seminal miscellany studies on Iberian dragonflies. As an acknowledged specialist of the Lestidae, his magnum opus is the monograph: *Die Binsenjungfern und Winterlibellen Europas (Lestidae)* (1997). Among his many other works he co-wrote: *The Odonata of Tunisia* (2000) and co-edited: *Guardians of the watershed: global status of dragonflies: critical species, threat and conservation* (2004). He has described one new species and two new subspecies of *Sympetrum*. The etymology reads: "...*named in honour of Reinhard Jödicke for his dedication to the International Journal of Odonatology during his 10-year editorship...*"

Reinhardt

Clubtail sp. *Asiagomphus reinhardti* **Kosterin** & **Yokoi,** 2016

Dr Klaus Reinhardt (b.1968) is a biologist, entomologist and odonatologist, particularly focussed on reproduction. His PhD was awarded by the University of Jena (2000) and he undertook various postdoctoral posts (2001–2010) at Illinois State University (2001), the University of Leeds (2002) as well as Sheffield University (2003–2005) when he was a research fellow (2005–2010) in the Department of Animal & Plant Sciences. He was a scientist at the Eduard-Karls University, Tübingen (2010-2014) and then became a professor of biology at Dresden Technical University, where he started the applied zoology laboratory (2014). He is a member of the Evolutionary & Ecological Entomology Group at Sheffield. His publications include: *Evolutionary consequences of sperm ageing* (2007) and *Bacteriolytic activity in the ejaculate of an insect* (2009). The etymology reads: *"The new species is named in honour of Klaus Reinhardt, a German odonatologist who contributed greatly to our knowledge of Odonata by taking an active part in establishing the International Dragonfly Fund, including personal donations, editing of Odonatological Abstract Service and numerous papers in IDF-Report".* Commenting on the honour he said: *"Having a newly discovered animal or plant species named after oneself is one of the most beautiful awards for a biologist,*

I'm extraordinarily honoured, for sure, but there are a lot of other dragonfly researchers who would have deserved this award before me."

Reinhold

Tropical Rockmaster *Diphlebia reinholdi* **Förster**, 1910

[JS. *Diphlebia euphaeoides* Tillyard, 1907]

Reinhold Förster (1906–c.1944) was the son of the author Johann Friedrich Nepomuk Förster (1865–1918) and godson of E B Williamson (q.v.). He emigrated to the US (late 1920s) but, not faring well, returned to Germany (1930s). The family tried to help Jewish people, using their American passports, but were caught and Reinhold was incarcerated in a mental home from where his ashes were later sent to his mother. No etymology was given for the name of this Australian species, which was erroneously labelled as having been collected in New Guinea. However, it is clearly named after the third son of the author. (Also see **Förster**)

Reinwardt

Hooktail sp. *Paragomphus reinwardtii* **Selys**, 1854

[Orig. *Onychogomphus reinwardtii*]

Caspar Georg Carl Reinwardt (1773–1854) was a German-born Dutch ornithologist who collected in Java (1817–1822). At 14 he went to study botany and chemistry in Amsterdam, where he also studied mathematics, classical and modern languages, and history. He gained a doctorate in natural philosophy and medical science (1801). That year he also became Professor of Chemistry and Natural History at the University of Harderwijk, where he was responsible for the University's botanical garden. Louis Napoleon made him director of his menagerie in Amsterdam and he became professor of natural history there. He was made responsible for all matters concerning agriculture, arts and sciences in Java (1816–1821) and contributed greatly to education and public health there. He founded the Botanical Gardens of Buitenzorg (Bogor) and travelled extensively. He laid the foundation for two important museum collections in Leiden: The National Museum of Natural History and The National Museum of Ethnography. When collecting he was meticulous and detailed in the documentation of his specimens. He was very concerned with the systematic development of collections and their documentation, and he was always pleased to extend opportunities for research to others. It was through this that many fellow naturalists honoured him in the names of various plants, an amphibian and nine birds. The Hooktail was found in Java.

Remartin

The Genus of Malachite Darners *Remartinia* **Navás**, 1911

Two-spined Darner sp. *Gynacantha remartinia* Navás, 1934

[A. Duskhawker sp.]

(See **Martin, R** & **René**)

Renaud

Blue-nosed Sprite *Pseudagrion renaudi* **Fraser**, 1953

Dr Renaud Paulian (1913–2003). He collected the type series in Madagascar (September 1952). (See **Paulian**)

Rendall

Blue Basker *Urothemis rendalli* **Kirby**, 1898
[JS *Urothemis edwardsii* (Selys, 1849)]

Dr Percy Rendall (1861–1948) was an itinerant zoologist who collected over much of Africa and in Trinidad and other Caribbean locations in the late 19th century. He made a collection including new fish species from the Upper Shiré River, British Central Africa, which was presented (1896) to the British Museum by Sir Harry Johnston. He collected the basker holotype in Nyasaland. He wrote: *Notes on the Ornithology of the Gambia* (1892). Two birds, a fish and a mammal are also named after him.

René

Western Talontail *Crenigomphus renei* **Fraser**, 1936
[S. *Dentigomphus rubrithora*]

(See **Martin, R** & **Remartin**)

Rentz

Swift Tigertail *Eusynthemis rentziana* **Theischinger**, 1998

Dr David Charles Foster Rentz (b.1942) is an Australian entomologist. He received his PhD from the University of California, Berkeley, then was appointed Curator of Entomology at the Academy of Natural Sciences of Philadelphia (until 1975). He then became a Curator at California Academy of Sciences, San Francisco. He was Curator of Orthopteroid Insects in the Australian National Insect Collection, Canberra (CSIRO) (1976–2001). He is an Adjunct Professor at the James Cook University and an Honorary Fellow of the California Academy of Sciences. His main interest is in *Tettigoniidae* (katydids). Among other works he wrote: *Grasshopper Country: The Abundant Orthopteroid Insects of Australia* (1996), the three volume *A Monograph of the Tettigoniidae of Australia* (1998, 1993 & 2001) and *A Guide to the Katydids of Australia* (2010). CSIRO recently published his *Guide to Australian Cockroaches* (2014). Rentz was awarded Member of the Order of Australia (2013). Now retired and living in Kuranda, Queensland, he was the first to record the tigertail species from south of the Hunter River.

Resla

Shadowdamsel sp. *Protosticta reslae* **Van Tol**, 2000

This is an acronym – Royal Entomological Society – The name is a combination of the characters in the abbreviation of the Royal Entomological Society (RESL), which organized the Project Wallace Expedition to northern Sulawesi (1985) in which the author participated. This expedition was the immediate reason for the author to start a study of the dragonflies of Sulawesi.

Reventazón

Shadowdamsel sp. *Palaemnema reventazoni* **Calvert**, 1931

This is a toponym; it is named for the type locality, a brook in woods on the left bank of the Rio Reventazón, Costa Rica.

Rhoads

Golden-Winged Dancer *Argia rhoadsi* **Calvert**, 1902

Samuel Nicholson Rhoads (1862–1952) was a professional collector and later (1902) a bookstore owner for twenty years, mainly dealing in natural history books. He collected the dancer holotype in Mexico (April 1899). Aged 18 he was given a farm, but spent all his leisure time studying birds. He undertook work for Harvard (1892–1895) and was taken on as a staff member at the Carnegie Museum, Pittsburgh. Nevertheless, he spent much of his time at the Philadelphia Academy of Science and contributed papers on birds and mammals. He made several trips to Mexico, including one with his wife (1899) where he collected mammals as well as the eponymous type. At least one such trip was cut short when he was recalled to Philadelphia (1893). He also collected in Ecuador, Nicaragua, Guatemala and the US. He wrote: *The Value of the Mole to Agriculture* (1898) and *The Mammals of Pennsylvania and New Jersey: A Biographic, Historic and Descriptive Account of the Furred Animals of Land and Sea, Both Living and Extinct, Known to Have Existed in These States* (1903). He suffered a mental breakdown aged 64 (1926) and spent the rest of his life in an institution

Ribeiro

Pond Damselfly sp. *Minagrion ribeiroi* **Santos**, 1962
[Orig. *Telagrion ribeiroi*]

José Ribeiro was a driver employed by the Museu Nacional, Rio de Janeiro, Brazil, who collected the type. The etymology says that: "…*José Ribeiro, functionario de Servico de Endemias Rurairs, collected the holotype in Rio de Janeiro.*" (September 1960)"

Ricci

Emerald Spreadwing sp. *Megalestes riccii* **Navás**, 1935

Matteo Ricci (1552–1610) was an Italian Jesuit priest who went as a missionary to China. After entering the order (1571), he later applied for a missionary expedition to India (1577). His journey began (March 1578) in Lisbon and he arrived in Goa in September 1578. Four years later, he was dispatched to China. Once in Macau, Ricci started learning the Chinese language and customs. This was the beginning of a long project that made him one of the first Western scholars to master Chinese script and classical Chinese. He travelled in a number of states before establishing the mission in Macau, but moved on to Zhaoqing (1583–1589) during which time he composed the first European-style map of the world in Chinese (1584). He compiled a Portuguese-Chinese dictionary, the first in any European language, developing a system for transcribing Chinese words in to the Latin alphabet; it was misplaced in the archives in Rome not rediscovered until 1934 and published in 2001. Expelled from Zhaoqing (1589), Ricci was allowed to relocate to Shaoguan. He was appointed head of the Chinese mission overall and continued to travel throughout China for the rest of his life. In his etymology Navás wrote: "…*Je l'ai appelé Riccii en l'honneur du premier missionnaire s. j. en Chine.*"

Richard

Skimmer sp. *Lanthanusa richardi* **Lieftinck**, 1942

Richard A Archbold (1907–1976) is most famous for the expeditions he funded (see **Archbold, R A**).

Richardson

Muskeg Emerald *Cordulia richardsoni* **Hagen**, 1861
(*nomen nudum*)
[JS *Somatochlora septentrionalis* (Hagen, 1861)]

Sir John Richardson (1787–1865) was a knighted (1846) Scottish naval surgeon and Arctic explorer, who assisted Swainson. He was a friend of Sir John Franklin, to whom he was also related by marriage, and took part in Franklin's expeditions (1819–1822 & 1825–1827). He also participated in the vain search for Franklin and his colleagues (1848–1949); their fate was not discovered until Rae's expedition (1853–1854). During these expeditions, he also collected a few dragonfly specimens. The Richardson Mountains in Canada are also named after him, as are a shark, eight birds, five mammals and four reptiles. *Although this name is listed here in use it is not available (nomen nudum), since the species was not formally described using this name.*

Ricker

Western River Cruiser *Macromia rickeri* **Walker**, 1937
[JS *Macromia magnifica* McLachlan, 1874]

Dr William Edwin Ricker (1908–2001) was a Canadian entomologist and fisheries biologist. He spent twenty-seven years working for the Fisheries Research Board of Canada, eleven years at Indiana University and a year at the Pacific Salmon Commission. He collected the type in British Columbia (1934). He produced nearly three hundred scientific papers and books, including 38 papers on stoneflies (his particular interest), which described about 88 new species, including some from outside North America, as well as one fossil species. He was aware of the accomplishments of the Russian fisheries scientist, Feodor Baranov and he taught himself Russian while at Indiana University so that he could read Baranov's original papers. He raised awareness of Russian fishery science in the English-speaking community by translating his 428-page *Russian–English dictionary for students of fisheries and aquatic biology* (1973). He received many awards and honorary degrees, but the recognition that he particularly cherished was the naming of the Canadian Fisheries Research Vessel the *W.E. Ricker* (1986). Walker wrote: *"While stationed at the Dominion Biological Laboratory at Cultus Lake, B. C., and conducting investigations on the sockeye salmon, Dr. W. E. Ricker spent some of his leisure time in collecting certain groups of aquatic insects and has generously presented the bulk of the material to the Royal Ontario Museum of Zoology."*

Ridley

Great Spreadwing *Lestes ridleyi* **Laidlaw**, 1902
[JS *Orolestes wallacei* (Kirby, 1889)]

Henry Nicholas Ridley (1855–1956) was a British botanist and collector on the island of Fernando de Noronha, Brazil (1887). He was known as 'Mad Ridley' or 'Rubber Ridley', as he was keen to ensure that the rubber tree was transplanted to British territory, thus freeing Britain from dependency on supplies of latex from Brazil. He was Superintendent of the Tropical Gardens in Singapore (1888–1912), where early experiments in growing rubber trees outside Brazil took place, the first successful growth having been achieved at Kew and plants then shipped to Singapore. He wrote: *The natural history of the island of Fernando de Noronha based on the collections made by the British Museum Expedition* (1887) and *The habits of Malay reptiles* (1889). Three reptiles, two mammals and a bird are also named after him.

Riel

Dash-winged Piedface *Thermochoria aequivocata rieli* **Navás**, 1914
[JS *Thermochoria equivocata* Kirby, 1889]

Dr Philibert Riel (1862–1943) was a physician and naturalist. He collected mainly in France in the first two decades of the twentieth century and sometimes collected butterflies and other insects for or with naturalist and museum curator Henri Gelin. He wrote a number of essays on the systematics of butterflies under the title: *Essai d'une classification des lépidoptères producteurs de soie* (1895–1904) as well as some on medicine such as: *De la Pneumonie tuberculeuse lobaire* (1888). There is also a book written about his fungi collection.

Riley

Riley's Hawker *Pinheyschna rileyi* **Calvert**, 1892
[A Bullseye Hawker, Orig. *Aeshna rileyi*]

Dr Charles Valentine Riley (1843–1895) was a British-born American entomologist and artist. Privately educated in France, Germany and England, he left for the US at the age of seventeen when his mother re-married. He became a farm labourer in Illinois on a farm owned by a fellow British ex-patriot. He took a job (1864) as a reporter, artist, and editor of the entomological department at the 'Prairie Farmer'. He was appointed as the first State Entomologist for the State of Missouri (1868) and studied the plague of grasshoppers that invaded many western states (1873–1877) and managed to convince the US Congress to establish the US Entomological Commission, which included a Grasshopper Commission, to which Riley was appointed chairman. He was appointed to the post of entomologist to the US Department of Agriculture, but resigned after a disagreement a year later. However, he was reappointed (1881–1894). Subsequently, he became the first Curator of Insects at the Smithsonian. (1885). He wrote over 2,400 papers and books and also published two journals, the 'American Entomologist' (1868–1880) and 'Insect Life' (1889–1894). The hawker species was collected on the Dr W L Abbott expedition to East Africa and lodged at the museum and Riley sent it to Calvert to study. As he was riding rapidly down a hill (1895), the bicycle wheel struck a granite paving block dropped by a wagon. He catapulted to the pavement and fractured his skull. He was carried home on a wagon and never regained consciousness. He died at his home the same day, leaving his wife with six children.

Rimanella (Rima)

Genus *Rimanella* **Needham**, 1934

[substitute name for preoccupied name *Rima* Needham, 1933]

Rima the Jungle Girl was the fictional heroine of W H Hudson's (1904) novel *Green Mansions: A Romance of the Tropical Forest*; she was later adapted (1974) for DC Comics and a TV cartoon.

Ringuelet

Tropical Dasher sp. *Micrathyria ringueleti* Rodrigues Capitulo, 1988

Dr Raúl Adolfo Ringuelet (1914–1982) was an Argentine zoologist and limnologist. The Institute of Museum, Universidad Nacional de La Plata awarded his doctorate (1939) and he held several professorships there: Adjunct Professor of General Zoology (1944–1948), Acting Professor (1946–1947) and Head (1947–1955) Invertebrate Zoology, Professor Acting Zoogeography (1958), Vertebrate Zoology Professor (1957–1966), Professor of Ecology and Zoogeography (1960 & 1972) and Professor of Limnology (1969–1978). He was also Professor of Systematic Zoology in the Faculty of Natural Sciences, University of Buenos Aires (1956–1964). He published more than 100 papers particularly on conservation of Latin American freshwater species. The etymology says: *"...The species is named in homage to Dr Raúl Adolfo Ringuelet (1914–1982), eminent limnologist of South America and founder of the "Instituto de Limnologia de La Plata".* The author worked at this institute.

Ris

The Genus *Risioneura* Munz, 1919

[JS *Nososticta* Selys, 1860]

The Genus *Risiolestes* **Fraser**, 1926

[JS *Austroargiolestes* Kennedy, 1925]

The Genus *Risiocnemis* Cowley, 1934

The Genus *Risiophlebia* Cowley, 1934

Glider sp. *Tauriphila risi* **Martin**, 1896

Pale Hunter *Austrogomphus risi* Martin, 1901

[JS *Austrogomphus amphiclitus* Selys, 1873]

Reedling sp. *Indolestes risi* **Van der Weele**, 1909

[Orig. *Lestes risi*]

Common Archtail *Nannophlebia risi* **Tillyard**, 1913

Pretty Relict *Chorismagrion risi* **Morton**, 1914

Darter sp. *Sympetrum risi* **Bartenev**, 1914

Skimmer sp. *Akrothemis risi* **Campion**, 1915

[Orig. *Oda risi*]

Sanddragon sp. *Progomphus risi* **Williamson**, 1920

Pincertail sp. *Onychogomphus risi* Fraser, 1922

[Orig. *Gomphus risi*]

Hooktail sp. *Paragomphus risi* Fraser, 1924

[Orig. *Mesogomphus risi*]
Threadtail sp. *Caconeura risi* Fraser, 1931
[Orig. *Indoneura risi*]
Duskhawker sp. *Gynacantha risi* **Laidlaw**, 1931
Slendertail sp. *Leptogomphus risi* Laidlaw, 1932
Black Bambootail *Disparoneura risi* **Schmidt**, 1934
[JS *Prodasineura verticalis delia* (Karsch, 1891)]
Cameroon Sprite *Pseudagrion risi* Schmidt, 1936
Navy Dropwing *Trithemis risi* **Longfield**, 1936
[Syn. of *Trithemis furva* Karsch, 1899]
Brown-winged Goldenring
Chlorogomphus risi Chen, 1950
Highland Pricklyleg *Porpax risi* **Pinhey**, 1958
Bluet sp. *Enallagma risi* Schmidt, 1961
Phantomhawker sp. *Planaeschna risi* **Asahina**, 1964
Wedgetail sp. *Acanthagrion risi* **Leonard**, 1977
[JS *Acanthagrion vidua* Selys, 1876]
Duskhawker sp. *Cephalaeschna risi* Asahina, 1981
Skimmer sp. *Oligoclada risi* **Geijskes**, 1984
Pond Damselfly sp. *Mesamphiagrion risi* **De Marmels**, 1997
[Orig. *Cyanallagma risi*]

Darner ssp. *Aeschna diffinis risi* Enderlein, 1912
[JS *Rhionaeschna variegata* (Fabricius, 1775)]
Jewelwing ssp. *Calopteryx orientalis risi* Schmidt, 1954
[JS *Calopteryx orientalis* Selys, 1887]

Dr Friedrich Ris (1867–1931) was a Swiss physician and entomologist who specialised in Odonata. He studied (1885) at the University of Zürich, graduating with a medical doctorate (1890), his thesis being: *Klinischer Beitrag zur Nierenchirurgie nach Erfahrungen aus der chirurgischen Klinik zu Zürich, 1881 bis 1890.* As a boy, he collected butterflies, and while at college, beetles, amassing 1,200 species of Coleoptera; but his main interest soon became Neuroptera, in the old sense (i.e. before the group was split), and particularly Odonata, which he collected around Zürich and in the Alps. He was then a house surgeon at Zürich Hospital for a while before becoming a ship's doctor taking four voyages, one to North America, two to South America and one to the Far East as far as Shanghai; on his shore leaves he made entomological excursions. On his return, he practised privately but very soon took an assistant physician position at a hospital in Zürich (1892–1894). He became (1894) assistant physician at the psychiatric sanatorium in Rheinau, Switzerland and was then director of a psychiatric clinic in Tessin canton, before returning to Rheinau as director for the rest of his career (1898–1931). At this time, the Rheinau Sanatorium was the largest mental hospital in Switzerland. All the while he was pursuing his interest in entomology in general and odonatology in particular. Most of his 125+ publications were on dragonflies, of which he described nearly 300 new species and named 27 new genera. He was the last odonate taxonomist who mastered the fauna of all continents. His major publication was the monograph of the libellulid dragonflies of the world: *Libellulinen*, which appeared in 9 volumes (1909–1919) in the series *Collections Zoologiques du Baron Edm. de Selys Longchamps Catalogue systématique et descriptif*. This monograph, sometimes called the 'Green Monster', being 1287 pages, is literally the heaviest (8 kg!) taxonomic publication ever published on dragonflies. It treated 556 species placed in 120 genera. His other major papers include: *Libellen (Odonata) aus der Region der amerikanischen Kordilleren von Costarica bis Catamarca* (1918) and *The Odonata or dragonflies of South Africa* (1921). His large Odonata collection of over 40,000 specimens is deposited at Senckenberg Museum in Frankfurt am Main.

Rita

Apache Dancer *Argia rita* **Kennedy**, 1919
[JS *Argia munda* Calvert, 1902]

This is a toponym; the species is named after the the Santa Rita Mountains in Arizona, the type locality.

Rita

Threadtail sp. *Nososticta rita* **Kovács** & **Theischinger**, 2016

Rita Kovács (b.2001) is the daughter of the author Tibor Kovács (b.1965).

Roberts

Robert's Forestwatcher sp. *Notiothemis robertsi* **Fraser**, 1944
[A Western Forestwatcher]

James Ernest Helme Roberts MB, OBE, FRCS, FRES (1881–1948) was an eminent British Harley Street physician and amateur entomologist and keen member of the Alpine Gardener Society. He studied medicine at St Bartholomew's Hospital qualifying (1906) and taking a further degree in surgery (1908). He was a houseman at Great Ormond Street (1913) and was a pioneer surgeon during WW1 during which he served (1914–1919) with distinction. Afterwards he was a surgeon at Queen Mary's Hospital, Roehampton, before returning to St Bart's (1919) until retiring (1946). The etymology says: "...*named after Mr J.E.H. Roberts in acknowledgement of much helpful criticism and loans of material and literature.*" G D Hale Carpenter collected the type specimens in Uganda.

Robinson, A

Cammaerts's Glyphtail *Isomma robinsoni* **Cammaerts**, 1987
[Orig. *Malgassogomphus robinsoni*]

Andrea Robinson was a local hunter, working for Professor Jaques Millot (q.v.), who collected the holotype in Madagascar (March 1960).

Robinson, P

Robinson's Rockstar *Tatocnemis robinsoni* **Schmidt**, 1951

Pierre Robinson assisted the collector (Madagascar 1934) of the holotype, Gregor von Olsufieff (q.v.).

Rocha

Straw-coloured Sylph *Macrothemis rochai* **Navás**, 1918
[JS *Macrothemis inacuta* Calvert, 1898]

Dr. Francisco Dias da Rocha (1869–1960) was a Brazilian naturalist and pharmacist who travelled to Europe with his father (1886). Francisco wanted to stay in Portugal to train as a physician, but instead entered his father's business, a store selling imported goods. However, in his spare time he studied science and collected natural history specimens. The collections were housed in the Museu Rocha that he founded (1887) at his home in Fortaleza and of which he was owner and director thereafter. He gave up business (1898) to concentrate on collecting and studying Ceará's flora and fauna. The museum had sections for archaeology, botany, minerology and zoology and the gardens were for local native plants, including cacti. He published the 'Boletim do Museu Rocha' (1908–1950). The journal included his own articles, mostly concerning insects that he collected in his home state such as: *Subsídio para o estudo da fauna cearense (catálogo das espécies animais por mim coligidas e anotadas)* (1950). The museum was later nationalised (1947). He also founded the School of Pharmacy and the School of Agriculture that grew into the Federal University of Ceará. He collected the type specimen in Ceará State, Brazil (1916).

Roderick

Thursday Island Mosquitohawk *Micromidia rodericki* **Fraser**, 1959

Roderick Dobson (1907–1979) (See **Dobson**).

Rogers, H

Dancer sp. *Argia rogersi* **Calvert**, 1902

H Rogers was a British amateur naturalist. He began collecting birds, insects – particularly butterflies, beetles and Odonata – and other natural history specimens, including shells, when he first went to Costa Rica (1877) and continuing well into the early twentieth century (1908). He collected this species in Caché, Costa Rica (1901) where he seems to have been based.

Rogers, JS

Sable Clubtail *Gomphus rogersi* Gloyd, 1936

Dr James Speed Rogers (1891–1955) was a biologist and entomologist who was Director, Museum of Zoology, University of Michigan (1946–1955). His interest in natural history was strengthened when he went on an expedition with the museum (1915) following which he wrote his first scientific paper on the craneflies he had collected. He had been a professor, Grinnell College, Iowa and was (1923) at Gainsville, Florida forming a Department of Biology and Geology for the University of Florida. He co-wrote: *Man and the biological world* (1952). He collected the Clubtail in Tennessee (1924). An amphibian is also named after him.

Roger Taylor

Flatwing sp. *Heteragrion rogertaylori* **Lencioni**, 2013

Roger Meddows Taylor (b.1949) is a multi-instrumentalist, best known as a songwriter and the drummer in the rock band Queen, of whom the author wrote: "*...whose powerful sound,* *wonderful lyrics and raspy voice have enchanted the world for over four decades"*. Four new species were described in tribute to the 40th anniversary of Queen – one for each of its original members.

Rokitansky

Clubtail sp. *Zonophora rokitanskyi* St. Quentin, 1973
[JS *Diaphlebia angustipennis* Selys, 1854]

Dr Gerth Rokitansky (1906–1987) was an Austrian ornithologist. He studied law at the University of Graz being awarded a doctorate of law (1930), but went on to study zoology at Munich and Vienna, where he was awarded his doctorate (1936). He was conscripted into military service (1940–1945) and taken prisoner on the Russian front. After the war, he was very ill for two years, suffering the effects of his imprisonment. He began work at the Vienna Museum of Natural History (1947) and remained there for two decades. He was Curator of Ornithology rising to Director of the museum's Department of Zoology (1969) until retiring (1971). His writing ranged from cataloguing collections, through scientific papers to popular articles calling for bird conservation.

Roland Müller

Duskhawker sp. *Gynacantha rolandmuelleri* **Hämäläinen**, 1991
Damselfly sp. *Risiocnemis rolandmuelleri* **Hämäläinen**, 1991

Roland Albert Müller (1936–2016) was an amateur Swiss zoologist and entomologist. Even as a schoolboy he was very interested in wildlife, particularly insects, making collections of butterflies and moths. He worked first as a locksmith and then for some years as a lift maintenance technician. He took a three-month course in taxidermy and in 1971 was appointed as a preparator at the Naturmuseum, St. Gallen, restoring exhibits that had been damaged by a leaky roof. After this (1986), he took a post as an archivist at Staatsarchiv in St. Gallen making microfilms, photographs for publications etc., until retiring for health reasons (1999). He made collecting trips to Colombia and Ecuador (1974), New Guinea (1979), the USA (1980 & 1988) as well as the Philippines (1985 & 1986). The first Philippines trip was for two and a half months and was a proper zoological expedition. He had intended to collect Lepidoptera, but Kiauta (q.v.) persuaded him to also collect Odonata. He and his co-workers collected the material examined by Hämäläinen (q.v.) during their seven further Philippine Expeditions (1985–1997), which (since 1987) focused on odonates. The National Museum of Natural History, Leiden, acquired (1998) The Roland Müller Collection which among much other material contains over 35,000 Odonata specimens. Among a few other papers he co-wrote with Hämäläinen: *Synopsis of the Philippine Odonata, with lists of species recorded from forty islands.* (1997). He had planned another trip (1998) but had a heart attack the month before he should have left. In the last decade of his life, he took up breeding exotic beetles. A Philippine Tiger Beetle *Thopeutica roalndmuelleri* and a caddisfly, both of which he collected, were named after him. He died at the age of eighty after a long illness. (See also **Müller, RA** & **Pistor**)

Rollinat

Demoiselle sp. *Ormenophlebia rollinati* **R Martin**, 1897
[Orig. *Lais rollinati*]

Pierre André Marie Raymond Rollinat (1859–1931) was a French herpetologist and zoologist. While still at school he trained under a local taxidermist. He spent a period in the military but thereafter spent his time in the laboratory that he kept adjacent to his large garden. There he established vivaria so he could observe the species, including European Pond Turtles, and study their breeding, hibernation and even embryonic development. He was a friend of Rene Martin (q.v.) and with him he wrote: *Description et moeurs des mammiferes, oiseaux, reptiles, batraciens et poissons de la France centrale*, 1914. His most important work was *La Vie des reptiles de la France centrale* (1934), which was re-issued no less than three times.

Roman

Neotropical Skimmer sp. *Micrathyria romani* **Sjöstedt**, 1918
Flatwing sp. *Heteragrion romani* Sjöstedt, 1918

Dr Per Abraham Roman (1872–1943) was a Swedish entomologist whose specialism was ichneumon wasps. He worked as an assistant at Stockholm Natural History Museum where his collection is housed. He twice collected in Amazonas, including the type (1914–1915 & 1923–1924). His publications include: *Entomologische Ergebnisse der schwedischen Kamtschatka-Expedition 1920–1922* (Entomological results of the Swedish Kamchatka Expedition 1920–1922).

Romana

Romana's Threadtail *Protoneura romanae* **Meurgey**, 2006

Romana Meurgey is the daughter of the author François Meurgey (q.v.). (Also see **Meurgey**)

Roppa

The Genus *Roppaneura* **Santos**, 1966

Olmiro Antônio Roppa (b.1925) of Santo Augusto, Rio Grande do Sul, Brazil, worked for many years as a preparator, collector and field assistant for the Museu National/UFRJ. He was the co-collector of *Roppaneura beckeri* in Minais Gerais, Brazil (1963), along with Professor Johann Becker (q.v.), who is honoured in the species name.

Rösel

Ruddy Darter *Libellula roeselii* **Curtis**, 1838
[JS *Sympetrum sanguineum* (Müller, 1764)]

August Johann Rösel von Rosenhof (1705–1759) was a German miniaturist, naturalist and entomologist. He is most famed for his incredibly detailed scientific drawings and paintings. Born into the nobility, he was privately educated then studied art at the Nuremberg Academy (1724–1726). A gifted portrait artist, he joined the royal court (1726–1728). After an illness, he decided to write and illustrate a book on German fauna. He spent much of his time studying insects, amphibians and reptiles resulting in two books: *Insecten-Belustigung* (1740) on insects and invertebrates and *Historia naturalis Ranarum nostratium* on frogs, published in parts (1753–1758). He was working on a volume on salamanders and lizards when he had a fatal stroke. A bush-cricket is also named after him.

Rosenberg

Glider sp. *Tramea rosenbergi* **Brauer**, 1866
Grey Duskhawker *Gynacantha rosenbergi* Brauer, 1867

Baron Carl (originally Karl) Benjamin Hermann von Rosenberg (1817–1888) was a German naturalist and geographer who collected in the East Indies, and later took Dutch nationality (1865). He enlisted in the Dutch colonial army and was in the Malay Archipelago for three decades, as cartographer and surveyor on the island of Sumatra (1840–1866), then as a civil servant in the Moluccas and around New Guinea (travelling there in the Dutch warship, *Etna*, during which he met Wallace (q.v.)). He went to Europe (1866–1868) but returned to the East Indies, finally returning to Europe (1871) in poor health. All the while he pursued his interest in ornithology, including writing a series of articles. He wrote: *Reistochten naar de Geelvinkbaai op Nieuw-Guinea in de jaren 1869 en 1870,* an important zoological and ethnographical study on New Guinea, and *Die Malayische Archipel, Land and Leute* (1878). He died in The Hague, but was buried in the family vault in Darmstadt. Kaup (q.v.) recognized several new dragonfly species among Rosenberg's material and proposed the name *rosenbergi* for them. However, since Friedrich Brauer alone wrote and published the description of the species, according to the rules of zoological nomenclature, Brauer, not Kaup, is the author of the species. A mammal and seven birds are also named after him.

Ross

Angola Claspertail *Onychogomphus rossii* **Pinhey**, 1966
Sylvan sp. *Coeliccia rossi* **Asahina**, 1985

Dark Yellowwing *Chlorocnemis nubilipennis rossii* Pinhey, 1969
[JS *Allocnemis flavipennis* (Selys, 1863)]

Dr Edward Shearman Ross (1915–2016) was (1980–2016) Curator of Entomology Emeritus at California Academy of Sciences. The University of California, Berkeley awarded both his BS in entomology (1937) and his PhD (1941). Shortly after this he entered military service (1942–1946) as Major in the Army Sanitary Corps. He became commanding officer of a malaria survey unit in New Guinea and Philippine combat zones, where he specialised in mosquito-borne diseases and parasitology. He was Assistant in Charge to Curator and Chairman, Department of Entomology, California Academy of Sciences (1939–1980). He was also a very fine pioneer 'macro' photographer. He was able to drive into remote areas in a four-wheel-drive safari truck he developed with a tent on the roof. It was fitted with other amenities that allowed him and his colleagues to live in the wilds for extended periods of time. He undertook fieldwork all over the world including New Guinea, Philippines, Mexico, Madagascar, Australia, Japan, China, Venezuela, Tunisia, Turkey, etc., hunting what he jokingly called 'small game' – insects. He once said: *"Like the man who climbed the mountain because it's there, I have to study embiidina simply because they exist."* He collected the holotype of the *Coeliccia* in Assam (October 1961). Ross also provided Pinhey with the dragonfly specimens he collected during his African expedition (1957–1958) for him to study. The yellowwing was collected by J C Thompson in Sierra Leone (1907) and the gomphid by Fr. Eduardo in Angola (1951). Although they were not collected during Ross' expedition, Pinhey named the dragonflies after him. He has published widely, including: *Lifelong safari: the story of a 93-year-old peripatetic insect hunter* (2009).

Rosser

Dancer sp. *Argia rosseri* **Tennessen**, 2002
Pond Damselfly sp. *Mesamphiagrion rosseri* Bota-Sierra, 2013

Dr Rosser William Garrison (b.1948) (See **Garrison**).

Rothschild

Spreadwing sp. *Lestes rothschildi* **Martin**, 1907
[JS *Lestes virgatus* (Burmeister, 1839)]

Baron Maurice Edmond Charles (Karl) de Rothschild (1881–1957) was a member of the French branch of the famous banking family. In his youth, he was a well-known playboy and quarrelled with his relations over an investment they regarded as risky – but which turned out to be highly profitable, as did many of his subsequent ventures. He became a politician and was a member of the Chamber of Deputies (1919–1929) and a senator (1929–1945), one of the few to vote against giving Marshall Pétain full powers (1940). He was instrumental in helping de Gaulle become the leader of the Free French in exile in England during WW2, but he later upset him and was virtually banished from France to the Bahamas. He travelled in East Africa where collected the type in Kenya (1904–1905). He also collected art and was a patron, philanthropist and big game hunter. Seven birds and a reptile are also named after him.

Rougeot

Cascader sp. *Zygonyx rougeoti* **Pinhey**, 1960
[JS *Zygonyx speciosus* Karsch, 1891]

Pierre-Claude Rougeot (1920–2002) was a French entomologist who was a Lecturer and Deputy Director General of the Laboratoire d'Entomologie du Muséum National d'Histoire Naturelle, Paris (1960). He collected part of the type series in Gabon (1957). He wrote many papers, particularly on butterflies and a number of the sections of: *Guide des papillons d'Europe et d'Afrique du Nord* (1975–1978) and longer works such as *Entomological Missions in Ethiopia* (1977).

Roux

Flatwing sp. *Argiolestes rouxi* **Ris**, 1915
[JS *Eoargiolestes ochraceus* Montrousier, 1864]

Dr Jean B Roux (1876–1939) was a Swiss zoologist. Geneva University awarded his doctorate (1899). He studied protozoa in Berlin (1899–1902), then was Curator, Natural History Museum, Basel, Switzerland (1902–1930). He was in New Guinea and Australia (1907–1908) and in New Caledonia and the Loyalty Islands (1911–1912), where he collected the type with Fritz Sarasin (q.v.). He wrote: *Les Reptiles de la NouvelleCalédonie et des Îles Loyalty* (1913). Six reptiles, an amphibian and a bird are also named after him.

Roy

Clubbed Cruiser *Macromia royi* **Legrand**, 1982
[Possibly JS *Phyllomacromia overlaeti* Schouteden, 1934]

Dr Roger Roy (b.1929) is an entomologist. He researched at the Institut Fondamental d'Afrique Noire at Dakar, Senegal (1958–1992), where he was in charge of Orthoptera and Dictyoptera, his particular focus being mantids. He collected the holotype in Senegal (November 1965) and invited the author to visit there (1981). He was President of the French Entomological Society (2008–2009) and is currently associated with the entomology collection at MNHN, Paris.

Rozendaal

Shadowdamsel sp. *Protosticta rozendalorum* **Van Tol**, 2000

Frank Gerard Rozendaal (1957–2013) and Caroline Rozendaal née Kortekaas were Dutch *"...ornithologists with a keen interest in Odonata"*. Frank was a founding member of the Dutch Birding Association and designed their logo; his passion was Southeast Asian birds. He wrote a number of papers on bats and birds such as: *Notes on macroglossine bats from Sulawesi and the Moluccas, Indonesia, with the description of a new species of Syconycteris Matschie, 1899 from Halmahera (Mammalia: Megachiroptera)* (1984) and the co-written: *Territorial songs and species-level taxonomy of nightjars of the Caprimulgus macrurus complex, with the description of a new species* (2004). Caroline is a biologist and helped her husband with his fieldwork. Frank was also keen on bats and named one after her – the Halmahera Blossom Bat *Syconycteris carolinae* – as well as a bird, the Tanimbar Bush Warbler *Cettia carolinae*. They collected the type specimens in Sangihe Island, off Sulawesi (May 1985). (See also **Caroline**)

Rua

Bluet sp. *Enallagma rua* **Donnelly**, 1968

Ruth Wills Birkhoff (b.c.1940), participant of the collecting trip to Guatemala (1964) during which the new species was found.

Ruby

Coraltail sp. *Ceriagrion rubiae* **Laidlaw**, 1916

Ruby Laidlaw was the daughter of the author F F Laidlaw. Although no etymology was given, it is certain that this Indian species was named after her. Previously he had named two Odonata species after his eldest daughter, Mary Louisa, and his son Eric was later credited with an eponym. Ruby and Eric were twins. (Also see **Louisa**, **Mary** (**Laidlaw**), **Laidlaw** & **Eric**)

Rudolph

Dancer sp. *Argia rudolphi* **Garrison** & von **Ellenrieder**, 2017

Rainer Rudolph (b.1943) is Professor Emeritus at the Westfälische Wilhelms-Universität Münster, Institut für Didaktik der Biologie (Institute for Didactics in Biology), North Rhine-Westphalia, Germany. He is one of the founders of the Gesellschaft deutschsprachiger Odonatologen (GdO) (1979), was Editor-in-Chief of the journal 'Libellula' (1981–1982) and an editorial board member of 'Libellula' (1982–1991). The main topics of his odonatological publications are regional faunistics, but especially Odonata in poetry and arts. He is an accomplished artist, illustrating his own work and that of others. The etymology records that the species is named: *"…in honour of our colleague Rainer Rudolph, in recognition for his generous support of the International Dragonfly Funds, which has helped further the knowledge of Odonata worldwide over the past twenty years".*

Rupasinghe

Rupasinghe's Shadowdamsel *Ceylonosticta rupasinghe* Priyadarshana & Wijewardana, 2016

Mahinda Sisirakumara Rupasinghe is Professor of Applied Sciences and Head of the Department of Natural Resources at the Sabaragamuwa University, Sri Lanka. His doctorate was awarded by Mainz University. Among his scientific papers is: *Patterns of crop raiding by Asian elephants in a human-dominated landscape in southeastern Sri Lanka* (2011).

Rüppell

Rüppell's Longleg *Notogomphus ruppeli* **Selys**, 1858
[Orig. *Gomphus ruppeli*]

Wilhelm Peter Eduard Simon Rüppell (1794–1884) was a German collector. He went to Egypt and ascended the Nile as far as Aswan (1817), and later made two extended expeditions to Northern and Eastern Africa, Sudan (1821–1827) and Ethiopia (1830–1834), where he collected the type. Abdim Bey helped him in Egypt. Although he brought back large zoological and ethnographical collections, his expeditions impoverished him. He wrote: *Reisen in Nubien, Kordofan und dem Petraischen Arabien* (1829), *Systematische Übersicht der Vögel Nord-ost-Afrikas* (1845) and *Reise in Abyssynien* (1838–1840). He also collected in the broadest sense and presented his collection of coins and rare manuscripts to the Historical Museum in Frankfurt (his home town). Eleven birds, five mammals, an amphibian and two reptiles are also named after him.

Rutherford

Feeblewing *Neophya rutherfordi* **Selys**, 1881
Western Orange Emperor *Anax rutherfordi* **McLachlan**, 1883
[JS *Anax speratus* Hagen, 1867]

David Greig Rutherford FLS, FRGS, FZS (d.1881) was a British entomologist and anthropologist. He was a member (1876), like McLachlan, of the Royal Entomological Society of London in whose proceedings he published a number of papers. These included: *Notes regarding some rare Papiliones* (1878) and *Description of a New Goliath Beetle from Tropical West Africa* (1879). He exhibited (1878) at the London Entomological Society two specimens of an orthopterous insect *Palophus centaurus*, West, from Old Calabar (Nigeria) and, on another occasion, exhibited and communicated a description of a new species of *Cetoniidæ*, from Mount Cameroon. McLachlan was a fellow member and attended the same meeting. He collected (1870s & 1880s) at a number of locations in western Africa such as Calabar (Nigeria), Mount Cameroon and Sierra Leone. He died in West Africa during a natural history expedition.

S

St Serapia

Green Skimmer *Orthetrum serapia* **Watson**, 1984

St Serapia was a slave and martyr; she was the servant of St Sabina and was responsible for the Roman noblewoman's conversion to Christianity. Both Sabina and Serapia were subsequently beheaded during the persecutions of Emperor Hadrian.

Saalas

Aztec Dancer *Argia saalasi* **Valle**, 1942
[JS *Argia nahuana* Calvert, 1902]

Professor Uunio Saalas (Unio Sahlberg before 1906) (1882–1969) was a Finnish zoologist and entomologist, the son of Professor John Reinhold Sahlberg (q.v.). He gained his masters (1908) and doctorate (1919) at Helsinki University and later worked there as Professor of Agricultural and Forest Zoology (1923–1952). He was a very prolific author publishing several hundred of papers and many books, mainly on forest entomology, beetles and the history of entomology. His major publications include: *Die Fichtenkäfer Finnlands* (two volumes 1917, 1923) and *Über die Flügelgeäder und die phylogenetische Entwicklung der Cerambyciden* (1936). He also wrote extensive biographies of his father (1960), grandfather (1958) and great-grandfather (1956), who were all renowned entomologists, and an autobiography covering his first 37 years of life. He collected the type of the damselfly during a visit to California (1928). (Also see **Sahlberg**)

Sachiyo

Skydragon sp. *Chlorogomphus sachiyoae* **Karube**, 1995

Sachiyo Karube (née Nirasawa) (b.1970) is an aquatic coleopterologist and is married to the author, Haruki Karube. She was a graduate of the Tokyo University of Agriculture. He said in an email: *"This species was dedicated for my wife who collected the first specimens. (She) studied systematic and life history of Mordellid beetles in her student days."* She co-wrote a number of papers including: *Notes on Cicada Shells Observed at Ueki, Kamakura from 1998 to 2001, 2002 to 2005 & 2006 to 2009.* The skydragon was known only from Tam Dao National Park, Vinh Phuc Province, Vietnam, but the author has since found it (2014) in Xuan Son National Park in Phu Tho Province. (Also see **Karube**)

Saeger

Eastern Mushroom Skimmer *Orthetrum saegeri* **Pinhey**, 1966

Professor Henri Jules de Saeger (1901–1994) was an entomologist who was Secretary to the Board of Directors of the Institut des Parcs Nationaux du Congo. He led a mission to

Garamba National Park, of the then Belgian Congo (1949–1952), during which he made an extensive collection, including the type, much of which formed the basis of study over a number of years. He wrote a number of papers including species descriptions and he also wrote the report of the ecological mission: *Exploration du Parc National de la Garamba* (1954) and later: *Les parcs nationaux du Congo Belge, où, La nature est men* (1956).

SAFE

Griptail sp. *Phaenandrogomphus safei* **Dow** & Luke, 2015

This is an acronym standing for the Stability of Altered Forest Ecosystems (SAFE) Project. The etymology reads: "*safei, a noun in the genitive case, formed from the acronym of the Stability of Altered Forest Ecosystems (SAFE) Project, where the type specimen was collected*" (in Kalabakan Forest Reserve, Sabah, Malaysian Borneo).

Sahlberg

Treeline Emerald *Somatochlora sahlbergi* Trybom, 1889

Professor Johan 'John' Reinhold Sahlberg (1845–1920) was a Finnish entomologist, inspiring teacher and collector. He collected insects from his school days, later inheriting collections made by his father and grandfather. He studied at the University of Helsinki, graduated MSc (1869) becoming a docent (1871) and gaining his doctorate (1880). He became Extraordinary Professor of Entomology at the University of Helsinki until retiring (1883–1918) and was also Curator of the Natural History Museum of the University of Helsinki during which time he united many previously separate collections, thus facilitating their use. He made expeditions, always assiduously collecting insects, to various parts of Finland, Lapland (1867), Russian Karelia (1869), Kola Peninsula (1870), Siberia (1876), the Mediterranean area (1895–1896, 1898-1899, 1903–1904 & 1906) and with his wife Mimmi and other family members, to Turkestan (1896). His foreign journeys amounted to three years in total. The Emerald was collected on the arduos Siberia expedition in which both Sahlberg and Philip Trybom participated as did Théel (q.v.). Sahlberg estimated that during this six month journey he travelled a total of 17,600 km; 3,200 km by railway, 6,950 km by ship along rivers, 1,600 km by various small craft (rowing, drifting on the current or drawn by dogs) and 5,900 km in horse-drawn vehicles. He described c.500 new insects, particularly (c.400) Coleoptera and Homoptera. He was a prolific writer penning over 350 works mainly on Coleoptera, Hemiptera and other insects. His major works include several catalogues of Finnish Coleoptera in the series *Enumeratio Coleopterorum Fenniae*. He was a modest, humble man, a devout Christian who could not accept Darwinism. His son, Uunio Saalas, wrote a 620-page biography of him. (Also see **Saalas**)

Saint Johann

Granite Ghost *Bradinopyga saintjohanni* Baijal & Agarwal, 1956
[JS *Bradinopyga geminata* Rambur, 1842, A. Rock Dweller]

This is a toponym; St. John's College was where the authors worked.

Sakai

Goldenring sp. *Anotogaster sakaii* **Zhou**, 1988

Professor Dr Seiroku Sakai (1924–2004) was a Japanese entomologist with a particular interest in Dermaptera (earwigs). He studied insect population and toxicology at Kyoto University, graduating in 1948 then completing his doctorate (1952), and becoming assistant professor, then full professor (1953) at Nagoya University. He set up his own company for pest control (1955) and also worked for Yashima Chemical Industry as a technical adviser in order to support his late brother's family. He became Professor in General Education, teaching biology and chemistry at Daito Bunka University (1966–1996) and continuing his Dermaptera research. He sought to gather all the original descriptions of new species of Dermaptera and to take photographs of all the existing types. To fulfil this goal, he travelled all over the world to visit museums, entomological laboratories and private collections assisted by his wife Toshiko. His Dermaptera collection is now housed by Osaka Museum of Natural History. He was elected President of Bio-geographical Society of Japan (1994–2003). He compiled the results of his survey and published them as a long series of books under the title *Dermapterorum Catalogus: A Basic Survey for Integrated Taxonomy of the Dermaptera of the World* (1970–1996); he followed this with the six volume *Forficula* (1997–2000). He gave himself a posthumous name while he was alive (although such a name is usually given by a Buddhist priest after death); the name translates as 'Pure and Undefiled Man Who Studies Dermaptera and Enjoys Drinking Sake'. An earwig, *Anisolabis seirokui,* that he found and a worm are also named after him.

Sakeji

Cream-sided Citril *Ceriagrion sakejii* **Pinhey**, 1963

This is a toponym; it commemorates Sakeji School, which is a Christian boarding school overlooking the Sakeji River near Kalene Hill in the Ikelenge District of Zambia.

Saliceti

Pond Damselfly sp. *Andinagrion saliceti* **Ris**, 1904
[Orig. *Oxyagrion saliceti*]

The name is not an eponym but refers to the Weeping Willow (*Salix babylonica*), which occurs on the coast around Buenos Aires, where the species was found.

Sallé

Masked Clubskimmer *Libellula sallaei* **Selys**, 1868
[JS *Brechmorhoga p pertinax* Hagen, 1861]

Auguste Sallé (1820–1896) was a French veterinarian, taxonomist, malacologist, entomologist and traveller. He moved to Mexico with his mother (1932) and she often accompanied him on expeditions. He collected, mostly birds and insects, especially Coleoptera, in the US, West Indies, Venezuela and Central America, including Mexico (1846–1856). He became a very successful natural history dealer in Paris, specialising in insects. He collected the type in Mexico (c.1867). His writings include descriptions of new species of birds and insects, as well as his journeys and a memoire. Many other taxa, including two reptiles, two fish and seven birds are also named after him as well as a great many insects.

Sandra

Tennessee Clubtail *Gomphus sandrius* **Tennessen**, 1983

Sandra Faye Tennessen née Erdman (b.1947) is the wife of the author, Dr Kenneth Joseph Tennessen (b.1946). (Also see **Tennessen**)

Santos

The Emerald genus *Santosia* **Costa** & Santos, 1992
[Homonym of *Santosia* Stål, 1858 in Heteroptera]
The Emerald genus *Cordulisantosia* **Fleck** & Costa, 2007
[replacement name for *Santosia* Costa & Santos, 1992]

Skimmer sp. *Elga santosi* **Machado**, 1954
[JS *Elga leptostyla* Ris, 1911]
Pond Damselfly sp. *Oxyagrion santosi* **Martins**, 1967

Dr Newton Dias dos Santos (1916–1989) was a Brazilian odonatologist and promotor of science education. He graduated at Escola de Ciências da Universidade do Distrito Federal, Rio de Janeiro (1938), took a medical qualification at the Faculdade Nacional de Medicina (1940) and was awarded a doctorate by Faculdade Nacional de Filosofia (1950). After receiving his medical degree he worked as a physician for two years, maintaining a laboratory of clinical analysis. Then he worked at the Experimental Fishery Station of Pirassununga, where his interest in Odonata started. He became Director of the Zoologist Section of Museu Nacional of the Federal University of Rio de Janeiro (1944–1945), then held different posts in the museum, becoming director (1961–1963). After retiring (1987) he was appointed as Professor Emeritus of Museu Nacional. He also worked as a teacher for more than 40 years, teaching various biological topics in several institutions, ranging from high school to university. His book: *Praticás de Ciências* (1955) on the practices in science opened a new era in the methodology of science teaching in schools in Brazil. He collected dragonflies all over Brazil. During his 600th collecting trip (1986), he had a heart attack on a muddy road in Amazonas, from which he never completely recovered. Of his 132 publications, 127 are devoted to Odonata. Although basically on taxonomy his papers, most of them only a few pages long, include notes on habitats, ecology and behaviour. His major paper was his doctoral thesis (1950) on the genus *Nephepeltia*. Many of his papers were on species in the genera *Micrathyria* and *Leptagrion*. He described 53 new species of Odonata and named four new genera, all from Brazil. He was also honoured in the name of at least one fish for his contributions to the museum's fish collection. (Also see **Newton**, **Newton Santos**)

Sapho

Demoiselle Genus *Sapho* **Selys**, 1853

Sappho (c.630–c.570BC) was an ancient Greek poet from the Isle of Lesbos, best known for her poems about love. The genus name was an emendation.

Sara Lisa

Congo Riverjack *Mesocnemis saralisa* Dijkstra, 2008

Sara Elisabeth 'Lisa' von Linné née Moraea (1716–1806) was the wife of Carl von Linné (Linnaeus). (Also see **Linnaeus**)

Sarasin

Flatwing sp. *Caledopteryx sarasini* **Ris**, 1915
[Orig. *Argiolestes sarasini*]

Dr Karl Friedrich 'Fritz' Sarasin (1859–1942) was a Swiss naturalist and anthropologist who was (1896–1942) head of the Basel Museum of Ethnology and head (1899–1919) of the Natural History Museum there too. He was also head of zoology (1919–1936) and remained (1936–1942) in charge of the ornithology department. He made a number of exploratory trips with his cousin Paul Sarasin (1856–1929) and others to Ceylon (Sri Lanka) (1883–1886 & 1907), Egypt and Sinai (1889) and Celebes (Sulawesi) (1893–1896 & 1902–1903), during which they collected flora and fauna as well as artefacts. He collected the eponymous type in New Caledonia (1911) during an expedition there and to the Loyalty Islands with his younger sister and the then head of the Basel Natural History Museum, Jean Roux. He co-edited and co-wrote with Roux the four-volume: *Nova Caledonia. Forschungen in Neu-Caledonien und auf den Loyalty-Inseln. Recherches scientifiques en Nouvelle-Calédonie et aux îles Loyalty* (1913–1926), among other books and papers. His final trip was to Siam (Thailand) at the age of seventy-two. An amphibian, five birds, a mammal and five reptiles are also named after him.

Sarep

Sarep Sprite *Pseudagrion sarepi* Kipping & Dijkstra, 2015

This is an acronym standing for 'Southern African Regional Environmental Program'.

Sasamoto

Sylvan sp. *Coeliccia sasamotoi* **Do**, 2011

Dr Akihiko Sasamoto (b.1977) is a Japanese psychiatrist whose PhD in medicine was awarded by Kyoto University (2014). He is also an amateur entomologist and editorial secretary of 'Tombo – Acta Odonatologica Japonica', the journal of the Japanese Society of Odonatalogy (since 2007). His main interest in odonatology is systematic classification and phylogeny in east Asian Odonata. He is a friend of Do Manh Cuong (q.v.) and cooperated with him in the study of dragonflies in northern Vietnam (2003). He has published widely, often with others, including the papers: *Description of a new Nososticta species from Biak Island, Indonesia (Zygoptera: Protoneuridae)* (2007) and *Taxonomic revision of the status of Orthetrum triangulare and melania group (Anisoptera: Libellulidae) based on molecular phylogenetic analyses and morphological comparisons, with a description of three new subspecies of melania* (2013). He has described 13 new species, 3 new subspecies and one new genus of Odonata. He collected the eponymous type in Laos (2009).

Satô

Sylvan sp. *Coeliccia satoi* **Asahina**, 1997
Shadowdamsel sp. *Protosticta satoi* Asahina, 1997
[Orig. *Protosticta khaosoidaoensis satoi*]

Skydragon ssp. *Chlorogomphus nasutus satoi* Asahina, 1995

Professor Masataka Satô (1937–2006) was a Japanese entomologist most interested in Coleoptera. He graduated from the Agricultural Department of Ehime University and was awarded his PhD by Kyoto University defending the thesis: *A revisional study on the superfamily Dryopoidea (Colepotera) of Japan*. He became Professor of Domestic Economy at Nagoya Women's University and then Honorary Professor. He published widely (1959–2006), mostly on beetles, from: *Notes on Japanese Hydraena (Coleoptera: Limnebiidae)*, (1959) to *A review of the genus Cucujus Fabricius* (2007) and including books such as: *Beetles and Nature* (1995).

Saunders

Bannerwing sp. *Thore saundersii* **Selys**, 1853
[JS *Polythore picta* Rambur, 1842]
Pincertail sp. *Onychogomphus saundersii* Selys, 1854

Captain William Wilson Saunders FRS FLS (1809–1879) was a British insurance broker, entomologist, botanist, horticulturalist and 'accurate artist in natural history subjects'. He attended the East India Company military academy (1827–1829) and was commissioned in the Royal Engineers. He spent a year in India (1830–1831) but then resigned his commission. However, while there he made a large collection of plants and insects, which he brought back and which formed the basis of his later collections. He joined his uncle's firm as an underwriter at Lloyds and married his cousin (1832), so his uncle became his father-in-law. He remained an enthusiastic amateur natural historian throughout his life. He was a founding member and later President of the Entomological Society (1841–1842 & 1856–1857), Treasurer of the Linnean Society of London (1861–1873) and a Fellow of the Royal Society (1853). Later he served in various offices of the Royal Horticultural Society. His entomological interests centred on Lepidoptera and Hymenoptera, but his collection contained insects from all orders. His Diptera collection held many new species and was described by F Walker, and several other entomologists across Europe, who wrote the scientific descriptions of the species. Selys visited Saunders (1851), who showed him the 'beautiful' collection to facilitate his monograph of 'des Caloptérygines'. He wrote more than thirty-five scientific papers (1831–1877) including his first entomology paper: *On the habits of some Indian insects* (1834) and *Descriptions of two hymenopterous insects from northern India* (1842). He also edited 'Insecta Saundersiana' (1850–1869), which contained descriptions of many of the insects in his collection by Francis Walker, Henri Jekel, and Edward Saunders, his son. Furthermore, some of the illustrations featured in Hewitson's Exotic Butterflies were drawn from specimens in Saunders' collection. When his shipping insurance business failed (1873), he was almost financially ruined and had to sell his insect collection to the British Museum and the Hope Museum, Oxford. Thereafter he retired and devoted himself to horticulture. He collected in Britain but also in Corfu and Albania.

Sauter

Slendertail sp. *Leptogomphus sauteri* **Ris**, 1912
Lillysquatter sp. *Agrion sauteri* Ris, 1916
[JS *Paracercion sieboldii* Selys, 1876]

Dr Hans Sauter (1871–1943) was a German entomologist who also had an interest in herpetology. He studied biology at Ludwig-Maximilians-Universität München and

Eberhard Karls Universität Tübingen. He collected insects in Formosa (Taiwan, then a Japanese 'possession') (1902–1904). He was in Tokyo (1905) and then returned to Taiwan for the rest of his life where he collected the types (1908 & 1911). He worked for a British trading company but spent as much time as he could on entomology. Though Japan and Germany were enemies during WW1, he kept his job and continued collecting, although he was kept under observation. He was the first person to offer private piano lessons in Taiwan and gave German and English lessons too.

Sawano

Clubtail ssp. *Davidius moiwanus sawanoi* **Asahina** & **Inoue**, 1973

Professor Dr Juzo Sawano (1913–2007) was a Japanese teacher of human anatomy and an amateur entomologist. He held various posts at Hiroshima University, including Dean of the School of Medicine (1974–1977) and after retirement was professor emeritus. All his papers on dragonflies were in Japanese and were about Japanese fauna. He collected the type in Hiroshima Prefecture, Japan (June 1962).

Say

Say's Spiketail *Cordulegaster sayi* **Selys**, 1854

Thomas Say (1787–1834) was an American naturalist whose primary interests were conchology, herpetology, ornithology and entomology, who has been called 'the Linnaeus of the New World'. He was an apothecary who as a naturalist was largely self-taught. He became a charter member and founder of the Academy of Natural Sciences of Philadelphia (1812) and was appointed chief zoologist with Major Stephen H Long's expeditions, which explored the Rocky Mountains (1819–1820). He was also chief naturalist on Long's next expedition (1823) to the headwaters of the Mississippi. He went to the utopian village of 'New Harmony' in Indiana, married artist Lucy Way Sistatre and settled there (1826–1834). Say wrote the three volume: *American Entomology, or Descriptions of the Insects of North America* (1824–1828) and the six-part *American Conchology* (1830–1834), which his wife illustrated. He described over 1,000 new species of beetles and over 400 new insects of other orders, including 37 Odonata (1840) published posthumously. His wife was the first woman admitted to membership of the Academy of Natural Science. He died in New Harmony, apparently from typhoid fever. Two birds, a reptile, a mammal and at least one fish are named after him as are other taxa.

Schaus

Ringtail sp. *Erpetogomphus schausi* **Calvert**, 1919

William Schaus (1858–1942) was an American entomologist. He was educated at Exeter Academy then in France and Germany. He began travelling (1881) to Mexico, Costa Rica, Guatemala, Panama, Cuba, Jamaica, the Guianas, Colombia and Brazil, as well as other parts of the Caribbean, amassing a collection of over 200,000 specimens of Lepidoptera. He added to his own collection by buying the Dognin collection of 26,000 tropical moths and around 5,000 butterflies from the old world. He collected with John T Barnes particularly in Mexico, Guiana and Costa Rica, and collected the eponymous holotype during an extended collecting trip with Barnes to Guatemala (1915–1918). He later donated that collection to the AMNH (1901 & 1905) and donated his own collection to the Smithsonian. He joined the Bureau of Entomology of the US Department of Agriculture (1919) and also became Honorary Curator of Insects at the Smithsonian (1921). He was awarded an honorary MA by the University of Wisconsin (1921) and honorary doctorate by the University of Pittsburgh (1925). He described no fewer than 329 new genera and over 5,000 species of Lepidoptera, mostly from tropical America. Among his written works was: *Descriptions of new American butterflies* (1902).

Schiel

Sprite sp. *Pseudagrion schieli* Villanueva, 2010

Dr Franz-Josef 'Jupp' Schiel (b.1968) is co-founder (1995) of INULA (Institut für Naturschutz und Landsschaftsanalyse). He studied biology at the University of Freiburg (1989–1996) and he then studied for his PhD as an external candidate at the University of Oldenburg, with a thesis entitled: *Preimaginal Ecology of Dragonfly Species of Astatic Waters*. He is also a founding member and member of the board of the Schutzgemeinschaft Libellen in Baden-Württemberg (since 2001). He funded the trip that collected the sprite, by sponsoring the International Dragonfly Fund. Villanueva dedicated the species to Schiel through information provided to him by Martin Schorr of the IDF. The etymology states that the species was named after: "...*Jupp Schiel of Sasbach, Baden-Württemberg, Germany for his contribution to this study and to German odonatology.*"

Schiess

Flatwing sp. *Teinopodagrion schiessi* **De Marmels**, 2001

Dr Heinrich Schiess (b.1952) is a Swiss entomologist. Among his published work is the book: *Schmetterlinge und Libellen in der Schwantenau* (2005) and the papers *Zur Insektenfauna der Umgebung der Vogelwarte Sempach, Kanton Luzern* (1982) and the co-written: *Onychogomphus uncatus in Deutschland: die historischen Funde am Hochrhein (Odonata: Gomphidae)* (2008).

Schmid

Velveteen sp. *Schmidtiphaea schmidi* **Asahina**, 1978

Skydragon ssp. *Chlorogomphus preciosus fernandi* Asahina, 1986
Clubtail ssp. *Davidius aberrans schmidi* Asahina, 1994

Dr Fernand Schmid (1924–1998) was a Swiss-Canadian entomologist; a world authority on Trichoptera (caddisflies). The University of Lausanne awarded his License in Science (1951) and DSc (1953). His early research involved describing new species of Trichoptera from Switzerland, Spain and the neotropics and reviewing the collection of them made by Navás (q.v.). He made collecting expeditions to the Indian sub-continent, including Pakistan (1953) and Sri Lanka (1954) then visiting Iran (1955) and mainland India (1958–1962). He took the position (1963) of Trichopterist at the Entomological Research Institute, in Ottawa (later known as the Biological Research Institute of Agriculture Canada). He continued to conduct research until the end of his life and wrote 111 papers on Trichoptera describing an amazing 1,424 new species and held the record for the most Trichoptera described by any one person. He also studied other taxa including butterflies, had a passion for orchids and loved Mozart. Among his most important papers and books was: *Some New Trends in Trichopterology* (1979), but he wrote widely from: *Notes sur l'armature céphalique des Sericostomatinae (Trichoptera)* (1947) to *Genera of the Trichoptera of Canada and adjoining or adjacent United States. The insects and arachnids of Canada Part 7* (1998).

Schmidt

The genus *Schmidtiphaea* **Asahina** 1978

Flatwing sp. *Philogenia schmidti* **Ris**, 1918
Sprite sp. *Pseudagrion schmidtianum* **Lieftinck**, 1936
Pincertail sp. *Onychogomphus schmidti* **Fraser**, 1937
Tropical King Skimmer sp. *Orthemis schmidti* **Buchholz**, 1950
Clubtail sp. *Shaogomphus schmidti* Asahina, 1956
[Orig. *Gomphus schmidti*]
Albertine Jewel *Chlorocypha schmidti* **Pinhey**, 1967
Goldenring sp. *Neallogaster schmidti* Asahina, 1982
Sylvan sp. *Coeliccia schmidti* Asahina, 1984
Skydragon sp. *Chlorogomphus schmidti* Asahina, 1986
Angola Bluet *Enallagma schmidti* Steinmann, 1997
[Unnecessary replacement name for *Enallagma risi* Pinhey, 1962; JS *Pinheyagrion angolicum* (Pinhey, 1966)]

Dr Erich Walther Schmidt (1890–1969) was a German entomologist and odonatologist. He studied mathematics and natural sciences at Bonn, Freiburg and München Universities (from 1909) and was awarded his doctorate by Bonn University (1915) defending his thesis on dragonfly morphology: *Vergleichende Morphologie des zweiten und dritten Abdominalsegments bei männlichen Libellen.* He fought as an NCO in WW1 (1914–1918), and was slightly wounded. After the war, he studied further to qualify as a teacher, teaching in a high school in Bonn (1920–1921). Then he worked as an assistant at the Horticulture College in Geisenheim am Rhein (1921–1926); as assistant at Keizer-Wilhelm-Gesellschaf Entomological Institute in Berlin (1927–1934); then at the Zoological Research Unit of Museum Alexander Koenig in Bonn (1934–1936) and as a local civil servant in Bonn (1936–1941). Thereafter he worked independently receiving grants from various foundations, before being granted a pension (1955). He suffered a bad head wound during a bombing raid on Bonn (late 1944) and had to spend a month in hospital. He was an avid collector of dragonflies, even doing some collecting while fighting in France during WW1. He made numerous collecting trips to countries in Central and Southern Europe, the Middle-East and Algeria. He bequeathed his Odonata collection (over 30,000 specimens of c.1,200 species) to Syoziro Asahina (q.v.), who received the collection in Tokyo. He wrote 93 publications (1915–1968), mostly on Odonata, covering fauna from several continents including the book: *Libellen, Odonata* in the series *Die Tierwelt Mitteleuropas* (1929) and research papers, such as *Odonata der Deutschen Limnologischen Sunda-Expedition* (1934), *Odonata, nebst Bemerkungen über die Anonisma und Chalcopteryx des Amazonas-Gebiet* (1943), *Libellen aus Portugiesisch Guinea, mit Bemerkungen über andere aethiopische Odonaten* (1951), *The Odonata of Madagascar, Zygoptera* (1951), *Revision der Gattung Microstigma Rambur*

(Ordnung Odonata, Zygoptera) (1958) and *Libellen aus Burma, gesammelt von Dr. R. Malaise, Stockholm* (1964). He also compiled a bibliography of odonatological publications *Bibliographia odonatologica*, but unfortunately it remained incomplete and only one part was published (1933). He described 104 new species and subspecies of Odonata. Many of his numerous subspecies have been rejected by later researchers. He was an eccentric, who wrote some unusual papers, such as: *Dürfen Entomologen heiraten?* (1955) (Should entomologists get married?), which he ended as follows: *Glücklich leben die Zikaden, den sie haben stumme Weiber!* (The cicadas have a happy life, since they have silent wives!) Needless to say, he remained a bachelor. While walking on a road near his home at Kippenhohn, on his way to study dragonflies, he was hit by a car and fatally injured.

Schneider, G

Skimmer sp. *Orthetrum schneideri* **Förster**, 1903

Gustav Schneider (1867–1948) was a Swiss zoologist, botanist, collector and trader. He trained at, and became the custodian of, the Zoological Institute, Basel where he lived. He travelled quite widely in Sumatra (1897–1899) and later to the USA, Singapore, Malaysia and Bermuda Islands, as well in other parts of Indonesia, to make zoological collections. He visited Sumatra for the first time (1888) in the company of the geologist and director of the Zoological Museum Zürich, Professor C Mösch. He was later custodian of the museums at Kolmar and Mülhausen. He also traded in natural curiosities. Among other works he wrote: *Conférence sur Sumatra* (1900) and *Die Orang Mamma auf Sumatra* (1958). He discovered this species, and collected the type in Sumatra (1899).

Schneider, W

Clubtail sp. *Burmagomphus schneideri* **Do**, 2011
Flatwing sp. *Rhinagrion schneideri* **Kalkman** & Villanueva, 2011
Dancer sp. *Argia schneideri* **Garrison** & von **Ellenrieder**, 2017

Dr Wolfgang Schneider (b.1953) is (since 2010) Honorary Staff Member at the Senckenberg Research Institute and Museum in Frankfurt. He studied biology, chemistry, and biochemistry at the Johannes Gutenberg-University in Mainz, Germany, where he graduated in biology and was awarded his PhD (1986) in zoology. From 1983 to 1988 he was a research assistant at the University of Tübingen, Special Research Unit 19 of the DFG (1983–1988) and then (1988–1991) worked as a taxonomist in the Fisheries Department of the FAO/UN in Rome. He became (1993–2008) Head of the Zoology Department at the Hessisches Landesmuseum in Darmstadt and deputy director there (2000–2005). He is a charter member of Worldwide Dragonfly Association and served as its president (2011–2013). He has made many contributions to the study of Middle Eastern zoology with a focus on Odonata, on which he has published 62 papers such as the co-written: *Dragonflies from mainland Yemen and the Socotra Archipelago - additional records and novelties* (2013). He has described four new species and subspecies of Odonata. A reptile is also named after him.

Schneider, WG

Turkish Clubtail *Gomphus schneiderii* **Selys**, 1850

Dr Wilhelm Gottlieb Schneider (1814–1889) was a German biologist, botanist, mycologist and entomologist, specialising in Neuroptera, from Breslau. His publications include: *Symbolae ad monographiam generis Chrysopae, Leach* (1851). He described nine new dragonfly species, five of which are currently recognised.

Schorr

Shadowdamsel sp. *Drepanosticta schorri* Villlanueva, 2011
Pincertail sp. *Nihonogomphus schorri* **Do** & **Karube**, 2011
Clubtail sp. *Euthygomphus schorri* **Kosterin**, 2016
Damselfly sp. *Idiocnemis schorri* Gassmann, Richards & **Polhemus**, 2016
Dancer sp. *Argia schorri* **Garrison** & von **Ellenrieder**, 2017

Martin Schorr (b.1958) is a German freelance ecologist and entomologist mostly involved in studies of wind turbine impact on avifauna. His uncle was Karl Schorr (q.v.). With Dennis Paulson, he maintains the World Odonata List website, is also a leading light of the International Dragonfly Fund (IDF founded 1996) that supports the work of other odonatologists, and he edits the fund's reports and journals and abstracts service. His diploma was written (1988) on Odonata conservation in Germany: *Grundlagen zu einem Artenhilfsprogramm Libellen der Bundesrepublik Deurschland*, which later (1990) was published by Ursus Scientific Publisher, Bilthoven. For the Societas Internationalis Odonatologica, he edited the journal 'Hagenia' (1991–1995), a newsletter directed at the German speaking members of SIO (Germany, Austria, Luxemburg, Switzerland), which was later the newsletter of the Gesellschaft Deutschprachiger Odonatologen (GDO), named 'Libellennachrichten' (since 1999), which he co-edited (2007–2010). He was Secretary General of SIO (1994–1996). Villanueva said the species was "*...named after Martin Schorr to appreciate his support and help and making the Odonata survey in Siargao and Bucas Grande (Philippines) possible*". Among his written works are: the bibliography of odonatological literature published in Germany, *Bibliografie der für Deutschland publizierten Libellenliteratur (Odonata)* (2012) and such papers as *Ecology and distribution of Lindenia tetraphylla (Insecta, Odonata, Gomphidae): a review* (1998), *Verzeichnis der Libellen (Odonata) Deutschlands* (2001) and *Libellula virgo Linnaeus, 1758 auf Grönland – Eine Neubewertung der Beobachtung von Fabricius (1780)* (2012). He co-authored a paper describing two new damselfly species from the Philippines (2011). (Also see **Martin Schorr**)

Schouteden

Congo Scissortail *Microgomphus schoutedeni* **Fraser**, 1949
Toothed Cruiser *Phyllomacromia schoutedeni* Fraser, 1954
[Orig. *Macromia schoutedeni*]
Crowned Leaftail *Phyllogomphus schoutedeni* Fraser, 1957

Professor Dr Henri Eugène Alphonse Hubert Schouteden (1881–1972) was a Belgian zoologist and explorer. The University of Brussels awarded his doctorate (1905). He worked at the Natural History Museum in Brussels (until 1910) and thereafter at the Museum of the Belgian Congo, Tervuren. He undertook many expeditions to the Congo, the first was a major collecting expedition (1921–1923) and the second was even more productive (1923–1926). He taught at the Colonial University and then taught medical entomology at the Institute of Tropical Medicine (1927–1952), both in Antwerp. He was also appointed Director of the Royal Museum for Central Africa in Tervuren (1927–1947). He wrote on both ornithology and entomology, including: *Faune du Congo Belge et du Ruanda-Urundi*. Thirteen birds, three reptiles, two amphibians and two mammals are also named after him, as are a number of fish.

Schreiber

Threadtail sp. *Neoneura schreiberi* **Machado**, 1975

Professor Dr Giorgio Schreiber (1905–1977) was an Italian biologist, zoologist, cytologist and geneticist. He graduated from Padua University (1927) and then worked in Italy (1927–1939). He fled fascist persecution and went to Brazil (1940). He undertook research at the Butantan Institute in Sao Paulo and then the University of Belo Horizonte, Minas Gerais and was invited to help construct the natural history course at its Faculty of Philosophy. Being the only person with the right education and attainment, he was soon supervising postgraduates as well as teaching. He was initially a contract teacher, then a full-time teacher of biology (1948–1953), an associate professor (1953–1968) and then full professor of zoology (1968–1974). He published widely (1928–1974). He was a good friend of the author.

Schröder

Clubtail sp. *Ebegomphus schroederi* **Belle**, 1970
[Orig. *Cyanogomphus schroederi*]

Dr Heinz Günther Schröder (b.1930) is an honorary researcher, who was formerly curator and a section head, responsible for Odonata, Auchenorrhyncha (Cicadas etc.) and butterflies of the Philippine Islands at the Natur-Museum Senckenberg, Frankfurt am Main. He has written widely on Lepidoptera, in particular with Hatashi and Treadway (q.v.) on the butterflies of the Philippines from: *Neue Lepidoptera von den Philippinen* (1976) to, with Treadaway, *Revised checklist of the butterflies of the Philippine Islands (Lepidoptera: Rhopalocera)* (2012). He has authored nine Philippine Lepidoptera species.

Schubart

Skimmer sp. *Elasmothemis schubarti* **Santos**, 1945
[Orig. *Dythemis schubarti*]

Dr Otto Rudolf Julius Schubart (1900–1962) was a German biologist who emigrated to Brazil (1934). He was regarded as the grand old man of Brazilian diplopodology (the study of millipedes and centipedes). He published widely on myriapods (1923–1962) and some are named after him, such as *Ommatoiulus schubarti*. He worked at the Museu Paulista in São Paulo, where a street is named after him as are a reptile and two amphibians.

Schultze, A

Duskhawker sp. *Gynacantha schultzei* **Le Roi**, 1915
[JS *Gynacantha sextans* McLachlan, 1895]

Dr Arnold Schultze (1875–1948) was a German soldier, geographer and entomologist specialising in *Lepidoptera*. He was a cadet in the Brandenburg Field Artillery Regiment

(1885), rising to lieutenant (1896), who transferred to the Prussian Foreign Office (1902–1904) and helped delineate the border between Cameroon and Nigeria. He was assigned to the Railway regiment (1904) but shortly afterwards retired (1905) and joined the Imperial Protection Force in Cameroon. However, ill health forced him to retire from active service that same year. After his retirement from the army, he took up the study of geography, natural history and political science in Bonn and was awarded his PhD (1910). As a geographer, he participated in the central African expedition of Duke Adolf Friedrich of Mecklenburg and published the geographical part of the scientific results of the expedition. He collected the hawker holotype in Spanish Guinea (1910). Whilst overseas, he collected in the Congo (1907–1910 & 1929–1932), Colombia (1926–1928) and Ecuador (1935–1939). On the return trip from Ecuador (September 1939), just southwest of the Canaries, his ship was sunk by the British cruiser *Neptune*. He and his wife were sent to Dakar and interned for some weeks, but were soon released to neutral Portugal because of his poor health. They remained there for the rest of his life. His collections are now in the museums of Stuttgart and Berlin. He published on a number of subjects including the paper: *Die afrikanischen Seidenspinner und ihre wirtschaftliche Bedeutung* (1914). Two birds are also named after him.

Schultze, L

Swamp Bluet *Enallagma schultzei* **Ris**, 1908

[JS *Africallagma glaucum* (**Burmeister**, 1839)]

Dr Leonhard Sigmund Friedrich Kuno Klaus Schultze-Jena (1872–1955) was a geographer, zoologist, botanist, philologist, and ethnographer. He studied medicine and natural sciences in Lausanne, Kiel and Jena (1891–1895), the latter awarding his PhD (1896) and continued his study of zoology at Naples and Messina (1896–1897). He took part in an expedition to South Africa (1903–1905). Among his written works is: *Zoologische und anthropologische Ergebnisse einer Forschungsreise im westlichen und zentralen Südafrika, ausgeführt in den Jahren 1903–1905 mit Unterstützung der Kgl. Preussischen Akademie der Wissenschaften zu Berlin* (1908). He was professor of zoology in Jena (1907) and then led an expedition that collected in New Guinea (1910–1911). He was then professor of geography at Kiel (1911–1912) and at Marburg (1913–1937). He also travelled in Mexico and Guatemala studying Mayan and Aztec culture and languages, and published translations from, and dictionaries of, these languages (1930s). An amphibian and four reptiles are also named after him.

Schumann

Dasher sp. *Micrathyria schumanni* **Calvert**, 1906

Schumann was a naturalist who made collections of Odonata and other insects all over Mexico (1880s–1900s) where he collected the type of this species in Vera Cruz. Unhelpfully, neither Calvert's paper, nor many collectors' labels give any further information, not even an initial!

Scissorhands

Pond Damselfly sp. *Archboldargia scissorhandsi* **Kalkman**, 2007

Edward Scissorhands was a movie character. Kalkman wrote: *"The inferior appendages resemble pairs of scissors and therefore the species is named after the character in Tim Burton's movie, Edward Scissorhands".*

Scortecci

Violet Dropwing *Trithemis annulata scorteccii* **Nielsen**, 1935
[JS *Trithemis annulata* (Palisot de Beauvois, 1805)]

Professor Dr Giuseppe Scortecci (1898–1973) was an Italian zoologist and an expert on desert herpetology. He took his doctorate at the University of Florence (1921) after military service in WW1, and then joined the staff of the Institute of Comparative Anatomy there. He moved to Milan as Curator of Lower Vertebrates at the Museum of Natural History and became Professor of Zoology, University of Genoa (1942). Before WW2 he explored on numerous occasions in the Sahara, Libya, Italian Somaliland (Somalia), Ethiopia and to Arabia. He wrote around 50 publications on herpetology, particularly about the fauna of desert regions. He died after a road traffic accident. Twelve reptiles, a bird and an amphibian are also named after him.

Scudder

Zebra Clubtail *Stylurus scudderi* **Selys**, 1873
[Orig. *Gomphus scudderi*]

Samuel Hubbard Scudder (1837–1911) was an American entomologist and palaeontologist, but is probably best known for his essay on the importance of first-hand, careful observation in the natural sciences; the treatise on inductive reasoning that he entitled *The Student, the Fish and Agassiz*. Williams College awarded his first degree (1857) and Harvard his second (1862). He was the first North American to study the palaeontology of insects. He wrote profusely publishing 791 papers (1858–1902) on every aspect of entomology, as well as geology, geography, zoography and ethnology. His *Nomenclator Zoologicus* (1882–1884) was a seminal and comprehensive list of all generic and family names in zoology and among his other longer works was: *Catalogue of the Described Orthoptera of the United States and Canada* (1900). He was also Curator, Librarian, Custodian, and President of the Boston Society of Natural History (1864–1870 & 1880–1887), a co-founder of the Cambridge Entomological Club and its journal 'Psyche' (1874) and much more. He described thirteen new Odonata species, seven of which remain valid.

Seabra

Bannerwing sp. *Chalcopteryx seabrai* **Santos** & **Machado**, 1961

Dr Carlos Alberto Campos Seabra (d.2001) was a Brazilian physician, biologist, zoologist and entomologist at the National Museum, Rio de Janeiro (1920–1958). His written work (1940s–1980s) covers a wide range of insects, including bees and beetles, such as the co-written: *Três espécies novas de Centris (Ptilotopus) Klug, 1810 (Hymenoptera – Apoidea)* (1962) and *Tavakilian Gérard. Un nouveau genre, une nouvelle espèce de Torneutini: Gnathopraxithea sarryi nov. sp. (Coleoptera, Cerambycidae)* (1986). He allowed Machado access to his extensive collection as acknowledged in the etymology: *"...que ofereceu a um dos autores (Machado) abundante colecao de Odonatas."* He collected the type of a bee which is also named after him *Dicranthidium seabrai.*

Seidenschwarz

Cebu Frill-wing *Risiocnemis seidenschwarzi* **Hämäläinen**, 2000

Dr Franz Gerhard Seidenschwarz (b.1954) is a German botanist and honorary consul (since 2009) living in Cebu, Philippines, where he collected the type (1998). He studied chemistry and biology in Munich, then worked as a lecturer at the Faculty of Forestry at the Ludwig Maximilian University (1984–1987). He moved to Cebu and worked (1987–2007) as professor at the University of San Carlos. He became an honorary citizen of Cebu (1997) and honorary consul. Among his written work is the book: *Plant World of the Philippines* (1994). The author acknowledged Dr Seidenschwarz thus: *I am grateful to Dr Franz Seidenschwarz for sending the valuable material for study and for providing detailed information on the collecting site.* It also honours Seidenschwarz's part in promoting the restoration of the last remaining virgin forest on Cebu. IUCN and Zoological Society of London included *Risiocnemis seidenschwarzi* among the 100 most threatened animal and plant species of the World in a book: *Priceless or Worthless?* (2012).

Seimund

Clubtail sp. *Burmagomphus seimundi* **Laidlaw**, 1932
[JS *Burmagomphus williamsoni* Förster, 1914]

Eibert Carl Henry Seimund (1878–1942) was a British taxidermist employed in the Zoology Department, BMNH (1897–1906). He collected in South Africa (1899–1903) whilst fighting in the South African War (1899–1902), in Fernando Po (1904) on a British Museum expedition, Thailand (1913) and Malaya (1916), having become assistant curator at the Selangor State Museum, Kuala Lumpur (1906). The Raffles Museum in Singapore holds specimens from his last expedition. He collected the type in the Federated Malay States (1921). A mammal and four birds are also named after him.

Sélika

Crimson Dropwing *Trithemis selika* Selys, 1869

Sélika was the heroine of the opera *L'Africaine* (The African Woman) by the German composer Giacomo Meyerbeer (1791–1864). An opera which Selys loved and saw at least five times.

Selys

The genus *Selysiothemis* **Ris**, 1897
The genus *Selysioneura* **Förster**, 1900
The genus *Selysiophlebia* Förster, 1904
[JS *Gynacantha* Rambur, 1842]
River Clubtail *Petalura selysii* **Guerin-Meneville**, 1837
[JS *Gomphus flavipes* (Charpentier, 1825)]
Selys' Sundragon *Helocordulia selysii* **Hagen**, 1878
[Orig. *Cordulia? selysii*]

Tropical Sprite *Nehalennia minuta selysi* **Kirby**, 1890
[Orig. *Nehalennia selysi*]
Selys's Glyphtail *Isomma hieroglyphicum* Selys, 1892
Green Spreadwing *Orolestes selysii* **McLachlan**, 1895
Threadtail sp. *Nososticta selysii* Förster, 1896
[Orig. *Caconeura selysii*]
Jewel sp. *Rhinocypha selysi* **Krüger**, 1898
Blue-faced Jewel *Chlorocypha selysi* **Karsch**, 1899
Flatwing sp. *Podopteryx selysi* Förster, 1899
[Orig. *Argiolestes selysi*]
Emperor sp. *Anax selysii* Förster, 1900
Western Bronze Cruiser *Macromia selysi* Kirby, 1900
[JS *Phyllomacromia aeneothorax* (Nunney, 1895)]
Forest Needle *Synlestes selysi* **Tillyard**, 1917
Leaftail sp. *Phyllogomphoides selysi* **Navás**, 1924
[Orig. *Gomphoides selysi*]
Pond Damselfly sp. *Nesobasis selysi* Tillyard, 1924
Grappletail sp. *Heliogomphus selysi* **Fraser**, 1925
Shadowdancer sp. *Idionyx selysi* Fraser, 1926
Skydragon sp. *Chloropetalia selysi* Fraser, 1929
[Orig. *Chlorogomphus selysi*]
Bold Leaftail *Phyllogomphus selysi* **Schouteden**, 1933
Common Hoetail *Diastatomma selysi* Schouteden, 1934
Pond Damselfly sp. *Metaleptobasis selysi* **Santos**, 1956

Helicopter Damselfly sp. *Mecistogaster marchali selysia* Navás, 1923
[JS *Mecistogaster modesta* Selys, 1860]

Baron Michel Edmond de Selys Longchamps (1813–1900) was a Belgian politician, entomologist and naturalist. As a liberal politician, (1846 on) he was a senator (1855–1900) serving as vice-president (1879) and then as president of the Senate (1880–1884). In the latter position, he was second only to the King in the State order of precedence in Belgium. Reflecting his liberal views, he avoided using his title of baron from the age of twenty-five, later always using the title of senator. As a self-taught scientist, he laid the foundation for the classification of Odonata and has often been called 'The father of Odonatology'. His wealth and social standing enabled him to amass – by collecting, exchanging and purchasing (including buying complete collections) – the finest collection of Odonata in the world at that time (containing c.1500 of the then known 2000 species) and extensive collections of other insects as well as European birds and mammals, the bird collection being almost complete. Since his death, the collections have been housed in the Royal Belgian Institute of Natural Science, Brussels. He travelled in many European countries and was appointed an honorary member of almost all European entomological societies, including the Entomological Society of London (1871). His first publications were mainly on the European species of Odonata (1831–1851), including the standard works: *Monographie des Libellulidées d'Europe* (1840) and *Revue des Odonates ou Libellules d'Europe* (1850) co-authored by H A Hagen (q.v.). He then began publishing synopses and monographs covering the whole world of odonate fauna. His synopses and later additions to them (1853–1886) covered Caloptérygines, Gomphines, Agrionines, Cordulines and Aeschnines. For Caloptérygines (1854) and Gomphines (1857) he also published more comprehensive monographs, both co-authored by Hagen. He published many faunal papers on Odonata of different countries, such as Belgium (1837, 1840), Great Britain (1845), Algeria (1849, 1866, 1871, 1902), Cuba (1857), Madagascar (1867), Seychelles (1869), Central Asia and Asia Minor (1872, 1887), New Guinea (1878, 1879), Central Africa (1881), the Philippines (1882, 1891), Japan (1883), Sumatra (1889) and Burma (1891). He described over 930 new species or 'varieties' (subspecies) of Odonata and named c.180 genera or subgenera. Of these, over 700 full species and over 130 genera are presently ranked as valid. Although only a few of his publications included illustrations, he left the world with a huge collection of c.1250 watercolour plates of Odonata, many of them drawn by himself. So far, only a few of them have been published. His c.250 publications also treated European species of other insect orders (Orthoptera, Neuroptera and Lepidoptera) and also birds and mammals. His great achievements as a scientist were recognized by the issue of a Belgian stamp in his honour (1986). The stamp shows his portrait, his signature and a few dragonflies and plants. Selys kept a diary almost daily from the age of ten to the end of his life. These diaries, which were published (2008) in two massive tomes (totalling 1750 pages), offer an incredible source of detailed information on his life with its joys and sorrows. They also show the extraordinary scope of his acquaintance, from servants to kings. The diaries reveal Selys as a warm and emotional man, loving father and husband, and a man who liked dancing, watching opera, riding a horse, hunting, smoking and sometimes gambling. He was a man who drank wine for the first time aged 40 and who learned to ride a bicycle when over 80 years old. The diaries also reveal that his interest in insects started (July 1824) when his brother-in-law brought him living individuals of *Libellula depressa* and *Calopteryx virgo* to look after; he tried to feed them. In addition to the eponymous odonate taxa, he was honoured with many other eponyms including at least 17 insect species of various orders, as well as a spider, a fossil mollusc, a fossil bird, a plant genus *Selysia* and mammal subgenus. (See also **Longchamps** & **Edmond**)

Sembe

Nighthawker sp. *Heliaeschna sembe* **Pinhey**, 1962

This is a toponym; the Sembe Forest in Congo is the type locality.

Semper

Slendertail sp. *Leptogomphus semperi* **Selys**, 1878
Satinwing sp. *Euphaea semperi* Selys, 1879
[JS *Euphaea refulgens* **Hagen**, 1853]

Carl Gottfried Semper (1832–1893) was a German ethnologist and zoologist. He studied at the Hanover Polytechnic (1851–1854) and was awarded his PhD by the University of Würzburg (1856). He lived in the Philippines and Palau (1858–1865), associated with the Museum Godeffroy. He collected (1858) around Manila and later (1859) travelled south, working mainly near Zamboanga in Mindanao and Basilan before returning (1860) to Manila. Almost immediately he went to the northeastern part of Luzon travelling through

Bulacan and Nueva Ecija to Baler in Aurora province and along the coast to Palanan in Isabela province and further across the Sierra Madre mountain chain to Cagayan, where a Malaria attack forced him to return to Manila. He left again (1861) for north Luzon going through Nueva Ecija to San Nicolas and further to Benquet province. From there he visited La Union province before arriving at Mankayan at the northern corner of Benquet again returning to Manila (November 1861), this time with dysentery. In order to recover he took a long voyage in the Pacific. He visited (1863) Bohol, Cebu, Leyte and Mindanao and (1864) central and eastern Mindanao, again returning to Manila and on to Europe (May 1865). The collection of insects he made there represents the beginning of knowledge on Philippine odonata. He was noted for his humane and unbiased attitude towards the indigenous cultures, and published several works about them including being one of the authors of part of the five-volume: *Reisen im Archipel der Philippinen* (1868–1916). He had a major stroke (1887), never fully recovering. Three reptiles and a bird are also named after him as is an anatomical structure that he discovered in the head of certain gastropods (1856).

Septima

Septima's Clubtail *Gomphurus septima* **Westfall**, 1956

Dr Septima Cecilia Smith (1891–1984). (See **Smith, S C**)

Servilia

Oriental Scarlet *Crocothemis servilia* Drury, 1773
[Orig. *Libellula servillia*]

Servilia was reputedly Brutus' Mother and Julius Caesar's mistress.

Serville

Stream Cruiser *Didymops servillii* **Rambur**, 1842
[JS *Didymops transversa* (Say, 1840)]

Jean Guillaume Audinet-Serville (previously Audinet de Serville) (1775–1858) was a French entomologist, specialising in Orthoptera. He wrote a number of major works including co-writing: *Dictionnaire des Insectes de l'Encyclopédie méthodique*; *Faune française* (1830) and the seminal *Histoire naturelle des Insectes Orthoptères* (1839) as well as contributing sections to such as *Histoire naturelle*.

Sevastopoulos

Yellow-legged Duskhawker *Acanthagyna sevastopuloi* **Pinhey**, 1961
[JS *Gynacantha nigeriensis* (Gambles, 1956)]

Demetrius George Sevastopoulos (1903–1987) was a London-born merchant and entomologist who spent some time in India (1940s) and later lived in Mombasa (1950s on). He was appointed to the board of East African Cargo Handling Services (1966). He published more than 300 papers, mostly on East African Lepidoptera and deposited a major part of his collection with the BMNH a few years before his death. He collected the type in Uganda (1952).

Séverin

Green-Striped Darner *Dromaeschna severini* **Förster**, 1908
[JS *Dromaeschna forcipata* Tillyard, 1907]
Flutterer sp. *Rhyothemis severini* **Ris**, 1913

Wilhelm Peter Robert (aka 'Guillaume') Séverin (1862–1938) was Curator (1899–1927) of the Musée Royal D'Histoire Naturelle de Belgique, Brussels. He was born in The Hague to German parents who took Dutch citizenship. He was not a graduate, but was trained in Liège as an artist and industrial designer; he was also a gifted musician. When his health weakened, his doctor instructed him to walk in the country air. There, in his free time, he developed a deep interest in insects, particularly Coleoptera. He visited Selys (1888) and his collection. Selys was very taken with the young entomologist, taking him on as a student and later (1890) he was appointed 'aide-naturalist' at the Royal Museum in Brussels and then promoted (1899) to the post of Curator of Arthropods. There he painted the larger part of the plates of the odonates. Part of his time was spent at the School of Tropical Medicine where he taught doctors who were preparing to go to the Congo about medical entomology necessary for their mission. He visited a number of US museums (1907) when attending an international zoology conference. The darner was named after him honouring his assiduous work of cataloguing Selys' Odonata collections.

Seydel

Southern Dark Claspertail *Onychogomphus seydeli* **Schouteden**, 1934
[Orig. *Tragogomphus seydeli*]
Black Cruiser *Phyllomacromia seydeli* **Fraser**, 1954
[Orig. *Macromia seydeli*]
Southern Red-tipped Jewel *Chlorocypha seydeli* Fraser, 1958

Charles Henri Victor Seydel (1873–1960) was a lepidopterist and botanical collector for Jardin Botanique National de Belgique, notably in the Belgian Congo (1924–1938), when Schouteden was also collecting there. He was the only member of the Lepidopterist Society living (1951) in Elisabethville, Congo. He advertised that he had specimens for sale, including the words, 'Prices are low'. A reptile is also named after him.

Seyrig

Cerulean Sprite *Pseudagrion seyrigi* **Schmidt**, 1951

André Seyrig (1897–1945) was a French civil engineer, miner, colonist and amateur entomologist. He volunteered for active service at just seventeen years of age during WW1 and was a lieutenant awarded the Legion of Honour at demobilisation (becoming a captain in the reserves). He entered the mining school at Nancy, qualifying as an engineer and taking a post in Spain for five years. Although offered promotion, he preferred to be in a less responsible position so he could continue with his entomology hobby. He went to Madagascar as an inspector for the General Company of Madagascar, travelling all over the country and taking frequent trips to Europe, US, Canada, Africa and Syria. These trips allowed him to collect and continue to study entomological collections. He became director of the mica mines near Ampandrandava, living there for fifteen years. He was a prolific collector of many faunal and floral taxa throughout Madagascar (1928–1944), including the eponymous type (1938–1939). His widow gave Seyrig's insect collection and manuscripts to the Muséum National d'Histoire Naturelle, Paris. Among other works he wrote: *Les ichneumonides de Madagascar* (1952). He was interred by the Vichy administration and while there a fellow internee was said to have 'gone mad and killed him'.

Sharp, D

Shadowdamsel sp. *Drepanosticta sharpi* **Laidlaw**, 1907
[Orig. *Platysticta sharpi*]

Dr David Sharp (1840–1922) was an English physician and entomologist whose primary focus was Coleoptera. He left school at 17 to help his father, who was a leather merchant, and about this time he began to collect beetles. As he did not like being in business it was decided he should become a physician and he studied for two years at St Bartholomew's Hospital, London and then attended Edinburgh University where he was awarded his bachelor of medicine (1866). He assisted a friend in his London practice for several years but was then offered a post as medical officer in the Crichton Asylum in Dumfries. While there he joined the Dumfriesshire and Galloway Scientific, Natural History, and Antiquarian Society writing his first entomology papers at that time. He moved back to England, first to Southampton (1882) and two years later to Dartford, Kent. He was offered, and accepted the post of Curator of the Cambridge University Museum of Zoology (1885–1909). Thereafter he retired moving to Brockenhurst on the edge of the New Forest for the rest of his life. He became (1862) a Fellow of the Entomological Society of London and was president (1887 & 1888), as well as joining and holding office in the Zoological Society, the Linnean Society and the Royal Society. He wrote a great many papers such as: *A revision of the British species of Homalota* (1869) and *The Staphylinidae of Japan* (1888). His longer works included: *Insecta: Coleoptera Vol. 1, Part* 2 (1882–1887) and *Insecta: Coleoptera Vol. 2, Part 1* (1887–1905) in the monumental *Fauna Centrali-Americana* series. He travelled a little to Switzerland, France and Spain, but mainly collected around his homes. Sharp and Laidlaw were fellow members of both the London Zoological Society and the British Association for the Advancement of Science. In *Anthropological and Zoological Results of an Expedition to Perak and the Siamese Malay States of 1899–1900* (1907), Laidlaw described the eponymous species among the other odonata and Sharp described the beetles.

Sharp, J E D

Orange Jewel *Chlorocypha sharpae* **Pinhey**, 1972
[JS *Chlorocypha luminosa* Karsch, 1893]

Mrs Jane E Sharp née (Jennie) Davis was an African American missionary and teacher who was born in Missouri and grew up in Boston. She was a teacher at Webster Grove, Missouri. She became a missionary who went to Liberia (1883) to take charge of the girls' department of Monrovian College and, after marrying Jesse Sharp, a local coffee planter and businessman, set up a school for girls at their home on the Mount Coffee estate thereafter known as 'Mrs Jane E Sharp's School for African Girls' (although the majority of pupils were 'African Americans' who emigrated to Liberia, but local girls were also pupils). She had formerly been the head of a high school in St Louis for 'coloured children'. She returned to the US (1902) to raise funds to build a dedicated school because her home was overflowing with pupils. Her aim was to establish a Hampton Institute of Liberia, similar to the Black University in Virginia; several of their graduates went to Liberia with her to teach. She collected the eponymous type and many other Odonata on Mount Coffee, Liberia (1896).

Shieh

Shadow-emerald sp. *Macromidia shiehae* Jiang, Li & Yu, 2008

Mrs Shwu-Feng Shieh (1936–2012) was the wife of the Taiwanese entomologist, Professor Chung-Chich Kuan.

Shirozu

Lyretail sp. *Stylogomphus shirozui* **Asahina**, 1966

Professor Dr Takashi Shirozu (1917–2004) was a Japanese entomologist, specialising in Lepidoptera. He graduated from the Faculty of Agriculture, Kyushu Imperial University (1941), almost immediately (1942) joining the Entomology Laboratory there as a research assistant. He rose to associate professor (1949) and full professor (1963). He was professor emeritus after retirement (1981) and president emeritus of its Museum of Natural History, as well as President Emeritus of The Lepidopterological Society of Japan of which he had been president (1970–1987). He wrote many scientific papers describing new species of butterflies and some longer works including: *Butterflies of Formosa in Colour* (1960) and *Butterflies of Japan Illustrated in Colour* (1964). A number of butterflies and other insects are also named after him. (Also see **Takashi**)

Shozo

Flatwing sp. *Rhipidolestes shozoi* K Ishida, 2005

Shozo Ishida (b.1930) is the father of the author, entomologist and odonatologist Katsuyoshi Ishida. Together they have written a number of papers and were among the authors of: *Illustrated Guide for Identification of the Japanese Odonata* (1988) and *Dragonflies of the Japanese Archipelago in Colour* (2001). In dedicating the species to his father Katsuyoshi wrote: "*The species is named in honour of Mr. Shozo ISHIDA, an amateur odonatologist and author's father, who gave him valuable guidance through his life and odonatological work.*" (See **Ishida**)

Shurtleff

American Emerald *Cordulia shurtleffi* **Scudder**, 1866

Carleton Alwood Shurtleff (1840–1864) was a friend of the author and fellow (amateur) entomologist and botanist, as well as a Civil War physician. He collected plants and insects from the age of ten. He was educated in the Brookline schools, and graduated from Harvard (1861), having studied under Professor Agassiz, who took an interested in him as he was very fond of botany and entomology. He studied medicine in the Harvard medical school and with his father, Dr S A Shurtleff. He enlisted at the very start of the American Civil War into a corps of medical cadets, a division of the regular army, and served during the siege of Vicksburg on a floating hospital on the Mississippi River. He contracted chills and fever there and returned home on a short furlough. As soon as his health was restored he returned and for three months after the battle of Gettysburg, served in the Cotton Factory Hospital at Harrisburg; he was then transferred to Philadelphia. At the time of his enlistment he had been nearly ready to graduate from the medical school; he therefore obtained (1864) a discharge, in order to go home and take his degree. However shortly after his return he contracted diphtheria and died June 26th, 1864. Among his papers was: *The Army Worm* (1862).

Sibylla

Skimmer sp. *Orthemis sibylla* **Ris**, 1919
[JS *Orthemis ambirufa* Calvert, 1909]

Maria Sibylla Merian (1647–1717) was a German-born, Swiss naturalist and illustrator. Her stepfather, an artist himself, taught her. She lived in Frankfurt, Nuremberg and Amsterdam. She published several books of her illustrations of nature including: *Neues Blumenbuch* (1675) and her major work: *Metamorphosis insectorum Surinamensium* (1705). Her observations of butterfly metamorphosis were a considerable contribution to entomology. She also worked in Suriname for two years sketching the colony there, as well as plants and animals. After a stroke (1715), she was unable to work and died in poverty two years later.

Siebers

Shadowdamsel sp. *Drepanosticta siebersi* **Fraser**, 1926

Hendrik Cornelis Siebers (1890–1949) was a Dutch zoologist and ornithologist in the Dutch East Indies (Indonesia). He worked for the Amsterdam University Zoological Museum (1920–1947) and took part in their central-east Borneo expedition (1925). He wrote: *Fauna Buruana, Aves* (1930). A mammal, three birds and four reptiles are also named after him. Fraser did not provide any etymology for this species collected in Java, but no doubt it was named after Hendrik Siebers.

Siebold

Genus *Sieboldius* **Selys**, 1854

Siebold's Dragonfly *Anotogaster sieboldii* Selys, 1854
[Orig. *Cordulegaster sieboldii*, A. Golden-winged/Jumbo Dragonfly]
Lilysquatter sp. *Paracercion sieboldii* Selys, 1876
[Orig. *Agrion sieboldii*]

Dr Philipp Franz Balthazar von Siebold (1796–1866) was a German physician, biologist and botanist. He was medical officer to the Dutch East Indian Army in Batavia, and at the Dutch Trading Post on Dejima Island, Nagasaki, Japan. His encyclopaedic knowledge soon attracted many Japanese to study under him, and the following year he established a boarding school on the outskirts of Nagasaki. He taught Western medicine and treated Japanese patients, accepting assorted ethnographic and art objects in payment. He collected in Japan, sending local collectors into the interior (1823–1829). It became known that he was copying a map of the northern (Ezo) regions of Japan, helped by the Imperial librarian and astronomer. The Government, suspecting the intention was to put the map to some use harmful to Japan, imprisoned all his Japanese students and friends, searched his house repeatedly, confiscated religious objects he might have exported illegally and told him that he would not be allowed to leave the country. He packed all of his manuscripts, maps and books in a lead-lined chest and hid it. He was banished forever from Japan (1829) leaving behind his Japanese mistress and two-year-old daughter. Returning to the Netherlands he was appointed (1831) advisor on Japanese affairs by King William I. He edited, and made many contributions to, *Fauna Japonica* (1833–1850), as well as a number of other books on Japan, its natural history, language and ethnography. He and his eldest son, Alexander von Siebold (1846–1911), were invited back to the Japanese court by the emperor (1859). The mission was a failure and he was pensioned off (1863), returning home by way of Java. His ethnographical collections were bought by the Bavarian government and exhibited in Munich. He continued to write scientific papers until his death. His botanical and zoological collections are preserved in Leiden, where he established a nursery for the introduction of Japanese plants to Europe. The Academy of St. Petersburg purchased eight volumes of his original drawings and dried Japanese plants from his German widow. The Siebold Memorial Museum was founded to honour the contributions made by him to the modernization of Japan, and built by the city of Nagasaki adjacent to Siebold's former residence. The form of blue hydrangea now planted next to the museum was named *Hydrangea otaksa* by Siebold; it is the city flower of Nagasaki. Siebold University was founded in Nagasaki (1999). An amphibian, two reptiles and three birds are also named after him.

Siemens

Skimmer sp. *Uracis siemensi* **Kirby**, 1897

Carl (Charles) Wilhelm (William) Siemens (later Sir William Siemens) (1823–1883) was a German-born engineer and entrepreneur who settled in Great Britain and took British citizenship. He studied at Göttingen University but did not take a degree, instead becoming an apprentice engineer (1842). With his brothers, he established a company with various inventions; they became a leading cable-laying company, which went on to lay cable across the Atlantic and around the world. It is through this that he became honoured when Messrs Siemens Bros, Cable SS *Faraday* undertook an expedition up the Amazon, during which E E Austen collected the eponymous holotype.

Silberglied

Jewel sp. *Indocypha silbergliedi* **Asahina**, 1988

Professor Dr Robert Elliot Silberglied (1946–1982) was an American entomologist. He developed an interest in entomology while still at school where he wrote his first research paper aged sixteen, on *Drosophila melanogaster* (1961). He went to Cornell University School of Agriculture (1963), which awarded his BSc (1967) and masters (1968). He began his doctoral work at Harvard (1968), which awarded his PhD (1973) throughout which (1968–1973) he was a teaching fellow. He was appointed Assistant Professor of Biology at Harvard University (1973) and Assistant Curator of Lepidoptera at the Museum of Comparative Zoology (MCZ). He remained at Harvard (until 1981) being promoted to Associate Professor of Biology and Associate Curator of Lepidoptera. He was also appointed (1976) as a biologist at the Smithsonian Tropical Research Institute (STRI). The position divided his research, teaching and administrative responsibilities between Cambridge and Panama, as he began spending half of each year in the American tropics in residence at Barro Colorado Island. He remained on the STRI staff as a research entomologist and scientist-in-charge of the research station until his death. He also did field work in the Galapagos and in Florida. His special area of interest was in matching larvae to adult tropical butterflies by raising them, but also included studies of ultraviolet reflectance patterns of butterflies and flowers, insect vision and insect behaviour, especially with regard to courtship, mating, and reproductive isolation. He died, aged just 36 in the Air Florida accident in Washington DC, when the Boeing 727 plunged into the Potomac.

Silke

Silke's Fairytail *Lestinogomphus silkeae* Kipping, 2010

Silke Kipping (1968–2003) was the wife of the author Jens Kipping (b.1965), who wrote that she *"...always encouraged me in my studies of African Odonata."*

Silsby

Jill's Shadowcruiser *Idomacromia jillianae* Dijkstra & Kisakye, 2004

Jillian 'Jill' Dorothy Silsby (b.1925) served as Honorary Secretary of the British Dragonfly Society for eight years (1989–1996), supported by her husband as treasurer. She was a founding member of the Worldwide Dragonfly Association and, since its foundation (1997), has been its honorary secretary and treasurer. She also edited 'Kimminisia', the newsletter of the UK National Office of the International Odonatological Society (1990–1997) and 'The Newsletter of the British Dragonfly Society' (1987–1997). She and her husband Ronnie (Ronald Ian Silsby, 1921–2002) lived in India for several years and for periods in Kenya and Saudi Arabia, as well as travelling all over the world looking for birds and dragonflies. She is author of: *Inland Birds of Saudi Arabia* (1980) and *Dragonflies of the World* (2001), which was written for interested amateurs as well as more experienced professionals. In their etymology, the authors wrote: *"This species is named in honour of a unique lady in odonatology, Jill Silsby, who has done so much for this field and has produced a handbook, which is of great value to all workers."* (Also see **Jillian**)

Simmonds

Pond Damselfly sp. *Melanesobasis simmondsi* **Tillyard**, 1924

Hubert Walter Simmonds (1877–1966) was a prominent English-born entomologist, who lived in Fiji (1919–1966) and contributed greatly to the economy of the country through his extensive travels in search of beneficial insects for the control of a number of serious pests there. He was completely self-taught and accepted a job in New Zealand (1902) during which time he made a large insect collection (1902–1909). He was offered a job as overseer on a rubber plantation in Fiji (April 1919). At the time when he collected the holotype (1919), he was already acting government entomologist – his early recognition as an entomologist led to his formal appointment to the post (1920–1937) which he held until retirement, but he continued with his work long after. He died in New Zealand.

Simone (Kelner-Pillault)

Proto sp. *Protolestes simonei* **Aguesse**, 1967
Damselfly sp. *Tanymecosticta simonae* **Lieftinck**, 1969
Wide-striped Sprite *Pseudagrion simonae* **Legrand**, 1987

Simone Kelner-Pillault (1925–1985) was a French biologist, taxonomist, entomologist, curator and teacher. She was an intern at the Agriculture School in Orléans (1941) becoming a teacher (1944) of mathematics and the sciences (1945). She passed her baccalaureate (1950) in Paris and was a science intern (1956) at CNRS, where she later defended her thesis (1967). She was then curator (1957) of Hymenoptera etc., at MNHN and helped create the diploma course in applied entomology at the Faculty of Science there (1962). She died in an accident at home two days before her retirement was due. Jean Legrand wrote her obituary in the 'French Entomological Review'.

Simone (Soares dos Santos)

Skimmer sp. *Carajathemis simone* **Machado**, 2012

Simone Soares dos Santos is a Brazilian biologist and the wife of Haroldo Lapertosa Jr (together they collected the eponymous type). Machado wrote: *"...I dedicate this species to the biologist Simone Soares dos Santos who, together with her husband Haroldo Lapertosa Junior, collected the material of this outstanding species."* The genus is a toponym referring to the type locality, Serra dos Carajás, a mountain chain in the Amazon forest of the Brazilian state of Pará.

Simpson

Bluebolt *Cyanothemis simpsoni* **Ris**, 1915

Dr James Jenkins Simpson (1881–1936) was a Scottish zoologist and entomologist. The University of Aberdeen awarded his MA (1904) and BSc (1906) and he was then Carnegie Research Scholar and Fellow at Aberdeen where his DSc thesis was on the natural history of the eel. He worked at their Natural History Department and then worked for the Colonial Office's Entomological Research Committee (Tropical Africa) (1909–1915). During WW1, he joined the West African Frontier Force to work on tropical diseases and their carriers (1915–1919). He was then Keeper of Zoology at the National Museum of Wales (1919–1925) and Director of the Public Museums of Liverpool (1926–1928). His final museum post was in Turkey, where he went to organize the Department of Oceanography and Marine Biological Research, in order to help advance the fishing industry in that country. He wrote a number of reports on various western African states he visited on behalf of the Entomological Research Committee, such as: *Entomological Research in British West Africa* (1918) and *Bionomics of Tsetse and Other Parasitological Notes in the Gold Coast* (1918). He collected in several West African countries including Nigeria, The Gambia and Ghana, as well as specimens of the type series in Sierra Leone (September 1912). He briefly collected in Portuguese East Africa (1909). He was reported lost at sea from the steamer *Kyrenia* (November 1936) while travelling from Greece to Turkey.

Simun

Shadowdamsel sp. *Drepanosticta simuni* **Dow** & **Orr**, 2012

Jeffry Simun (b.1972) is a member of staff at Gunung Mulu National Park and a friend of the first author. He collected (2005) the holotype at the foot of Gunung Mulu in the park, which is situated in Sarawak, Malaysia.

Sinclair

Common Redcoat Damselfly *Xanthocnemis sinclairi* Rowe, 1987

Dr Andrew Sinclair (1794–1861) was a Scottish surgeon, colonial administrator and pioneer amateur scientist who was the original collector of most of the New Zealand endemic dragonfly species. He studied medicine at Glasgow University (1814–1816), then took a surgical course at L'Hôpital de la Charité in Paris (1817). He continued to study at Edinburgh University (1818–1819), which awarded his MD. He joined the Navy (1822) as an assistant surgeon, and served at Cape of Good Hope and in the Mediterranean (1823–1833). Throughout this time, he collected botanical specimens for BMNH and was aboard HMS *Sulphur* (1835) accompanying Beecher on his survey of the Pacific coasts of both North and South America where he continued to collect. His health became compromised and he returned to England (1839) but when recovered he went to Australia (1841) and to the Bay of Islands, New Zealand, continuing to collect on trips ashore. He returned to Scotland and at that time donated his shell and insect collection to BMNH. He again went to Australia (1843) as surgeon-superintendent aboard a convict ship *Asiatic* bound for Tasmania. He took discharge in Sydney and accompanied Governor FitzRoy to New Zealand where he spent the rest of his life. After time exploring New Zealand's natural resources he was persuaded to take the post of colonial secretary (1844–1856). During his time in office and after retirement he continued to collect, sending specimens to Kew from all over the country and also making several trips back to Europe. He drowned while exploring the headwaters of the Rangitata River with Julius von Haast. The spot where Sinclair met his death is 15km from the type locality.

Sini

Damselfly sp. *Igneocnemis siniae* **Hämäläinen**, 1991
[Orig. *Risiocnemis siniae*]

Sini Wallenius (b.1957) is the former wife of the author. The University of Helsinki awarded her MSc (1986) and she now works as a senior officer at the Finnish Ministry of Agriculture and Forestry. Hämäläinen wrote: *"Named after my wife Sini Wallenius in appreciation of her patience for my dragonfly work."*

Siqueira

Pond Damselfly sp. *Leptagrion siqueirai* **Santos**, 1968

Dr Luis Manoel Paes Siqueira is Professor of Parasitology at the Faculty of Medicine, Federal University of Pernambuco.

Sjöstedt

Variable Sprite *Pseudagrion sjoestedti* **Förster**, 1906
[Orig. *Pseudagrion sjöstedti*]

Quarré's Tiger *Diastatomma sjöstedti* **Schouteden**, 1934
[JS *Gomphidia quarrei* (Schouteden, 1934)]

Helicopter Damselfly sp. *Microstigma anomalum sjöstedti* **Schmidt**, 1958
[JS *Microstigma anomalum* Rambur, 1842]
Variable Sprite *Pseudagrion sjöstedti pseudosjöstedti* **Pinhey**, 1964
[JS *Pseudagrion sjoestedti* Föster, 1906]

Bror Yngve Sjöstedt (1866–1948) was a Swedish entomologist and ornithologist. He collected natural history specimens even before he started at school, keeping them in his family's summerhouse. He studied at the University of Uppsala (1886), which awarded his BA (1890). He was then commissioned by the Swedish Academy of Science to lead a two-year expedition in Cameroon (1890–1891), collecting for Uppsala University Zoological Department and for the State Natural History Museum, but he stipulated the right to collect for himself too. He was awarded his PhD (1896) for a thesis he wrote on the birds he collected on that expedition, and he produced his first paper the same year and was subsequently appointed as assistant at the newly established Entomological State Institute (1897). He went to the USA and Canada to visit entomological stations and to study their methods (1898). He also worked on the entomological collections at the Naturhistoriska Riksmuseet where he later became professor and curator. He was part of the Swedish Zoological Expedition to Mount Kilimanjaro (1905–1906) collecting 4,300 species, a third of which were new and he published *Akaziengallen und Ameisen auf den Ostafrikanischen Steppen* and edited three volumes on the expedition results (1910). He married (1918) the opera singer Rosa Grünberg (1878–1960). He was President of the Board of the Natural History Museum (1922–1933) and Vice-President of the Vassijaure Institute of Natural Science (1927–1933) and produced an entomological inventory of the park in three volumes. In all, he wrote over 120 publications and described 72 species (more than 30 of which are synonyms) and nine genera (two still in use) of odonates. His major odonatological publications include: *Odonaten aus Kamerun West Afrika* (1900) and *Wissenschaftliche Ergebnisse ders schwedischen entomologischen Reise des Herr Dr. A. Roman in Amazonas 1914–1915. 1. Odonata* (1918). He also wrote a small handbook on Swedish dragonflies (1902).

Sjupp

Fineliner sp. *Teinobasis sjupp* Kalkman, 2008

Almost an eponym in that the fineliner was named after a pet of Carl von Linné, a racoon given to him by The Crown Prince of Sweden. Sjupp escaped through the fence of the Uppsala Botanical Garden (June 1747) and was killed by a dog.

Skinner

Bannerwing sp. *Cora skinneri* **Calvert**, 1907

Dr Henry Skinner (1861–1926) was an American entomologist and outstanding lepidopterist who became Professor of Entomology and Curator at the Philadelphia Academy of Science. He graduated from the University of Pennsylvania BSc (1881) and MD (1884) and then entered medical practice. He eventually gave up medicine for full time entomology (1891). Like Calvert, he edited the 'Entomological News' (1890–1910) for seven years whilst Calvert was associate editor and, like Calvert, he was President of The Entomological Society of Philadelphia (1908), which became The American Entomological Society.

Smedley

Limniad sp. *Amphicnemis smedleyi* **Laidlaw**, 1926
[Orig. *Amphicnemis louisae smedleyi*]

Norman Smedley (1900–1980) was a British zoologist. He joined the Durham Light Infantry when just sixteen, seeing active service almost immediately in France during WW1 (1916–1918). He was Assistant Curator at the Raffles Museum, Singapore (1920s–1930s) and took part in at least one collecting expedition in the Malay States (1927), as well as representing the Straits Settlements at the Fourth Pacific Congress in Java (1928). Ill health brought an end to his colonial service and he became Curator of Doncaster Museum. During WW2, he worked for the Ministry of Information and organised the collecting of medicinal herbs in northeast England. He moved to Suffolk (1952) and became Curator of Ipswich Museum (1953–1964) where his interest turned away from the natural world and more to archaeology. For many years, he had deplored the destruction of the relics of our rural past; he had accordingly made a collection of farm and craft tools which he stored at Beccles, hoping that one day he would be able to found a rural life museum. Someone donated a barn and some land and the Museum of East Anglian Life was opened at Stowmarket (1967), with Smedley as director, until he retired again (1974). He wrote a number of papers, such as: *Amphibians and reptiles from the Cameron Highlands, Malay Peninsula* (1931) and *An Ocean Sunfish in Malaysian Waters* (1932) and books and articles about East Anglian life, such as: *Life and Tradition in Suffolk and Northeast Essex* (1976) and *East Anglian Crafts* (1977) and was President of the Suffolk Institute of Archaeology (1968). He collected part of the type series in Siberut island, west of Sumatra (1924).

Smith, A J

Smith's Dragonfly *Procordulia smithii* **A White**, 1846
[P. Ranger Dragonfly]

Lieutenant Alexander John Smith (1812–1872) was a British naval officer. He discovered a skink and presented the holotype that was collected in New Zealand during Ross' expedition with HMS *Erebus* and HMS *Terror* to the Southern Ocean and Antarctic (1839–1843). He later became a politician in Victoria, Australia. It seems likely he was also the collector of the dragonflies in New Zealand. The type was almost certainly in Andrew Sinclair's collection forwarded to BMNH and Smith may have been part of the collecting team.

Smith, F

Hooktail sp. *Megalogomphus smithii* **Selys**, 1854
[Orig. *Heterogomphus smithii*]

Frederick Smith (1805–1879) was an English entomologist. After school, he was apprenticed to a landscape engraver, but developed an interest in insects and became (1841) Curator of the Collections and Library of The Entomological Society of London, although engraving continued to be how he made a living. However, he took on (1849) arranging the BMNH

Hymenoptera collection and was then employed by them. He engraved illustrations for the Hymenoptera catalogue as well as some of A R Wallace's monographs and his own increasingly prolific papers (1837–1879). He was President of the Entomological Society (1862–1863) and repeatedly vice-president. He was appointed as Assistant Keeper of Zoology at the Museum, published more than 150 entomological papers and described more than 700 new species and subspecies of ant.

Smith, H H

Red-tailed Pennant *Cannacria smithii* **Kirby**, 1894
[JS *Brachymesia furcata* Hagen, 1861]
Dancer sp. *Argia smithiana* Calvert, 1909

Herbert Huntingdon Smith (1851–1919) was an American naturalist who mainly worked on the natural history of Brazil. He first went to Brazil (1870) on the Morgan expedition and returned to stay (1874–1876), spending a year exploring the Amazon. Back in the USA, he began working for 'Scribner's Magazine', writing on Brazil and frequently returning there. He married Amelia 'Daisy' Woolworth (1880), also a naturalist. They lived in Brazil (until 1886), travelling widely and visiting Paraguay, but spending most time at Chapada dos Guimarães, where intensive collecting (especially of insects) resulted in the discovery of many new species. After a few months in Rio de Janeiro, they returned to the USA where he sold his collection. He collected in Mexico (1889) for Godman, the results appearing in: *Biologia Centrali-Americanum*. He was then commissioned by the Royal Society to collect in the West Indies (1889–1895) including St Vincent and the Grenadines where he collected nineteen species of dragonfly including the holotype *Cannacria smithii*. He became curator at the Carnegie Museum and collected in Colombia (1898–1902) for the AMNH before resuming curatorship of the Carnegie Museum on his return. He collected for a 'Shell Syndicate' in Alabama before being appointed curator at the Alabama Museum of Natural History (1910). He wrote: *Brazil, the Amazons and the coast* (1879) and *Do Rio de Janeiro a Cuyabá: Notas de um naturalista* (published posthumously, 1922). He suffered a bout of influenza and became deaf and tragically, while collecting snails on his way to work along a railway, he didn't hear an approaching train and was fatally hit. The spot on the University of Alabama campus was known for many years as 'Smith's Crossing'. He collected the damselfly type in Chapada, Brazil. An amphibian is also named after him. (Also see **Herbert** & **Thisma**)

Smith, M A

Spine-tufted Skimmer *Crocorthetrum smithi* **Fraser**, 1921
[JS *Orthetrum chrysis* (Selys, 1891)]

Dr Malcolm Arthur Smith (1875–1958) was an English physician and herpetologist who was in Bangkok (1902–1924) and collected the types (1920–1921) during which time visited French Indochina (1918). He practised medicine in Siam (now Thailand) including five years as court physician, publishing his memoirs under the title: *A Physician at the Court of Siam* (1947). He was President of the British Herpetological Society (1949–1954). Three amphibians, a bird, a fish and seven reptiles are also named after him. (Also see **Malcom (Smith)**)

Smith, S C

Clubtail sp. *Gomphus septima* **Westfall**, 1956

Dr Septima Cecilia Smith (1891–1984) was an American zoologist. The University of Texas, Austin, awarded both her BSc and MSc and Johns Hopkins her doctorate (1927). She was an instructor in medicine at the University of Alabama, where she became Associate Professor of Zoology and eventually Professor of Parasitology and Marine Biology (1927–1962) then Professor Emeritus. She undertook a number of collecting trips such as to Florida (1934). She collected the type in Alabama (1939). (Also see **Septima**)

Smith, W A

Sioux Snaketail *Ophiogomphus smithi* **Tennessen** & Vogt, 2004

William Arthur Smith (b.1947) is an American entomologist who worked at the Wisconsin Department of Natural Resources (2001). He compiled a: *Checklist of Wisconsin Dragonflies* (1993) with the second author and collected the type there (1994). He provided specimens and assistance to the authors and has co-written a paper with Vogt, describing a new dragonfly species: *Ophiogomphus susbehcha* (1993).

Snelleman

Jewel sp. *Melanocypha snellemanni* Albarda, 1879
[Orig. *Micromerus snellemanni*]

Johannes François Snelleman (1852–1938) was a Dutch zoologist, ethnographer and orientalist, who became a museum director. He took part as zoologist in the Royal Netherlands Geographical Society scientific expedition to central Sumatra mapping the Hari River (1887–1889), where he collected the eponymous type. He wrote about the expedition in the two volume: *Bijdragen tot de kennis der fauna van Midden-Sumatra* (1887 & 1892). He also edited the four-volume: *Encyclopaedie van Nederlandsch-Indië* (1899–1905). He was Curator of the National Museum of Ethnology in Leiden (late 1880s–c.1900) and became Director (1901–1915) of the Ethnological Museum, Rotterdam and the maritime museum there until taking early retirement because of ill health.

Snellen

Yellow-striped Flutterer ssp. *Rhyothemis phyllis snelleni* **Selys**, 1878
[Orig. *Rhyothemis snelleni*]

Pieter Cornelis Tobias Snellen (1832–1911) was a Dutch merchant in Rotterdam and an amateur entomologist. Selys made a diary entry (April 1867) that he received specimens from Snellen collected in Celebes. Selys met Snellen in person in Maestricht over twenty years later.

Sollaart

Lieutenant sp. *Brachydiplax sollaarti* **Lieftinck**, 1953

A Sollaart was a naturalist and collector who was a member of a Dutch colonial family in Indonesia. He collected insects (1950–1954), particularly Lepidoptera, Coleoptera, Odonata and Hymenoptera in Indonesia (Sumatra) for Leiden, Rotterdam and other Dutch museums. Most were taken on Medang Ara Estate, Kwala Simpang, Sumatra Timur. He sent specimens to various institutions including: Zoological Museum Amsterdam, RMNH (Leiden), BPBM (Honolulu) and Rotterdam Natural History Museum.

Sombu

Cruiser sp. *Macromia sombui* **Vick**, 1988

Sombu Paryar (b.1956) worked for British entomologist Colin Smith (b.1933) (who was known as 'Putali Baje', which translates from Nepali to 'butterfly grandpa') who was Curator of the Annapurna Regional Museum in Pokhara, Nepal. Sombu was collecting dragonflies in the tropical forests in the vicinity of the Royal Chitwan National Park (November 1986), when he collected the holotype, which is now in the author's collection.

Sommer

Hooktail sp. *Megalogomphus sommeri* **Selys**, 1854
[Orig. *Heterogomphus sommeri*]

Michael Christian Sommer (1785–1868) was a German financier and trader in insects, the son of a spice merchant. He collected insects from childhood, which he greatly added to as an adult, being sent specimens by his overseas contacts. He established an import-export business with trading partners in the Americas and around the world, which in time turned into a business trading in insects. He was a 'Kommissionär' and raised money to enable people to emigrate, undertake overseas explorations or, importantly for him, collecting trips. This credit had to be paid off by 'products of nature' i.e. botanical or zoological specimens, most often insects, given in turn directly to the investors or sold on to collectors. He kept a very large collection himself, which was equivalent to a catalogue used for buyers to select their future purchases. Through this trade, he maintained contacts with all the important collectors and entomologists of his era.

Somnuk

Cinchtail sp. *Amphigomphus somnuki* **Hämäläinen**, 1996

Somnuk Panpichit (b.1935) was one of Bro Amnuay Pinratana's major helpers while Pinratana accumulated his extensive collections of Thai butterflies, moths, beetles and dragonflies, (c.1980-c.2003). He collected for Pinratana all over Thailand, the holotype of the eponymous species being found at Doi Suthep, near Chiang Mai City. Hämäläinen participated in many of these joint trips, including one four-day trip to Ranong alone with Somnuk (1988). They did not have any common language to communicate, but managed very well despite this. Well over a decade ago, Pinratana stopped increasing his collections, which are now housed in Bangkok. Several other species are named after Somnuk including a beetle and some Lepidoptera.

Sonehara

Migrant Hawker ssp. *Aeshna mixta soneharai* **Asahina**, 1988

Imato Sonehara (1921–2000) was a Japanese odonatologist. He collected the type series in Japan (1963 & 1965). He entered teacher training after WW2 at Hokkaido University where he graduated (1948), thereafter working as a science teacher in junior and then high schools until retirement (1981). He began studying Odonata near Saku City (1961) and studied them further under Asahina, resulting in his first two papers being published; he went on to publish another 61 papers. One of his last was co-authored with his son-in-law Hidenori Ubukata. He wrote: *The Life History of Epitheca bimaculata sibirica at Mt. Yatsugatake*. He died as a result of a road traffic accident.

Sônia

Pond Damselfly sp. *Franciscobasis sonia* **Machado** & **Bedê**, 2016

Sônia Rigueira is a biologist who is President of the Terra Brasilis Institute, Aimorés, Minas Gerais, Brazil, where the second author also works. The authors honoured her "… *in recognition for her important contribution to the conservation of odonates in the State of Minas Gerais*".

Sonja

Genus *Sonjagaster* Lohmann, 1992
[JS *Cordulegaster* Leach, 1815]

Sonja Meury née Lohmann (b.1975) is the daughter of the author Heinrich Lohmann (b.1945).

Sooretama

Pond Damselfly sp. *Metaleptobasis sooretamae* **Santos**, 1957
[JS *Telagrion longum* Selys, 1876]

This is a toponym; Sooretama National Park in Espirito Santo, Brazil was the type locality.

Sophie

Sedgling *Agrion sophia* **Selys**, 1840
[JS *Nehalennia speciosum* (Charpentier, 1840)]
Ebony Cruiser *Phyllomacromia sophia* Selys, 1871
[Orig. *Macromia sophia*]
Sprite sp. *Nehalennia sophia* Selys, 1876
[Homonym; JS *Nehalennia minuta selysi* Kirby, 1890]

Sophie Caroline de Selys Longchamps née d'Omalius d'Halloy (1818–1869) became the wife (1838) of the author. Selys met Sophie for the first time on 27th January 1835 and his diary reveals that by the time he saw her again a few weeks later (3rd March 1835) he was in love. She died of cancer and Selys wrote a long diary account of her last days. He concluded with: "…*At half past eight, she expired. We were crying while kneeling near the bed, me kissing her with my hands in hers, my head on her pillow, against her shoulder. It's all over. R.I.P. my poor Sophie*."

Souter

Scissortail sp. *Microgomphus souteri* **Fraser**, 1924
[Orig. *Microgomphus torquatus souteri*]
Threadtail sp. *Elattoneura souteri* Fraser, 1924
[Orig. *Disparoneura souteri*]

Sir Charles Alexander Souter KCIE CSI (1878–1958) was a tax collector and later the Governor's Council under the British Raj. After leaving the University of Edinburgh and Caius College, Cambridge, he joined the Indian Civil Service as an assistant collector (1901) and later became an administrator in Coorg. He was then Secretary to the Public Works Department in Madras and was appointed (1930) a member of the Board of Revenue.

Southwell

Sylvan sp. *Coeliccia southwelli* **Dow** & **Reels**, 2011

Lukut 'Luke' Southwell (b.1950) and his son John Hudson Southwell (1981–2011). The former was a collector on, and the manager of, the odonatological expedition to the Hose Mountains in central Sarawak, Malaysian Borneo (April 2011), which collected the type. According to Brian Row Mcnamee in his book: *With Pythons & Head-Hunters in Borneo* (2009), Luke (who guided him in 2008 in the Niah Caves National Park) has an interesting history. He is apparently a member of the Kayan people brought up in a longhouse in the Bario Highlands. His birth name was Lukut, but he was a sickly twin given away as a baby to be brought up by Australian missionaries. He later attended university in Australia. The species is also named in memory of his son John as the etymology states: "*Named for our very good friend Luke Southwell from Miri in Sarawak, who has been a vital figure in all of our trips to Mount Dulit, and without whose initiative the trip on which the holotype and two paratypes were collected would not have taken place; and in memory of his son, John Hudson Southwell, who tragically passed away on 31 January 2011.*"

Sowerby

Clubtail sp. *Burmagomphus sowerbyi* **Needham**, 1930
[Orig. *Gomphus sowerbyi*]

Arthur de Carle Sowerby (1885–1954) was a naturalist, explorer and artist, born in China, where his father was a Baptist Missionary. He went to Bristol University but only stayed a short time before returning to China, where he began to collect specimens for the BMNH in Tai-yuan Fu. He collected mammals (1907) for the BMNH during an expedition to the Ordos Desert in Mongolia, and (1908) was part of the Clark Expedition to Shansi and Kansu (Shaanxi & Gansu) provinces. He wrote an account *Through Shên Kan, the Account of the Clark Expedition in North China 1908–09* (1912). There was a revolution in China (1911) and Sowerby led an expedition to evacuate foreign missionaries from Shensi and Sianfu provinces. During WW1, he was a technical officer in the Chinese Labour Corps and saw service in France. After the war, he settled in Shanghai and established 'The China Journal of Science and Arts', which he edited until the Japanese occupied Shanghai during WW2. The Japanese Army interned him for the duration, but despite this he appears to have been able to go on writing and publishing, as evidenced by *Birds recorded from or known to inhabit the Shanghai* area (1943). He emigrated to the USA (1949) and lived the rest of his life in Washington DC. He spent his time in genealogical research, which resulted in a family history: *The Sowerby Saga*. Three birds, an amphibian and a reptile are also named after him.

Spaeter

Bannerwing sp. *Polythore spaeteri* **Burmeister** & Börzsöny, 2003

Hartmut Spaeter (1922–2007) was a German businessman, philanthropist, nature lover and traveller. He inherited some money and invested wisely enough for it to grow considerably and this, coupled with his modest lifestyle, allowed him to follow his passion for overseas travel. In his will, he created the Hartmut Spaeter Foundation under the aegis of the Environment Foundation Greenpeace, to donate to environmental projects. The etymology simply says: *"Named to honour a person for his sponsorship of biological investigations at the biological observation center of Panguana (Huanuco, Peru)."*

Spegazzini

Springwater Dancer *Argia Spegazzinii* **Navas**, 1933
[JS *Argia plana* Calvert, 1902]

Carlo Luigi (Carlos Luis) Spegazzini (1858–1926) was an Italian-born Argentinean mycologist and naturalist. He was trained in oenology but from the outset his main interest was fungi. He travelled from Italy to Brazil (1879), but quickly moved from there to Argentina to escape an epidemic of yellow fever. He was a member (1881) of the Italo-Argentine expedition to Patagonia and Tierra del Fuego. The expedition was shipwrecked and Spegazzini had to swim for it, bearing all his notes on his shoulder to keep them from the sea. He later took up permanent residence in Argentina (1884). He became a professor at the University of La Plata (1887–1912). In the same period, he was also Curator of the National Department of Agriculture Herbarium, first head of the Herbarium of the Museo de La Plata and founder of an arboretum and an Institute of Mycology in La Plata city. He is most remembered for his study of mycological and vascular plants, but he travelled widely and collected natural history specimens wherever he went. He published about 100 papers on vascular plants, mostly in Argentinean journals, and described around 1,000 new taxa. A mammal is also named after him.

Spence

Sprite sp. *Pseudagrion spencei* **Fraser**, 1922

William Spence (1783–1860) was a British economist and entomologist and the eldest of four children of farmer Robert Spence. He was apprenticed to Russian merchants and ship-owners Carhill, Greenwood & Co. but little else is known about his early life except that at the age of ten he was in the care of a clergyman who taught him botany. He married Elizabeth Blundell (1804) and very soon supported her brother Henry to set up the highly successful oil and colour company, Blundell Spence. He was a close friend of the Reverend William Kirby, a fellow entomologist, with whom he wrote and published the 4-volume: *Introduction to Entomology* (1815–1826). He was the founder of the Entomological Society (1833), later becoming president (1847). The type specimens were collected by Thomas Bainbrigge Fletcher (1878–1950) (q.v.). Fraser did not provide any etymology for this name and the entry is the most likely candidate. A bird is also named after him.

Stahl

Red Relic *Pentaphlebia stahli* **Förster**, 1909

Rev Gottlieb Heinrich Stahl (b.1875) was a missionary in Africa. He was at (1893) the Mission in Nyasoso, Cameroon, where he collected some insect specimens, including the type series (1901–1909). The mission was abandoned during WW1.

Stainberger

Blue Sprite ssp. *Pseudagrion microcephalum stainbergerorum* Marinov, 2012

Daisy Blanche Stainton (b.1984) and Simona Joop Kraberger (b.1983) are honoured; at the time, both were PhD students at the School of Biological Studies, University of Canterbury, New Zealand. According to Marinov's etymology, they were both passionate about finding a new species and suggested the name that he used. The Marsden Fund of New Zealand and a grant from Canterbury University funded much of the work undertaken by Marinov on the Odonata of Tonga. Daisy and Simona were among those who undertook field sampling as part of the 'Banana Bunchy Top Group'. Both have already published papers (10 and 21 respectively) (2014).

Starmühlner

Starmuehlner's Shadowdamsel *Drepanosticta starmuehlneri* St. Quentin, 1972
[JS *Drepanosticta lankanensis* (Fraser, 1931), P Drooping Shadowdamsel]

Professor Dr Ferdinand Starmühlner (1927–2006) was an Austrian zoologist, limnologist marine biologist and professor at the Zoological Institute at the University of Vienna. He was particularly interested in marine invertebrates. He participated in two research trips to Afghanistan (1949) and Iran (1952) and led others including to Icelandic waters (1955), Madagascar (1958), the Indian Ocean (1974) and to Tonga and Samoa (1985). He wrote a number of papers such as: *Checklist of the fauna of mountain streams of tropical Indopacific Islands* (1985) and contributions to longer works such as: *Ecology and Biogeography in Sri Lanka* (1984), as well as full length works including: *Between the Arctic Circle and the Equator* (1999). He was also a populariser of natural history appearing in many TV programmes. He collected the type in Sri Lanka (1970).

Starre

Marshal sp. *Pornothemis starrei* **Lieftinck**, 1948
Sprite sp. *Pseudagrion starreanum* Lieftinck, 1949

Captain J J van der Starre, who collected the types in Sumatra (February 1938), was a friend of the author (See **Van der Starre**). I assume the same man was decorated with the equivalent of the Distinguished Service Cross when he saved his Dutch merchant navy ship and crew attack off Java by Japanese submarines (1942).

Staudinger

Rubyspot sp. *Hetaerina staudingeri* **Förster**
[*This is an unpublished manuscript name only*]

Otto Staudinger (1830–1900) was a German entomologist and insect dealer who never went to South America, but (1865–1935) much of the material that was collected in Peru was sent to the Staudinger & Bang-Haas company where Otto Staudinger and afterwards his son and partner Andreas Bang-Haas worked on Lepidoptera, Orthoptera, Coleoptera and Hymenoptera.

Stawiarski

Tropical Dasher sp. *Micrathyria stawiarskii* **Santos**, 1953

Professor Vitor (Victor) Stawiarsky (b.1903) taught biology at a Buenos Aires College (1929) and was at Museu Nacional, Rio de Janeiro, Brazil (1930–1945). He collected the holotyope of this new dragonfly (December 1943). He also collected the holotype of a toad that is named after him (1948).

Steele

Swamp Slim *Aciagrion steeleae* **Kimmins**, 1955

Mary Steele (d.c.1960) was a collector of insects and plants. She collected the holotype at Lake Bangweulu in Rhodesia (October 1946).

Stella

Florida Baskettail *Epitheca stella* **Williamson**, 1911
[Orig. *Tetragoneuria stella*]

Stella Ann Deam née Mullin (1870–1953) was an American collector from Indiana who collected (1904) a great deal of material from Florida; the botanical specimens being for her husband (m.1893), Charles Clemon Deam (1865–1953) (q.v.), the first state forester of Indiana. Williamson wrote: "...*I am indebted to her for this new species as well as for many other dragonflies from Florida.*" (Also see **Deam**)

Stevens

Clubtail sp. *Perissogomphus stevensi* **Laidlaw**, 1922
Shadowdamsel sp. *Protosticta stevensi* **Fraser**, 1922
[JS *Protosticta gravelyi* Laidlaw, 1915]
Shadowdancer sp. *Idionyx stevensi* Fraser, 1924

Featherleg sp. *Copera annulata stevensi* Laidlaw, 1914
[JS *Pseudocopera annulata* (Selys, 1863)]

Herbert Stevens (1877–1964) was a tea planter in Sikkim (India) and an ornithologist and zoologist also interested in entomology. He took part in a number of expeditions, normally as a museum collector. He was in Tonkin (1923–1924 & 1929) and was ornithologist on the Kelly-Roosevelt expedition of the Field Museum, Chicago, to Indochina. He was in New Guinea (1932–1933), collecting for the Museum of Comparative Zoology, Harvard. He wrote: *Notes on the birds of the Sikkim Himalaya* and his autobiography: *Through deep defiles to Tibetan uplands* (1934). He was a great benefactor of the British Ornithologists' Club and bequeathed to them his house in Tring (Hertfordshire), which was their headquarters for many years. Eight birds are also named after him as are a number of insects.

Stevenson

Duskhawker sp. *Gynacantha stevensoni* **Fraser**, 1927
[A. Duskhawker sp.]

Robert Louis Balfour Stevenson (1850–1894) was a well-known novelist, poet, essayist and travel writer, who had spent time in Samoa. Fraser did not provide any etymology of the name. The specimen was collected in Tonga by G H Hopkins. Since the description was published in the series 'Insects of Samoa', it is most likely that the species was named after the novelist.

Strachan

Red Rockdweller *Bradinopyga strachani* **Kirby**, 1900
[Orig. *Apeleutherus strachani*]

Dr William Henry Williams Strachan (1859–1921) was a British physician born in the Bahamas. He was educated at Guy's Hospital (1877–1882) receiving his medical degrees there (MRCS 1881 & LRCP 1882) and he became Junior, then Senior Medical Officer, to the Kingston Hospital, Jamaica (1882–1897). He was Principal Medical Officer of Lagos (1897–1899, where he collected the holotype) responsible for combatting malaria and was one of those who recognized the mosquito as the vector for the disease. He was acting Colonial Secretary (1899–1900) then Chief Medical Officer (1902, when he was awarded a CMG). He retired to Jamaica (1911) although he later moved to London where he died of chronic Bright's disease and uraemia. Most of his publications were on public health such as: *Guide to the Preservation of Health in West Africa* (1910). He was described once as "*...a middle-sized, rather dark man from the West Indies, a good talker and fine performer on the piano*".

Strohm

Pond Damselfly sp. *Acanthallagma strohmi* **Williamson** & Williamson, 1924

John Winnett Strohm (1880–1952) was a retired captain from the United States Army who was a member of the University of Michigan 'Williamson Expedition' to Amazonas and Mato Grosso, Brazil (1922), where he collected the first specimens of this "*...handsome and interesting species*".

Strouhal

Bluet sp. *Enallagma strouhali* St. Quentin, 1962
[JS *Enallagma risi* Schmidt, 1961]

Professor Dr Hans Strouhal (1897–1969) was an Austrian zoologist and speleologist. He studied mathematics, botany and later zoology in Vienna, receiving his PhD there (1926). Whilst an assistant at the Zoological Department of the University and Natural History Museum of Vienna (1927–1938) he was summarily dismissed by the Nazis. In WW2, he was drafted (1940) and rose to the rank of major until captured by American forces after which, he was able to resume his teaching (1945). He was appointed as director (1951) and became director both of the department and museum until retirement (1963). Thereafter he continued with his work whenever his health was good enough. He wrote his last paper: *Die Landisopoden der Insel Zypern* in 1968.

Stüber

Bluetail sp. *Ischnura stueberi* **Lieftinck**, 1932
Offshore Emerald *Anacordulia stüberi* Lieftinck, 1938
[JS *Metaphya tillyardi* Ris, 1913]

Pond Damselfly sp. *Papuargia stueberi* Lieftinck, 1938
[Orig. *Papuargia stüberi*]

Wilhelm Carl Julius Stüber (1877–1942) was a German planter in Dutch Government service in Dutch New Guinea, as well as a professional collector, particularly of insects, orchids and other botanical material. He collected in Irian Jaya (1930–1939) and (1936) with British entomologist Lucy Evelyn Cheesman (1881–1969). At least one orchid is named after him. She reported "*I accompanied Herr Stuber, a German planter in the Dutch Government service who was making a reconnaissance with a view to opening up a road leading to gold-bearing country farther inland. As he is a collector of orchids and butterflies, Herr Stuber is friendly with the Papuans of that district.*" Stüber presented himself (1930) to the Zoological Museum and Laboratory of the Botanical Gardens in Buitenzorg, Java, where Lieftinck had recently (1929) taken the post of zoologist; he offered to collect specimens and a financial agreement was made. They exchanged many letters (1930–1939), often describing the behaviour of Odonata species that Stüber collected for the museum. In total, Stüber sent Lieftinck 13,800 Odonata specimens of 165 species! To quote a recent article in 'Agrion' by Matti Hämäläinen and Albert G Orr (2016): *Although Stüber was already advanced in years, he was a diligent and active collector and his contribution to the knowledge of New Guinean Odonata was phenomenal. It is fair to suggest that his prodigious collecting activity enabled Lieftinck, who was still at the beginning of his career, to specialize in the study of New Guinean Odonata and rapidly become the regional expert on this group.* He was among the over 400 German-born internees who lost their lives when the freighter *van Imhoff*, which was on its way from Sumatra to India, was attacked and sunk by a Japanese bomber near Nias Island on 19th January 1942.

Stuckenberg

Stuckenberg's Sprite *Pseudagrion stuckenbergi* **Pinhey**, 1964

Dr Brian Roy Stuckenberg (1930–2009) was a South African entomologist. He studied at Rhodes University, Grahamstown, which awarded his MSc. He was Assistant Professional Officer at the Natal Museum in Pietermaritzburg (1953) whilst working toward his PhD, which the University of Natal awarded (1972). He became Director of the Natal Museum (1976–1994). His special interest and expertise was in Diptera. After retirement, he continued to work as Director Emeritus; Honorary Keeper of Entomology. He collected the holotype of this damselfly in Madagascar (November 1957).

Stuhlmann

Red-veined Dropwing *Libellula stuhlmanni* **Gerstäcker**, 1891
[JS *Trithemis arteriosa* Burmeister, 1839]
Eastern Horntail *Nepogomphoides stuhlmanni* **Karsch**, 1899
[A Congo Scissortail, Orig. *Notogomphus stuhlmanni*]

Professor Dr Franz Ludwig Stuhlmann (1863–1928) was a German zoologist and naturalist who collected in East Africa (1888–1900). He made his career in the German Colonial Forces and Civil Service. He did not confine himself to zoological specimens and a number of the artefacts he collected in Africa are in anthropological exhibits. Stuhlmann travelled with Emin Pasha and, after Emin's murder, he and others who had survived an outbreak of smallpox went back from the area of Lake Albert with a large collection and a lot of cartographic material from which the first comprehensive map of German East Africa (Tanzania) was made. The German Government published a monograph by Stuhlmann: *Mit Emin Pasha ins Herz von Africa Eine Reiseberich* (1894). Five birds, two mammals, a reptile and an amphibian are also named after him.

Sudha

Oread sp. *Calicnemia sudhaae* Mitra, 1994

Sudha Rani Mitra was the mother of the author Dr Tridib Ranjan Mitra (1942–2012).

Suenson

Clubtail sp. *Sinogomphus suensoni* **Lieftinck**, 1939
[Orig. *Gomphus suensoni*]
Emerald Spreadwing sp. *Megalestes suensoni* **Asahina**, 1956
[JS *Megalestes heros* Needham, 1930]

Erik Eigin Suenson (1887–1966) was a Danish engineer who lived and worked in China (1917–1946). He collected Coleoptera (particularly Scarabs), Odonata and other material there over a number of decades (1919–1937) and his 1936–1937 collection was passed to Lieftinck who identified the Odonata species including the clubtail holotype. The bulk of his collection is now housed in the Zoologisk Museum København.

Susana

Threadtail sp. *Epipleoneura susanae* Pessacq, 2014

Susana Ringuelet (b.1942) is the mother of the author Pablo Pessacq (b.1973).

Susanna

Pond Damselfly sp. *Palaiargia susannae* Kovács & **Theischinger**, 2015

Zsuzsanna Benkó Kovácsné (b.1965) is the wife of the author Tibor Kovács (b.1965).

Sutter

Reedling sp. *Indolestes sutteri* **Lieftinck**, 1953
[Orig. *Lestes sutteri*]

Dr Ernst Sutter (1914–1999) was a Swiss zoologist, particularly interested in ornithology. His PhD (1943) was about the growth of brains in certain birds. He became Curator of the Bird Collection at the Naturhistorisches Museum of Basel (1945–1980) and continued his work as Curator Emeritus thereafter. He organised a nine-month long field expedition to the Lesser Sunda Islands, his only such expedition. He stayed in Sumba from May to October (1949) and made extensive collections there (65,000 specimens including 770 birds), which included the eponymous type. He co-wrote the report on the expedition (1951). He was also a pioneer in using radar to study bird movements.

Suzuki

Skydragon sp. *Chlorogomphus suzukii* Oguma, 1926
[Orig. *Orogomphus suzukii*]
Lyretail sp. *Stylogomphus suzukii* Oguma, 1926
[Orig. *Gomphus suzukii*]

Motojiro Suzuki (d.1942) was a Japanese insect dealer. He collected in Kyoto, Honshu, Japan, and may have taken the types.

Sven Hedin

Clubtail sp. *Trigomphus svenhedini* **Sjöstedt**, 1932
[Orig. *Gomphus sven-hedini*]
Jewel sp. *Indocypha svenhedini* Sjöstedt, 1932
[Orig. *Rhinocypha sven-hedini*]

Dr Sven Anders Hedin (1865–1952) was a Swedish geographer, explorer, photographer and map maker who also wrote travelogues, which he illustrated himself. He participated in three expeditions to central Asia (1894–1908). Later, he organized and led a Sino-Swedish Expedition (1927–1935). The participating scientists worked almost independently, while Sven Hedin, as a manager on site, negotiated with the authorities, organized, raised funds and charted routes etc. Archaeologists, astronomers, botanists, anthropologists, geographers, geologists, meteorologists and zoologists from Sweden, Germany and China studied Mongolia, Xinjiang and Tibet. The culmination of his life's work was a posthumous publication: *Central Asia Atlas*. Other works included: *The scientific results of my travels in Central Asia 1894–1897* (1900) and the 12-volume *Scientific results of a journey in Central Asia* (1904–1907).

Svihla

Sylvan sp. *Coeliccia svihleri* **Asahina**, 1970
Grappletail sp. *Heliogomphus svihleri* Asahina, 1970
[Orig. *Leptogomphus svihleri*]

Dr Arthur Svihla (1902–1996) was an American zoologist, professor of zoology (1936–1943) and museum curator at the University of Washington where he became Chairman of the Zoology and Physiology Department (1943). He collected in Louisiana (1925–1927) with his botanist wife Dr Ruth Dowell Svihla and in Washington (1930), Oregon (1931) and Idaho (1932). He was made Secretary of the Ichthyologist Society (1933) and President of the Pacific Northwest Bird and Mammal Society. He studied *Tanypteryx hageni* for five seasons (1954–1959), culminating in a detailed description of its life history: *Life history of Tanypteryx hageni Selys (Odonata)* (1959). Among his other written works are: *The Hawaian Rat* (1936) and *Another record of the larva of Epiophlebia laidlawi Tillyard, (Odonata: Anisozygoptera)* (1961). The Ford Foundation sponsored a University of Florida expedition to Burma during which he collected the holotypes that Asahina described in: *Burmese Odonata collected by Dr Arthur Svihla with supplementary notes on Asiatic Ceriagrion species* (1970).

Sykina

Ruddy Darter ssp. *Sympetrum sanguineum sykinia* **Belyshev**, 1955

Zykina was the surname of an unspecified Russian lady after whom the taxon was named.

Symoens

Katanga Sprite *Pseudagrion symoensii* **Pinhey**, 1967

Southern Leaftail *Phyllogomphus symoensi* **Lieftinck**, 1969

[JS *Phyllogomphus selysi* Schouteden, 1933]

Professor Jean-Jacques André Symoens (1927–2014) was a Belgian ecologist and botanist. The Free University of Brussels awarded both his first degree in chemical sciences and his PhD in botanical science. He then became a researcher at the Lake Tanganyika station and afterwards professor at l'Université Officielle du Congo Belge et du Ruanda-Urundi in Elisabethville (now Lumumbashi), where he taught botany and plant ecology and where his wife taught pharmaceutical chemistry. He returned to Belgium to become professor at Vrije University, Brussels (1972), teaching botany and hydrobiology as well as the history of biological sciences at the University of Mons-Hainaut. His hobby was numismatics (the collection and study of coins). He continued to teach and write long after retirement age. His last book was: *Vegetation of Inland Waters,* part of the Handbook of Vegetation Science Series (2013).

T

Tabe

Clubtail ssp. *Trigomphus citimus tabei* **Asahina**, 1949

Professor Masachika Tabe worked at the Zoology Department of Tokyo Imperial University. He collected the type in Japan (1946) where the subspecies is endemic.

Takakuwa

Skydragon sp. *Chlorogomphus takakuwai* **Karube**, 1995

Dr Masatoshi Takakuwa (1947–2016) was an entomologist who was at the Kanagawa Prefectural Museum of Natural History, Japan, until retirement (2008). He was a specialist in the taxonomy of *Mordellidae* (Tumbling Flower-beetles). He collected quite widely and had a number of insects named after him, some of which he was the first to collect. He collected with Tatsuya Niisato (q.v.) in Vietnam during the mid-1990s. Among his publications are longer works such as the co-written: *The Longicorn-Beetles of Japan in Color* (1984) and many papers such as: *Two new cerambycid beetles from Lutao Island off Southwest Taiwan* (1991).

Takasaki

Longleg sp. *Anisogomphus takasakii* **Yamamoto**, 1954
[JS *Anisogomphus maacki* (Selys, 1872)].

Yasuo Takasaki is a Japanese Public Health professional who lives in Nagoya city, Aichi Prefecture, central Japan. He is an amateur odonatologist active in the Japanese Society for Odonatology. He collected the type in Japan (1951). Among his published papers is: *Dragonflies of the projected site of the exposition at Seto City, Aichi* (1998).

Takashi

Clubtail sp. *Stylurus takashii* **Asahina**, 1966
[Orig. *Gomphus takashii*]

Dr Takashi Shirozu (1917–2004) (See **Shirozu**)

Tamoyo

Dancer sp. *Argia tamoyo* **Calvert**, 1909

The Tamoyo are a people native to the type locality in Chapada, Brazil.

Tao

Demoiselle sp. *Matrona taoi* Phan & **Hämäläinen**, 2011

Dr Nguyen Thien Tao (b.1982) is a Vietnamese herpetologist. His PhD was awarded by Kyoto University, Japan, when he defended his thesis on the molecular and morphological systematics and distribution pattern of various rhacophorid frog species. He is Curator of Herpetology and head of the Department of Nature Conservation, Vietnam National Museum of Nature, of the Vietnam National Academy of Natural Science, Hanoi. He is (since 2016) Visiting Associate Professor at Kyoto University. His research interests are in the taxonomy, evolutionary origin, and diversification of amphibians and reptiles, as well the practical elucidation of the phylogeny of various amphibian and reptile groups. He has extensive experience in taxonomy and ecology of amphibians and reptiles throughout Vietnam. The authors wrote: "*The species is named after Mr Nguyen Thien Tao (Biology Department, Vietnam National Museum of Nature) as a token of gratitude for his friendly support of the first author's field work and research activity.*" He has published (since 2007) more than 70 papers on herpetology, among them 31 new species of amphibians and reptiles from Vietnam, Laos and Cambodia, such as: *A new species of Theloderma (Amphibia: Anura: Rhacophoridae) from Vietnam* (2016). He recently discovered a new salamander, a new crocodile newt and a new white-headed viper species in northern Vietnam. A reptile is named after him, as is a frog jointly with his brother Dr. Nguyen Quang Truong.

Taracumbi

Melville Island Threadtail *Nososticta taracumbi* **Watson** & **Theischinger**, 1984

This is a toponym; the Taracumbi Falls on Melville Island is where the species can be found.

Tarascos

Dancer sp. *Argia tarascana* **Calvert**, 1902

This Mexican species is named after a people; the Tarascos Indian tribe of Gerrero, Mexico.

Tarboton

Threadtail sp. *Elattoneura tarbotonorum* Dijkstra, 2015

Warwick Rowe Tarboton (b.1943) and Michèle Tarboton (b.1949) are a South African husband and wife. Warwick was awarded his BSc (1965), MSc (1979) and PhD in zoology by the University of the Witwatersrand (1991), as well as an MSc by Sheffield University (UK). He started working life (1966) as a geologist but switched to ornithology (1974), a passion since he was eleven years old, working for 22 years as an ornithologist with the former Transvaal Directorate of Nature Conservation. In 1996 he changed to freelance work. Michèle was a biology schoolteacher; both are retired and continue to pursue their interest in birds and dragonflies where they live in the northern part of South Africa. They were honoured for their advance of the identification of African Odonata through their guides. They have written three handbooks of South African Odonata, the latest being: *A Guide to the Dragonflies & Damselflies of South Africa* (2015). Warwick has published extensively including 40 papers and 200 articles on birds and has authored or co-authored more than a dozen books, including the various editions of: *Sasol Fieldguide to the Birds of Southern Africa* (1993) and *Roberts' Nests and Eggs of southern African birds* (2011). He collected the holotype.

Tariana

Threadtail sp. *Epipleoneura tariana* **Machado**, 1985

Named after the indigenous people who live in the area of Brazil where the species was found.

Tarry

Common Spreadwing *Lestes tarryi* **Pinhey**, 1962

[JS *Lestes plagiatus* (Burmeister, 1839), A. Highland Spreadwing, Stream Spreadwing]

David W Tarry was a veterinarian entomologist at the Ministry of Agriculture, Fisheries and Food, Central Veterinary Laboratory, UK, who collected a few dragonflies in Nigeria, including this species (April 1960). He visited Denmark (1975) to study 'measures against headfly injury to sheep and cattle'. He published more than twenty papers (1960s–1990s) on veterinary entomology such as: *Observations on the ecology of Glossina morsitans submorsitans Newst. in the Guinea-Sudan transition savanna of Northern Nigeria* (1968) to *Transmission of bovine virus diarrhoea virus by blood feeding flies* (1991).

Tarui

Clubtail ssp. *Davidius moiwanus taruii* **Asahina** & **Inoue**, 1973

Yoshitsugu Tarui (1931–1978) was a Japanese high school teacher at Karaku Middle School, Kyoto and an amateur entomologist. He collected the type in Japan (1955) and was co-author of a R*eport on the Biological Survey of Freshwater Planarians in Kyoto City and its Adjacent District, Honshu* (1967).

Taurepan

Sylph sp. *Macrothemis taurepan* **De Marmels**, 2008

The Taurepan are an indigenous people living in areas of Venezuela where the sylph was discovered.

Telesford

Dancer sp. *Argia telesfordi* **Meurgey**, 2009

John Telesford is a consultant entomologist and ecologist for the Caribbean Community Climate Change Centre. At the time the damselfly was discovered he was an entomologist for the Pest Management Service at the Grenada Ministry of Agriculture. The author says: "*This species is dedicated to John Telesford, entomologist at the Grenada Ministry of Agriculture who led us to aquatic habitats of Grenada and for his help on the field.*" According to the description article John Telesford drove the author all over the island during his two-week collecting trip (2009).

Telosticta

Damselfly genus *Telosticta* (**Dow** & **Orr**, 2012)

(See **Tol**)

Tennessen

Sanddragon sp. *Progomphus tennesseni* Daigle, 1996

Pond Damselfly sp. *Oxyagrion tennesseni* **Mauffray**, 1999
Threadtail sp. *Drepanoneura tennesseni* von **Ellenrieder** & **Garrison**, 2008

Dr Kenneth (Ken) Joseph Tennessen (b.1946) is an American entomologist who also describes himself as a poet, Vietnam veteran, whittler and daydreamer. The University of Florida awarded his PhD (1975). One of his primary goals as a biologist has been to blend the science of biology with the study and appreciation of nature. Although retired, his research focus is still Odonata, which he has studied all over the world, resulting in the largest collections of Odonata specimens and photos in western North America. He has written or co-written more than seventy publications, including scientific papers such as: *Description of the Nymph of Enallagma minusculum (Odonata: Coenagrionidae)* (1975) and *Psaironeura angeloi, a new species of damselfly (Zygoptera: Coenagrionidae) from Central and South America* (2016) and also a booklet *Dragonflies of Washington*, as well as field guides to North American dragonflies and damselflies. He has also co-authored a book of poems: *Dragonfly Haiku* (2016) and wrote a novel *Utterly Bugged* (2013) about an entomologist transported back in time. He has described twenty-two new odonate species. (Also see **Sandra**)

Teodoro

Ringed Forceptail *Aphylla theodorina* **Navás**, 1933
[Orig. *Gomphoides theodorina*]
Firetail sp. *Telebasis theodori* Navás, 1934
[Orig. *Argia theodori*]

H Teodoro of the Christian School in Caxias, Rio Grande do Sul, Brazil, collected the type in Porto Alegre, Brazil (1931). Described as 'a local person' he sent the type to Navás who labelled the specimens as being collected by H Teodoro.

Tepeaca

Clubskimmer sp. *Brechmorphoga tepeaca* **Calvert**, 1907

This Mexican species is named after a people; the Tepeaca are an Indian tribe in the type locality.

Teramoto

Clubtail sp. *Borneogomphus teramotoi* **Karube** & **Sasamoto**, 2014
Emerald sp. *Hermicordula teramotoi* Yokoi, 2015

Toshiyuki Teramoto (b.1955) is an amateur entomologist from Tokyo. He has collected dragonflies in Laos with Naoto Yokoi (q.v.) and with him and Sasamoto (q.v.) co-authored: *Description of a new Sinorogomphus from Northern Laos (Odonata: Chlorogomphidae)* (2011).

Terira

Dancer sp. *Argia terira* **Calvert**, 1907

The Costa Rican species is named after a people – the Terira are a local Indian tribe.

Terue

Bluet sp. *Coenagrion terue* **Asahina**, 1949
[Orig. *Agrion terue*]

Terue Asahina née Yoshioka was the wife (1940) of the author Dr Syoziro Asahina (1913–2010). (Also see **Asahina**).

Tezpi

Tezpi Dancer *Argia tezpi* **Calvert**, 1901

Tezpi was the equivalent of the biblical Noah in Michoacan legend. The legend is related thus: *"When the floodwaters began to rise, a man named Tezpi entered into a great vessel, taking with him his wife and children and diverse seeds and animals. When the waters abated, the man sent out a vulture, but the bird found plenty of corpses to eat and didn't return. Other birds also flew away and didn't return. Finally, he sent out a hummingbird, which returned with a green bough in its beak."*

Thaís

Flatwing sp. *Heteragrion thais* **Machado**, 2015

Thaís Fernanda Costa e Silva Magalhães de Souza (b.1980) is the wife of Professor Marcos Magalhães de Souza who collected the type in Minas Gerais, Brazil.

Theá

Pond Damselfly sp. *Tepuibasis thea* **De Marmels**, 2007

The etymology says: *"Theá in Greek means 'goddess'. This new species is named in allusion to the sacred status (to the local Amerindian tribes) of the mountains of Pantepui."*

Theebaw

Threadtail sp. *Prodasineura theebawi* **Fraser**, 1922
[Orig. *Caconeura theebawi*]

King Thibaw (or Theebaw) Min of Burma (1859–1916). No etymology is given for this name, but J R Elton Bott collected the specimen in King Island, Mergui, Burma, so it was almost certainly named after the recently deceased king, or an allusion to the type locality.

Théel

Treeline Emerald *Somatochlora theeli* Trybom, 1889
[JS *Somatochlora sahlbergi* Trybom, 1889]

Professor Dr Johan Hjalmar Théel (1848–1937) was a Swedish zoologist, who started his career in Uppsala as a disciple of Lilljeborg, then Tullberg, culminating in his doctoral thesis (1872). His first paper (1875) resulted in his appointment as lecturer in Uppsala where he was inspired to work on marine life. He succeeded Tullberg (1883) and later (1889) was appointed Extraordinary Professor of Comparative Anatomy. Three years later (1892) he became Curator and Professor of Invertebrate Zoology at the Swedish Museum of Natural History, Stockholm and Director of Kristineberg Marine Zoological Station, until retiring from the museum (1916) and devoting all his energy to Kristineberg (1916–1924).

As an excellent artist, he illustrated his own works. He took part (1875 & 1876) in two of the Nordenskiöld's expeditions to the mouth of the Yenisei River. The latter expedition, which he led, was participated in by the Swedish entomologist Dr Arvid Filip Trybom (1850–1913) (who later became chief of the fishery department of the Board of Agriculture in Sweden) and Finn John Salhberg (q.v.). He published mainly on Holothuridae (sea cucumbers) and other marine invertebrates, his major work being *Report on the Scientific Results of the Exploring Voyage of HMS Challenger 1873–76*, appearing in two parts (1881 & 1885). His habit of wearing his leather briefcase over his shoulder led his students to name it 'epiThéel' (pronounced like the Swedish word 'epitel', meaning epithelium). A number of marine organisms are also named after him.

Theischinger

Rainforest Vicetail *Hemigomphus theischingeri* **Watson**, 1991

Dr Günther Theischinger (b.1940) is an Austrian-born Australian entomologist who works for the Department of Environment, Climate Change and Water, New South Wales. His love of aquatic insects started on his daily walk home from school by the River Danube for twelve years. On leaving school he worked for Austrian Railways which gave him the opportunity to collect moths attracted to railway lights and enabled him to all over Austria trapping moths. He switched careers, working for Linz City Council as an educator, social worker and environmental advisor. After marrying (1966) the couple moved temporarily to Australia where he worked as a quality controller for ICI in Sydney and began collecting odonates. Able to afford a house back in Austria, they returned (1970) and, after several jobs, Günther was hired as Collection Manager and Curator of Invertebrates at the Oberösterreichisches Landmuseum, a job he relinquished after the death of his parents and parents-in-law (1979) whereupon he returned to Australia. He worked in private industry for nearly 20 years before becoming curator of all aquatic insects at CSIRO and taxonomist with the Environment Protection Authority in New South Wales. He has been a visiting scientist at the Australian National Insect Collection Canberra, ACT, and is a Research Associate of the Australian Museum Sydney, NSW and Visiting Fellow at the Smithsonian Institution, US. He has described more than 60 new species and has collected all over the world and written more than 200 papers (1965–2012) and longer works. Among other works he co-wrote: *Complete Field Guide to Dragonflies of Australia* (2006). Watson merely says: *"Named for my colleague, Günther Theischinger"*. Theischinger collected the eponymous type in Queensland (1976). He has described 132 new species of Odonata, most from New Guinea and Australia, and named 10 new genera.

Thelma

Bluetail sp. *Ischnura thelmae* **Lieftinck**, 1966

Thelma Blanche Miesen Clarke (1906–1988), with her husband Dr John Frederick Gates Clarke (1905–1990), collected the holotype on Rapa Island (1963). They were keen amateur entomologists.

Theodor

(See **Teodoro**)

Therese

Spotted Anatya *Anatya theresiae* **Selys**, 1900

[JS *Anatya guttata* (Erichson, 1848)]

Princess Therese Charlotte Marianne Auguste von Bayern (Princess Theresa of Bavaria) (1850–1925) was a zoologist, botanist, anthropologist and explorer in Tunisia, Russia, the Arctic, Mexico, Brazil and western South America. Her father was Luitpold, Prince Regent of Bavaria. She was the first woman to receive an honorary degree from Ludwig Maximilian University, Munich (1897). She wrote several books such as: *Ausflug nach Tunis* (1880) and *Meine Reise in den Brasilianischen Tropen* (1897). Her South American anthropological collection (1898) is held by the State Museum of Ethnology, Munich, where all her collections were transferred at her death. She sometimes sent specimens direct to Selys. A reptile and two birds are also named after her.

Théry

Mediterranean Bluet *Agrion caerulescens theryi* **Schmidt**, 1959

[JS *Coenagrion caerulescens* (Fonscolombe, 1838)]

André Théry (1864–1947) was a French zoologist and entomologist who spent a large part of his life in Rabat, Morocco, and founded the Society of Natural History of North Africa (1909) and the Society of Natural Sciences in Morocco (1920). His collection and archive are at the MNHN, Paris with which he was associated (1942) after he had returned to France (1929). He was later President of the Entomological Society of France (1939). He was particularly interested in Coleoptera and among at least 237 entomological papers (1891–1947) wrote: *Revision des buprestides de Madagascar* (1905) about Jewel Beetles. He also published on non-entomological topics, including general zoology, botany, geology and meteorology.

Thespis

Dancer sp. *Argia thespis* **Hagen**, 1865

Thespis (c. 6th Century BC) was a singer, poet and possibly playright in ancient Greece who was apparently the first person ever to appear on stage playing a role as another person.

Thienemann

Pincertail sp. *Onychogomphus thienemanni* **Schmidt**, 1934

Professor Dr August Friedrich Thienemann (1882–1960) was a German limnologist, ecologist and zoologist. He studied (1901–1905) science and philosophy at the universities of Greifswald, Innsbruck and Heidelberg, culminating with his doctorate in Heidelberg with Robert Lauterborn. He was then (1907) head of the Biological Department of Fisheries and wastewater issues at the Zoological Institute of the Westfälische Wilhelm-University in Münster, where he conducted his famous research on the lakes of the Eifel. During his military service, he sustained a serious shrapnel wound at Reims (1914) and when he recovered was transferred to military government service (1915). He was (1917) Professor of Zoology at the University of Keil. He took part in the Deutschen Limnologischen Sunda-Expedition (1928–1929) where the type was taken in Java. He then became Director of the former Hydrobiologische Anstalt der Kaiser-Wilhelm-Gesellschaft (now the Max-Planck-

Institut für Limnologie) at Plön until retirement (1960). He is best known for his work on the biology of *Chironomidae* (non-biting midges) and his contributions to the field of lake topology. He published a great many papers over half a century (1909–1959).

Thisma

Dancer sp. *Argia thisma* **Calvert**, 1909
[JS *Argia lilacina* Selys, 1865]

Thisma is an anagram of Smith with an 'a' "...*added for euphony*". (See **Smith, HH**)

Thomas

Desert Basker *Urothemis thomasi* **Longfield**, 1932

Bertram Sidney Thomas (1892–1950) was an English civil servant and the first westerner to cross the Arabian 'Empty Quarter'. He was educated at Trinity College, Cambridge then began work at the Post Office. He served during WW1 in Belgium and was posted to Mesopotamia (Iraq) (1916–1918) as assistant political officer (1918–1922). He went on to become Assistant British Representative in Transjordan (Jordan) (1922–1924) and became Finance Minister and Wazir (Advisor) to the Sultan of Muscat & Oman (Oman) (1925–1932). He made several expeditions during this time, including the historic crossing (1930–1931), during which he described the animals, culture and people of the area. Just before making his historic crossing of the Rub'al Khali Desert, Thomas collected 64 specimens of Odonata in the Qara Mountains (1930), including the holotype of this dragonfly. He wrote a number of books including his account of the journey *Arabia Felix* (1932), He returned to England and lived in the house he was born in until he died.

Thomasson

Pincertail sp. *Nihonogomphus thomassoni* **Kirby**, 1900
(Orig. *Aeshna thomassoni*)

John Thomas Thomasson (b.1869) from Bolton, Lancashire, UK, was one of the patrons of the Whitehead Expedition to Hainan. He presented a collection of Odonata to the Natural History Museum that had been collected by John Whitehead (q.v.) and that Kirby studied and described in: *On a small collection of Odonata (Dragonflies) from Hainan, collected by the late John Whitehead* (1900).

Thorey

Thorey's Grayback *Tachopteryx thoreyi* **Hagen**, 1858
[Orig. *Uropetala thoreyi*, A. Gray Petaltail]

Georg Christoph Gottlieb Thorey (1790–1884) of Hamburg, Germany, was a professional collector and trader. He managed the ornithology collection at the Hamburg Museum (1843–1846). He seems to have visited southern Africa and Zanzibar. He sent the holotype (originating from the USA) to Hagen.

Tillyard

Australian Cruiser *Macromia tillyardi* **Martin**, 1906
Adytum Swamp Damselfly *Agrion tillyardi* Perkins, 1910

[JS *Megalagrion adytum* (Perkins, 1899)]
Evader sp. *Metaphya tillyardi* **Ris**, 1913
Tropical Evening Darner *Telephlebia tillyardi* **Campion**, 1916
Slim sp. *Aciagrion tillyardi* **Laidlaw**, 1919
Wiretail sp. *Isosticta tillyardi* Campion, 1921
Skimmer sp. *Palaeothemis tillyardi* **Fraser**, 1923
Common Glider *Tramea tillyardi* **Lieftinck**, 1942
[JS *Tramea loewii* Brauer, 1866]
Mountain Tigertail *Eusynthemis tillyardi* **Theischinger**, 1995

Bronze Needle *Synlestes weyersii tillyardi* Fraser, 1948
[Orig. *Synlestes tillyardi*]
Whitewater Rockmaster ssp. *Diphlebia lestoides tillyardi* Fraser, 1956

Robert 'Robin' John Tillyard (1881–1937) was an English-born Australian based entomologist described as a 'giant among odonatologists'. He passed exams to enter the Royal Military Academy at Woolwich but was rejected on medical grounds. He won scholarships to study classics at Oxford and to read mathematics at Cambridge, taking up the latter, but, after graduating BA (1903) and plagued by rheumatism he moved to Australia for his health where his interest in dragonflies developed. There he taught mathematics at Sydney Grammar School. Wanting to pursue entomology, he gave up his teaching job and was admitted as a research student in the University of Sydney, being awarded a scholarship. He married (1909) his friend from Cambridge, Patricia Cruske (1880–1971), who later illustrated some of his work. He was awarded his BSc (1914), the first time the university had awarded a degree for research. He was badly injured in a railway accident (1914) and had a slow recovery, but (1915–1920) became Linnean Macleay Fellow in Zoology at the University of Sydney, during which time his seminal book: *The Biology of Dragonflies (Odonata or Paraneuroptera)* was published (1917) and Sydney University conferred a DSc on him. He gave advice to the NZ government on problems at trout fisheries (1919) and while there became (1920) Head of Biology at the Cawthorn Institute in New Zealand for eight years, during which time he wrote his second major book: *The Insects of Australia and New Zealand* (1926). He then became the Australian Government's Chief Entomologist in the Division of Economic Entomology at the Commonwealth Council for Science (1928) and spent time recruiting entomologists from England, including Herbert Womersley (1929). He had been awarded a ScD by Cambridge (1920). He was unhappy in Nelson and this, coupled with his continuing back pain, contributed to him suffering a breakdown (1933) and his resignation (1934). He died as a result of a road traffic accident; he had also previously broken an arm in a cycling accident in New Zealand and received broken ribs in another car accident in California. He wrote 213 publications, including five books on insects (three being on fossil insects) around seventy of which were on his first love, dragonflies. He described c.125 new species or subspecies and named 24 new genera of extant Odonata, most of them from Australia. In addition, he described several fossil species and genera. His personal Odonata collection was donated after his death to BMNH and the Australian National Collection, as well as several other museums in Australia and New Zealand. Philip S Corbet wrote (2003): "*...No scientist has contributed so much to odonatology, across so broad a front, as Robin John Tillyard, ... He contributed extensively to basic knowledge through his own research, primarily in the fields of* *functional morphology, taxonomy, embryology, ontogeny and phylogeny. He contributed to the science of odonatology and to odonatologists through his book,* The Biology of Dragonflies *(1917), the first book to treat the biology of the order Odonata as a whole, and a masterly and inspiring synthesis of contemporary knowledge. The debt owed by odonatologists to Tillyard is incalculable."* (Also see **Patricia (Tillyard)**)

Timotocuica

Darner sp. *Andaeschna timotocuica* **De Marmels**, 1994

The species is named after a people; a native tribe in Venezuela, where the species was found.

Tindale

Small Reedling *Austrolestes albicauda tindalei* **Tillyard**, 1925
[JS *Indolestes alleni* Tillyard, 1913]

Dr Norman Barnett 'Tinny' Tindale (1900–1993) was an Australian anthropologist, archaeologist and entomologist. He worked at the South Australian Museum (1919–1968) starting as an entomologist's assistant, apart from WW2 service as a Wing Commander (1942) despite having lost an eye while helping his father process photographs (1917). He had published thirty-one papers on entomological, ornithological and anthropological subjects before the University of Adelaide awarded his BSc (1933). He collected prolifically across Australia, but is most remembered for his work mapping aboriginal tribal groupings during his anthropological surveys (1938–39 & 1952–54). After retirement, he took a teaching post at the University of Colorado, which awarded his honorary doctorate (1967) and he remained in the US until his death. He collected the type on Groote Eylandt Island (1921–1922).

Tirio

Pond Damselfly sp. *Tuberculobasis tirio* **Machado**, 2009

The Tirio are an indigenous people of Brazil, in whose village the author spent a month, during which time they helped him with his collecting.

Tisi

Sino Riverjack *Mesocnemis tisi* **Lempert**, 1992

Matthias ('Tisi') Lempert (b.1960) is the brother of the author Jochen Lempert (b.1958).

Titschack

Forceptail sp. *Phyllocycla titschacki* **Schmidt**, 1942
[Orig. *Gomphoides titschacki*]

Hans Erich Woldemar Titschack (1892–1978) was a German zoologist principally interested in insects and birds. He studied biology, geology and chemistry in Jena, Berlin and Bonn (1912–1914), but volunteered to serve in WW1 when he was seriously injured on the eastern front and discharged. He returned to study in Bonn (1916) and completed his PhD (1919). He worked as a zoologist (1919–1924) studying the clothes moth and developed the first mothproofing for wool. He became head of the Entomology Department of the Zoological Institute and State Zoological Museum in Hamburg (1924), later becoming professor (1934). He collected in the Canary Islands (1931) and France (1934). He went to southern Peru on the Hamburg Expedition (1936) where he collected extensively (taking 38,000 specimens, mostly insects) from the Pacific coast to Rio Apurimac in the eastern Andes. Erich Schmidt (q.v.) worked on the Odonata that he had collected virtually immediately, but the results were not properly published until some years later after the disruption of WW2. Titschack was bombed out of his home and job in Hamburg (1945) and took a post at Poznan University. He returned to Hamburg after the war (1951) and began to re-build the Altona Museum collection (1951–1957). After retirement, he continued to work on entomology, turning his attention to Thysanoptera.

Todd

Epaulet Skimmer *Orthetrum chrysostigma toddii* **Pinhey**, 1970
[JS *Orthetrum chrysostigma* (Burmeister, 1839)]

Alexander Todd was at Fourah Bay College, University of Sierra Leone. He collected the type in Sierra Leone (1965).

Tol

Damselfly genus *Telosticta* **Dow & Orr**, 2012
Flatwing sp. *Celebargiolestes toli* **Kalkman**, 2016

Dr Jan van Tol (b.1951). *Telosticta* is a genus of damselfly currently encompassing 17 species. It is a feminine noun, compounded from the Greek τ λος, meaning toll. It is named for Jan van Tol, whose surname originates from the Dutch word of the same meaning, in honour of his contributions to our knowledge of the *Platystictidae*, and the suffix –sticta, a common generic ending in the family. (see **Van Tol &** also see **Jan (van Tol)**)

Tomokuni

Phantomhawker sp. *Planaeschna tomokunii* **Asahina**, 1996
Sylvan sp. *Coeliccia tomokunii* Asahina, 1997
[JS *Coeliccia scutellum* Laidlaw, 1932]
Spectre sp. *Petaliaeschna tomokunii* **Karube**, 2000

Dr Masaaki Tomokuni (b.1946) is a Japanese zoologist and entomologist who was Director of the Department of Zoology, National Museum of Science & Nature, Tokyo. He has undertaken a number of collecting expeditions for the museum, including to Vietnam, where he collected the types (1995 & 1998), and Nepal. He has a particular interest in Hemiptera (Shield Bugs). Among his publications, which are mostly on Hemiptera and Heteroptera (1972–2016) are several books including: *A field guide to Japanese bugs – Terrestrial heteropterans* (1993) and *The Encyclopaedia of Animals in Japan* (1996).

Tonto

Dancer sp. *Argia tonto* **Calvert**, 1902

This species from Arizona was named after a people; the Tonto Apache Indians are a tribe in Arizona.

Tony (Watson)

The Genus *Tonyosynthemis* **Theischinger**, 1995
Alpine Redspot *Austropetalia tonyana* Theischinger, 1995

Dr (Tony) John Anthony Linthorne Watson (1935-1993). The species etymology reads: "*A tribute to the late Dr J.A.L. (Tony) Watson, great friend and odonatologist.*" (See **Watson**)

Tony (O'Farrell)

The Genus *Tonyosynthemis* **Theischinger**, 1995

Anthony 'Tony' Frederick Louis O'Farrell (1917-1997). The etymology reads: "*...in memory of Prof. A.F.L. (Tony) O'Farrell (1917-1997) and Dr J.A.L. (Tony) Watson (1935-1993) two unforgettable friends and outstanding odonatologists.*" (See **O'Farrell**, also see above & **Watson**)

Toto

Wisp sp. *Agriocnemis toto* Dijkstra, 2015

Alvaro Bruno Toto Nienguesso (b.1967) is an engineer and management consultant currently (since 2016) Deputy CEO of H+W Art & Bau GmbH, Technische Universität Dresden, Germany. His special area of expertise is change management. He was founder and CEO (2010-2016) of Albrutoni Angola Lda. He trained in geotechnical and mining engineering (1987-1992) at the Technische Universität Bergakademie Freiberg and took an associate's degree in processing technology and engineering at Technische Universität Dresden (1995-1998). The etymology describes him as "*...the driving force behind biodiversity research in the VII Academic Region Uige Province and kwanza norte province, Angola.*" He co-authored: *Relicts of a forested past: Southernmost distribution of the hairy frog genus Trichobatrachus Boulenger, 1900 (Anura: Arthroleptidae) in the Serra do Pingano region of Angola with comments on its taxonomic status* (2014) and *Plants from disturbed savannah vegetation and their use by Bakongo tribes in Uíge, Northern Angola* (2016).

Tournier

Tournier's Hooktail *Paragomphus tournieri* **Legrand**, 1992

Jean-Luc Tournier (1907-1985) was a French pharmacist based in Saigon who later was an ethnographer during a mission to collect medicinal plants in French West Africa. He was the founder Director of the Centre IFAN (1942) that became the Musée d'Abidjan, Côte d'Ivoire (1947) and another IFAN base in Ziéla. His collection was bequeathed to the Natural History Museum of Besançon.

Townes

Townes's Clubtail *Stylurus townesi* Gloyd, 1936

Dr Henry Keith Townes (1913-1990) was an American entomologist who collected the eponymous type in South Carolina. He was at Furman University, Greenville aged just 16! He graduated with a BSc in biology (1933) and a BA in language. His PhD was obtained at Cornell University (1933-1937). His focus was on Hymenoptera systematics, particularly *Ichneumonidae*. A number of jobs followed graduation, including a fellowship at the

Academy of Natural Sciences of Philadelphia (1940–1941) to work on a catalogue of Nearctic *Ichneumonidae*. He was employed by the US Department of Agriculture in Washington, DC (1941–1949). Except for a two-year interlude as an advisor to the Philippine government on pests of rice and corn (1952–1954), he worked at the North Carolina State University (1949–1956). Thereafter he worked under the auspices of University of Michigan with research grants from National Science Foundation and the National Institutes of Health, allowing him to publish three volumes of *Bulletin of the U.S. National Museum* and *Memoirs of the American Entomological Institute* (including a four-volume revision of the genera of *Ichneumonidae*) (1959–1978). From early days (1933) he built a collection of over 700,000 specimens, which was eventually transferred to the American Entomological Institute to Gainesville, Florida.

Toxopeus

Emerald sp. *Hemicordulia toxopei* **Lieftinck**, 1926

Professor Lambertus Johannes Toxopeus (1894–1951) was a Dutch zoologist, entomologist and collector, born in Java. He attended University in Amsterdam and began a teaching career (1923), continuing his study until his PhD was awarded (1930) when he returned to Java. He taught in schools for some years, but during WW2, was a reserve officer monitoring an airport at Buitenzorg before being interned by Japanese forces. After the war, he lectured at the University of Batavia, but went to the Netherlands (1947–1949) to recover from illness brought on by his time as a POW. He returned to become a professor at Bandoeng University (1949) where he died in a road traffic accident. His main interest was Lepidoptera, specifically *Lycaenidae*, and other insects, but he also collected plants on the Royal Dutch Geographical and Treub Society expedition to the western Moluccas (Buru, Maluku) (1921–1922), where he collected (1921) the eponymous type, and in Sumatra (1934). He was also on the Third Archbold Expedition to Dutch New Guinea (1938–1939) collecting over 100,000 insects! Three birds and a mammal, as well as a number of plants and butterflies, including the sedge moth genus *Toxopeia*, are also named after him.

Travassos

Clubskimmer sp. *Brechmorhoga travassosi* **Santos**, 1946

Professor Lauro Pereira Travassos (1890–1970) was a Brazilian zoologist, helminthologist and entomologist who travelled widely in Brazil. He studied medicine in Rio, graduating from the Faculty of Medicine (1913) and his doctoral thesis translates as: *About the Brazilian species of the subfamily Heterakinea*. He worked briefly at the Institute of Experimental Medicine, and went on to take the Chair of Parasitology, Faculty of Medicine of São Paulo (1926). He lectured in Europe and took various academic positions in Brazil, establishing himself as a teacher, but also as a researcher going on many fieldtrips and describing hundreds of new species. He was Santos' teacher and introduced him to entomology. A school in São Paulo is also named after him. He remained dedicated to scientific endeavour his whole life and never stopped working until the day he died.

Treadaway

Griptail sp. *Phaenandrogomphus treadawayi* **Müller** & **Hämäläinen**, 1993
[Orig *Onychogomphus treadawayi*]

Colin Guy (Trig) Treadaway Hoare (b.1923) is an entomologist. His fascination with butterflies started when he was a child and developed into a lifelong passion; he became a world expert on the Lepidoptera of the Philippines. He was sent to Canada (1940) by his guardian to escape the blitz. He made his own way to his uncle's farm where he was treated as an indentured servant working 6½ days a week, so he ran away (1941). He worked as a labourer, then as an assistant to a chemical engineer. He joined the RCAF (1942) becoming a navigator to a British squadron serving in Europe until demobilisation (1946) when he entered McGill University, Montreal to study for his degree in chemical engineering (1947–1951). He worked for Procter and Gamble first in USA and then (1953) in Manila and studied Philippine butterflies in his spare time. Further postings took him to Peru and Morocco (1958) and Germany (1960), where he married his second wife (1972) and eventually became a board member (1978). He was appointed honorary member of the Senckenberg Museum Research Institute, Germany, to which he has dedicated his very large private collection of Lepidoptera from the Philippines. Since retiring (1987), he has been even more deeply involved with Philippine Lepidoptera. In association with the Senckenberg, he has made research trips almost every year to the Philippines, camping for weeks in jungles. A male specimen of the odonate was collected in Busuanga Island, Philippines, during a butterfly collecting trip led by Treadway (1991). He has participated in the publishing of over 100 journal articles in Japan, Germany, Holland, England and the Philippines, mostly on butterflies and moths of the Philippines, covering over 300 holotypes, and has written or co-written several books, including: *Checklist of the butterflies of the Philippine Islands* (1995). He is a Fellow of the Royal Entomological Society, London and an honorary staff member of the Malaysian University, Sarawak, North Borneo (2003).

Trebbau

White-banded Shinywing *Iridictyon trebbaui* **Rácenis**, 1968

Dr Pedro (Peter) Trebbau Milovich (b.1929) is a German-born Venezuelan veterinarian, zoologist and conservationist. He grew up in Germany but left for Venezuela (1952) where he collected the type in Bolivar, Venezuela (April 1956). He was President of the Fundacion Nacional de Parques Zoológicos y Acuarios, Venezuela and became Director of Caracas' El Pinar Zoo (1962). Among other publications he co-wrote: *The Turtles of Venezuela* (1984).

Tristán

Ringtail sp. *Erpetogomphus tristani* **Calvert**, 1912

Professor José Fidel Tristán (1874–1932) was a naturalist at the Museo Nacional in San Jose, Costa Rica where he became President of the Board (1926) and Director (1930–1932). He collected the type series in Costa Rica (July 1911) and often helped visiting scientists with collecting. The Lyceum of Costa Rica awarded his BSc (1894) and he then worked in the National Museum (1894–1897). He obtained a scholarship to study at the Pedagogical Institute of Santiago de Chile; where he received the title of Professor of Physics and Chemistry, before returning to the Costa Rica (1900). He then taught in a number of different schools. During this time, he also organized several field expeditions. He wrote: *Insectos de Costa Rica, pequeña colección arreglada* (1897) and *Las Libélulas* (1909).

Troch

Darting Cruiser *Macromia trochi* Grünberg, 1911
[JS *Phyllomacromia picta* Hagen, 1871]

The epithet *trochi* is a misspelling present in some catalogues. (See **Koch**)

Trotter

Black-kneed Featherleg *Pseudocopera trotteri* **Fraser**, 1922
[JS *Pseudocopera ciliata* (Selys, 1863)]

Lieutenant General Ernest Woodburn Trotter (1871–1935) was a British colonial police officer. He joined the police in Burma (1889) as an inspector, and became Assistant Superintendent (1894) and then District Superintendent. He was Adviser to the Siamese Gendarmerie and Deputy Commissioner of Police at Bangkok and was awarded the Order of the White Elephant (fourth class 1909, third class 1915) and Order of the Crown of Siam (second class 1916) by the King of Siam in recognition of the valuable service rendered by him. He was awarded the Police Medal by the Indian Government (1918) for his part in tackling sedition in Burma. He served in Siam (now Thailand) (1901–c.1926). He was a keen natural historian collecting botanical and zoological specimens during his various postings. He collected the holotype among around 100 dragonfly specimens from Bangkok (September 1921). He remained in Siam after retiring and is buried in Bangkok.

Tryon

Coastal Evening Darner *Telephlebia tryoni* **Tillyard**, 1917

Henry Tryon (1856–1943) was an English-born Australian scientist. He started medical studies at London University but gave them up to pursue natural history. He travelled to Sweden, then to New Zealand where he managed a grazing property for his father, and then (1882) to Queensland where he collected in the cane fields, giving his specimens to Queensland museum. In light of this he was given a voluntary position (1882) then became an employee (1883) being promoted assistant curator (1885). He worked for the Department of Agriculture (1887) and went on to become Government Entomologist for Queensland (1894) and lent Tillyard specimens to study. He became a vegetable pathologist (1901). His made great efforts to rear South American Cactus Moths in order to control Prickly Pear, a very invasive species in Australia, but he failed. He undertook an expedition to New Guinea (1895–1896) where he acquired many new species of sugar cane. He was founding secretary of the Royal Society of Queensland (1883–1888). He was also a keen ornithologist. He retired from public service (1925) but continued in temporary employment (1925–1929). A fruit fly and two reptiles are also named after him.

Tschugunov

White-faced Darter *Leucorrhinia dubia tschugunovi* **Bartenev**, 1909
[JS *Leucorrhinia dubia* (Vander Linden, 1825).

Sergey Mikhaylovich Tschugunov (Chugunov) (1854–1920) was a Russian entomologist. He collected the type in Siberia as well as other insects, such as a butterfly (1903). He is also honoured in the name of a flatworm *Aponurus tschugunovi* Issatschikov, 1928.

Tsiama

Rusty Threadtail *Elattoneura tsiamae* **Aguesse**, 1966
[JS *Elattoneura vrijdaghi* Fraser, 1954]

This is a toponym; the type locality was the Tsiama Forest, in Congo-Brazzaville.

Tuanui

Redcoat sp. *Xanthocnemis tuanuii* Rowe, 1981

Manuel Tuanui (1927–1984) was a New Zealand conservationist and farmer. He and his family were extremely helpful to biologists working on Chatham Island from 'Tuku Base', which was situated on Manuel Tuanui's farm. The Tuku Nature Reserve was created on 1,238 hectares of land still largely covered with dense native forest, donated by Manuel and Evelyn Tuanui for the conservation management of the Taiko (a species of tube-nosed bird), with adjacent land covenanted for the same purpose. The etymology reads: "*The name tuanuii is proposed for this species in honour of Mr Manuel Tuanui of Chatham Island.*"

Tullia

Parasol sp. *Neurothemis tullia* Drury, 1773
[Orig. *Libellula tullia*]
Helicopter sp. *Agrion tullia* **Burmeister**, 1839
[Syn. of *Mecistogaster lucretia* (Drury, 1773)]

Tullia Minor is a semi-legendary figure in Roman history; the last queen of Rome. She was the youngest daughter of Servius Tullius, Rome's sixth king. She married Lucius Tarquinius and they arranged the overthrow and murder of Servius, securing the throne for her husband and making her an infamous figure in ancient Roman culture.

Tümpel

Banded Demoiselle ssp. *Calopteryx splendens tümpeli* Scholz, 1908
[JS *Calopteryx splendens ancilla* Hagen, 1853]

Dr Rudolf Johannes Tümpel (1864–1938) was a German teacher and natural historian. His doctoral thesis (1894) was published as: *Naturwissenschaftliche Hypothesen im Schulunterricht.* He also wrote: *Die Geradflügler Mitteleuropas* (1901).

Tunti

Skydragon sp. *Chlorogomphus tunti* **Needham**, 1930

This species is named after a people, as the author wrote: *Tunti is the name of a tribe of aborigines of Southwest China.*

Tupi

Dancer sp. *Argia tupi* **Calvert**, 1909

The Tupi are an indigenous people in the type locality of Chapada, Brazil.

Turconi

Jewel sp. *Rhinocypha turconii* **Selys**, 1891

Turconi was a very distinguished Italian engineer and amateur entomologist who sent some newly collected specimens to Selys for his collection, including the eponymous holotype that he collected (1887) in Cebu, Philippines.

Turner

Jewel Flutterer *Rhyothemis turneri* **Kirby**, 1894
[JS *Rhyothemis resplendens* Selys, 1878]
Flame-tipped Hunter *Austroepigomphus turneri* **Martin**, 1901
[Orig. *Austrogomphus turneri*]

Gilbert Turner was a farmer at MacKay, Queensland. When he retired, he took up entomology as a hobby and collected insects sending several collections, including the flutterer type specimen, to Kirby at the BMNH (1894). He wrote at least one paper: *Notes upon the Formicidae of Mackay, Queensland* (1897) and has other insect taxa named after him, including ant species such as *Notoncus gilberti* and the ant genre *Turneria*.

Tydecks-Jürging

Damselfly sp. *Palaiargia tydecksjuerging* **Orr, Kalkman** & Richards, 2014
Pond Damselfly sp. *Papuagrion tydecksjuerging* **Orr** & Richards, 2016

Anke Tydecks-Jürging (b.1955) and Michael Jürging (b.1959) were jointly honoured in recognition of their generous support for Odonata research in New Guinea through the International Dragonfly Fund (IDF).

Tyler

Black-winged Dragonlet *Trithemis tyleri* **Kirby**, 1899
[JS *Erythrodiplax funerea* (Hagen, 1861)]

Charles Henry Dolby-Tyler (1863–1900) was British Vice-Consul (1898) and then Consul (1899) in Panama. He was previously in Guayaquil, Ecuador, where he wrote an account of an earthquake (1896) and a book *The River Napo* (1894), as well as some entomological papers. He collected the holotype 20 miles north of Panama City and sent it to Kirby (1898). He led a travelling life, having been born in Gibraltar, and is said to have died in Canada.

U

Ubadschi

(see **Oubaji**)

Uchida

Striped Emerald sp. *Somatochlora uchidai* **Förster**, 1909

Dr Seinosuke Uchida (1884–1975) was a Japanese ornithologist and entomologist. He graduated in veterinary medicine at Tokyo Imperial University and gained his PhD with the thesis: *Studies on amblycerous mallophaga of Japan*. He worked as an engineer at the Ministry of Agriculture and Commerce and was President of the Ornithology Congress of Japan. He wrote widely on Mallophaga (bird lice), including: *Mallophaga from birds of Formosa* (1917). He collected the type (1907). Five birds are also named after him.

Uemura

Leaftail sp. *Oligoaeschna uemurai* **Asahina**, 1990

Yoshinobu Uemura (b.1950) is a Japanese entomologist and curator at the Toyosato Museum of Entomology. He collected the type in Mindanao (1987). He has written or co-written a number of books, such as the two volume: *The Butterflies of Thailand* (2011 & 2014) and papers such as: *Efficient display of insect specimens* (1994).

Ueno

Jewel sp. *Rhinocypha uenoi* **Asahina**, 1964
Skydragon sp. *Watanabeopetalia uenoi* Asahina, 1995
[Orig. *Chlorogomphus uenoi*]
Slendertail sp. *Leptogomphus uenoi* Asahina, 1996
Oread sp. *Calicnemia uenoi* Asahina, 1997
Sylvan sp. *Coeliccia uenoi* Asahina, 1997

Dr Shun-Ichi Ueno (b.1930) is a Japanese entomologist who collected the holotype of the *Coeliccia* species in Taiwan. He works at the Department of Zoology, the National Science Museum, Tokyo. He has undertaken a number of collecting expeditions in China, Taiwan (1962), Nepal (1979) and the Japanese islands. He has published widely, particularly on beetles.

Uhler

Uhler's Sundragon *Helocordulia uhleri* **Selys**, 1871
[Orig. *Cordulia uhleri*]

Philip Reese Uhler (1835–1913) was an American librarian and entomologist specialising in Heteroptera. He was educated at Harvard and taught by Louis Agassiz. He became Head of the Museum of Comparative Zoology Library & Insect Department there (1864–1867) and then an associate in natural sciences at Johns Hopkins University (1876). He was also librarian at the Peabody Library in Baltimore. He made a number of collecting trips including to Haiti and the Western US as well as to Europe (1888) to study insect collections in European museums. He was Secretary and then President (1873) of the Maryland Academy of Science. He wrote many papers on entomology and geology and also: *Check-list of the Hemiptera Heteroptera of North America* (1886).

Ulmec

Dancer sp. *Argia ulmeca* **Calvert**, 1902

The Mexican species is named after a people; the Ulmec (Olmec) Indians.

Uluna

Damselfly sp. *Lieftinckia ulunorum* Marinov, 2016

The Uluna are a tribe of people in Guadalcanal in the Solomon Islands. The holotype was collected within the area owned by this tribe.

Umbarga

(See **Mbarga**)

Underwood

Dancer sp. *Argia underwoodi* **Calvert**, 1907

Cecil Frank Underwood (1867–1943) left London for Costa Rica (1889) to collect natural history specimens for a living, staying for the rest of his life. He was an all-round naturalist who collected for a number of overseas museums, and was a taxidermist at Costa Rica's National Museum, San José. He described many new mammals from Central America, often with George Goodwin, and collected the damselfly holotype. Four mammals, two birds and two amphibians are also named after him.

Untamo

Genus *Untamo* **Kirby**, 1889
[Syn. of *Neurothemis* Brauer, 1867]

Untamo is a character from the Finnish national epic poem *Kalevala*, which W F Kirby translated into English – *Kalevala: The Land of Heroes* (1907). In the epic, Untamo was a mean and rapacious man, who started a war against his brother's (Kalervo) family. From the family, only Kalervo's son Kullervo survived. Later Kullervo revenges and kills Untamo and his whole family. (See also **Aino**).

Ursula

Beech Tigertail *Eusynthemis ursula* **Theischinger**, 1998

Ursula Jones (b.1996) is Günther Theischinger's granddaughter.

Usuda

Hainan Skydragon *Chlorogomphus usudai* **Ishida**, 1996

Akimasa Usuda (1933–2006) was a Japanese odonatologist at Nagoya Insect Museum, Japan. He wrote the book: *The Insects of Nagoya* (1989). He collected in South Carolina (1994) and the type in Hainan that same year.

V

Vadon

Dark-tipped Claspertail *Onychogomphus vadoni* **Paulian**, 1961

Jean Pierre Léopold Vadon (1904–1970) was a naturalist. He served in the French army in Morocco (1924–1926) then became a schoolmaster (1926–1930). Here he developed an interest in entomology as evidenced by his sending a collection of insects to Paris (1931). He obtained (1931) a post in Cameroon, teaching with the health service, as part of which he began to study the haematophagous Diptera, establishing a map of the distribution of the tsetse fly. He returned to Paris on 'convalescent leave' (1932) but then transferred (1933) to teach at the European School, Tananarive, Madagascar. However, wanting to study insects, he moved to a more rural post and taught for thirty-six years (1934–1970) at the Marine Training School at Maroantsetra, being also French Consul (1963). Throughout his teaching career he collected insects and got his pupils to collect for him too. The school needed timber, which they took from the forest at Ambodivohangy which became his prime source of specimens. Throughout he sent his material to Lebis in France who preserved, arranged and catalogued the specimens. Eventually they came to the notice of Professor Jeanne, who moved the collection to the Muséum National d'Histoire Naturelle as the Vadon-Lebis collection, and Vadon became a corresponding member of the Entomological Department (1945), as did Lebis, on whose death (1963) the collection was donated to the museum. In later years, he developed techniques with R Paulian (q.v.). Eleven genera of insects are named after him, as is a reptile. He collected the Odonata type in Madagascar (1958). He died in Paris of an unknown disease.

Vakoana

Sprite sp. *Pseudagrion vakoanae* **Aguesse**, 1968

This is a toponym; named after the type locality, the Vakoana Forest in Madagascar.

Valle

Kinkedwing sp. *Anisopleura vallei* St. Quentin, 1937

Professor Kaarlo Johannes Valle (1887–1956) was a Finnish entomologist, zoologist and limnologist with an interest in botany. He gained his MSc (1914) at Helsinki University and DPhil (1927) at Turku University. After graduation, he first worked as a biology teacher in a school in Kotka (1915–1919) and then held various posts in Helsinki, including an assistantship at the Zoology Department of Helsinki University (1919–1922). He moved (1922) to work at the Zoology Department of Turku University, becoming Docent of Zoology (1928) and later Professor of Zoology (1943–1956). He was acting professor of Zoology at Helsinki University (1938–1941), where he had taught as Docent of Zoology (1933–1943). His numerous publications include two handbooks on the dragonflies of

Finland (1922, 1952), two books on the fish of Finland (1934, 1941) and a series of four handbooks of butterflies and moths of Finland: *Suurperhoset Macrolepidoptera* (1935-1946). He published several local faunistic papers and synopses of Finnish Odonata (his major study interest), the last one: *Die Verbreitungsverhältnisse der ostfennoskandischen Odonaten* (1952), having distribution maps. He also published odonate results from several collecting trips made by other entomologists to Siberia, Southern Russia, Azores, Madeira, Cap Verde Islands, Spain, Marocco, Cyprus, Newfoundland and the United States. His Odonata and Lepidoptera handbooks greatly increased the interest in these insect groups in Finland. His limnological publications include the two-part *Ökologisch-limnologische Untersuchungen über die Boden- und Tieffauna in einigen Seen nördlich vom Ladoga-See* (1927-1928), the first being his thesis. He died one month after retiring.

Vallis

Large Wiretail *Labidiosticta vallisi* **Fraser**, 1955
[Orig. *Phasmosticta vallisi*]

Eliah Close 'Closie' Vallis (1890–1965) was an Australian amateur entomologist. He and his elder brother, Robert Eliah Vallis (1887–1954), grew up loving nature and collected birds' eggs then butterflies and beetles during which they learned how to mount and preserve their specimens. Together they helped to found the Rockhampton and District Field Naturalists Club (1948). He worked as a maintenance carpenter in the Lakes Creek meatworks, spending his holidays collecting insects in northern Queensland where he discovered and collected the type series. The year he died his local authority named a park after him. Both brothers' collections were donated to the Queensland Museum and the Rockhampton Botanic Gardens. He is also commemorated in the name of a beetle, which he collected, *Stigmodera (Castiarina) vallisii* (C M Deuquet, 1964).

Van Brink

Bluet sp. *Coenagrion vanbrinkae* **Lohmann**, 1993
[Orig. *Coenagrion vanbrinki*]
Goldenring sp. *Cordulegaster vanbrinkae* Lohmann, 1993
[Orig. *Cordulegaster vanbrinki*]
Clubtail sp. *Archaeogomphus vanbrinkae* **Machado**, 1994
[Orig. *Archaeogomphus vanbrinki*]

Professor Dr Janny Margaretha van Brink (1923–1993) was a Dutch geneticist. Her doctorate was awarded (1959) when she defended her thesis: *L'Expression de la morphologique digamétie chez les sauropsidés et les monotremes*. She was professor emeritus of cytogenetics at Utrecht, having been professor (1980–1986) and a lecturer there (1965–1980). The publication (1994) of the proceedings of the Eleventh International Symposium of Odonatology, Trevi, 1991 was dedicated to her memory; she had been a stalwart of the International Odonatological Society, and principal initiator. Much of her work on genetics was about Odonata and she published papers such as: *Notes on chromosome behaviour in the spermatogenesis of the damselflyEnallagma cyathigerum (Charp.) (Odonata: Coenagrionidae)* (1964). She died after a long illness at her home, known by a whole generation of odonatologists as 'Dragonfly Lodge'. (See also **Janny**)

Van der Starre

Shadowdamsel sp. *Protosticta vanderstarrei* **Van Tol**, 2000

J J van der Starre was a Dutch entomologist who collected in Indonesia (Sumatra and Sulawesi) in the 1940s and collected the holotype there (1940). Van Tol's etymology reads: *"... after Mr. J.J. van der Starre, who travelled extensively along the coasts of Sulawesi in the early 1940s, collecting many interesting Odonata in areas previously unexplored and even poorly investigated today. Mr. Van der Starre also collected the holotype of this species".* (Also see **Starre**)

Van der Weele

Nighthawker sp. *Heliaeschna vanderweelei* **Martin**, 1907
[JS *Heliaeschna simplicia* (Karsch, 1891)]

Dr Herman Willem van der Weele (1879–1910) was a Dutch entomologist principally interested in Neuroptera. He studied at Beren University, eventually leading to a doctorate with his published dissertation entitled: *Morphologie und entwicklung der gonapophysen der Odonaten* (1906). He became Second Conservator of Insects at RMNH, Leiden, and went on an expedition to Java where he died of cholera. In his short academic life, he published twenty-seven papers in English, French, German and Dutch, the most important being a world revision (in German) of the owlflies *Ascalaphiden* in the series: *Collection zoologiques du Baron Edm. De Selys Longchamps. Catalogue Systematique et decriptif.*

Van Dyke

Spineleg sp. *Merogomphus vandykei* **Needham**, 1930

Dr Edwin Cooper Van Dyke (1869–1952) was an American entomologist, particularly a coleopterist, who started collecting as a boy. He took a bachelor's degree (1889–1893) at the University of California and then qualified as a physician (1895) at the Cooper Medical School, Stanford, graduating MD (1895). He practised medicine in San Francisco (1895–1913), but then abandoned general practice and became an instructor in entomology at the University of California. He married (1915) and his wife often accompanied and helped him on collecting trips in the western USA, China (1923–1924), Europe (1933) and the southern states and Florida (1939–1940), shortly after which she died. He stayed as a member of the faculty, becoming Professor of Entomology (1927–1939), retiring as Professor Emeritus. Throughout this time, he strived to increase the entomology collection. He described over 430 species, mostly beetles and particularly his first love, Carabidae, and he wrote over 150 papers with his last major one being: *The Coleoptera of the Galapagos Islands* (1953) describing 84 new species. An amphibian is also named after him.

Van Mastrigt

Damselfly sp. *Hylaeargia vanmastrigti* **Theischinger** & Richards, 2013

Brother Henricus 'Henk' Jacobus Gerardus van Mastrigt (1946–2015) was a Dutch Franciscan lay brother and entomologist. As a boy, he collected butterflies and wanted to work as a forester. He had no formal biology training but when he became a brother at the Franciscan Mission in Jayapura, Papua, Indonesia, he studied the insects, particularly the butterflies of western New Guinea, for several decades. He studied theology and philosophy (1966–1969) and economics (1969–1973) at the School of Economics in Rotterdam. After six months language training, he worked as one of the last Dutch missionaries (1974–2015) in Papua province, Indonesia, all the while keeping his entomological collections going. He returned to the Netherlands briefly (1994) for treatment for chronic leukaemia. He is particularly known for his forays into the Foja Mountains, organized by Conservation International (2005 & 2008), and has undertaken several other surveys over the years. He was editor of *Suara Serangga Papua*, the journal of the workgroup on insects of Papua. Among his many published papers is: *A new species of Hypolycaena (C. & R. Felder, 1862) from Papua, Indonesia (Lepidoptera: Lycaenidae)* (2013) as well as three fieldguides to Papuan butterflies. At least one butterfly is also named after him.

Van Someren

Tiny Bluet *Azuragrion vansomereni* **Pinhey**, 1956
Blue-sided Sprite *Pseudagrion vansomereni* Pinhey, 1961
[JS *Pseudagrion sudanicum* Le Roi, 1915]

Dr Victor Gurney Logan van Someren (1886–1976) was born in Australia but qualified at Edinburgh University in both medicine and dentistry. However, he was most famed for his 'magnificent collection of the butterflies of East Africa'. He was appointed Medical Officer of British East Africa (1912) (Kenya) and spent 40 years practising there, during which time he studied its natural history. He and his brother Robert started a survey of the birds of Kenya and Uganda (1906) and their collection ultimately exceeded 25,000 specimens. He was Honorary Curator of the Natural History Museum in Nairobi (1914–1938), and a Fellow of both the Linnean Society of London and the Royal Entomological Society. Part of his collection of insects was donated to the Natural History Museum in Tring and his collection of African birds to the Field Museum, Chicago. Most of his collection of butterflies (c.23,000 specimens) was left to the AMNH. He collected the damselfly types in Uganda (1952). He wrote, among other works, *Days with Birds – Studies of Habits of some East African Species* (1956), but the body of his work was about butterflies such as his ten part: *Revisional notes on African Charaxes (Lepidoptera: Nymphalidae)* (1963–1975). Two birds are also named after him.

Van Tol

Shadowdamsel sp. *Sulcosticta vantoli* Villanueva & **Schorr**, 2011
Jewel sp. *Heliocypha vantoli* **Hämäläinen**, 2016
Flatwing sp. *Celebargiolestes toli* **Kalkman**, 2016

Dr Jan van Tol (b.1951) is a Dutch entomologist who has concentrated on Odonata since the early 1980s. He has been working at Naturalis Biodiversity Center (formerly National Museum of Natural History), Leiden, for four decades, first as Co-ordinator of Invertebrate Mapping Schemes in the Netherlands (1977), then as Curator of the Department of Hemiptera, Orthoptera and Odonata (1986) and as Head of the Department of Entomology (1999). He became Director of Research at the museum (2008) until retiring (2016), becoming an Honorary Research Associate. He was President of the Netherlands Entomological Society for ten years (1999–2008). He was a Commissioner of the

International Commission on Zoological Nomenclature (President 2009–2016) and is on the editorial board of the 'International Journal of Odonatology' and various other journals. He has written many papers on the taxonomy of South-East Asian Odonata, including a series of papers (1987–2007) entitled: *The Odonata of Sulawesi and adjacent Islands*, largely based on his own collecting results. He has contributed to longer works as well as writing a number of books including: *De libellen van Nederland (Odonata)* (1983) as co-author and *Phylogeny and biogeography of the Platystictidae (Odonata)* (2009), his doctoral thesis at the University of Leiden. He published: *An annotated index to names of Odonata used in publications by M.A. Lieftinck* (1992) and has also contributed the Odonata contents to the *Catalogue of Life* website, which is an online database of the world's known species of animals and plants. He has collected extensively in Indonesia, Malaysia and Vietnam. He has described a total of 66 new species and named two new genera in Odonata. (Also see **Tol** & **Jan (Van Tol)** & **Telosticta**)

Vander

Genus *Vanderia* Kirby, 1890
[JS *Lindenia* De Haan, 1826]

Pierre Léonard Vander Linden (1797–1831) (See **Linden**).

Vanida

Velvetwing sp. *Dysphaea vanida* **Hämäläinen**, **Dow** & Stokvis, 2015

The species epithet is based on the common Thai female name Vanida which means 'girl'. The name is a noun in apposition and is not named after any particular person.

Varrall

Midget sp. *Mortonagrion varralli* **Fraser**, 1920

Ethel Grace Fraser née Varrall (1881–1960) married (1905) the author Dr Frederic Charles Fraser (1880–1963). The dedication may refer to her or her father. No etymology was given. (Also see **Ethel**)

Vazques

Darner sp. *Rhionaeschna vazquezae* González Soriano, 1986
[S. *Aeshna vazquezae*]

Dr Leonila Vázques Garcia (1914–1995) was a Mexican lepidopterist. She graduated (1934) from the Institute of Biology at National Autonomous University of Mexico, which also awarded her masters (1936) and doctorate (1946), her theses being on Mexican Lepidoptera. She worked there at the entomology laboratory, starting as assistant zoologist (1937). She began teaching (1941) biology and was Assistant Professor of Biology (1944–1954). One of her most important books was a textbook on arthropods: *Zoology of the Phylum Arthropoda* (1941–1987), used by Mexican students over many years. She undertook fieldwork in various parts of Mexico and visited museum collections in the US. She described numerous species of Lepidoptera. She was honoured in the name as the author's teacher. Around 15 taxa are named in her honour.

Verschueren

Longleg sp. *Podogomphus verschuereni* **Schouteden**, 1934
[JS *Notogomphus spinosus* (Karsh, 1890)]

René Charles Marie Verschueren (b.1883) was a collector working in the Congo (from c.1912) principally of botanical, but also entomological specimens for the Musée Royal du Congo Belge and other institutions. Schoutenden (q.v.) earlier (1917) dedicated a Heteroptera species to him, *Crollius verschuereni*. A number of plant names also honour him, such as *Cleistopholis verschuereni* De Wild, 1920 and *Uragoga verschuereni* De Wild, 1930.

Vervoort

Jewel sp. *Watuwila vervoorti* **Van Tol**, 1998

Professor Dr Willem 'Wim' Vervoort (1917–2010) was an eminent Dutch marine zoologist and museum director. His interest in marine biology was sparked by his father's business as a wholesale dealer in fish. He studied biology at Leiden University (1936–1941), but his study was interrupted when mobilised (1939) and he saw service as a sergeant in the infantry. After the German invasion, he returned to Leiden to complete his study, being awarded his master's degree (1941). The occupying forces closed the university and he joined the staff of the Rijksmuseum van Natuurlijke Historie in Leiden as a scientific assistant. The director was Boschma (q.v.) who supervised his PhD on Copepoda (1946) at which time he became Curator of Invertebrates. He undertook two Antarctic trips aboard a whaler (1946–1947 & 1947–1948). He took a post as junior lecturer (later lecturer) at the university (1950–1959), but returned to the museum (1959) in the same post until becoming Director (1972–1982) until he retired, being Emeritus Professor thereafter. Concurrently with being Director, he was Professor of Systematic Zoology at the University. He was a distinguished authority on the taxonomy of Hydrozoa and Copepoda. When the Emperor Hirohito of Japan (a keen marine biologist himself) visited the Netherlands (1971), he wanted to meet his research colleagues, Lipke Holthuis (q.v.) and Verwoort at RMNH, Leiden. Vervoort undertook a number of field trips to Scandinavia, the Mediterranean, Suriname and the Caribbean. He participated as a senior zoologist in the last great colonial exploration in Netherlands New Guinea; the Star Mountains Expedition (1959). He also visited many prominent museums visiting the BMNH annually. He published 170 papers and books (1941–2009) and described 167 new taxa. He is commemorated in the names of 34 marine invertebrates, two spiders, a butterfly and another insect as well as the dragonfly.

Vick

Blue-shouldered Yellowwing *Allocnemis vicki* **Dijkstra** & Schütte, 2015

Graham Spencer Vick (b.1947) is an English amateur odonatologist and retired mathematics teacher. His late wife, Mary Christine Vick (q.v.), was a fellow scientist and teacher. He graduated from Nottingham University (1965–1968) and later took his master's at Reading University (1980–1983). He taught in a number of schools and colleges including Shiplake College, Henley-on-Thames until retirement (1983–2007). His entomological studies have concentrated on the dragonfly faunas of Nepal, Cameroon and New Caledonia, but he has also published on Venezuela, Malaysia and Brazil. He made two expeditions to Nepal

(1980s) and was also much aided by resident entomologist Colin Smith. He made several expeditions to Cameroon, helped by the local efforts of Otto Mesumbei, a resident of Nyasoso (near Mount Kupe), with whom he, Mary Vick and David Chelmick set up the Cameroon Dragonfly Project (1996–2005). For his New Caledonia work he collaborated in the two species' descriptions with D Allen Davies, who made many visits to the island; and Vick himself visited (2013) after Allen's death. He was honoured for his considerable contribution to the study of African dragonflies. He has described ten new species (including four in jointly authored papers): two from New Caledonia, one from Nepal, and seven from Africa. One is now in synonymy. Among other works he wrote: *List of the Dragonflies from Nepal with a summary of their altitudinal distribution* (1989) and *A checklist of the Odonata of the South-West Province of Cameroon, with a description of Phyllogomphus corbetae* (1999). He has described several larvae from Africa and ten new species, including one named after his wife. (See also **Mary**)

Victor

Shadowdancer sp. *Idionyx victor* **Hämäläinen**, 1991

The name of this species, described from Hong Kong, is not an eponym. It refers to the Latin word 'victor' meaning 'a winner'. Symbolically, this Hong Kong taxon, which had earlier been identified as same species as the Bornean *Idionyx yolanda*, 'won' its own specific status.

Victor

Sanddragon sp. *Progomphus victor* St. Quentin, 1973

Victor is a masculine forename, but not a specified one. The author gave no etymology, so it is not possible to know for sure if this was named after a particular person. However, the lack of a Latin declension points to it not being an eponym.

Victoria (Falls)

Elf sp. *Tetrathemis victoriae* Pinhey, 1963
[Orig. *Archaeophlebia victoriae*]

This is a toponym; Victoria Falls is an area where this species is found. David Livingstone, who discovered the falls, named them in honour of Queen Victoria.

Victoria (Lake)

Victoria's Jewel *Chlorocypha victoriae* **Förster**, 1914)
[Orig. *Libellago rubida victoriae*]
Victoria's Duskhawker *Gynacantha victoriae* **Pinhey**, 1961

These are toponyms referring to Lake Victoria. The find locality of the jewel was given as 'Entebbe, Victoria Nyanza, Uganda'. The hawker was found in several places in Uganda, also in Entebbe. David Livingstone named the lake in honour of Queen Victoria.

Victoria

Bannerwing sp. *Polythore victoria* **McLachlan**, 1869
[Orig. *Thore victoria*]

Thylacine Darner *Acanthaeschna victoria* **Martin**, 1901
Lesser Pincer-tailed Wisp *Agriocnemis victoria* **Fraser**, 1928
Alpine Redspot *Austropetalia victoria* Carle, 1996
[JS *Austropetalia tonyana* Theischinger, 1995]

Queen Victoria (Alexandrina Victoria) (1819–1901) of Great Britain.

Vikhrev

Emerald ssp. *Hemicordulia tenera vikhrevi* **Kosterin, Karube** & Futahashi, 2015

Nikita Evgenyevich Vikhrev (b.1959) is a Russian biologist, businessmen, photographer and traveller from Moscow. While still a schoolboy he was interested in ornithology and has co-authored three papers in this field. He graduated from the Biology Faculty of Moscow State University (1981) and then entered the Institute of Molecular Biology of Russian Academy of Sciences. He defended (1987) his PhD Thesis entitled: *Investigation of influenzia virus with monoclonal antibodies*. After this he became a schoolteacher (1987–1990) then a successful businessman trading medical and laboratory equipment. He took a sabbatical year (1999) to pursue his hobby of photography in Madagascar, and upon his return to Moscow he founded the Rosfoto Photographic Agency. He gave up (2005) all his businesses interests and his ornithological hobby in favour of entomology and became a prolific taxonomist of Diptera based in the Zoological Museum of Moscow State University, specialising mostly in certain genera of Muscidae. For the last decade, he has travelled all over the world to collect flies and other, mostly 'inconspicuous', insects and to visit many entomological collections. He has also spent time supporting the collecting expedition of a number of his colleagues. He has published c.40 entomological papers so far and described around forty fly species new to science. He has been married four times and has three children; several fly species have been named after some of his wives, children and colleagues. In private life, he enjoys being a heavy smoker. Once he noticed a dragonfly hovering over a pond in the Cambodian rainforest and told his friend Oleg Kosterin about it. Kosterin went and collected it and found it to be a new sub-species which he named it after Vikhrev.

Villiers

White-shouldered Threadtail *Elattoneura villiersi* **Fraser**, 1948
[Orig. *Prodasineura villiersi*]
Abbott's Skimmer *Oxythemis villiersi* Fraser, 1949
[JS *Orthetrum abbotti* Calvert, 1892]
Blue-breasted Waxtail *Ceriagrion villiersi* Fraser, 1951
[JS *Ceriagrion bakeri* Fraser, 1941]
Comoro Knifetail *Nesocordulia villiersi* **Legrand**, 1984
Congo Double-spined Cruiser *Phyllomacromia villiersi* Legrand, 1992
[Orig. *Macromia villiersi*]

Dr André Villiers (1915–1983) was a French entomologist and herpetologist. When just 13 he started collecting insects, which he took to the MNHN for identification and, after military service in Algeria he began work there (1937) as a technical assistant in entomology. He became attaché (1944), having been awarded his PhD by the University of Paris (1943), following which he took a post as the first Curator of Entomology at the French Institute at Dakar, Senegal (1945). He collected widely, including Ivory Coast (1946), Portuguese Guinea (1947) and Togo (1950) where the first three odonate types were taken. He returned to the MNHN (1956) and became professor (1976). During his time there, he continued to make collecting trips to West Africa and also to Iran, Madagascar and the Caribbean. He had more than 660 publications, the majority being on entomology, but also some (c.35) on herpetology including his first book: *Les Serpents de l'Ouest Africain* (1950).

Vilmi

Mosaic Darner sp. *Gynacantha vilma* Steinmann, 1997

[JS *Aeschna subviridis* Selys, 1850 (a replacement name of the preoccupied *Aeschna viridis* Rambur, 1842) all are *nomina oblita*]

Vilmi Steinmann was the wife (1958) of the author Henrik Steinmann (1932–2009).

Viloria

Pond Damselfly sp. *Teinopodagrion vilorianum* **De Marmels**, 2001

Dr Angel Luís Viloria Petit (b.1968) is a Venezuelan biologist. He graduated from the Museo de Biología, Universidad del Zulia, Maracaibo, Venezuela (1991) and gained his PhD at London University (1998). He has been Professor of Zoology and Entomology at the Faculty of Sciences, University of Zulia for eleven years, and Director of the Museum of Biology (MBLUZ) (1994–1995). He began working (2000) at the Centre for Ecology of the Venezuelan Institute for Scientific Research (IVIC), first as research associate, then Deputy Director (2005–2007 and currently Research Associate and Director. He has a particular interest in the historical biogeography of butterflies and has over 100 publications to his name mostly on Lepidoptera. The etymology says: "*I dedicate this species to my friend, Dr. Angel Luís Viloria... ...He has collected many precious dragonflies in the Sierra de Perijá, a most dangerous frontier region between Venezuela and Colombia, infested by contrabandists, drug smugglers and guerrillas.*"

Vinson

Mauritius Bluetail *Ischnura vinsoni* **Fraser**, 1949

Joseph Lucien Jean Vinson (1906–1966) was a Mauritian zoologist. He was Director of the Mauritius Institute and was awarded an OBE in 1963. He collected the type in Mauritius (1947). Among his papers and longer publications is: *Catalogue of the Coleoptera of Mauritius and Rodriguez* published by the Institute (1956) and, co-written with his son, Jean-Michel: *The Saurian Fauna of the Mascarene Islands*, which was published after his death (1969). At least one beetle, a gecko and a mammal are also named after him.

Virginie

Virginie's Rockstar *Tatocnemis virginiae* **Legrand**, 1992

Virginie Legrand (b.1971) is the oldest daughter of the author, Dr Jean Legrand (b.1944). (Also see **Legrand**)

Vnukovsky

Siberian Hawker *Aeshna crenata wnukoswskii* **Belyshev**, 1973

[JS *Aeshna crenata* Hagen, 1856]

Dr V V Vnukovsky was a Russian zoologist working in Omsk and Tomsk at the beginning of the 20th century. Among his published papers (1920s and 1930s) is: *Contribution to the Fauna of Western Siberian Odonata* (1928). His collection is housed in the Zoological Museum of Tomsk University.

Volxem

Emerald sp. *Neocordulia volxemi* **Selys**, 1874
[Orig. *Gomphomacromia volxemi*]

Camille Van Volxem (1848–1875) was a Belgian entomologist. He travelled in Central Spain, Portugal, Morocco and Brazil and collected entomological specimens including many water-associated species. He co-wrote: *Liste des Criocérides recueillies au Brésil par feu C. van Volxem, suivie de la description de douze nouvelles espèces américaines de cette tribu* (1881). He gave the type specimen of the emerald, which he had collected during his trip through Minas Gerais, Brazil (1872), to Selys. A number of other insects are named after him.

Vrydagh

Golden-winged Jungleskimmer *Hadrothemis vrijdaghi* **Schouteden**, 1934
Ochre Threadtail *Elattoneura vrijdaghi* **Fraser**, 1954
Hintz's Skimmer *Orthetrum vrydaghi* Fraser, 1954
[JS *Orthetrum hintzi* Schmidt, 1951]

Jean-Marie Vrydagh (1905–1962) was a biologist and entomologist who later became a teacher working in the Congo (c.1930–c.1957). He collected the types of the three odonate species in the (then) Belgian Congo (March 1938). He is recorded as collecting earlier (1930) at Mbwa in Uele-Itimbiri, Congo (1932) and at Bambesa in the same area later (1934–1939). He wrote a number of articles about insects associated with crops including: *Tableau systematique des insects nuisibles aux plantes cultive'es au Congo Belge* (1930) as well as other subjects, including small mammals and birds, such as: *Observations ornithologiques en region occidentale du Lac Albert et principalement de la plaine d'Ishwa.* (1949). The binomials of two species refers to the same person but are spelled incorrectly.

Vulcano

Firetail sp. *Telebasis vulcanoae* **Machado**, 1980
[Orig. *Helveciagrion vulcanoae*]

Maria Aparecida Vulcano d'Andretta (b.1921) is a Brazilian entomologist, specialising in Diptera and Coleoptera, and a collector of insect fossils. At one time, she was a researcher at Museu Paulista. She wrote a great many papers (1960–1980), such as: *Two new species of the genus Scatonomus* (1973), as well as a number of longer works and popular science articles. She is also a fine illustrator. She even chose a VW Beetle as her car because Coleoptera inspired it!

W

Wahlberg

Pallid Spreadwing *Lestes wahlbergi* **Ris**, 1921
[JS *Lestes pallidus* Rambur, 1842]

Johan August Wahlberg (1810–1856) was a Swedish naturalist and collector. He studied chemistry and pharmacy at Uppsala University (1829) and later studied forestry, agronomy and natural science and graduated from the Skogsinstitutet (Swedish Forestry Institute) (1834), supporting himself by working in a chemist's shop in Stockholm. His travelling started when he joined an entomology collecting trip in Norway (1832), followed by forestry projects in Sweden (1833) and Germany (1834). He joined the Office of Land Survey becoming engineer (1836) and then an instructor at the Land Survey College. He travelled and collected widely in southern Africa (1838–1856), sending thousands of specimens back to Sweden. He returned briefly to Sweden (1853) but was soon back in Africa where he lived in Walvis Bay, Namibia (1854). He was exploring the Thamalakane River (headwaters of the Limpopo) in what is now Botswana, when a charging wounded elephant killed him. Four birds are also named after him. He collected the damselfly in Caffraria, Eastern Cape, South Africa.

Wahnes

Genus *Wahnesia* **Förster**, 1900

Skimmer sp. *Protorthemis wahnesi* Förster, 1897
[JS *Protorthemis coronata* (Brauer, 1866)]

Carl Wahnes (1835–1910) was a German naturalist who collected in New Guinea for a wealthy patron, Wolf Von Schonberg, and for Walter Rothschild. He was over fifty before his first expedition and seventy-five by his last. He made four extended collecting trips, the first to Borneo (1886–1890) and New Guinea (1890–1896). He returned briefly to Germany but set out again the following year back to New Guinea (1897–1901) again returning to Germany. His third (1902–1906) and fourth (1907–1909) were also to New Guinea, with a brief return to Germany in between. He died just six months after his final return. Two birds and at least one butterfly are also named after him.

Wai-wai

Threadtail sp. *Epipleoneura waiwaiana* Machado, 1985

The Wai-wai are an indigenous people who live in the area in Brazil where the species was found.

Walker

Taiga Bluet *Enallagma walkeri* Muttkowski, 1911
[JS *Coenagrion resolutum* Hagen, 1876]

Walker's Darner *Aeshna walkeri* **Kennedy**, 1917
Treeline Emerald *Somatochlora walkeri* Kennedy, 1917
[JS *Somatochlora sahlbergi* Trybom, 1889]
Skimmer sp. *Oligoclada walkeri* **Geijskes**, 1931
Sooty Saddlebags *Tramea walkeri* **Whitehouse**, 1943
[JS *Tramea binotata* Rambur, 1842, A. Striped Saddlebags]

Dr Edmund Murton Walker (1877–1969) was a Canadian entomologist interested in animals from the age of seven and insects when he was twelve. He therefore studied natural science at the University of Toronto and graduated in 1900 but then, at the request of his father, he studied for his MD (1903). He worked as a physician at Toronto General Hospital for a few months and preferring a career in biology he returned to study at the Department of Biology, Toronto University, then (1905) Berlin University for postgraduate work in zoology. He returned to Toronto taking up the post of lecturer of invertebrate zoology (1906), later becoming Professor of Entomology and staying on the staff his whole career, becoming Head of Zoology (1934) until retirement (1948) and Professor Emeritus thereafter. He founded the invertebrate collection at the Royal Ontario Museum (1914) and was Assistant Director (1918–1931), later Honorary Director (1931–1969) there. He was editor of the 'Canadian Entomologist' (1910–1920). His early research work focussed on Orthoptera, but he became interested in dragonflies (early 1900s) which became his major focus. Throughout this time, he travelled widely across Canada, studying and collecting Odonata and other insects. His major zoological discovery was finding (with T B Kurata) a peculiar new insect species at Sulphur Mountain in Alberta (1913). Next year he described it as *Grylloblatta campodeiformis*, a 'living fossil', presently often placed in the order Notoptera. He published c.200 research papers, notes and book reviews many of which included detailed information on the habitats, ecology and behaviour of the local species. His major publications on Odonata are: *The North American dragonflies of the genus Aeshna* (1912), *The North American dragonflies of the genus Somatochlora* (1925) and the three-volume: *The Odonata of Canada and Alaska* (1953, 1958 & 1975). He described 24 new species or subspecies of Odonata. A book: *Centennial of Entomology in Canada 1863–1963: A tribute to Edmund M. Walker* was published in his honour (1966).

Wall

Claspertail sp. *Nepogomphus walli* **Fraser**, 1924
[Orig. *Onychogomphus walli*]
Wall's Grappletail *Heliogomphus walli* Fraser, 1925
Velvetwing sp. *Dysphaea walli* Fraser, 1927
Wall's Shadowdamsel *Ceylonosticta walli* Fraser, 1931

Colonel Dr Frank Wall (1868–1950) was a physician and herpetologist, who was born in Ceylon (Sri Lanka) where his father initiated the study of natural history on the island. He qualified as a physician in the UK and worked for the Indian Medical Service (1893–1925). He was a member of the Bombay Natural History Society and a recognised expert on Indian snakes. He wrote more than two hundred papers, such as: *A Popular Treatise on the Common Indian Snakes* (1905–1919), and the books: *The monograph of sea snakes* (1909) and *Ophidia taprobanica, or the snakes of Ceylon* (1921). Eleven reptiles are also named after him. He retired to England and died in Bournemouth.

Wallace, AR

Limniad sp. *Amphicnemis wallacii* **Selys**, 1863
Threadtail sp. *Nososticta wallacii* Selys, 1886
[Orig. *Alloneura wallacii*]
Great Spreadwing *Orolestes wallacei* **Kirby**, 1889
[Orig. *Lestes wallacei*]
Fineliner sp. *Teinobasis wallacei* **Campion**, 1924

Alfred Russel Wallace (1823–1913) was an English naturalist, evolutionary scientist, geographer and anthropologist; the father of zoogeography. He was also a social critic and theorist, a follower of the utopian socialist Robert Owen. His interest in natural history began through attending public lectures, while working as an apprentice surveyor. He went to Brazil, (1848), on a self-sustaining natural history collecting expedition. Even then he was very interested in how geography limited or facilitated the extension of species' ranges. He not only collected but also mapped, using his surveying skills. His return to England (1852) was a near disaster; his ship, the brig *Helen*, caught fire and sank with all his specimens, and he was lucky to be rescued by a passing vessel. He spent the next two years writing and organising another collecting expedition to the Indonesian archipelago. He managed to get a grant to cover his passage to Singapore (1862) and had the benefit of letters of introduction and the like prepared for him by representatives of the British and Dutch governments. He spent nearly eight years there, during which he undertook about 70 different expeditions involving a total of around 14,000 miles of travel visiting every important island in the archipelago at least once, some many times. He collected a remarkable 125,660 specimens, including more than 1,000 new species. He wrote: *The Malay Archipelago* (1869), which is the most celebrated of all writings on Indonesia and ranks as one of the 19th century's best scientific travel books. He also published: *Contributions to the History of Natural Selection* (1870) and *Island Life* (1880). His essay, *On the law which has regulated the introduction of new species*, which encapsulated his most profound theories on evolution, was sent to Darwin. He later sent Darwin his essay: *On the tendency of varieties to depart indefinitely from the original type*, presenting the theory of 'survival of the fittest'. Darwin and Lyell presented this essay, together with Darwin's own work, to the Linnean Society. Wallace's thinking spurred Darwin to encapsulate these ideas in *The Origin of Species*; the rest is history. Wallace developed the theory of natural selection, based on the differential survival of variable individuals, halfway through his stay in Indonesia. He remained for four more years, during which he continued his systematic exploration and recording of the region's fauna, flora and people. For the rest of his life he was known as the greatest living authority on the region and its zoogeography, including his discovery and description of the faunal discontinuity that now bears his name, the Wallace's Line, a natural boundary running between the islands of Bali and Lombok in the south and Borneo and Sulawesi in the north, separating the Oriental and Australasian faunal regions. Two mammals, a reptile, an amphibian and at least twelve birds, among other flora and fauna are also named after him.

Walsh, B D

Brush-tipped Emerald *Somatochlora walshii* **Scudder**, 1866
[Orig. *Cordulia walshii*]

Handsome Clubtail *Gomphus fraternus walshii* **Kellicott**, 1899
[JS *Gomphus crassus* Hagen, 1878]

Benjamin Dann Walsh (1808–1869) was an English-born American entomologist who championed the application of scientific methods to control agricultural pests, particularly biological controls. He graduated with a BA from Trinity College, Cambridge (1831) promoted to MA (1834). He became a fellow of Trinity (1833), where he resided for twelve years with the intention of becoming an Anglican minister, but he became disillusioned with the clergy and Cambridge University (1837) and left. He married (1838) and emigrated to the US where they lived for the next twelve years in a log cabin 20 miles from the nearest settlement farming 300 acres. Eventually they moved from the area after an outbreak of malaria settling in Rock Island, Illinois and starting a successful lumber business (1851–1858). He became involved in politics as a radical republican opposed to slavery and was elected (1858) to the local council purely to expose their corruption, which he did and then resigned. He sold his lumber business (1858) and bought property which he rented out devoting himself to full time entomology, an interest he had developed at Cambridge. He lectured (1859) and exhibited his insect collection, including at the State Horticulture Society (1860). He also began publishing entomological papers and joined various learned societies. He was appointed as the first official State Entomologist (1867–1869) for Illinois (only the second state to have such a position). He wrote 385 papers and co-wrote another 478 and was co-founder and editor of 'American Entomologist' with C V Riley (q.v.) as junior editor. He was an early supporter of, and correspondent with, Darwin. He was run over and killed by a locomotive while out catching butterflies.

Walsh, M E

Walsh's Clubtail *Acrogomphus walshae* **Lieftinck**, 1935
[Orig. *Acrogomphus walshi*]

Mrs Maria Ernestine Walsh-Held (1881–1973) managed a tea plantation on Java. She was born in Switzerland, married in Australia and settled in Soekaboemi, Java (1911). She was widowed (1913) and thereafter devoted her time to studying and collecting insects. She began selling specimens and employed Indonesian collectors to help her. She travelled across Timor (1928–1929), south Sumatra, the type locality (1935) and Borneo (1937), collecting both insects and plants. She eventually returned to Switzerland (1958) where she lived for the rest of her life. Other insects she collected, such as the beetle *Oberea walshae*, also honour her. Buitenzorg Museum, Bogor, Java, Indonesia acquired many specimens, but they were also taken by BMNH and other European museums.

Walsingham

Giant Darner *Anax walsinghami* **McLachlan**, 1883

Thomas de Grey, 6th Baron Walsingham (1843–1919), was an English Conservative MP (1865–1870), amateur entomologist and lepidopterist. He was educated at Eton and Trinity

College, Cambridge (BA 1865, MA 1870, Hon LLD 1891). He was Chief Whip in Disraeli's second administration (1874–1875). He succeeded to the title (1870) so went to the House of Lords in addition to managing the estate and serving as a trustee of the British Museum. He collected butterflies and moths from an early age and was particularly keen on Microlepidoptera; his butterfly collection was one of the most important ever made with c.260,000 specimens which he donated to the BMNH, along with 2,600 books. He was twice President of the Entomological Society of London and was elected a Fellow of the Royal Society (1887). His other claims to fame (or notoriety depending on your point of view) were having married three times, having played first-class cricket (1862–1866) and having shot (1888) 1070 grouse in one day on Blubberhouse Moor, Yorkshire, England.

Walter

Scoop-tailed Slim Sprite *Aciagrion walteri* Carfi & D'Andrea, 1994
[JS *Pseudagrion cyathiforme* Pinhey, 1973]

Professor Dr Valter (Walter) Rossi (b.1946) is a biologist and mycologist at the Department of Life, Health & Environmental Sciences, University of L'Aquila, Italy. He has a special interest in parasitic fungi on insects and for this reason he collects insects on his expeditions although not his area of expertise, so he can make them available to entomologists to study. He has published widely with descriptions of new fungi and plants, including: *A new contribution to the knowledge of the Laboulbeniales (Ascomycetes) from Sierra Leone* (1994) in the same edition of the journal in which Cari & D'Andrea's description of the odonate appears. He has participated in several collecting trips, notably to Brazil, Ecuador, Costa Rica and numerous expeditions to Sierra Leone (including 1980, 1984–1985, 1986–1987, 1991–1992, 1993–1994). A number of insects are named after him, such as the assassin bug *Katanga walterrossii* Dioli, 1994 collected (1991) by Rossi in Sierra Leone.

Walthère

Clubtail sp. *Cyanogomphus waltheri* **Selys**, 1873
Big Red Damselfly sp. *Minagrion waltheri* Selys, 1876
Threadtail sp. *Neoneura waltheri* Selys, 1886
Amberwing sp. *Perithemis waltheri* **Ris**, 1910
[JS *Perithemis icteroptera* (Selys, 1857)]

Baron Charles Michel Edgard Walthère de Selys Longchamps (1846–1912) was the youngest son of the author Baron Michel Edmond de Selys Longchamps (1813–1900). He collected the type specimens of all four of the new species in Brazil (1872). The 'black sheep' of the family, Walthère was an atheist, non-conformist and socialist who had a relationship with Joséphine Davignon, the family cook, who became pregnant. They started a self-imposed exile in Paris (1875) to protect the family name. While this embarrassed Selys, his relationship with his son remained warm. The couple had two more children before marrying (1881) and another three afterwards. In his diaries Selys doesn't mention any of the grandchildren until the day Walthère married and even in the official records of Belgian nobility their marriage date was given as a decade earlier (1871). (Also see **Selys**)

Watanabe, K

Lyretail ssp. *Stylogomphus shirozui watanabei* **Asahina**, 1984

Ken-ichi Watanabe (b.1951) is a Japanese odonatologist who worked as a high school science teacher in Ishigaki-jima island, Japan. He is currently (since 2014) President of the Japanese Society for Odonatology and Editor of 'Tombo – Acta Odonatologica Japonica' (since 2015). He has authored two books in Japanese on the Odonata of Okinawa (1986 & 2007) and studied the geographical variation of *Coeliccia* species in the Ryukyu islands, publishing (1988): *Mesepisternal pattern of Coeliccia ryukyensis (Platycnemididae)*. He collected the type series in Iriomote Island, Japan (1981 & 1983).

Watanabe, Y

Genus *Watanobeopetalia* **Karube**, 2002

Dr Yasuaki Watanabe (b.1932) is a Japanese entomologist who was at the Laboratory of Entomology, Tokyo University of Agriculture. His focus is Coleoptera. Among his many written works (1960–2015) are the co-authored: *Description of a new species of genus Syntomium in Japan (Col. Staphylinidae)* (1960) to *New or Little Known Species of the Genus Stenus (Col. Staphylinidae) from Japan* (2015). A number of beetles are also named after him.

Waterhouse

Riverhawker sp. *Tetracanthagyna waterhousei* **McLachlan**, 1898

Charles Owen Waterhouse (1843–1917) was an English entomologist specializing in Coleoptera. He became assistant keeper at the BMNH describing hundreds of new species and writing many papers on the beetles at the museum. He also contributed *Buprestidae* to Godman & Salvin's *Biologia Centrali-Americana* (1889). He was President of the Royal Entomological Society (1907–1908). He described (1877) a new giant aeshnid species from Borneo as *Gynacantha plagiata*. The generic combination was McLachlan's suggestion. Later Selys (1883) erected a new subgenus *Tetracantagyna* for this species and McLachlan (1898) upgraded it to a full genus and named one of the new species after Waterhouse, following Selys' manuscript name.

Waterston

Ethiopian Hawker *Pinheyschna waterstoni* **Peters** & **Theischinger**, 2011

Banded Demoiselle ssp. *Calopteryx splendens waterstoni* **W Schneider**, 1984
[Orig. *Calopteryx waterstoni*]

Andrew Rodger Waterston OBE FRS (1912–1996) was a Scottish naturalist, zoologist, entomologist and malacologist, son of the entomologist James Waterston. His BSc was awarded by Edinburgh University (1934) and he took part in their expedition (1935) to Barra, Hebrides, an island that became his summer home for the next thirty years. In 1938 he married a fellow participant in that expedition, Marie Campbell. He became Assistant Keeper in the Department of Natural History at the Royal Scottish Museum (1935). He was seconded to the Ministry of War (1939–1942) joining the Royal Scots during WW2 (1943–1944). He transferred to the Colonial Office (1943) becoming Locust Officer in the Middle East, based in Cairo, Egypt, but travelling to Saudi Arabia, Yemen, Ethiopia and Palestine. He was then Chief Locust Officer and subsequently Entomological Advisor (1947–1952) in charge of the Middle East, Eritrea and Ethiopia. He returned to the museum (1952) becoming

Keeper of Natural History (1958–1973) until he retired and then Keeper Emeritus (1972–1977). He returned to the Middle East to study eastern dragonflies and conducted the first systematic research of Odonata in the Arabian Peninsula (1980s–1990s), work later continued by Wolfgang Schneider. Waterston was also one of the founders of the Scottish Natural History Library. He published widely, especially on conchology and entomology, specialising in Odonata. He described four new species and one new subspecies of Odonata.

Watson

Eastern Billabongfly *Austroagrion watsoni* **Lieftinck**, 1982
Tropical Cascade Darner *Spinaeschna watsoni* Theischinger, 1982
Jewel sp. *Rhinocypha watsoni* **Van Tol** & **Rozendaal**, 1995
Firetail sp. *Telebasis watsoni* **Bick** & **Bick**, 1995

Dasher ssp. *Micrathyria mengeri watsoni* **Dunkle**, 1995

Dr John Anthony Linthorne 'Tony' Watson (1935–1993) was an Australian entomologist. Before university he spent a year (1953) travelling in Britain and western Europe with his parents. He graduated with BSc (zoology) (1956) and with 1st class honours (1957) from the University of Western Australia, then undertook postgraduate research on silverfish in the laboratory of Sir Vincent Wigglesworth at Cambridge University which awarded his PhD (1962). He was a postdoctoral fellow (1962–1963) and later Visiting Professor (1964–1965) at Western Reserve University, Cleveland, Ohio, before returning to Australia again as a Queen Elizabeth II Fellow to the Division of Entomology, CSIRO, studying silverfish (1965–1967). In 1967 he joined the Division's Termite Section becoming its leader (1970) and was promoted to Chief Research Scientist (1981). He co-wrote *Atlas of Australian Termites* (1993), having already contributed a number of chapters to *The Insects of Australia* and published 68 papers, including books and book chapters on termites. He was also assigned responsibility for dragonflies in the Australian National Insect Collection and allowed part of his official time for this work. But his love of dragonflies, which had developed at an early age, was always to the fore and he became the leading expert of Australian Odonata on which he published c.75 papers (1956–1994). These include the books: *The dragonflies (Odonata) of South-western Australia* (1962) and *The Australian Dragonflies* (with Theisschinger (q.v.) and Abbey, 1993). His other major publications include: *The Australian Gomphidae* (1991) and co-writing the *Odonata* chapter in *Zoological Catalogue of Australia Vol. 6* (1988). He named eleven new odonate genera and subgenera and described 55 new odonate species, many of them jointly. "*I also wish to express my special gratitude to my friends … Dr J.A.L. Watson (Canberra) who supported my work in many ways.*" (Also see **Tony (Watson)**)

Watuliki

Black Flasher *Aethiothemis watulikii* **Pinhey**, 1962
[JS *Aethiothemis basilewskyi* Fraser, 1954]

B K Watuliki collected the holotype in the Democratic Republic of the Congo (1960). Pinhey said of him (1962): "*The author is particularly indebted to those entomologists who have presented material to this Museum. Firstly, Col. T. H. E. Jackson, of Kenya, whose assistant Mr*

B. K. Watuliki has sent many consignments from Eastern Nigeria and the Congo Republic (the former French Moyen Congo)."

Wayana

Darner sp. *Staurophlebia wayana* **Geijskes**, 1959

Named after a people; the Wayana (also spelled Oajana or Oayana) who live in the area in Surinam where the species was found.

Weibezahn

Pond Damselfly sp. *Metaleptobasis weibezahni* **Rácenis**, 1955
[JS *Metaleptobasis brysonima* Williamson, 1915]

Professor Franz Herbert Weibezahn Massiani (1925–2004) was a Venezuelan limnologist and ecologist. He was Professor Emeritus at Universidad Central de Venezuela (VCU) and Consulting Professor for the Universidad Simon Bolivar. Liceo Aplicación awarded his BSc in biological sciences (1945) and he went on to medical school at VCU, but left after a year. He attended the University of California at Berkeley (1946) obtaining a BSc in zoology. He worked for the Ministry of Agriculture and then became Associate Professor at VCU. He became head of the research group at the Limnology School of Biology (1969–1979) before retiring (1980). He had very long-term research associations with the Orinoco River and the Guri Reservoir. Among many other works he wrote: *El Río Orinoco como ecosistema* (1990).

Weiske

Ochre-tipped Darner *Dromaeschna weiskei* **Förster**, 1908

Emil Weiske (1867–1950) was a German traveller and specimen preparator who collected zoological specimens and ethnological objects and founded a private museum. He was in California (1890–1892), unsuccessfully prospecting for gold, but successfully learning English. He went to Hawaii (1892) to collect birds and insects, then moved on to the Fiji Islands (1894) and continued collecting there until he moved to North Queensland (1895–1897) for two years where he collected the type, also visiting Papua (New Guinea) (1897). He had the habit of fishing with dynamite when in 1900 the stick of dynamite he was about to use exploded and shattered his right hand. His companion, a local inhabitant, saved him from bleeding to death. After this, he returned to Germany (1900) and established a mobile museum (1904), which he took touring. He went to Lake Baikal (1908) in Siberia to collect seals and other animals. He had an assistant, Otto Taschmann, to make up for his one-handedness. His final overseas trip was to Argentina (1911–1913). Back again in Germany he bought a house and re-established his private museum. A number of butterflies and four birds are also named after him.

Werner, FG

Sylvan sp. *Coeliccia werneri* **Lieftinck**, 1961

Professor Dr Floyd Gerald Werner (1921–1992) was an American entomologist. Harvard awarded his BSc in biology (1943) and, during his student years, he served as assistant in the *Coleoptera* section of the Museum of Comparative Zoology. He served (WW2) as an entomologist in the US Army in the South Pacific where his medical unit worked with mosquitos and malaria in Okinawa and Korea. He was a member of the Philippine Zoological Expedition (1946–1947) and collected many new species, including the above. After demobilisation, he continued (1947–1950) with his PhD at Harvard. He collected in the USA, Canada and Mexico (1948 & 1949) then became Assistant Professor of Zoology at University of Vermont (1950). He returned to Okinawa (1951) studying sweet potato pests. He undertook fieldwork for the Entomology Department, Arizona University (1954) as part of the Arizona Economic Insect Survey and was promoted to associate professor (1958) then professor (1962). He took a sabbatical (1963) to collect in Brazil and Argentina. He retired (1989) and took on the task of editing 'Coleopterist Bulletin'.

Werner, FJM

Elegant Dropwing *Trithemis werneri* **Ris**, 1912

Professor Dr Franz Josef Maria Werner (1867–1939) was an Austrian explorer, zoologist and herpetologist. He graduated from the Gymnasium (1888), then studied zoology at the University of Vienna leading to his PhD (1890) and he continued his studies in Leipzig. He was appointed (1895) as assistant (instructor) at the Natural History Museum, Vienna, and Privatdozent (1909) and eventually professor (1919–1933). It is said that Steindachner who was director there, disliked him and forbade him access to the herpetological collection. Werner made a long series of expeditions (1894–1936) into southeastern Europe, Asia Minor and north and east Africa. He collected in Algiers and Morocco (1928 & 1930), Egypt (1899 & 1904), the Sudan (1905 & 1914), Uganda, where he collected the type (1905), and Syria and Cyprus (1935). He collected on fourteen islands in the Aegean (1936), as well as at other times covering forty islands. His many publications included some on a number of insect groups, including Orthoptera, but were mainly herpetological: *Amphibien und Reptilien* (1910) and a standard work on venomous snakes. Twenty-eight reptiles and six amphibians are named after him.

Westermann

Threadtail sp. *Phylloneura westermanni* **Hagen**, 1860
[Orig. *Alloneura westermanni*]

Bernt Wilhelm Westermann (1781–1868) was a wealthy Danish businessman who collected insects. He travelled on business to India (where the holotype was collected) and the Dutch East Indies (Indonesia) (via the Cape of Good Hope) and collected while there. He returned to Denmark (1817) where he became a ship-owner and sugar refiner. When he died, he had amassed over 45,000 species of insect, which is now housed in the collection of the University of Copenhagen.

Westfall

Pond Damselfly sp. *Metaleptobasis westfalli* Cumming, 1954
[JS *Metaleptobasis foreli* Ris, 1918]
Rubyspot sp. *Hetaerina westfalli* **Rácenis**, 1968
Westfall's Snaketail *Ophiogomphus westfalli* **Cook** & Daigle, 1985

Westfall's Knobtail *Epigomphus westfalli* **Donnelly**, 1986
Threadtail sp. *Epipleoneura westfalli* **Machado**, 1986
Dark Leaftipper *Malgassophlebia westfalli* **Legrand**, 1986
Westfall's Clubtail *Gomphus westfalli* Carle & May, 1987
Westfall's Dancer *Argia westfalli* **Garrison**, 1996
Pond Damselfly sp. *Amazoneura westfalli* Machado, 2001
[Orig. *Forcepsioneura westfalli*]

Western Slender Bluet *Enallagma traviatum westfalli* Donnelly, 1964
[Orig. *Enallagma westfalli* Selys, 1876]

Dr Minter Jackson Westfall Jr (1916–2003) was an American odonatologist. He was deputy game warden on Merrit Island, Florida, US (1936), then attended Rollins College where he received his BSc (1941). He left for graduate studies at Cornell University and after service in the US Army (1943–1944) returned to Cornell and earned his PhD (1947). Then (1947) he was hired as Assistant Professor of Zoology at the University of Florida (Gainesville), later becoming Professor of Zoology until retiring (1985). Many of his students became renowned odonatologists, who remember him as a friendly, inspiring and patient mentor with attention to detail. He worked actively in Societas Internationalis Odonatologica, becoming Member of Honour (1979), edited its newsletter 'Selysia' (1970–1986) and was the leading force in the establishment (1985) of the Society's International Odonata Research Institute (affiliated to the Florida State Collection of Arthropods in Gainesville, Florida), where he worked as Director (1985–1996). His wife's (Margaret Lucille Stepherd q.v.) failing health forced the couple to move (1996) to Gainesville in Georgia to be closer to receive support from their son. Work on dragonflies was both his research focus and dearest hobby. He travelled extensively in the US and across Central America, visiting Venezuela (1980). He also participated with his wife in numerous international odonatological symposia in different parts of the world. During his travels, he used every opportunity to collect dragonflies. It was said that '... *he could not live without dragonflies and that dragonflies could not live with him*'. He co-wrote: *A manual of the Dragonflies of North America (Anisoptera)* (1955) with J G Needham (q.v.) and *Damselflies of North America* (2006) with M L May (q.v.), the standard regional handbooks of Odonata. His publications include descriptions of two new genera, 15 new species and one subspecies of Odonata. (Also see **Minter**)

Westwood

Cruiser sp. *Macromia westwoodii* **Selys**, 1874

Professor John Obadiah Westwood (1805–1893) was an English entomologist, palaeographer and archaeologist, also noted for his artistic talents. He originally trained to be a lawyer, but instead pursued entomology and archaeology and is regarded as one of the foremost entomologists of his era. He was Curator and later Professor of Zoology in the University of Oxford and President (1852–1853), then Honorary Life President of the Entomological Society of London. He wrote many papers in their proceedings and longer works, including: *The Cabinet of Oriental Entomology* (1848), which features his own richly painted scenes of insect and plant life and *Thesaurus Entomologicus Oxoniensis: or illustrations of new, rare and interesting insects, for the most part coloured, in the collections presented to the University of Oxford by the Rev F W Hope* (1874). He is the author (1837) of North American *Libellula axilena* (Bar-winged Skimmer). He died after a stroke.

Weyers

Bronze Needle *Synlestes weyersii* **Selys**, 1868

Joseph Léopold de Weyers (fl.1842–1908) was a Belgian engineer, malacologist and entomologist. He was a founder member and Librarian of the Malacologist Society of Belgium (1863–1868), becoming secretary (1868–1872) and its administrator (1875–1877). He was also librarian at the Royal Entomological Society of Belgium until resigning (1870) (Selys was President during part of that time). Collectors in Victoria, Australia, sent specimens to him in Belgium via Port Denilson. Weyers gave Selys the Bronze Needle specimen that he had been sent, to describe. Being an engineer he travelled and collected whenever he could, sending specimens to zoologists to study, but also dealing in them for profit. He collected scarab beetles (1871–1872) at Hastière, Belgium, ants and grasshoppers in Aguilas, Spain (1882) and grasshoppers in Sumatra (1883) (which is when he started collecting there), as well as butterflies (1889) and spiders, mostly around Fort de Kock Padang where he was resident. He co-wrote: *Contribution à la faune malacologique de Sumatra (récoltes de M.J.-L. Weyers)* (1899).

Whedon

Horned Clubtail *Gomphus whedoni* Muttkowski, 1913
[JS *Arigomphus cornutus* (Tough, 1900)]

Professor Dr Arthur DeWitt Whedon (1879–1969) was an American zoologist. He collected the dragonfly in his birth state, Iowa (1908) and the University of Iowa awarded his BA (1907) and MSc (1912). He was an assistant in zoology at the University of Pennsylvania (1916–1918) where he took his PhD and subsequently Instructor in Zoology (c.1920) and (c.1925) Head of the Zoology Department, later becoming Professor of Zoology and Physiology, North Dakota State Agricultural College. He published two papers (1918, 1927) on the comparative morphology of dragonflies. R A Muttkowski, who wrote a catalogue of the Odonata of North America (1910), wrote a paper with A D Whedon: *On Gomphus cornutus Tough (Odonata)* (1915).

Wheeler

Shadowdamsel sp. *Drepanosticta wheeleri* **Fraser**, 1942
[JS *Drepanosticta fontinalis* Lieftinck, 1937]
Tiger sp. *Gomphidictinus wheeleri* Fraser, 1942
[JS *Gomphidictinus perakensis* Laidlaw, 1902]

Dr Raymond Wheeler, with others, collected (1932 & 1936) the type series amongst 104 species in Malaysia where he was resident in Butterworth, Penang. Fraser's etymology says: *"I am indebted to Dr. Raymond Wheeler for the opportunity of reporting on a collection of some 104 specimens of Odonata which he and others collected in the northern parts of the Federated Malay States."*

Whellan

Yellow-faced Citril *Ceriagrion whellani* **Longfield**, 1952
Swarthy Sprite *Pseudagrion whellani* **Pinhey**, 1956
[JS *Pseudagrion hamoni* Fraser, 1955]

James Arden Whellan BSc (1915–1995) was a British entomologist and botanist. He entered Liverpool University (1934), which awarded his BSc. He served as a lieutenant in the army during WW2 and then worked as an entomologist (1947–1967), later Chief Entomologist, for the department of Agriculture, Southern Rhodesia (Zimbabwe). He worked in Malawi (1967–1974) and then Malaysia (1974–1976) and settled in Australia after retiring. He and Pinhey worked together at the Department of Agriculture where he sparked Pinhey's interest in Odonata; Whellan was keen on them and specimens when he could, in addition to botanical specimens. He wrote a number of papers on economic entomology, such as: *The Red Locust, A Forgotten Menace* (1961) and *Agricultural Ethnology* (1971), as well as papers in the 'Journal of the Entomological Society of Southern Africa' such as: *A new species of the genus Mecostibus (Orthoptera: Lentulidae) from Rhodesia* (1975). At least one plant is also named after him, *Euphorbia whellanii*.

White

Cruiser sp. *Macromia whitei* **Selys**, 1871

Adam White (1817–1878) was a Scottish zoologist. At the age of eighteen, he obtained a post in the British Museum in the Zoology Department, having become acquainted with the then Keeper of Zoology, J E Gray. There he specialized in insects and crustaceans. He wrote: *List of the Specimens of Crustacea in the British Museum* (1847) and *A Popular History of Mammalia* (1850) and such papers as: *Descriptions of unrecorded species of Australian Coleoptera of the families Carabidae, Buprestidae, Lamellicornia, Longicornia, etc.* (1859). He was a member of the Entomological Society of London (1839–1863) and a Fellow of the Linnean Society (1846–1855). He wrote up the results of insects collected on the voyage of HMS *Erebus* and HMS *Terror*. J O Westwood (q.v.) named a beetle after him, *Taphroderes whitii*. Selys met White during his visit to the British Museum (July 1851).

Whitehead

Common Blue Jewel *Rhinocypha whiteheadi* **Kirby**, 1900
[JS *Heliocypha perforata* (Percheron, 1835)]

John Whitehead (1860–1899) was a British explorer and naturalist who collected in Borneo (1885–1888), the Philippines (1893–1896) and Hainan (1899), when he collected the type. He wrote: *Explorations of Mount Kina Balu, North Borneo* (1893) and he *may* have been the first European to reach the summit of the mountain. He died of fever in Hainan, at the age of 38. Thirteen birds, five mammals and an amphibian are also named after him.

Whitehouse

Whitehouse's Emerald *Somatochlora whitehousei* **Walker**, 1925

Francis Cecil Whitehouse (1879–1959) was a British-born Canadian amateur odonatologist. He became a bank manager and settled in Canada (1905), working there until retirement

(1934). He loved fishing, and during his many trips developed a passion for dragonflies. A close association with Edmund Walker of the Royal Ontario Museum and University of Toronto helped him greatly in his pursuit of dragonfly research and he published extensively, especially on the Odonata of Alberta, British Columbia and Jamaica. He made a significant collection, which was divided between the Royal British Colombia Museum and the Spencer Museum, University of British Colombia, although many specimens were sent to Walker. He collected the type in Alberta (1917). He also wrote a few papers on Odonata from Alberta and British Colombia, such as a 16 page booklet: *Dragonflies (Odonata) of Alberta* (1918). His other publications include: *A guide to the study of dragonflies of Jamaica* (1943) and *Catalogue of the Odonata of Canada, Newfoundland and Alaska* (1948). He also wrote a novel, a book of poetry and two books on sport fishing. His booklet on Albertan dragonflies starts with this couplet: *A bug designed a double debt to pay, to feed the fish and keep the flies away.* He described one new species and one new subspecies of dragonfly (both synonyms).

Wijaya

Wijaya's Scissortail *Microgomphus wijaya* **Lieftinck**, 1940

Prince Vijaya was a figure from Sri Lankan history after which various areas were named. He was the first King of Sri Lanka and his reign (543BC–505BC) marks the beginning of their history according to their chronicles.

Wildermuth

Shadowdamsel sp. *Drepanosticta wildermuthi* Villanueva & **Schorr**, 2011

Professor Dr Hansruedi Wildermuth (b.1941) is a Swiss entomologist. He studied zoology, botany and palaeontology, and has a diploma and PhD in developmental biology from the Institute of Zoology, University of Zurich, Switzerland. He was a biology teacher at high school and a senior lecturer for zoology and conservation biology at University of Zurich. He retired from teaching (2003), but is still active in conservation and odonatology and continues to publish. His main interests are education in environmental biology, applied conservation biology and entomology, especially odonatology. He is co-editor and board member of odonatological journals and a consultant in nature conservancy. His publications include: a biology book and educational documents for secondary schools, several books on nature conservation and field guides. His Odonata publications include: *Les Libellules de Suisse* (lead editor, 2005), *Die Falkenlibellen (Corduliidae) Europas* (2008), *Libellen schützen, Libellen fördern* (lead editor, 2009), *Taschenlexikon der Libellen Europas* (together with Andreas Martens, 2014) and numerous scientific papers, especially on ecology, behaviour, faunistics and conservation of European Odonata. He is a supporter of Martin Schorr's International Dragonfly Fund.

Wilkins

Clubtail sp. *Cyclogomphus wilkinsi* **Fraser**, 1926

Mr Wilkins was the author's collecting companion at Hunse in Mysore, India (1924). Typically, Fraser records nothing more to help us identify him further.

Williams, D

Powder-striped Sprite *Pseudagrion williamsi* **Pinhey**, 1964
[JS *Pseudagrion kersteni* (Gerstäcker, 1869)]

Donald Williams was the author's fieldwork assistant in Northern Rhodesia (1960).

Williams, FX

Sanddragon sp. *Progomphus williamsi* **Needham**, 1943
[JS *Progomphus clendoni* (Calvert, 1905)]

Dr Francis Xavier Williams (1882–1967) was an American entomologist of insect behaviour and ecology, taxonomist, embryologist, expert illustrator, explorer and world traveller, visiting the Galapagos Archipelago, Kansas, Massachusetts, the Hawaiian Islands, the Philippine Islands, Panama, Ecuador, British Guiana, Brazil, Missouri, Guatemala, New Caledonia and East Africa. St Ignatius College (now the University of San Francisco) awarded his bachelor's degree (1903), as did Stanford University. The University of Kansas awarded his master's (1912) and Harvard his doctorate (1915). His longest sojourn, other than to Hawaii, was to the Galápagos Islands for 17 months (1905–1906), where he collected about 4,000 insects. He worked for the Hawaiian Sugar Planters' Association (1917–1949), after which he retired back to California where he was appointed a research associate in the Department of Entomology of the California Academy of Sciences and later donated his library and collections to them. He wrote: *Handbook of the Insects and Other Invertebrates of Hawaiian Sugar Cane Fields* (1931) as well as other papers including the description of a new ant. He collected the holotype of the sanddragon in Guatamala (July 1934). An amphibian is also named after him.

Williamson

The Emerald genus *Williamsonia* Davis, 1913

Bannerwing sp. *Polythore williamsoni* **Förster**, 1903
[Orig. *Thore williamsoni*]
Williamson's Darner *Aeshna williamsoniana* **Calvert**, 1905
Williamson's Emerald *Somatochlora williamsoni* **Walker**, 1907
Clubtail sp. *Gomphus williamsoni* Muttkowski, 1910
[a hybrid specimen or an aberrant *Gomphus lividus* Selys, 1854]
Williamson's Hawaiian Damselfly *Megalagrion williamsoni* Perkins, 1910
[Orig. *Agrion williamsoni*]
Slender Baskettail *Tetragoneuria williamsoni* Muttkowski, 1911
[JS *Epitheca costalis* Selys, 1871]
Slendertail sp. *Leptogomphus williamsoni* **Laidlaw**, 1912
Clubtail sp. *Burmagomphus williamsoni* Förster, 1914
[Orig. *Burmagomphus vermicularis williamsoni*]
Skimmer sp. *Elasmothemis williamsoni* **Ris**, 1919
[Orig. *Dythemis williamsoni*]
Yellowwing sp. *Chlorocnemis williamsoni* **Martin**, 1921
[JS *Allocnemis nigripes* (Selys, 1886)]

Sprite sp. *Pseudagrion williamsoni* **Fraser**, 1922
Tiger sp. *Gomphidia williamsoni* Fraser, 1923
Two-Striped Forceptail *Aphylla williamsoni* Gloyd, 1936
[Orig. *Gomphoides williamsoni*]
Threadtail sp. *Epipleoneura williamsoni* **Santos**, 1957
Flatwing sp. *Oxystigma williamsoni* **Geijskes**, 1976
[JS *Oxystigma petiolatum* (Selys, 1862)]
Wedgetail sp. *Acanthagrion williamsoni* **Leonard**, 1977
Three-spined Darner *sp. Triacanthagyna williamsoni* von **Ellenrieder** & **Garrison**, 2003
Demoiselle sp. *Mnesarete williamsoni* Garrison, 2006
Firetail sp. *Telebasis williamsoni* Garrison, 2009
Pond Damsel sp. *Tuberculobasis williamsoni* **Machado**, 2009

Edward Bruce Williamson (1878–1933) was a famous American odonatologist, a widely-recognised authority of the New World odonate fauna, an ardent collector, keen observer and evocative writer. He graduated BSc from Ohio State University (1898) and became Assistant Curator at the Carnegie Museum, Pittsburgh (1898-1899), then held fellowship at Vanderbilt University (1900–1901). He went (1902) to work at Wells County Bank at Bluffton, Indiana, which was owned by his family with his father as director. He became assistant cashier (1903), cashier (1905) and president and director of the bank (1918–1928) which then closed (1928) following the stock market crash. In his remaining four and half years he managed to refund all the bank customers. While in banking, he continued his scientific research and fieldwork on dragonflies, usually in cooperation with the Museum of Zoology of the University of Michigan at Ann Arbor, where he had been appointed as Honorary Curator of Odonata (1916). The museum invited him (1929) to become research associate there and provided an office. He moved his collections and library to Ann Arbor (1929). He actively collected dragonflies (from 1896) in Indiana, elsewhere in the US and made expeditions to Guatemala and Honduras (1905, 1909), British and Dutch Guiana and Trinidad (1912), Colombia and Panama (1916–1917) and Venezuela (1920). He accumulated an unrivalled collection of Odonata of the Americas. He published over 100 papers and notes on Odonata. These include revisions of several American genera and many grippingly written travel reports with detailed information of local fauna. His major papers include: *The dragon-flies of Indiana* (1900) and *Notes on American species of Triacanthagyna and Gynacantha* (1923). He described 92 new species and named 14 new genera. In his early career, he also published on such diverse topics as crayfish, salamanders and birds. He was a keen cultivator of irises, breeding many new varieties and running a business selling them. He suffered poor health for most of his life. Three amphibians are also named after him. (See also **Bruce, Ethel (Merriman) & Jesse**)

Willink

Sylph sp. *Macrothemis willinki* **Fraser**, 1947
[JS *Elasmothemis constricta* (Calvert, 1898)]
Firetail sp. *Telebasis willinki* Fraser, 1948

Dr Abraham 'Bram' Willink (1920–1998) was a Dutch-Argentine entomologist, He was born in Friesland and died in Tucuman, his parents having moved to Argentina just after he was born, and he became an Argentine citizen (1940). He studied at Agustín Álvarez National College, Mendoza (1934–1938) for his bachelors' degree and at the Museum of La Plata (1939–1943), which awarded his doctorate (1946), his thesis on Hymenoptera being: *Contribución al estudio de los esfecoideos argentinos (Hym.: Bembicidae)*. He joined (1944–1946) the Facultad de Ciencias Naturales e Instituto Miguel Lillo, Universidad Nacional de Tucuman as a student assistant, helping to organize and manage their Entomology Section, working on systematics and the biogeography of Neotropical wasps. He rose to Director (1958–1966) and Dean of the Faculty of Natural Sciences (1975–1977 & 1978–1979). He became Professor of Entomology until retirement (1988) and Professor Emeritus thereafter. He also spent periods at Cornell, the Smithsonian, the University of London and Leiden. His main contributions were made on Hymenoptera and he collected across many parts of Argentina. His published papers (1969–1998) include: *Las especies del* género *Incodynerus* (1969) and the co-written: *Revisión del genero Pachodynerus Saussure (Hymenoptera: Vespidae, Eumeninae)* (1998), but perhaps his most significant work was the co-written book *Biogeography of Latin America* (1973). He, with Captain Hayward (q.v.), collected the type specimens in Argentina (1946). Three genera and around 50 species were named after him.

Wilson, C E

Wilson's Groundling *Brachythemis wilsoni* **Pinhey**, 1952
[A. Swamp Groundling]

C E Wilson was an American entomologist (fl.1900–fl.1950). He was a charter member of the Florida Entomological Society (1916) and Assistant Entomologist at the State Plant Board, Gainesville (1917) and then Instructor in Zoology at the University of Indiana whilst working toward his PhD. He then enlisted and was appointed Neural Histologist in the Brain Surgery Department of USA Medical Corps (1918). He was appointed (1919–1924) Entomologist at the Virgin Islands Experimental Agricultural Station. He collected this dragonfly in southern Sudan (December 1950) whilst collecting butterflies (1942–1953), working for the Research Division of the Sudan Ministry of Agriculture as an economic entomologist. He wrote a number of papers, including: *Some Florida Scale Insects* (1917), *Insect pests of cotton in St. Croix and means of combatting them* (1923) and *Butterflies of the Northern and Central Sudan* (1950).

Wilson, H

Dancer sp. *Argia wilsoni* **Calvert**, 1902
[JS *Argia calida* Hagen, 1861]

H Wilson collected for MCZ in Guatemala, including the type (1885).

Wilson, K D P

Sylvan sp. *Coeliccia wilsoni* **Zhang** & Yang, 2011

Keith Duncan Peter Wilson (b.1953) is a British hydrobiologist, environmentalist and conservationist who is currently Technical Director, Al Fahad Environmental Services, UAE. The University of Reading awarded his BSc (1975) and King's College London his MSc in applied hydrobiology (1977). He has extensive knowledge and experience of establishing and managing marine protection areas and wetlands, built up through a distinguished career in fisheries and marine environmental services. He was Assistant Biologist for the UK Southern Water Authority (1978–1983) then Fisheries and Conservation Officer (1983–1989). Subsequently he was Fisheries, Conservation and Recreation Manager, North West Region, National Rivers Authority (1989–1991), when he moved to Hong Kong as Fisheries Officer for their Agriculture and Fisheries Department (1991–1995), Marine Parks Officer (1995–1996) and Senior Fisheries Officer (1996–2003). He then became Ecological Consultant and Sole Proprietor of Dragonfly Ecological Services, Asia (2003–2008). He moved to UAE (2008) first taking up the post of Senior Manager (Environment), Nakheel, Waterfront and Palm Jebel Ali, Dubai (2008–2009) then becoming Director Marine Programme, Emirates Marine Environmental Group (2009–2013). He has also, on numerous occasions, been called upon to serve as an expert witness in connection with planning disputes related to development proposals. He is recognised as an international expert on the Odonata of China. He has written or co-written more than eighty papers on fisheries management, wetland conservation and particularly Odonata. He has authored several books such as *Hong Kong Dragonflies* (1995) and *Field Guide to the Dragonflies of Hong Kong* (2003). He has described c.50 new species of Odonata from China and named four genera.

Witte

Tanganyika Siphontail *Neurogomphus wittei* **Schouteden**, 1934
Shadow Hawker *Aeshna wittei* **Fraser**, 1955
[JS *Afroaeschna scotias* (Pinhey, 1952)]
Blue-spotted Yellowwing *Allocnemis wittei* Fraser, 1955
[Orig. *Chlorocnemis wittei*]
Katanga Jewel *Chlorocypha wittei* Fraser, 1955

Variable Sprite *Pseudagrion wittei* Fraser, 1949
[JS *Pseudagrion sjoestedti* Förster, 1906]

Dr Gaston-François de Witte (1897–1980) was a civil servant, naturalist and herpetologist. He was originally a colonial administrator, working in the Belgian Congo (Democratic Republic of Congo) (from c.1920), but worked as a naturalist, becoming a collector of zoological and botanical specimens for the Institut des Parcs Nationales Congo-Belge in Tervuren (1938). He was also honorary conservator at the Royal Belgian Institute for Natural Sciences. He directed the scientific exploration missions in the Belgian Congo's Albert National Park (1933–1935) and Upemba National Park (1946–1949). He authored the reports, in which Fraser described the Odonata with, in the first mission 2,569 specimens collected from 42 species in 22 genera including the eponymous sprite which Fraser described as a new species. Most of his published works (1930–1965) are on reptiles, such as: *Les caméléons de l'Afrique centrale : république démocratique du Congo, République du Rwanda et Royaume du Burundi* (1965). He described twenty-four new species of reptile and three birds, two reptiles and four amphibians are also named after him.

Wnukovskiy

(See **Vnukovskiy**)

Wollaston

Tigertail sp. *Palaeosynthemis wollastoni* **Campion**, 1915
[Orig. *Synthemis wollastoni*]

Dr Alexander 'Sandy' Frederick Richmond Wollaston (1875–1930) was a physician, surgeon, naturalist, ornithologist, botanist and explorer. He studied medicine at King's College, Cambridge (graduating 1896) and then qualifying as a surgeon (1903). He travelled widely including to Lapland, the Dolomites, Sudan and Japan. During 1905–1907) he went to the Ruwenzori Mountains in Uganda. He was on the BOU expedition to the Snow Mountains in Dutch New Guinea (1910–1911) to ascend its highest mountain, but it was deliberately misdirected by Dutch authorities. He led another expedition there (1912–1913), during when he collected the eponymous type. During WW1, he was a naval surgeon in East Africa where he met his future brother-in-law, Richard Meinertzhagen. He went on an Everest reconnaissance expedition (1921) as physician, ornithologist and botanist where he collected a new primula, later named after him. He married Mary 'Polly' Meinertzhagen (1923) and was in Colombia that year. Five years later he was appointed as Honorary Secretary of the Royal Geographical Society. He was invited to become a tutor at his alma mater and while teaching, was shot dead in his rooms by a deranged student (Douglas Potts), who then shot and killed the policeman who had come to arrest him and finally committed suicide (1930). He wrote a number of books based on his explorations, including: *From Ruwenzori to the Congo: a Naturalist's Journey Across Africa* (1908) and *An Expedition to Dutch New Guinea* (1914) as well as others, including a biography: *Life of Alfred Newton, Professor of Comparative Anatomy Cambridge University, 1866–1907* (1921). Two mammals, a bird and an amphibian are also named after him.

Woodford

Skimmer sp. *Protorthemis woodfordi* **Kirby**, 1889
[Orig. *Nesocria woodfordi*]

Charles Morris Woodford (1852–1927) was the Resident Commissioner in the Solomon Islands Protectorate (1896–1914) where he collected the type. He was an adventurer, a naturalist and also a philatelist. He established the first postal service in the islands and issued their first stamps, personally franking the envelopes. He wrote: *A Naturalist Among the Headhunters* (1890), which is referred to in a letter by his friend, the novelist Jack London, and *Notes on the Solomon Islands* (1926). Two mammals and five birds are also named after him.

Woytkowski

Threadtail sp. *Protoneura woytkowskii* LK Gloyd, 1939

Felix Woytkowski (1892–1966) was a Polish (born in what is now Ukraine) botanist and entomologist. He studied economics in Belgium and Oxford before serving in the army in WW1. He married and emigrated from Ukraine to Peru (1929), lured by promises of free passage and a homestead farm. The promises did not materialize, but he established himself as a collector, sending insects and plants to scientists in Europe and the US. There he collected more than one thousand insects new to science, including the damselfly holotype

(1936), as well as many hundreds of plants. He worked for seven years for a pharmaceutical company and also became Professor of French at La Molina College of Agriculture, Lima. However, he discovered that because he never became a citizen he was not entitled to a pension and so he returned to Poland (1965). He wrote: *Peru, My Unpromised Land* (1974). Many plants and insects are named after him.

Wright

Bold Skimmer ssp. *Orthetrum stemmale wrightii* **Selys**, 1869
[Orig. *Libellula wrightii,* A. Tough Skimmer]

Professor Dr Edward Perceval Wright (1834–1910) was an Irish ophthalmic surgeon, botanist and zoologist. He studied at Trinity College, Dublin (1852) graduating (1957) then becoming Curator of the University Museum (1857) and a Lecturer in Zoology (1858). He also lectured in botany at the medical school before studying for his MA at Dublin University (1859) and another MA from Oxford. He gained his MD (1862) and studied ophthalmic surgery in Vienna, Paris and Berlin before practising as a surgeon. He became Professor of Botany at Trinity College, Dublin (1869–1905) and Curator of the Herbarium. He spent most vacations travelling in Europe collecting natural history specimens; six months were spent in Seychelles (1867) where he collected the eponymous type. He wrote many papers and longer works from: *Catalogue of British Mollusca* (1855) to *Notes from the Botanical School of Trinity College, Dublin* (1896). His interests were always wide, from coral to fossil amphibians, and he is named in many other taxa including a reptile.

Wucherpfennig

Clubtail sp. *Zonophora wucherpfennigi* **Schmidt**, 1941

Ferdinand Wucherpfennig (1878–1958) was a German-born Brazilian entomologist and Lepidopterist. He moved to Sao Paulo (1923) and collected insects, particularly butterflies, across Brazil, including Amazonas (1930–1931), Rio Grande do Sul (1935) and Sao Paulo (1937). He collected the type series of the clubtail (1930, 1935) at Manicoré and Amazonas. The etymology reads thus: "*Die vorliegende Art hat Herr F. Wucherpfennig aus Dingelstädt (Eichsfeld), jetzt Sao Paulo, offenbar als einziger bisher gesammelt; sie sei ihm zu Ehren für seine erfolgreiche Sammeltätigkeit, die er auch auf andere Insektengruppen ausdehnte, benannt.*"

X

Xiao

Darter sp. *Sympetrum xiaoi* Han & **Zhu**, 1997

Cai-Yu Xiao (formerly Tsai Yu Hsiao) (1903–1978) was a Chinese entomologist whose focus was Hemiptera. He described no less than 56 Hemipteran genera. He collected in China and Korea. He wrote, among other books and papers: *Heteroptera insects Identification Manual Volume: Semi Hemiptera Heteroptera suborder* and *Epidemiology of the Diseases of Naval Importance in Korea* (1948). The etymology says: "*The species is dedicated to our teacher, Professor Dr Cai-yu Xiao.*"

Y

Yagasaki

Regal Pond Cruiser ssp. *Epophthalmia elegans yagasakii* **Eda**, 1986

Dr Yasushi Yagasaki (1920–2010) was a Japanese microbiologist at the Department of Microbiology, Tokyo Dental College. He was founder (1972) of the Matsumoto Dental University where he was professor and served as Chairman of the Board of Trustees (1977, 2003 & 2006). He collected the type in North Korea (1986).

Yamasaki

Sylvan sp. *Coeliccia yamasakii* **Asahina**, 1984

Dr Tsukane Yamasaki (b.1939) is a Japanese entomologist who is Professor Emeritus, Laboratory of Applied Entomology, at the University of Tokyo. He was formerly at the Department of Biology, Saitama University, Urawa. He collected the holotype in Thailand (1973). He has published many papers, particularly on Orthoptera, such as: *Some New or Little Known Species of the Meconematinae (Orthoptera, Tettigoniidae) from Japan* (1982) and some longer works such as a biography of Kano Tadao (1992) and *Grasshoppers and Crickets* (1979). He wrote Asahina's obituary (2011).

Yanagisawa

Longleg sp. *Anisogomphus yanagisawai* **Sasamoto**, 2015

Takashi Yanagisawa (b.1982) is an amateur entomologist. He collected the holotype male at Doi Inthanon in Thailand (2013) on a paved road alongside a stream (in the previous year some females had been collected at the same site). He studied law at Hosei University and currently works for the Hiroshima Prefecture Private Kindergartens Federation.

Yang, B

Flatwing sp. *Rhipidolestes yangbingi* **Davies**, 1998
Flatwing sp. *Sinocnemis yangbingi* KDP **Wilson** & **Zhou**, 2000

Bing Yang is a Chinese entomologist who works at the Kunming Institute of Zoology, Academia Sinica. He collected both of the types in Emeishan, Sichuan, China (1992) having guided Allen Davies (q.v.) during his visit to Yunnan (1991). Yang and Davies jointly described several new species of dragonflies.

Yang, Z-D

Pincertail sp. *Nychogomphus yangi* Zhang, 2014

Zu-De Yang (b.1943) is a Chinese biologist and entomologist at the Biology Department, Hanzhong Teachers' College. Hanzhong City, Shaanxi Province, China. Among his papers are: *A New Species of the Genus Burmagomphus from Northern Szechuan (Odonata: Gomphidae)* (1994) and *A New Species of Protoneurid Damselfly from Shaanxi, China (Odonata: Protoneuridae)* (1995).

Yanomami

Swampdamsel sp. *Tuberculobasis yanomami* **De Marmels**, 1992
[Orig. *Leptobasis yanomami*]

The Yanomami are an indigenous people who live in the same region on the Venezuela-Brazil border where the damselfly was found.

Yazhen

Spookhawker sp. *Periaeschna yazhenae* Xu, 2012

Yazhen Lin (1944-2010) was the mother of the author Qi-han Xu (b.1963).

Yerbury

Crimson Dropwing *Trithemis yerburii* **Kirby**, 1890
[JS *Trithemis aurora* (Burmeister, 1839)]
Yerbury's Elf *Tetrathemis yerburii* **Kirby**, 1894

Lieutenant-Colonel John William Yerbury (1847–1927) was an English army officer and amateur zoologist, entomologist and, in particular, dipterist. He collected in Aden (1884–1896) including reptiles and botanical specimens. When he was a major he collected in western India (1886–1887) and stayed at Trincomalee, Ceylon (1880s & 1890s). The type of the Elf was in his collection, and collected in Kandy, Ceylon (1892). He was in Rhodesia (Zambia/Zimbabwe) just before WW1. He wrote: *Seashore Diptera* (1919) as well as papers on rodents. A reptile is also named after him.

Yie

Basker sp. *Urothemis signata yiei* **Asahina**, 1972

Dr Shi-Tao Yie (1903-1975) was Professor of Entomology at National Taiwan University and an avid student of Mecoptera, at least one of which is also named after him. He was also Chairman of the Republic of China Plant Protection Society and the Research Institute of Plant Pathology and Entomology. He wrote many papers on Mecoptera and longer works, such as: *The biology of Formosan Panorpidae and morphology of eleven species of their immature stages* (1951).

Yokoi

Skydragon sp. *Chlorogomphus yokoii* **Karube**, 1994
Flatwing sp. *Rhinagrion yokoii* **Sasamoto**, 2003
[JS *Rhinagrion hainanense* Wilson & Reels, 2001]
Grisette sp. *Devadatta yokoii* Phan, **Sasamoto** & Hayashi, 2015

Dr Naoto Yokoi (b.1956) is a Japanese entomologist and odonatologist at the Fukushima Sericultural Experiment Station. He gained (1992) his PhD in agriculture at Hokkaido University with a thesis on the ecology and control of the cerambycid beetle *Psacothea hilaris,* a serious pest of mulberry trees. He has regularly visited and collected dragonflies in Laos (since 1994) discovering more than 100 Odonata species not previously known, including numerous undescribed species. Several new species have already been described from his material. He himself has authored or co-authored five of them. His over 20 faunistic and taxonomic papers on Laos Odonata includes the book: *A list of Lao dragonflies* (2014).

Yoshiko

Demoiselle sp. *Noguchiphaea yoshikoae* **Asahina**, 1976

Darter ssp. *Sympetrum risi yosico* Asahina, 1961

Yoshiko Noguchi (1926–1976). The demoiselle binomial is an example of a clever double eponym where both parts of the name honour one person. Asahina dedicated the demoiselle in memorium to his faithful collaborator who had just died. (**See Noguchi**)

Yoshihiro Hirose

Skydragon sp. *Chlorogomphus yoshihiroi* **Karube**, 1994

Dr Yoshihiro Hirose (b.1969) is a Japanese entomologist from Hokkaido. He graduated (1999) from the doctoral program of the Department of Bioindustry Graduate School at Tokyo University of Agriculture. His thesis treated the evolutionary genetics of Japanese Odonata, using biochemical methodology. He has published two guide books on dragonflies of Hokkaido (1993 & 2007) and several other research papers on Japanese dragonflies. Presently he is a teacher at Hokkaido Eco College of Wildlife and Nature.

Yoshitomi

Paddletail sp. *Sarasaeschna yoshitomii* Kiyoshi, Katatani, **Kompier** & Yeh, 2016

Dr Hiroyuki Yoshitomi (b.1972) is a Japanese entomologist at the Entomological Laboratory, Faculty of Agriculture, Ehime University. He has been consistently publishing for some years (2003–2017), primarily on Coleoptera, and has expertise in genetics, systematics (taxonomy) and evolutionary biology. The authors etymology reads: "*...the species is dedicated to Dr. Hiroyuki Yoshitomi who has provided his valuable materials for this study*".

Yu Lan

Threadtail sp. *Prodasineura yulan* **Dow** & Ngiam, 2013

Yu Lan Lu (1941–1985) was the mother of the second author, Robin Wen Jiang Ngiam (b.1974). She was born in Singapore during WW2 and, as an infant, was sent to stay with an aunt in the ancestral home in Hainan for safety. When ten years old, while doing farm chores, she fell and was trampled by cattle, becoming seriously ill with a heavy fever. She was thought to be dead, covered with a shroud and the villagers went to fetch a coffin. Fortunately, her aunt noticed faint breathing, and she was saved but sadly, the fever caused blindness in her left eye. Soon after the accident she returned to Singapore, and finally (1954) was able to join her parents in Johor. She moved back to Singapore (1960) to stay with her elder brother and later (1965) got a job in a school for the blind where she met her future husband. They married (1970) and had two children. She was diagnosed with breast cancer (1982) from which she eventually died.

Yunosuke

Demoiselle sp. *Mnais yunosukei* **Asahina**, 1990
[Orig. *Vestalis yunosukei*]

Yunosuke Kimura is a well-known Japanese amateur lepidopterologist. A graduate of Shimane University, his 'day job' was working for the Mitsubishi Corporation. In this capacity he worked in Bangkok (1970–1978). There he spent his free time (1974–1978) collecting butterflies, often with Aroon Samruadkit (q.v.) and Bro. Amnuay Pinratana (q.v.). Since then he has often returned to Thailand and has collected more than 1,100 butterfly species there. His books include the three-volume: *The Butterflies of Thailand* (2011, 2013 & 2015). He found the demoiselle at Doi Hom Pok (1989). (Also see **Kimura**)

Yu Tse-hong

Scarlet Pygmy *Nannodiplax yutsehongi* **Navás**, 1935
[JS *Nannophya pygmaea* Rambur, 1842]

Brother Yu Tse-hong was a Jesuit priest in Anhui, China, where he collected the type specimen in Anhui (1934).

Z

Zane

Flatwing sp. *Argiolestes zane* Kalniņš, 2014

Zane Pīpkalēja (b.1980) is married to the author Dr Mārtiņš Kalniņš (b.1978) and is mother of their two children, David and Peter. The University of Latvia, Faculty of Geography and Earth Science awarded (2004) her masters in environmental science. For many years, she worked as a project manager in Amata Municipality (Vidzeme, Latvia), where her main tasks were related to local community development and encouraging small entrepreneurship. As part of this, she identifies and encourages opportunities for NGOs to implement nature conservation and environmental management projects, giving advice, help and support. She took part (2009) in an expedition to Indonesia, with Dr Mārtiņš Kalniņš and other biologists. Currently, she works in an educational project that aims to improve natural sciences in schools. Recently (2016), she got funding for a project as part of which a Latvian dragonflies atlas will be published.

Zerny

Striped Longleg *Notogomphus zernyi* St. Quentin, 1942
[Orig. *Podogomphus zernyi*]

Blue-eye ssp. *Erythromma lindenii zernyi* **Schmidt**, 1938
[Orig. *Agrion lindeni zernyi*]
Hintz's Skimmer *Orthetrum hintzi zernyi* Schmidt, 1951
[JS *Orthetrum hintzi* Schmidt, 1951]

Dr Hans Anton Zerny (1887–1945) was an Austrian entomologist and lepidopterologist at the Natural History Museum, Vienna. As a schoolboy, he collected and prepared plants and animals and published his first note on insects. He studied at Vienna University (1908–1912), which awarded his doctorate, spending some of his time at the Zoological Station in Trieste. He took a post at the Vienna Natural History Museum (1912) where he worked for the rest of his life becoming Keeper of the Lepidoptera collection (1932–1945). His major foci were Neuroptera, Diptera and Odonata. He made a number of collecting expeditions to South America and east Africa. He wrote or co-wrote around 50 papers, mostly on Lepidoptera. His name is also associated with other taxa including butterflies.

Zetek

Flatwing sp. *Philogenia zeteki* **Westfall** & Cumming, 1956

Professor James Zetek (1886–1959) was an American entomologist. He graduated (1911) with a BA from the University of Illinois and then worked for the Isthmian Canal Commission in the Panama Canal Zone as their entomologist (1911–1913). He married

(1914) a local girl, and was the Republic of Panama's entomologist (1914–1915). He became Professor of Biology and Hygiene at the National Institute of Panama (1916–1918) and the first Director of the Smithsonian Tropical Research Institute on Barro Colorado Island, Gatun Lake, Panama Canal. Thereafter he was entomologist to the Panama Canal (1918–1956) and during this time, he gave 'enthusiastic assistance' to Robert Cumming, making possible the collection (1950) of the eponymous species. His primary research interest was the study of termites and termite control. He co-wrote: *The Black Fly of Citrus and Other Subtropical Plants (1920)*. A reptile and two amphibians are also named after him.

Zheng, L-Y

Shadowdamsel sp. *Protosticta zhengi* Xin Yu & Wenjun Bu, 2009

Professor Le-Yi Zheng (b.1931) was a teacher and researcher at the Institute of Entomology, Department of Biology at Nankai University, Tianjin. His main research topic is Heteroptera. He has collected Odonata and other insects in Yunnan including collecting the holotype of the Shadowdamsel at Menghun, Xishuangbanna, Yunnan, China (1958). Among his authored or co-authored papers are: *A New Species of the Genus Neoephemera McDunnough from China (Ephemeroptera: Neoephemeridae)* (2010) and *A new genus of Orthotylini (Heteroptera: Miridae) with descriptions of four new species* (2101). The etymology says: "… Species epithet is in honor of the collector Professor Le-yi Zheng."

Zheng, Z-M

Kinkedwing sp. *Anisopleura zhengi* Yang, 1996

Zhe-Min Zheng (b.1932) is a Chinese entomologist who studied Diptera (Syrphidae, etc.) and Orthoptera in particular, writing papers on them, such as: *A new genus and two new species of Syrphidae from China (Diptera, Syrphidae)* (2005). He named many new species in these groups. He worked at the Institute of Zoology, Shaanxi Normal University, Xi'an, China. The type was taken in Shaanxi. I believe Zu-de Yang was one of his students.

Zhou

Clubtail sp. *Davidius zhoui* **Chao**, 1995
Shadowdamsel sp. *Drepanosticta zhoui* **KDP Wilson** & **Reels**, 2001

Wen-Bao Zhou is an entomologist at the Zhejiang Museum of Natural History, Hangzhou, China. His papers (1984–2008) include: *A List of Butterflies in Zhejiand (China)* (1989) and *Macromia hamata* sp. nov. from Guizhou, China (Odonata: Corduliidae) (2003). He has also written papers with Wilson (q.v.) describing new species, such as: *Sinocnemis yangbingi gen. nov., sp. nov. and Sinocnemis dumonti sp. nov., new platycnemidids from southwest China (Odonata: Plactycnemididae)* (2000) and *Priscagrion kiautai gen. nov., spec. nov. and P. pinheyi spec. nov., new damselflies from southwestern China (Zygoptera: Megapodagrionidae)* (2001). His publications include descriptions of over 40 new odonate species from China. The etymology states that *"…this damselfly is named in honour of Zhou Wen-bao who first studied this species from Hainan."* (1986).

Zhu

Oread sp. *Calicnemia zhuae* Zhang & **Yang**, 2008

Paddletail sp. *Sarasaeschna zhuae* Xu, 2008

Hui-Qian Zhu (d.2009) was a Professor in the Department of Life Sciences at Shanxi University, Taiyuan. I am still hoping to find more information about her.

Zielma

Pond Damselfly sp. *Oxyagrion zielmae* JM **Costa**, LOI De Souza & J **Muzón**, 2006

Zielma de Andrade Lopes (b.1972) is the wife of the second author Luiz Onofre Irineu de Souza (1948–2010).

Zoe

Zoe Waterfall Damsel *Paraphlebia zoe* **Selys**, 1861

It seems unlikely that this is an eponym at all given the construction of the binomial… it might refer to the meaning of Zoe: being Greek for 'life'.